Recent Titles in This Series

152 **Christopher K. McCord, Editor,** Nielsen theory and dynamical systems, 1993

151 **Matatyahu Rubin,** The reconstruction of trees from their automorphism groups, 1993

150 **Carl-Friedrich Bödigheimer and Richard M. Hain, Editors,** Mapping class groups and moduli spaces of Riemann surfaces, 1993

149 **Harry Cohn, Editor,** Doeblin and modern probability, 1993

148 **Jeffrey Fox and Peter Haskell, Editors,** Index theory and operator algebras, 1993

147 **Neil Robertson and Paul Seymour, Editors,** Graph structure theory, 1993

146 **Martin C. Tangora, Editor,** Algebraic topology, 1993

145 **Jeffrey Adams, Rebecca Herb, Stephen Kudla, Jian-Shu Li, Ron Lipsman, Jonathan Rosenberg, Editors,** Representation theory of groups and algebras, 1993

144 **Bor-Luh Lin and William B. Johnson, Editors,** Banach spaces, 1993

143 **Marvin Knopp and Mark Sheingorn, Editors,** A tribute to Emil Grosswald: Number theory and related analysis, 1993

142 **Chung-Chun Yang and Sheng Gong, Editors,** Several complex variables in China, 1993

141 **A. Y. Cheer and C. P. van Dam, Editors,** Fluid dynamics in biology, 1993

140 **Eric L. Grinberg, Editor,** Geometric analysis, 1992

139 **Vinay Deodhar, Editor,** Kazhdan-Lusztig theory and related topics, 1992

138 **Donald St. P. Richards, Editor,** Hypergeometric functions on domains of positivity, Jack polynomials, and applications, 1992

137 **Alexander Nagel and Edgar Lee Stout, Editors,** The Madison symposium on complex analysis, 1992

136 **Ron Donagi, Editor,** Curves, Jacobians, and Abelian varieties, 1992

135 **Peter Walters, Editor,** Symbolic dynamics and its applications, 1992

134 **Murray Gerstenhaber and Jim Stasheff, Editors,** Deformation theory and quantum groups with applications to mathematical physics, 1992

133 **Alan Adolphson, Steven Sperber, and Marvin Tretkoff, Editors,** p-Adic methods in number theory and algebraic geometry, 1992

132 **Mark Gotay, Jerrold Marsden, and Vincent Moncrief, Editors,** Mathematical aspects of classical field theory, 1992

131 **L. A. Bokut′, Yu. L. Ershov, and A. I. Kostrikin, Editors,** Proceedings of the International Conference on Algebra Dedicated to the Memory of A. I. Mal′cev, Parts 1, 2, and 3, 1992

130 **L. Fuchs, K. R. Goodearl, J. T. Stafford, and C. Vinsonhaler, Editors,** Abelian groups and noncommutative rings, 1992

129 **John R. Graef and Jack K. Hale, Editors,** Oscillation and dynamics in delay equations, 1992

128 **Ridgley Lange and Shengwang Wang,** New approaches in spectral decomposition, 1992

127 **Vladimir Oliker and Andrejs Treibergs, Editors,** Geometry and nonlinear partial differential equations, 1992

126 **R. Keith Dennis, Claudio Pedrini, and Michael R. Stein, Editors,** Algebraic K-theory, commutative algebra, and algebraic geometry, 1992

125 **F. Thomas Bruss, Thomas S. Ferguson, and Stephen M. Samuels, Editors,** Strategies for sequential search and selection in real time, 1992

124 **Darrell Haile and James Osterburg, Editors,** Azumaya algebras, actions, and modules, 1992

123 **Steven L. Kleiman and Anders Thorup, Editors,** Enumerative algebraic geometry, 1991

122 **D. H. Sattinger, C. A. Tracy, and S. Venakides, Editors,** Inverse scattering and applications, 1991

(*Continued in the back of this publication*)

Nielsen Theory and Dynamical Systems

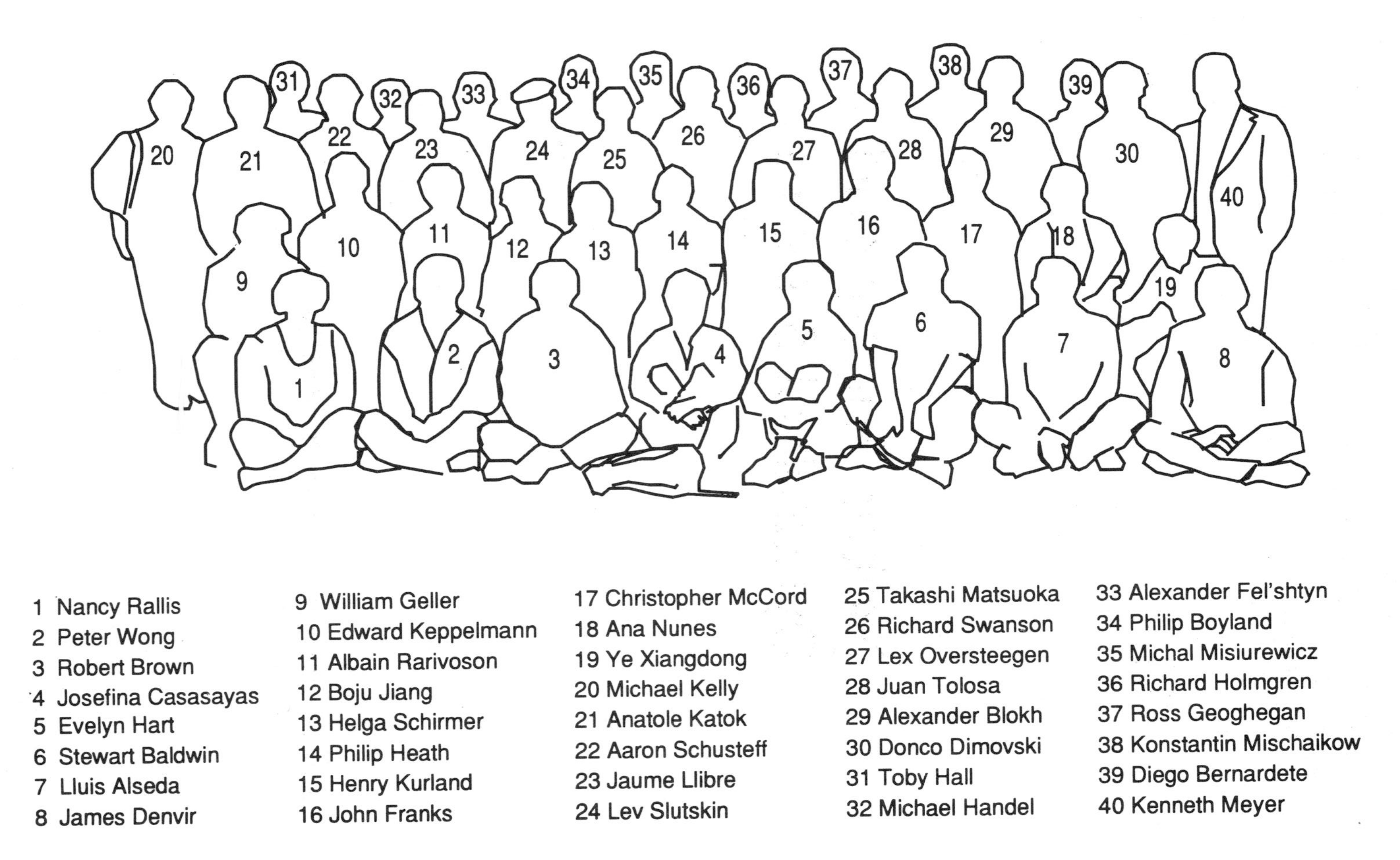

1 Nancy Rallis
2 Peter Wong
3 Robert Brown
4 Josefina Casasayas
5 Evelyn Hart
6 Stewart Baldwin
7 Lluis Alseda
8 James Denvir
9 William Geller
10 Edward Keppelmann
11 Albain Rarivoson
12 Boju Jiang
13 Helga Schirmer
14 Philip Heath
15 Henry Kurland
16 John Franks
17 Christopher McCord
18 Ana Nunes
19 Ye Xiangdong
20 Michael Kelly
21 Anatole Katok
22 Aaron Schusteff
23 Jaume Llibre
24 Lev Slutskin
25 Takashi Matsuoka
26 Richard Swanson
27 Lex Oversteegen
28 Juan Tolosa
29 Alexander Blokh
30 Donco Dimovski
31 Toby Hall
32 Michael Handel
33 Alexander Fel'shtyn
34 Philip Boyland
35 Michal Misiurewicz
36 Richard Holmgren
37 Ross Geoghegan
38 Konstantin Mischaikow
39 Diego Bernardete
40 Kenneth Meyer

Participants in the AMS-IMS-SIAM Joint Summer Research Conference on Nielsen Theory and Dynamical Systems

CONTEMPORARY MATHEMATICS

152

Nielsen Theory and Dynamical Systems

AMS-IMS-SIAM Summer Research Conference
on Nielsen Theory and Dynamical Systems
June 20–26, 1992
supported by the National Science Foundation

Christopher K. McCord
Editor

American Mathematical Society
Providence, Rhode Island

The AMS–IMS–SIAM Joint Summer Research Conference in the Mathematical Sciences on Nielsen Theory and Dynamical Systems was held at Mt. Holyoke College, South Hadley, Massachusetts, from June 20–26, 1992, with support from the National Science Foundation, Grant DMS-8918200 02.

1991 *Mathematics Subject Classification.* Primary 54H20, 54H25, 55M20;
Secondary 34C35, 58C30, 58F20.

Library of Congress Cataloging-in-Publication Data

AMS-IMS-SIAM Joint Summer Research Conference on Nielsen Theory and Dynamical Systems (1992: Mt. Holyoke College)
Nielsen theory and dynamical systems/Christopher K. McCord, editor.
p. cm.—(Contemporary mathematics, ISSN 0271-4132; v. 152)
"Proceedings of the AMS-IMS-SIAM Joint Summer Research Conference on Nielsen Theory and Dynamical Systems, held June 20–26, 1992, with support from the National Science Foundation."
ISBN 0-8218-5181-0 (acid-free)
1. Topological dynamics—Congresses. 2. Fixed point theory—Congresses. I. McCord, Christopher Keil. II. Title. III. Series: Contemporary mathematics (American Mathematical Society); v. 152.
QA611.5.A47 1992 93-26685
514—dc20 CIP

Printed in the United States of America.
The paper used in this book is acid-free and falls within the guidelines
established to ensure permanence and durability. ♾
Printed on recycled paper. ♻

All articles in this volume were printed from copy prepared by the authors.
Some articles were typeset using $\mathcal{AMS}$-TeX or $\mathcal{AMS}$-LaTeX,
the American Mathematical Society's TeX macro system.

10 9 8 7 6 5 4 3 2 1 98 97 96 95 94 93

Contents

Preface

In June, 1992, a one-week conference on Nielsen Theory and Dynamical Systems was held at Mt. Holyoke College in South Hadley, Massachusetts, as part of the AMS-IMS-SIAM Joint Summer Research Conference Series. The idea for this conference grew out of the special session for Nielsen fixed point theory at the Topology '90 conference, held at the University of Hawaii in August, 1990. At that session, I noticed that most speakers indicated a desire to develop dynamical applications of their work. But many seemed unsure just what form such applications should take, as they were not sufficiently conversant with dynamical systems to know what its issues and problems were. It was clear that Nielsen fixed point theory had reached a level of development where it had the opportunity to make some real contributions to dynamics; but it was equally clear that some input and guidance from the dynamics community was needed to make those contributions effective.

The Nielsen Theory and Dynamical Systems Conference was the obvious response to this opportunity/need. The conference organizing committee consisted of Joan Birman, Robert Brown, John Franks, Boju Jiang and Christopher McCord (chair). It brought together about 50 topologists and dynamicists. During the week, they heard talks on Nielsen theory and its potential for dynamics; on problems in dynamics that might be amenable to Nielsen theory; and on problems in dynamics that have been successfully analyzed via Nielsen theory. More importantly, the week spent together gave the two research communities the opportunity to meet, interact, and form the personal contacts that are so important for interdisciplinary research. These proceedings attest to the value of such interactions. I can cite one personal example: Konstantin Mischaikow's article contains a result whose proof was supplied the night before his talk. Parts of the proof were supplied at the conference by Phil Boyland, David Fried, Ross Geoghegan, Boju Jiang, Konstantin Mischaikow and myself.

The articles in this volume are contributions by speakers at the conference whose papers, after review, were viewed as conveying the conference's theme. The papers include surveys, expository papers, and technical presentations of new results. Collectively, they provide a good impression of the current status of the interface between Nielsen theory and dynamical systems. I hope that they will promote further interaction between the two fields.

I would like to thank the organizing committee, the conference speakers and the conference participants, whose participation was the real success of the conference. I would also like to thank the American Mathematical Society and the National Science Foundation for making the conference possible, and the AMS staff, particularly Carole Kohanski and Donna Harmon, for their assistance in the organization of the conference and this proceedings.

Christopher McCord
Cincinnati, Ohio

Contemporary Mathematics
Volume 152, 1993

Torus maps and Nielsen numbers

LL. ALSEDÀ, S. BALDWIN, J. LLIBRE, R. SWANSON, W. SZLENK

ABSTRACT. The main results in this paper concern the minimal sets of periods possible in a given homotopy class of torus maps. For maps on the 2−torus, we provide a complete description of these minimal sets. A number of results on higher dimensional tori are also proved; including criteria for every map in a given homotopy class to have all periods, or all but finitely many periods.

1. Introduction

In dynamical systems, it is often the case that topological information can be used to study qualitative properties of the system. This article deals with the problem of determining the set of periods (of the periodic orbits) of a torus map given the homotopy class of the map. To fix terminology, suppose f is a continuous self–map on the torus T^m. A *fixed point of* f is a point x in T^m such that $f(x) = x$. We will call x a *periodic point of period* n if x is a fixed point of f^n but is not fixed by any f^k, for $1 \leq k < n$.

Denote by $Per(f)$ the set of natural numbers corresponding to periods of periodic orbits of f.

Our aim is to provide a description of the minimal set of periods (see below) attained within the homotopy class of a given torus map $f : T^m \to T^m$. We also present a few results, described below, in a more general setting.

Toward this end, it is convenient to distinguish among several subsets of the natural numbers N. First, there is the set of periods, $Per(f)$, mentioned above. When the mapping $g : T^m \to T^m$ is homotopic to f, we shall write $g \simeq f$. Define the *minimal set of periods* of f to be the set

$$MPer(f) = \bigcap_{g \simeq f} Per(g).$$

1991 *Mathematics Subject Classification.* 54H20, 34C35.
Key words and phrases. Torus maps, Nielsen Numbers, periods.
The complete version of this paper will be submitted elsewhere.

For instance, if f is a circle map of degree d, then we have (see for example [2])

(1) For $d = 1$, $MPer(f) = \emptyset$.
(2) For $d = 0$ or $d = -1$, $MPer(f) = \{1\}$.
(3) For $d = -2$, $MPer(f) = N \setminus \{2\}$.
(4) For $d \geq 2$ or $d \leq -3$, $MPer(f) = N$.

In order to determine $MPer(f)$, we will use Nielsen fixed point theory (see [5] and [6]).

In the case of torus maps the Nielsen number is easily computable. To do it we proceed as follows. Given a map $f : T^m \to T^m$, the Nielsen number of f equals the absolute value of the Lefschetz number: $N(f) = |L(f)|$ by a theorem of Brooks, Brown, Pak and Taylor [3]. Recall that the Lefschetz number is given by

$$L(f) = \sum_{k=0}^{m} (-1)^k \operatorname{Trace}(f_{*k}),$$

where $f_{*k} : H_k(T^m, Z) \to H_k(T^m, Z)$ is the kth order homology endomorphism of f. Let A be the integer matrix corresponding to f_{*1}. Then the linear map $A : R^m \to R^m$ covers a unique algebraic endomorphism $f_A : T^m \to T^m$ with $f_A \simeq f$. One can show

$$L(f) = L(f_A) = \det(I - A) = \prod_{i=1}^{m} |1 - \lambda_i|$$

where λ_i for $i = 1, 2, \ldots, m$ are all the eigenvalues of A (see [4] and [7] for the hyperbolic case). Thus, for each $k \geq 1$

$$N(f^k) = \left|\det(I - A^k)\right|,$$

and if $N(f^k) \neq 0$, the mapping $f_A^k : T^m \to T^m$ has exactly the number of fixed points given by $N(f^k)$.

Let A be an $n \times n$ matrix. We set

$$T_A = \{n \in N : 1 \text{ is not an eigenvalue of } A^n\} = \{n \in N : N(f_A^n) \neq 0\}.$$

The set T_A can be thought of as the largest possibility for the set $MPer(f_A)$. We also set

$$S_A = Per(f_A) \cap T_A,$$

the expected $MPer(f_A)$.

The next result shows how to compute the minimal sets of periods of a torus map. Its proof uses a result from Simó, Swanson and Walker [8].

THEOREM 1.1. *For each integer matrix A there exists f homotopic to f_A such that $Per(f) = MPer(f_A) = S_A$.*

(t,d)	$MPer(f_A)$
$(-2,1)$	$\{1\}$
$(-2,2)$	$N\setminus\{2,3\}$
$(-1,0)$	$\{1\}$
$(-1,1)$	$\{1\}$
$(-1,2)$	$N\setminus\{3\}$
$(0,0)$	$\{1\}$
$(0,1)$	$\{1,2\}$
$(0,2)$	$N\setminus\{4\}$
$(1,1)$	$\{1,2,3\}$

TABLE 1. The set $MPer(f_A)$ in the nine exceptional cases.

Since $N(f)$ depends only on the eigenvalues of A, it turns out that $MPer(f)$ also depends only on the eigenvalues of A, or, equivalently, on the coefficients of the characteristic polynomial. The latter gives a more convenient parametrization to study the minimal sets of periods.

2. The 2-torus

Given a 2×2 matrix A let t be the trace of A and d be the determinant of A. The following result characterizes completely the minimal sets of periods for the 2-torus maps (see also Figure 1).

THEOREM 2.1. *Assume that*

$$(t,d)\notin\{(-2,1),(-2,2),(-1,0),(-1,1),(-1,2),(0,0),(0,1),(0,2),(1,1)\}.$$

Then the following four statements hold.

(1) If $t-d=1$, then $MPer(f)=\emptyset$.
(2) If $t+d=-1$, then $MPer(f)=\{n : n \text{ is odd}\}$.
(3) If $t+d=0$ or $t+d=-2$, then $MPer(f)=N\setminus\{2\}$.
(4) If (t,d) does not lie on one of the four lines in (1)–(3), then $Per(f)=N$.

Otherwise, the minimal sets of periods are given in Table 1.

If we look at Table 1 we can see that periods 3 and 4 play a special role in the characterization of the minimal sets of periods for 2–torus maps. Thus, to prove the above results we must find the points (t,d) where periods 3 and 4 are missing. They are given in Figure 2. We note that these points are distributed around

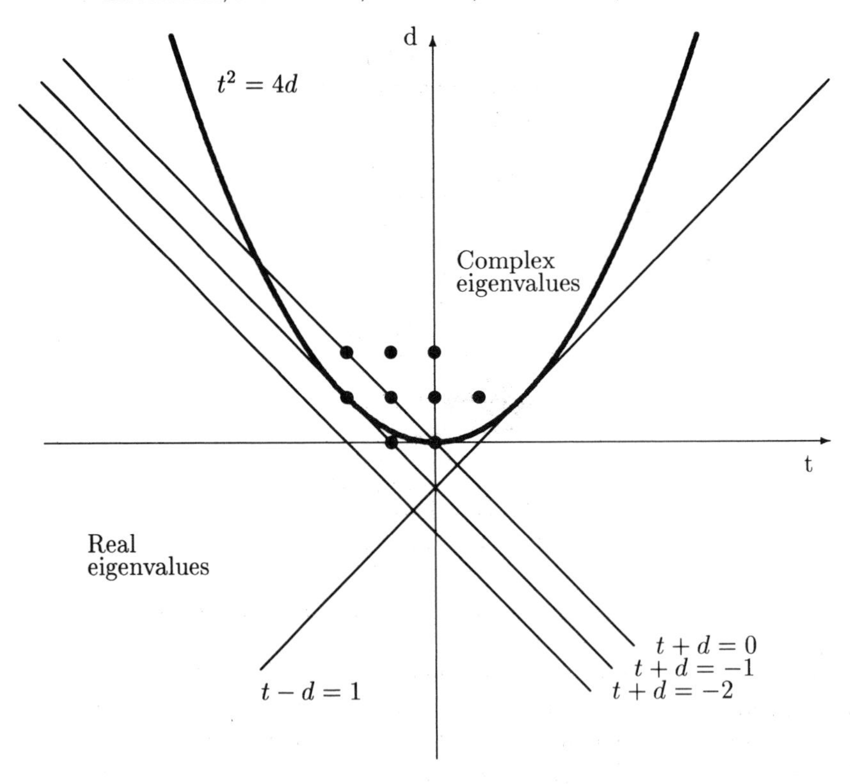

FIGURE 1. The bifurcation diagram in the (t, d)–plane of the minimal sets of periods of 2-torus maps.

the points $(-1, 1)$ and $(0, 1)$ which correspond to the case when the matrix has cube roots of unity and fourth roots of unity respectively.

3. The n-torus

One of the main results of this paper is the following.

THEOREM 3.1. *Let $f : T^m \to T^m$ be a torus map and let A be the integer matrix associated to f_{*1}. Then one of the following occurs:*

(1) All eigenvalues of A are equal to zero or a root of unity and, hence, $MPer(f)$ is finite.

(2) At least one eigenvalue of A is different from zero or a root of unity, and $MPer(f)$ is a cofinite subset of T_A.

We note that (2) of the above theorem is of interest only if 1 is not an eigenvalue of A (which would make $T_A = \emptyset$). It also improves Halpern's result [**4**], which got a similar result with "infinite" instead of "cofinite in T_A".

From the previous theorem it follows easily:

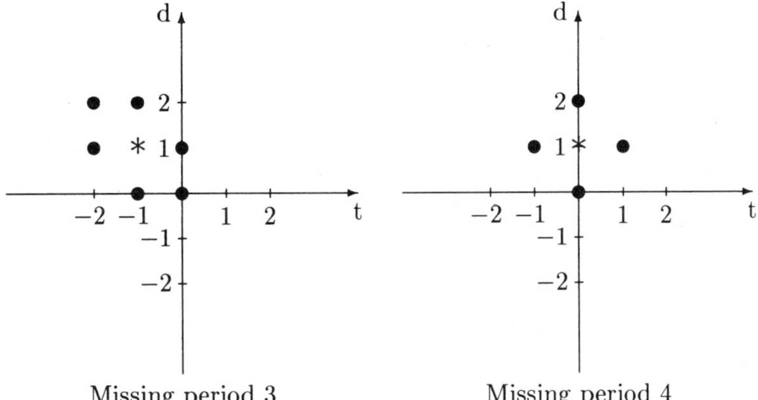

Missing period 3 Missing period 4

FIGURE 2. The points marked by an asterisk give the values where $N(A^n) = 0$ (and $N(A^m) \neq 0$ if and only if n does not divide m) for $n = 3$ or 4, respectively. The other points mark places where $N(A^n) \neq 0$, but f_A has no points of period n for $n = 3$ or 4, respectively.

COROLLARY 3.1. *Let f be an m-dimensional torus map, and let A be the integer matrix of f_{*1}. If A has no eigenvalue which is a root of unity and has some eigenvalue different from zero, then $Per(f)$ is cofinite in N.*

The finite set of missing periods in the previous corollary can be arbitrarily large. In particular, if $m > 1$ is any integer, and A is an $m \times m$ matrix of integers having $p(x) = x^m + 2\,x^{m-1} + 2\,x^{m-2} + \cdots + 2\,x + 2$ as its characteristic polynomial, then $Per(f_A)$ is a cofinite subset of $N \setminus \{k : k > 1 \text{ and } k|(m+1)\}$.

The following results are a useful tool for obtaining the set of periods from the sequence of Nielsen numbers of a torus map.

THEOREM 3.2. *If $f : T^m \to T^m$ is a torus map and $N(f^{n+1})/N(f^n)$ is well-defined and greater than τ for some $\tau > 1$ and all $n \geq n_0$. Then $Per(f)$ contains all integers greater than or equal to the maximum of $2n_0$ and $4\left(1 + \left|\frac{\log(\tau-1)}{\log \tau}\right|\right)$.*

PROPOSITION 3.1. *If A has no eigenvalues having modulus 1, then for all $n \geq k \geq 1$,*

$$\frac{N(f_A^{n+1})}{N(f_A^n)} \geq \prod_{i=1}^{m} \left| \frac{1 - |\lambda_i|^{k+1}}{1 + |\lambda_i|^k} \right|.$$

In particular, (letting $k = 1$),

$$\frac{N(f_A^{n+1})}{N(f_A^n)} \geq \prod_{i=1}^{m} \big|1 - |\lambda_i|\big|,$$

for all n.

Thus, if no eigenvalue of A has modulus 1, the numbers n_0 and τ from the above theorem are easily found, giving an effective algorithm for calculating $MPer(f_A)$.

The following two theorems go in the same direction of the previous one and, in particular, give sufficient conditions to assure that a torus map has periodic points of all periods.

THEOREM 3.3. *Let f be an m-dimensional torus map such that its sequence $\{N(f^n)\}$ of Nielsen numbers is strictly increasing. Then $Per(f) = \mathbb{N}$.*

THEOREM 3.4. *Let f be an m-dimensional torus map and let A be the integer matrix of f_{*1}. Suppose that all the eigenvalues of A are real, positive, different from 1 and* $\det A > 1$. *Then $Per(f) = \mathbb{N}$.*

If we are interested in knowing if a given natural number occurs as a period of a torus map we have the following proposition. In particular, this result helps in determining completely the minimal set of periods of a torus map once we know that it is cofinite in T_A.

PROPOSITION 3.2. *Assume that f is a torus map, and*

$$N(f^n) > \sum_{\frac{n}{k} prime} N(f^k).$$

Then f has a periodic point of period n.

REFERENCES

1. Ll. Alsedà, S. Baldwin, J. Llibre, R. Swanson and W. Szlenk, *Minimal sets of periods for torus maps via Nielsen numbers*, To appear in Pacific J. of Math.
2. Ll. Alsedà, J. Llibre and M. Misiurewicz, *Combinatorial dynamics and entropy in dimension one*, Advanced Series on Nonlinear Dynamics, Vol. **5**, World Scientific, Singapore, 1993.
3. R.B.S. Brooks, R.F. Brown, J. Pak and D.H. Taylor, *Nielsen numbers of maps of tori*, Proc. Amer. Math. Soc. **52** (1975), 398–400.
4. B. Halpern, *Periodic points on tori*, Pacific J. of Math. **83** (1979), 117–133.
5. B. Jiang, *Lectures on Nielsen fixed point theory*, Contemporary Mathematics **14**, Amer. Math. Soc., 1983.
6. T.H. Kiang, *The theory of fixed point classes*, Springer-Verlag, Berlin, 1989.
7. S. Smale, *Differentiable dynamical systems*, Bull. Amer. Math. Soc. **73** (1967), 747–817.
8. C. Simó, R. Swanson and R. Walker, *Torus maps with degenerate Nielsen numbers*, preprint.

Departament de Matemàtiques, Universitat Autònoma de Barcelona, 08193 – Bellaterra, Barcelona, SPAIN
E-mail address: imat0@cc.uab.es

Department of Mathematics, Parker Hall, Auburn University, AL 36849, U.S.A.
E-mail address: sbald@auducvax.bitnet

Departament de Matemàtiques, Universitat Autònoma de Barcelona, 08193 – Bellaterra, Barcelona, SPAIN
E-mail address: imat0@cc.uab.es

Department of Mathematical Sciences, Montana State University, Bozeman, MT 59717, U.S.A.
E-mail address: umsfdswa@waves.oscs.montana.edu

Institute of Applied Mathematics and Mechanics, University of Warsaw, Ul. Banacha 2, 02–097 Warsaw, POLAND
E-mail address: szlenk@plearn.bitnet

Contemporary Mathematics
Volume 152, 1993

Wecken Properties for Manifolds

ROBERT F. BROWN

The beginnings of topology are closely associated with the subject of fixed point theory. Subsequently, topology developed into a major branch of mathematics, connected in a multitude of ways with many of the other branches, while fixed point theory pursued its own, more specialized, ends. In recent years, however, fixed point theory has re-established close relations with topology, especially with the topology of manifolds. This report describes a part of fixed point theory that depends very much on manifold theory. I thank Boju Jiang, Mike Kelly and Helga Schirmer for their help in its preparation.

1. The Wecken Property

Topological fixed point theory is concerned with finite polyhedra and with (continuous) maps from a polyhedron to itself. Given a finite polyhedron X and a map $f : X \to X$, the set of fixed points of f is denoted by

$$Fix(f) = \{x \in X : f(x) = x\}$$

The principal object of study in the subject goes by the unimaginative name of the *minimum number* of the map f. It is denoted by $MF(f)$ and defined to be the minimum number of fixed points among all the maps $g : X \to X$ homotopic to f. That is,

$$MF(f) = \min \#\{Fix(g) : g \sim f\}$$

where $\sim$ means homotopic. Given a map f, the problem is to calculate $MF(f)$. For instance, $MF(f) = 0$ means that there is a fixed point free map g homotopic to f. The calculation of $MF(f)$ has to depend just on a knowledge of f since there is no way to look at all the maps in its homotopy class.

In a 1927 paper on the fixed point theory of maps of compact surfaces [15], Jakob Nielsen proposed a method for computing $MF(f)$ for such maps. We will not reproduce his approach here, but instead describe a technique equivalent to Nielsen's that is simpler to explain. It makes use of the following fundamental result in topological fixed point theory, published by Heinz Hopf in 1929, [3].

1991 *Mathematics Subject Classification.* 55M20.
This paper is in final form and no version of it will be submitted for publication elsewhere.

HOPF CONSTRUCTION THEOREM. *Given X, a finite polyhedron, and a map $f : X \to X$, there exists a map $\hat{f}$ homotopic to f such that $Fix(\hat{f})$ is finite and each fixed point of $\hat{f}$ is in a maximal simplex of X.*

One evident, and important, consequence of Hopf's theorem is that $MF(f)$ is always finite. The theorem tells us that, about each fixed point p of $\hat{f}$, we can find a neighborhood that is homeomorphic to a euclidean space and contains no other fixed points. Consequently, the fixed point index, one of the basic tools of fixed point theory, has to be understood only in this special and very convenient case, as follows. For $p \in Fix(\hat{f})$ we identify such a neighborhood with some euclidean space $\mathbf{R}^n$, with p corresponding to the origin $\mathbf{0}$. The dimension of the euclidean space may vary with the choice of the fixed point p. We denote by S^{n-1} a sphere about p chosen small enough so that $f(S^{n-1}) \subseteq \mathbf{R}^n$. Define the map $I - \hat{f} : S^{n-1} \to \mathbf{R}^n - \mathbf{0}$ by $(I - \hat{f})(x) = x - \hat{f}(x)$. See Figure 1. The *index* of $\hat{f}$ at p, denoted by $i(\hat{f}, p)$ is defined to be the degree of $I - \hat{f}$.

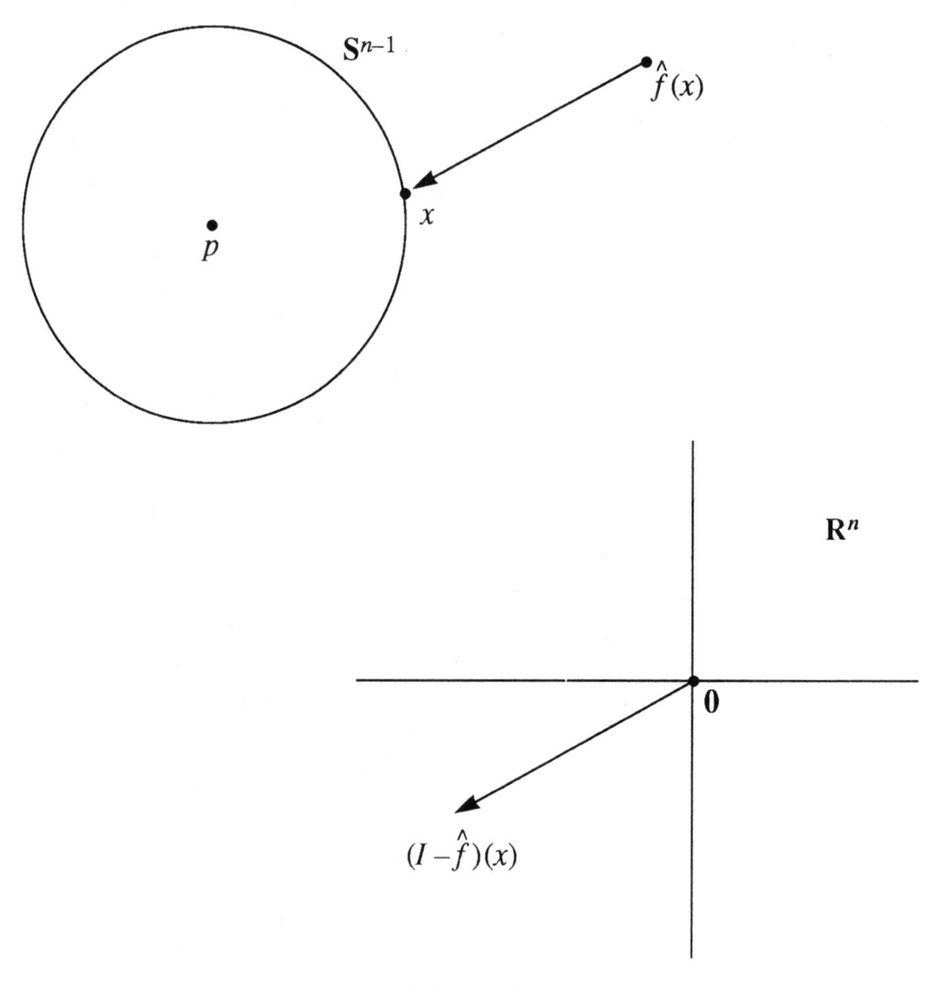

Figure 1

Intuitively, the degree measures how many times $\hat{f}$ wraps around p and whether the wrapping preserves or reverses orientation. To define the fixed point index formally, choose a generator μ of the infinite cyclic group $H_{n-1}(S^{n-1})$ and let $\mu' = i_*(\mu)$ where $i_* : H_{n-1}(S^{n-1}) \rightarrow H_{n-1}(\mathbf{R}^n - 0)$ is the inclusion-induced isomorphism. Then we use the induced homomorphism $(I-\hat{f})_* : H_{n-1}(S^{n-1}) \rightarrow H_{n-1}(\mathbf{R}^n - 0)$ to define the index $i(\hat{f}, p)$ by

$$(I - \hat{f})_*(\mu) = i(\hat{f}, p) \cdot \mu'$$

If $i(\hat{f}, p) = 0$, then $\hat{f}$ can be modified in a neighborhood of p to eliminate this fixed point. That is, there is a map g homotopic to $\hat{f}$ (in fact identical to it except on an arbitrarily small neighborhood of p) such that $Fix(g) = Fix(\hat{f}) - \{p\}$ (see [1]).

Next, define an equivalence relation on $Fix(\hat{f})$ in the following way. Fixed points p and q are *Nielsen equivalent* if there is a path α in X from p to q such that the paths α and $\hat{f}(\alpha)$ are homotopic by a homotopy keeping the end-points fixed (see Figure 2). An equivalence class is called a *fixed point class* of $\hat{f}$ and denoted by $\mathbf{F}$. The *index* $i(\mathbf{F})$ of a fixed point class $\mathbf{F}$ of $\hat{f}$ is the sum of the indices of the fixed points in it, that is,

$$i(\mathbf{F}) = \sum_{p \in \mathbf{F}} i(\hat{f}, p)$$

A fixed point class $\mathbf{F}$ is called *essential* if its index is nonzero. The *Nielsen number* $N(\hat{f})$ is defined to be the number of essential fixed point classes.

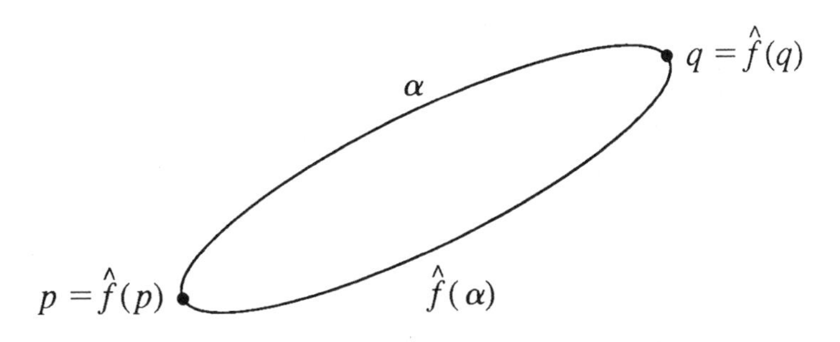

Figure 2

Now we are ready to define Nielsen's basic concept. Given any map $f : X \rightarrow X$ of a finite polyhedron, use Hopf's Theorem to homotope f to a map $\hat{f}$ with a nice fixed point set. The *Nielsen number* of f is just $N(f) = N(\hat{f})$, which can be proved independent of the choice of $\hat{f}$ so it is well-defined. The key property of the Nielsen number is that it is homotopy-invariant. That is, if $g : X \rightarrow X$ is a map homotopic to f, then $N(g) = N(f)$. Thus the map g has at least $N(f)$ essential fixed point classes which certainly means it has at least that many fixed points. Thus the homotopy invariance of the Nielsen number tells us

THEOREM. $N(f) \leq MF(f)$

(There are several detailed expositions of Nielsen theory: [1], [6], [14].)

Nielsen conjectured in his 1927 paper that, for maps of compact surfaces, the Nielsen number equals the minimum number. That would imply that he had solved the problem of calculating $MF(f)$ for maps of surfaces since the Nielsen number $N(f)$ can be calculated, in principle at least, just from information about the map f. Since this conjecture has a long and interesting history, let's state it formally.

NIELSEN CONJECTURE. *If M is a compact 2-manifold, then $MF(f) = N(f)$ for all maps $f : M \to M$.*

Nielsen knew that the conjecture was true for some surfaces: the disc, annulus, two-sphere, torus, Möbius band, projective plane, and Klein bottle. Although he didn't conjecture that the Nielsen number would calculate the minimum number for a map of any finite polyhedron other than a surface, it was natural to ask what polyhedra X would have the property $MF(f) = N(f)$ for all maps $f : X \to X$. The first successful attack on this problem was published in a series of papers in 1941-2 by Franz Wecken [18] and, for that reason, the property carries his name. That is, a finite polyhedron X is said to be *Wecken* if $MF(f) = N(f)$ for all maps $f : X \to X$. In this terminology, the Nielsen Conjecture states that all compact surfaces are Wecken. Wecken didn't prove anything about 2-manifolds, but he demonstrated that many finite polyhedra are Wecken and, in particular, he showed that, for manifolds of higher dimension, Nielsen's technique would calculate the minimum number, that is,

WECKEN'S MANIFOLD THEOREM. *If M is a compact n-manifold, $n \geq 3$, then M is Wecken.*

(Since we are discussing polyhedra in this exposition, we are restricted to triangulable manifolds, although in fact topological n-manifolds are also Wecken if $n \geq 3$.)

Wecken's theorem might be viewed as evidence in favor of the Nielsen Conjecture, but the next paper to be published on the subject claimed to present a counterexample to the Conjecture. In a brief note published in an obscure journal in 1956 [19], Josef Weier presented the following example.

WEIER'S EXAMPLE. *Let M be a disc with four discs removed and let $f : M \to M$ be the map defined by retracting M to a union of four loops: $\alpha \cup \beta \cup \gamma \cup \delta$ joined at a point x_0 (see Figure 3), and then mapping the loops according to the directions:*

$$\alpha \to \delta\alpha\delta^{-1}, \quad \beta \to \beta, \quad \gamma \to \delta\gamma\delta^{-1}, \quad \delta \to 1$$

The point x_0 is fixed in the construction. There is one fixed point x_1 on the loop α, which occurs when the middle third of α is stretched over itself, and it's not difficult to see that $i(f, x_1) = -1$. Similarly, there is a fixed point $x_2 \in \gamma$ with $i(f, x_2) = -1$ also. The loop β can be deformed slightly so that x_0 is the only fixed point on that loop whereas δ contains no other fixed point since f is defined to be constant on δ. Weier calculated that $i(f, x_0) = 0$ so, as we mentioned, f can be modified slightly to make $Fix(f) = \{x_1, x_2\}$. Weier demonstrated that x_1 and x_2 are Nielsen equivalent so f has a single fixed point class, of index -2, and thus $N(f) = 1$. On the other hand, Weier claimed that no map homotopic to f could have fewer than two fixed points, and therefore $MF(f) = 2$. However, he did not explain why he believed this to be true. Weier claimed that the Nielsen Conjecture is false because the surface consisting of the disc with four discs removed is not Wecken but, for a long time, though no one could produce a map homotopic to Weier's that had a single fixed point, neither could anyone verify Weier's claim.

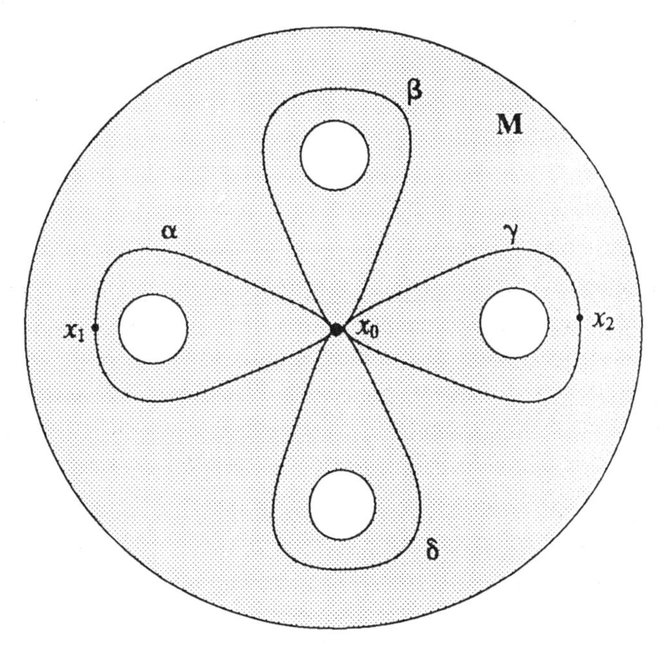

Figure 3

However, positive results concerning the use of the Nielsen number to calculate the minimum number were published. In particular, in 1980 Boju Jiang [4] obtained a refinement of Wecken's work that demonstrated the great effectiveness of Nielsen's technique, as long as the finite polyhedron avoided the one pathology that was known to cause trouble, that of local separating points. A point x of a space X is said to be a *local separating point* if there is a neighborhood U of x in X such that U is connected but $U - \{x\}$ is disconnected (see Figure 4).

WECKEN-JIANG THEOREM. *Suppose X is a finite polyhedron with no local separating points. If X is* not *a 2-manifold, then X is Wecken.*

We emphasize that the only finite polyhedra without local separating points that can possibly *fail* to be Wecken are surfaces, and the Wecken-Jiang Theorem says nothing about them.

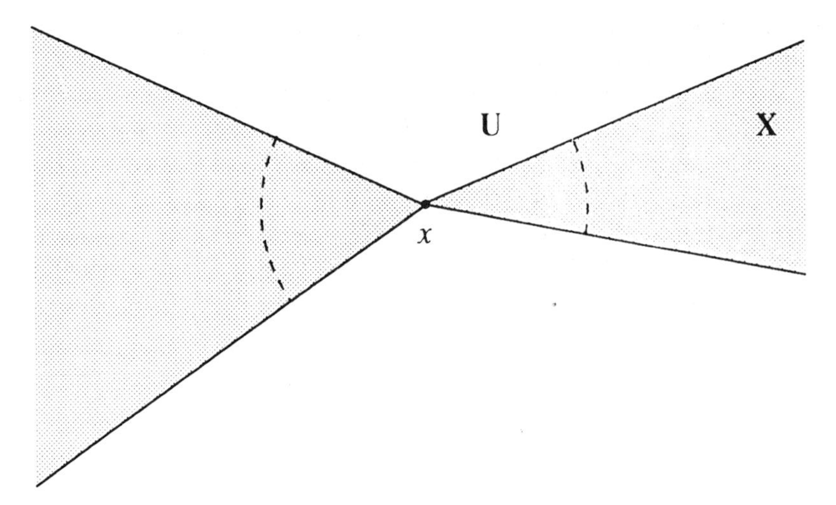

Figure 4

Then, in papers published in 1984 and 1985, Jiang demonstrated that the setting of maps of surfaces, for which Nielsen had defined his number in order to calculate the minimum number $MF(f)$, was the one setting (without local separating points) in which the Nielsen number *did* sometimes fail to do the job. His first example [7] demonstrated that the surface consisting of a disc with two holes, often called the *pants surface* (Figure 5 shows why), is not Wecken.

JIANG'S PANTS EXAMPLE. *Let P be the pants surface and let $f : P \to P$ be the map defined by retracting P to a union of two loops, written $\alpha \cup \beta$, meeting at a point x_0 (see Figure 5), and then mapping the loops according to the directions:*

$$\alpha \to \alpha^{-1}, \quad \beta \to \alpha^{-1}\beta^2$$

Jiang's example is certainly similar to Weier's, though somewhat simpler. In addition to x_0, there is again a fixed point x_1 in α. However, this time the map behaves somethat differently than before and it can be shown that $i(f, x_1) = +1$. On the other hand, near the fixed point $x_2 \in \beta$, the map is like Weier's and $i(f, x_1) = -1$. Jiang showed that $i(f, x_0) = 0$, so this fixed point can be eliminated, and that the two remaining fixed points are Nielsen equivalent. Adding the indices, we see that this time the single fixed point class is not essential and therefore $N(f) = 0$. Jiang proved that no map homotopic to f could be fixed point free and since a single fixed point would be of index zero, and thus removable, every map homotopic to f has at least two fixed points, that is, $MF(f) = 2$. Thus, 57 years after it was stated, the Nielsen Conjecture was proved false: there is a surface that is not Wecken, namely, the pants.

The next year, Jiang published a paper [8] which used a modification of the pants example to demonstrate that all surfaces with negative Euler characteristic are like the pants: they are not Wecken. Since the seven surfaces for which

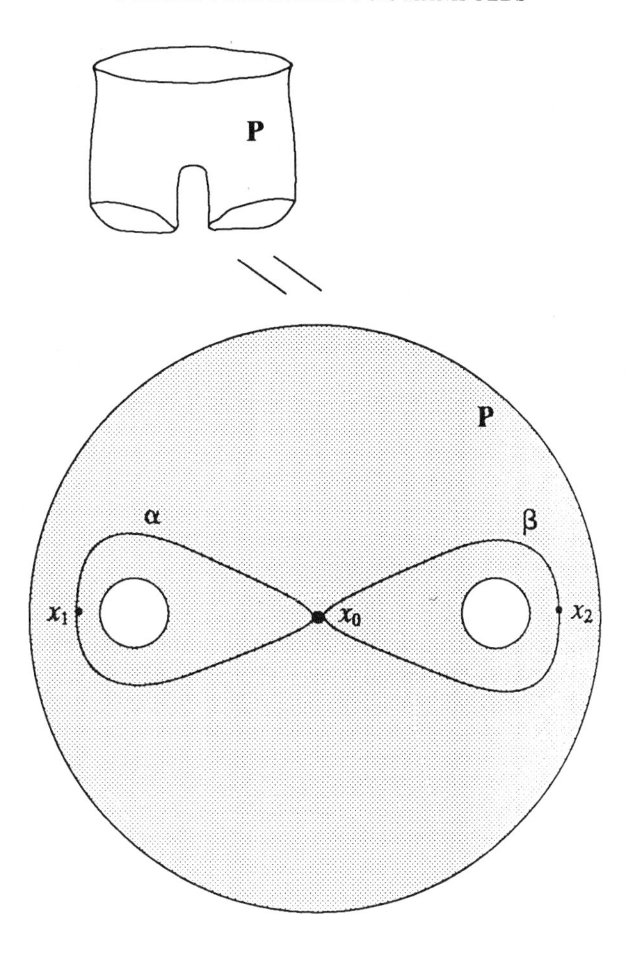

Figure 5

Nielsen knew his conjecture to be true are the only ones that have Euler characteristic non-negative, Jiang was able to characterize the surfaces with the Wecken property in terms of the Euler characteristic. Denote the Euler characteristic of a finite polyhedron X by $\chi(X)$.

JIANG'S 2-MANIFOLD CHARACTERIZATION THEOREM. *A compact 2-manifold M is Wecken if and only if $\chi(M) \geq 0$.*

Once we know that surfaces can fail to be Wecken, we can seek to measure the extent of that failure. If the Nielsen number were always close in value to the minimum number even when they are not equal, this might be a useful approximation result in some circumstances. On the other hand, the Nielsen number might be entirely unsuitable as an approximation. Such would certainly be the case for a finite polyhedron X that was *totally non-Wecken* which means that, given $m \geq 1$, there exists a map $f_m : X \to X$ with the property $MF(f_m) - N(f_m) \geq m$.

In independent papers of 1986 and 1987, respectively, Xingguo Zhang [20] and Michael Kelly [11] proved that the pants surface P is totally non-Wecken. Zhang's construction also applies to the disc with more than two discs deleted.

Kelly's examples are maps $k_m : P \to P$ defined by retracting P to a union of two loops, written $\alpha \cup \beta$, meeting at a point x_0, and then mapping the loops according to the directions:

$$\alpha \to (\beta\alpha\beta^{-1}\alpha^{-1})^m\beta\alpha, \quad \beta \to 1$$

which have the property $N(k_m) = 0$. Kelly uses techniques of geometric topology to show that $MF(k_m) = 2m$. In the case of the pants surface, Zhang's examples $z_m : P \to P$ are also defined by retraction to $\alpha \cup \beta$, but the mapping of the loops is given by

$$\alpha \to 1, \quad \beta \to \alpha^{-1}(\beta\alpha\beta^{-1}\alpha^{-1})^m$$

and $N(z_m) = 1$ for these maps. Zhang uses an algebraic method based on that of [7] and [8] to prove that $MF(z_m) \geq m$.

In [12], Kelly showed that examples such as his and Zhang's can be embedded in any surface of negative Euler characteristic that has a non-empty boundary, without changing either their Nielsen number or their minimum number, which proves that such surfaces are totally non-Wecken. Subsequently, Jiang [9] was able to remove the restriction to non-empty boundary, to obtain

KELLY-JIANG THEOREM. *If M is a compact 2-manifold with $\chi(M) < 0$, then M is totally non-Wecken.*

I can't leave the subject of the Wecken Property without mentioning, as a postscript, a paper of Deyu Tong published in 1989 [17]. Tong proved that Weier was right all along, that is, $MF(f) = 2$ for his example on the disc with four holes, just as he said it was. However, it's quite certain that, whatever reason Weier had, in 1956, for believing he had found a counterexample to the Nielsen Conjecture, the argument could not have been the one Tong gives. A crucial tool for Tong's proof, which is based on techniques of Jiang, is a calculation of braid groups by Joan Birman that was not published until 1968.

2. The Boundary-Wecken Property

For manifolds with nonempty boundary, there is a "relative" Wecken property that arises naturally in the context of relative Nielsen fixed point theory. We begin by sketching this relative theory in the more general setting of finite polyhedra.

Let X be a finite polyhedron, A a subpolyhedron and let $f : (X, A) \to (X, A)$ be a map of pairs, that is, a map from X to itself such that $f(A) \subseteq A$. The minimum number in this category is

$$MF_A(f) = \min \#\{Fix(g) : g \sim_A f\}$$

where $\sim_A$ means homotopic as maps of pairs. As before, the problem is to calculate $MF_A(f)$ just from information about the map of pairs f. For the case of $f : (M, \partial M) \to (M, \partial M)$, where $X = M$ is a compact (triangulated) manifold with nonempty boundary $\partial M = A$, we will denote the minimum number for the category of maps of pairs by $MF_\partial(f)$.

The Nielsen number for maps of pairs was introduced by Helga Schirmer in 1986 [16]. Let $f : (X, A) \to (X, A)$ be a map of pairs and denote the restriction of f to A by $\bar{f} : A \to A$. The Hopf construction can be applied to f as a map of pairs, to homotope it to a map of pairs that has a finite number of fixed points and the fixed points in A lie in maximal simplices of A. Thus the Nielsen number $N(\bar{f})$, can be defined as in Section 1. Schirmer pointed out that if $\mathbf{F}$ is a fixed point class of f as a map from X to itself, then $\mathbf{F} \cap A$ is a union of fixed point classes of $\bar{f}$. See Figure 6 where $p, q, r \in Fix(f)$ are in a single fixed point class of f and the points $p, q \in A$ are also Nielsen equivalent with respect to $\bar{f}$. Letting $N(f, \bar{f})$ denote the number of essential fixed point classes of $f : X \to X$ that contain essential fixed point classes of $\bar{f}$, she defined the *relative Nielsen number* $N(f; X, A)$ by

$$N(f; X, A) = N(\bar{f}) + [N(f) - N(f, \bar{f})]$$

Thus, to compute $N(f; X, A)$ it is necessary to calculate $N(\bar{f})$ and then add to it the number of essential fixed point classes of f that do *not* contain any essential fixed point classes of $\bar{f}$.

For the context we are working in, with X a compact (triangulated) manifold M with nonempty boundary and A that boundary, we will use the more compact notation $N_\partial(f)$ for the relative Nielsen number. We say that M, a compact manifold with nonempty boundary ∂M, is *boundary-Wecken* if $MF_\partial(f) = N_\partial(f)$ for all maps $f : (M, \partial M) \to (M, \partial M)$. In her 1986 paper, Schirmer proved the following relative version of Wecken's Manifold Theorem

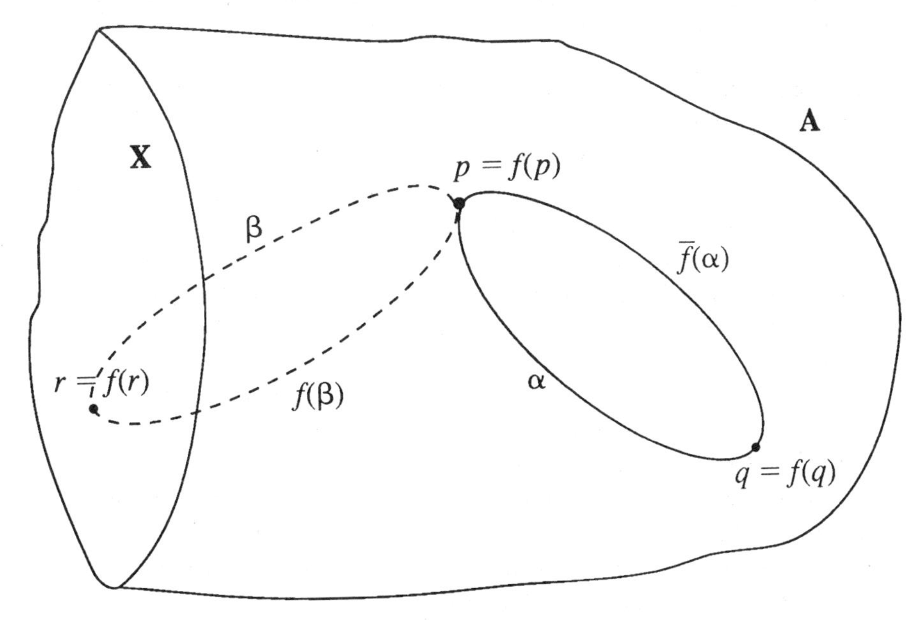

Figure 6

SCHIRMER'S MANIFOLD THEOREM. *Let M be a compact triangulated n-manifold with nonempty boundary ∂M. If $n \geq 4$, then M is boundary-Wecken. Furthermore, if $n = 3$ and $\chi(C) \geq 0$ for every component C of ∂M, then M is boundary-Wecken.*

On the other hand, if a 3-manifold M does have a boundary component C with negative Euler characteristic, its fixed point behaviour with respect to boundary-preserving maps $f : (M, \partial M) \to (M, \partial M)$ is entirely different, as Jiang showed in [9]. Call a manifold M with nonempty boundary *totally non-boundary-Wecken* if, given $m \geq 1$, there exits a map of pairs $f_m : (M, \partial M) \to (M, \partial M)$ such that $MF_\partial(f_m) - N_\partial(f_m) \geq m$.

JIANG'S 3-MANIFOLD THEOREM. *If M is a compact, orientable, triangulable 3-manifold such that $\chi(C) < 0$ for some component C of ∂M, then M is totally non-boundary-Wecken.*

It is not known whether non-orientable 3-manifolds exhibit the same behavior.

In contrast to the 3-manifold theory, the boundary-Wecken property for surfaces quite closely resembles the theory of the Wecken property. Recall that all surfaces of negative Euler characteristic are totally non-Wecken. A paper [2] of Brian Sanderson and myself, along with complementary results of Kelly [12], demonstrate that "most" surfaces of negative Euler characteristic that have nonempty boundary are also totally non-boundary-Wecken, as follows.

THEOREM. *If*

$$M = K - \{\textit{union of disjoint open discs}\}$$

where K is a closed surface that is not *the two-sphere S^2 or the real projective plane RP^2, then M is totally non-boundary-Wecken.*

The only surfaces with nonempty boundary that have non-negative Euler characteristic are the disc, annulus and Möbius band. We prove in [2] that they are boundary-Wecken as well as Wecken.

However, the Wecken and boundary-Wecken theories are not identical for surfaces; the difference shows up in the pants surface P which, we recall, is totally non-Wecken. Brian Sanderson and I prove in [2] that, for this surface, the relative Nielsen number $N_\partial(f)$ does approximate $MF_\partial(f)$; in fact $MF_\partial(f) - N_\partial(f) \leq 1$, and thus P is *almost boundary-Wecken*, that is, $MF_\partial(f) - N_\partial(f)$ is bounded for all $f : (P, \partial P) \to (P, \partial P)$.

Our analysis of maps $f : (P, \partial P) \to (P, \partial P)$ depends on their restriction $\bar{f} : \partial P \to \partial P$. The boundary ∂P consists of three simple closed curves and we can think of $\bar{f}$ as three maps, each from a simple closed curve to a simple closed curve. Orienting the boundary components by means of an orientation of P, each map of a boundary component has a well-defined degree. The image $\bar{f}(\partial P)$ can have one, two or three components; let $Im_\partial(f)$ be the number of boundary components in that image. In most cases, a map $f : (P, \partial P) \to (P, \partial P)$ has the property $MF_\partial(f) = N_\partial(f)$, as the following more precise statement explains.

THEOREM. *Let $f : (P, \partial P) \to (P, \partial P)$ be a map. If $Im_\partial(f) \geq 2$ then $MF_\partial(f) = N_\partial(f)$. If $Im_\partial(f) = 1$, let f^* denote the restriction of f to the component of ∂P that contains $f(\partial P)$. If the degree of f^* is nonzero, or if $\bar{f}$ is of degree zero on all components, then it is still true that $MF_\partial(f) = N_\partial(f)$. Otherwise, that is, if f^* is of degree zero but $\bar{f}$ is not of degree zero on the other components, then $MF_\partial(f) - N_\partial(f) \leq 1$.*

Kelly [12] has constructed a map $f : (P, \partial P) \to (P, \partial P)$ for which $N_\partial(f) = 1$ but $MF_\partial(f) = 2$. The map is the constant map on one boundary component and maps each of the other boundary components homeomorphically on to that one. In Wecken theory a surface is either Wecken or totally non-Wecken. The pants surface shows us that boundary-Wecken theory can behave quite differently:

THEOREM. *The pants surface P is almost boundary-Wecken but not boundary-Wecken.*

3. Homeomorphisms of Manifolds

Let M be a compact n-manifold, which may have empty boundary or not. Now we are not concerned with selfmaps of M but rather with homeomorphisms $f : M \to M$. There is a corresponding minimum number, as follows:

$$MF^h(f) = \min \#\{Fix(g) : g \simeq f\}$$

where $\simeq$ means isotopic. Since, if the boundary ∂M of M is non-empty, a homeomorphism must map the boundary to itself, we have the relative Nielsen number $N_\partial(f)$ of a homeomorphism $f : (M, \partial M) \to (M, \partial M)$, as in the previous section. The Wecken property for homeomorphisms states that the corresponding minimum number can be calculated by means of the appropriate relative Nielsen number, that is, M is *Wecken for homeomorphisms* if $MF^h(f) = N_\partial(f)$ for all homeomorphisms $f : M \to M$.

In 1981, Jiang [5] announced that the Nielsen-Thurston classification of homeomorphisms of surfaces could be used to prove that all compact 2-manifolds with empty boundary are Wecken for homeomorphisms (which, in this case, means $MF^h(f)$ equals the classical Nielsen number $N(f)$). Later, Jiang and Guo [10] proved that all surfaces have this property:

SURFACE HOMEOMORPHISM THEOREM. *Let M be a compact 2-manifold. Then M is Wecken for homeomorphsisms.*

Kelly [13] has obtained a result like Jiang's original statement, if the dimension of the manifold is high enough.

THEOREM. *If M is a compact n-manifold with empty boundary, $n \geq 5$, then M is Wecken for homeomorphisms.*

For M an n-manifold with non-empty boundary, $n \geq 5$, Kelly constructs a homeomorphisms of M into itself (that is, an embedding) that is isotopic to a given homeomorphisms f such that g has exactly $N(f)$ fixed points, but g maps the boundary of M into its interior, rather than back to the boundary as in the Jiang-Guo result. Thus the status of the Wecken for homeomorphisms property is unknown in this case. Nothing at all in known about the Wecken for homeomorphsims property for n-manifolds when $n = 3$ or 4.

4. Some Unsolved Problems

We can see from the table of Figure 7, which summarized the results described in this paper, that the Wecken properties of compact n-manifolds are quite well-understood. For $n \geq 4$, the Wecken properties for maps have been determined, although much less is known for isotopy classes of homeomorphisms, as we explained at the end of the previous section.

The remaining questions for manifolds of dimension $n = 2$ concern the boundary-Wecken property. Writing a surface M with nonempty boundary as

$$M = K - \{\text{union of disjoint open discs}\}$$

where K is a closed surface, we know what happens if K is not the two-sphere or projective plane: the surface is totally non-boundary-Wecken. Deleting one, two or three discs from the two-sphere produces the disc and annulus, which are boundary-Wecken, and the pants surface which is almost boundary-Wecken but not boundary-Wecken. Deleting one disc from the projective plane we have the Möbius band, a boundary-Wecken surface. That leaves the sphere with four or more discs removed and the projective plane with two or more discs removed.

	Dimension of the Manifold M			
Property	2	3	4	5 or more
Wecken	$\chi(M) \geq 0$	Yes	Yes	Yes
totally non-Wecken	$\chi(M) < 0$	No	No	No
boundary-Wecken	$\chi(M) \geq 0$	$\chi(C) \geq 0$	Yes	Yes
totally non-boundary-Wecken	$M = K - d\text{'s}$; $K \neq S^2, RP^2$	M orientable and some $\chi(C) < 0$	No	No
almost boundary-Wecken	pants	—	No	No
Wecken for homeomorphisms	Yes	—	—	$\partial M = \emptyset$

Figure 7

For each such surface we can ask: is it boundary-Wecken, totally non-boundary-Wecken or almost boundary-Wecken? If it is almost boundary-Wecken, what is the bound on $MF_{\partial}(f) - N_{\partial}(f)$ (like the bound of one for the pants surface)?

For manifolds of dimension $n = 3$, Wecken's classical result states that they are all Wecken, and now the boundary-Wecken properties have been characterized, when the manifold is orientable. The boundary-Wecken property, in the non-orientable case, and the Wecken for homeomorphisms property offer good opportunities for investigation, perhaps by taking advantage of the extensive literature of the geometric topology of 3-manifolds.

REFERENCES

1. R. Brown, *The Lefschetz Fixed Point Theorem*, Scott-Foresman, 1971.
2. R. Brown and B. Sanderson, *Fixed points of boundary-preserving maps of surfaces*, Pacific J. Math (to appear).
3. H. Hopf, *Über die algebraische Anzahl von Fixpunkten*, Math. Z. **29** (1929), 493–524.
4. B. Jiang, *On the least number of fixed points*, Amer. J. Math. **102** (1980), 749–763.
5. ———, *Fixed points of surface homeomorphisms*, Bull. Amer. Math. Soc. **5** (1981), 176–178.
6. ———, *Lectures on Nielsen Fixed Point Theory*, Contemp. Math. **14** (1983).
7. ———, *Fixed points and braids*, Invent. Math. **75** (1984), 69–74.
8. ———, *Fixed points and braids. II*, Math. Ann. **272** (1985), 249–256.
9. ———, *Commutativity and Wecken properties for fixed points of surfaces and 3-manifolds*, preprint.
10. B. Jiang and J. Guo, *Fixed points of surface diffeomorphisms*, Pacific J. Math. (to appear).
11. M. Kelly, *Minimizing the number of fixed points for self-maps of compact surfaces*, Pacific J. Math. **126** (1987), 81–123.
12. ———, *The relative Nielsen number and boundary-preserving surface maps*, preprint.
13. ———, *The Nielsen number as an isotopy invariant*, preprint.
14. T. Kiang, *The Theory of Fixed Point Classes*, Springer, 1989.
15. J. Nielsen, *Untersuchungen zur Topologie der geschlossenen zweiseitigen Flächen*, Acta Math. **50** (1927), 189–358.
16. H. Schirmer, *A relative Nielsen number*, Pacific J. Math. **122** (1986), 459–473.
17. D. Tong, *On an example of Weier*, Topological Fixed Point Theory and Applications; Springer Lecture Notes **1411** (1989), 171–181.
18. F. Wecken, *Fixpunktklassen. I*, Math. Ann. **117** (1941), 659–671; *II* **118** (1942), 216–234; *III* **118** (1942), 544–577.
19. J. Weier, *Über Probleme aus der Topologie der Ebene und der Fläche*, Math. Japon. **4** (1956), 101–105.
20. X. Zhang, *The least number of fixed points can be arbitrarily larger than the Nielsen number*, Acta Sci. Nat. Univ. Pekin. **1986**, 15–25.

UNIVERSITY OF CALIFORNIA, LOS ANGELES

E-mail address: rfb@math.ucla.edu

Contemporary Mathematics
Volume **152**, 1993

Lefschetz zeta functions and forced set of periods

JOSEFINA CASASAYAS, JAUME LLIBRE, AND ANA NUNES

ABSTRACT. We study C^1 self maps of compact manifolds whose periodic points are transversal and obtain information on the set of periods from the associated Lefschetz zeta function.

1. Introduction and basic definitions

Given a continuous self-map of a compact manifold M of dimension n, its *Lefschetz number* is defined as

$$L(f) = \sum_{k=0}^{n}(-1)^k \text{tr } (f_{*k}),$$

where $f_{*k} : H_k(M;\mathbb{Q}) \to H_k(M;\mathbb{Q})$ is the endomorphism induced by f on the k-th rational homology group of M.

One of the most useful results to prove the existence of fixed points or, more generally, periodic points for continuous maps $f : M \to M$ in a given homotopy class is Lefschetz fixed point theorem, which says that if $L(f) \neq 0$ then f has a fixed point.

When studying the periodic points of f, i.e., the set

$$\text{Per}(f) = \{m \in \mathbb{N} : f \text{ has a periodic orbit of least period } m\},$$

it is convenient to consider the *Lefschetz zeta function* of f,

$$Z_f(t) = \exp\left(\sum_{m=1}^{\infty} \frac{L(f^m)}{m} t^m\right),$$

1991 *Mathematics Subject Classification.* Primary 54H25, 58F20.

The first two authors were supported in part by DGICYT Grant PB90–0695.

The final version of this paper will be submitted for publication elsewhere.

which is a generating function for the Lefschetz numbers of all iterates of f.

The function $Z_f(t)$ can be computed (see [**4**]) from the homological endomorphisms f_{*k} through

$$Z_f(t) = \prod_{k=0}^{n} \det\left(I_{j_k} - tf_{*k}\right)^{(-1)^{k+1}}, \tag{1}$$

where $j_k = \dim_{\mathbb{Q}} H_k(M;\mathbb{Q})$, I_{j_k} is the $j_k \times j_k$ identity matrix and we take $\det\left(I_{j_k} - tf_{*k}\right) = 1$ when $j_k = 0$.

In this paper we shall restrict ourselves to maps of class C^1 which verify some additional assumptions. Denote by PO the set of periodic orbits of f and, given $\gamma \in PO$, let $p(\gamma)$ be the least period of γ. A C^1 map $f : M \to M$ of a compact manifold is called *transversal* if $f(M) \subset \operatorname{Int}(M)$ and for every $\gamma \in PO$, $x \in \gamma$ and $m \in \mathbb{N}$

$$\det\left(Df^{mp(\gamma)}(x) - I\right) \neq 0.$$

The Lefschetz numbers $L(f^m)$ and the Lefschetz zeta function $Z_f(t)$ associated to a transversal map f are related in a simple way with the periodic structure of f. Denote by $\operatorname{Fix}(f^m)$ the set of fixed points of f^m. Given $x \in \operatorname{Fix}(f^m)$, let $u_+(x)$ (respectively $u_-(x)$) be the number of real eigenvalues of $Df^m(x)$ which are strictly greater than one (respectively strictly smaller than one). Then, if f is transversal,

$$L(f^m) = \sum_{x \in \operatorname{Fix}(f^m)} (-1)^{u_+(x)}, \forall m \in \mathbb{N},$$

and the following proposition holds.

PROPOSITION 1 (FRANKS, [**5**]). *Let $f : M \to M$ be a transversal map. Then*

$$Z_f(t) = \prod_{\gamma \in PO} \left(1 - (-1)^{u_-(\gamma)} t^{p(\gamma)}\right)^{(-1)^{u(\gamma)}}, \tag{2}$$

where $u_+(\gamma) = u_+(x)$ and $u_-(\gamma) = u_-(x)$ for $x \in \gamma$, and $u(\gamma) = u_+(\gamma) + u_-(\gamma) + 1$.

Our results are based on the explicit relation between $Z_f(t)$ and the periodic orbits of f given by Proposition 1. Actually, they are based on Theorems 2 and 7 below, that we shall state without proof, and which, using (2), give a set of equalities envolving the numbers of orbits of each period. In these theorems, following the notation introduced by Matsuoka in [**9**], we shall distinguish four different types of periodic orbits, according to the parity of u_+ and u_-:

$$EE = \{\gamma \in PO : u_+(\gamma) \text{ and } u_-(\gamma) \text{ are even}\},$$
$$EO = \{\gamma \in PO : u_+(\gamma) \text{ is even and } u_-(\gamma) \text{ is odd}\},$$
$$OE = \{\gamma \in PO : u_+(\gamma) \text{ is odd and } u_-(\gamma) \text{ is even}\},$$
$$OO = \{\gamma \in PO : u_+(\gamma) \text{ and } u_-(\gamma) \text{ are odd}\}.$$

We shall also need the following definitions. Denote by $\mu : \mathbb{N} \to \{-1,0,1\}$ the *Möbius function* given by $\mu(1) = 1$ and by the following rule: if $n = p_1^{k_1} ... p_j^{k_j}$ is the prime decomposition of n, $\mu(n) = 0$ if $k_i > 1$ for some $i \in \{1, ..., j\}$, and $\mu(n) = (-1)^j$ otherwise. Given a non-negative integer d, denote by $\{\alpha_n(d)\}_{n \in \mathbb{N}}$ the sequence defined by

$$\alpha_n(d) = \frac{1}{n} \sum_{k|n} \mu\left(\frac{n}{k}\right) d^k.$$

For each positive odd r and each non-negative m, let

$$\beta_{2^m r}(d) = \sum_{k=0}^{m} \alpha_{2^k r}(d).$$

Our main purpose will be, given a transversal map f and its associated Lefschetz zeta function, to obtain what we shall call *forced periods* , i.e. we define the forced set of periods of the transversal map $f : M \to M$ to be the set

$$\mathrm{FPer}(f) = \cap_g \mathrm{Per}(g),$$

where g ranges over all transversal maps $g : M \to M$ homotopic to f.

2. Transversal maps of sphere type

Given a transversal map $f : M \to M$, we shall say that it is of *sphere type* if the associated Lefschetz zeta function is of the form

$$Z_f(t) = (1-t)^{-1}(1-\sigma_1 dt)^{\sigma_2} = Z_f(t; \sigma_1, \sigma_2, d),$$

where d is a non-negative integer and $\sigma_i \in \{-1, 1\}$, $i = 1, 2$. The reason for this notation is that in the case of the sphere S^n, the Lefschetz zeta function associated to any continuous map $f : S^n \to S^n$ is of sphere type with $\sigma_1 = D/|D|$, $\sigma_2 = (-1)^{n+1}$ and $d = |D|$, where D denotes the degree of f. This follows from (1), $H_0(S^n; \mathbb{Q}) \approx \mathbb{Q}$ and $f_{*0} = \mathrm{id}$, $H_k(S^n; \mathbb{Q}) \approx 0$ for $k = 1, ..., n-1$, $H_n(S^n; \mathbb{Q}) \approx \mathbb{Q}$ and $f_{*n}(1) = D$.

For each odd $r \geq 1$ and $n \geq 0$ let $PO(2^n r)$ be the set $\{\gamma \in PO : p(\gamma) = 2^n r\}$. Denote by $EE_{2^n r}$ (respectively $EO_{2^n r}$,$OE_{2^n r}$,$OO_{2^n r}$) the cardinal of $EE \cap PO(2^n r)$ (respectively $EO \cap PO(2^n r)$, $OE \cap PO(2^n r)$, $OO \cap PO(2^n r)$), and by δ_{ij} the Kronecker symbol. The results of this section are based on the following theorem, which provides a set of equalities relating not only the number of periodic orbits of each period but also their type with the numbers σ_1, σ_2 and d.

THEOREM 2 ([**1**]). *Let $f : M \to M$ be a transversal map of sphere type and $Z_f(t; \sigma_1, \sigma_2, d)$ its Lefschetz zeta function. For each odd $r \geq 1$, the following equalities hold:*

(3.a) $$EE_r + EO_r + \sigma_1\sigma_2\beta_r(d) = OO_r + OE_r + \delta_{1r},$$

$$
\begin{aligned}
&\sum_{k=0}^{m} EE_{2^k r} + EO_{2^m r} + \sigma_2\big(\beta_{2^m r}(d) + \frac{\sigma_1 - 1}{2}\beta_r(d)\big) = \\
(3.b) \qquad & OO_{2^m r} + \sum_{k=0}^{m} OE_{2^k r} + \delta_{1r}, \qquad \forall m \geq 1.
\end{aligned}
$$

In the case when $d = 1$, the above equalities may be obtained from Theorem 3 of [**9**]. In the case when $d > 1$ it is easy to obtain, using Theorem 2, the following result on the set of periods.

THEOREM 3. *Let $f : M \to M$ be a transversal map of sphere type with Lefschetz zeta function $Z_f(t; \sigma_1, \sigma_2, d)$. If $d > 1$, then $\mathbb{N} \setminus S \subset Per(f)$, where $S = \{n \in \mathbb{N} : n = 2r, r \text{ odd}\}$.*

PROOF. From the definition of the sequence $\{\beta_n(d)\}_{n \in \mathbb{N}}$ when $d > 1$ we have $\beta_1(d) = d$ and $\beta_r(d) > 1$ for every odd $r > 1$. Then, it follows immediately from equation (3.a) that $\mathrm{Per}(f)$ contains every odd number r.

Denote by g the $2^k r$-th iterate of f for $k \geq 1$. We claim that $2 \in \mathrm{Per}(g)$ and so the theorem follows because $2^{k+1} r \in \mathrm{Per}(f)$ for every $k \geq 1$.

To prove the claim, assume that $2 \notin \mathrm{Per}(g)$ and consider equations (3.a) and (3.b) for g. Subtracting equation (3.a) for $r = 1$ from equation (3.b) for $r = m = 1$ we obtain

$$
(4) \qquad EO_1 - \sigma_2 \left(\beta_2(d) - \left(\frac{\sigma_1 + 1}{2} \right) d \right) = OO_1.
$$

On the other hand, since g is an even iterate of f, both EO_1 and OO_1 are zero, so that (4) implies $\alpha_2(d) = (\sigma_1 - 1)d/2 \leq 0$, which is a contradiction. $\square$

We shall now consider some particular cases of self-maps of manifolds in which, using Theorem 3 together with additional properties of these maps, we obtain $\mathbb{N}$ as the forced set of periods.

THEOREM 4. *Let $f : S^2 \to S^2$ be a transversal map of the 2-sphere of degree D such that $\det\big(Df^{p(\gamma)}(x)\big) \geq 0$ for every periodic orbit γ and every $x \in \gamma$. If $D \geq 2$ then $Per(f) = \mathbb{N}$.*

PROOF. Using Theorem 3, it is enough to show that $2 \in \mathrm{Per}(f^r)$ for every odd r. As in the preceeding proof, we shall assume that $2 \notin \mathrm{Per}(f^r)$ and reach a contradiction. From equations (3) we obtain again equation (4) for the map $g = f^r$. Since $\sigma_2 = (-1)^{2+1} = -1$, (4) implies that g must have fixed points of type OO. But in dimension two det $Dg(x) < 0$ at every fixed point of type OO. $\square$

The map $A : S^n \to S^n$ which sends the point whose coordinates in $\mathbb{R}^{n+1}$ are $(x_1, ...x_{n+1})$ to the point $(-x_1, ..., -x_{n+1})$ is called the *antipodal map*. We shall say that $f : S^n \to S^n$ is *antipodal preserving* if it commutes with the antipodal map. Every antipodal preserving map $f : S^n \to S^n$ is the lift of the map $\tilde{f} : \mathbb{R}P^n \to \mathbb{R}P^n$ defined trough $p \circ f = \tilde{f} \circ p$, where p is the natural projection from S^n to $\mathbb{R}P^n = S^n/\sim$ with the equivalence relation $x \sim y$ if and only if $x = -y$.

THEOREM 5. *Let $f : S^n \to S^n$ be an antipodal preserving map of the n-sphere with n even and denote by D the degree of f.*

(a) *$S \subset Per(f)$, where $S = \{m \in \mathbb{N} : m = 2r, r\, odd\}$.*
(b) *If $|D| > 1$ then $\mathbb{N} = Per(f)$.*

PROOF. Statement (b) follows directly from Theorem 3 and statement (a).

To prove satetement (a) it is enough to show that $2 \in \text{Per}(f)$ because an odd iterate of an antipodal preserving map is also antipodal preserving .

Let $\tilde{f} : \mathbb{R}P^n \to \mathbb{R}P^n$ be the map induced by f. Writing down equations (3.a) with $r = 1$ for the maps f and $\tilde{f}$ we obtain

$$
\begin{aligned}
EE_1(f) + EO_1(f) &= OO_1(f) + OE_1(f) + 1 + D, \\
EE_1(\tilde{f}) + EO_1(\tilde{f}) &= OO_1(\tilde{f}) + OE_1(\tilde{f}) + 1,
\end{aligned}
\tag{5}
$$

where we have used that $H_n(\mathbb{R}P^n; \mathbb{Q}) \approx \{0\}$ for n even.

Suppose then that $2 \notin \text{Per}(f)$, and so $EE_1(f) = EE_1(\tilde{f})$, $EO_1(f) = EO_1(\tilde{f})$, $OE_1(f) = OE_1(\tilde{f})$ and $OO_1(f) = OO_1(\tilde{f})$. Subtracting equations (5) we obtain $D = 0$. On the other hand, the first of equations (5) may be written

$$EE_1(f) + EO_1(f) + OO_1(f) + OE_1(f) = 2\,(OO_1(f) + OE_1(f)) + 1 + D.$$

The number on the first term must be even because it is the total number of fixed points of an antipodal preserving map. Therefore, D must be odd, which contradicts $D = 0$. $\square$

Note that the preceeding theorem ensures that f has infinitely many periods even in the case when its degree D is ± 1. For arbitrary maps of sphere type, in the case when d is 0 or 1, the results which may be obtained from the equalities of Theorem 2 depend on the triple (σ_1, σ_2, d) and the forced set of periods is, as one should expect, much smaller. Some of the results of this type obtained in [**1**] are contained in the following proposition.

PROPOSITION 6. *Let $f : M \to M$ be a transversal map of sphere type with Lefschetz zeta function $Z_f(t; \sigma_1, \sigma_2, d)$. The following hold.*

(a) *If $\sigma_1\sigma_2 d \neq 1$ then $1 \in Per(f)$.*
(b) *If $\{1, 2\} \cap Per(f) \neq \emptyset$ then $\sigma_1 = \sigma_2 = d = 1$.*
(c) *If $(\sigma_1, \sigma_2, d) \notin \{(-1, 1, 2), (1, 1, 1)\}$ then either $\{n \in \mathbb{N} : n = 2^j, j \geq 1\} \subset Per(f)$ or for some $m \geq 0$ we have $EE_{2^m} + OE_{2^m} > 0$ and $\{n \in \mathbb{N} : n = 2^j, 1 \leq j\ m\} \subset Per(f)$ when $m \geq 1$.*

PROOF. Statement (a) is a direct consequence of equation (3.a) for $r = 1$.

In the hypothesis of statement (b), equations (3.a) for $r = 1$ and (3.b) for $r = m = 1$ become respectively

$$\sigma_1\sigma_2 d = 1, \qquad \sigma_2\left(\alpha_2(d) + \frac{\sigma_1+1}{2}d\right) = 1.$$

The first equality implies $d = 1$ and then from the second we obtain $\sigma_1 = \sigma_2 = 1$.

To prove statement (c) suppose first that

$$EE_{2^m} + OE_{2^m} = 0, \forall m \geq 0. \tag{6}$$

Then equations (3.b) for $r = 1$ become

$$EO_{2^m} - OO_{2^m}1 - \sigma_2\left(\beta_{2^m}(d) + \frac{\sigma_1-1}{2}d\right), \forall m \geq 1,$$

and it is easy to check that, under the hypothesis, the term on the right is always non-zero. Hence, if (6) holds, $\{n \in \mathbb{N} : n = 2^j, j \geq 1\} \subset \mathrm{Per}(f)$. Suppose that (6) does not hold and let m be the samllest non-negative integer such that $EE_{2^m} + OE_{2^m} > 0$. If $m \geq 1$ we may repeat the above argument and conclude that $EO_{2^k} - OO_{2^k} \neq 0$ for all $1 \leq k \leq m$. In particular, $\{n \in \mathbb{N} : n = 2^j, 1 \leq j \leq m\} \subset \mathrm{Per}(f)$. □

3. Lefschetz zeta functions whose zeros and poles are roots of unity

In this section we shall consider transversal maps $f : M \to M$ such that all the zeros and poles of $Z_f(t)$ are roots of unity. Lefschetz zeta functions of this form occur for some interesting general classes of maps. For instance, using a result of Fried (Theorem 6 of [**6**]), it is easy to see that this is the case whenever $\mathrm{Per}(f)$ is finite. The same happens when f is a map of a connected surface and the topological entropy $h(f)$ is zero (see [**2**], [**3**], [**7**] and [**8**]).

It is easy to check that if all the zeros and poles of $Z_f(t)$ are roots of unity then $Z_f(t)$ has a factorization of the form

$$Z_f(t) = \prod_{p\in P}\prod_{i=1}^{N_p}\left(1 - \sigma_1(i,p)t^p\right)^{\sigma_2(i,p)}, \tag{7}$$

where P is a finite subset of $\mathbb{N}$, the N_p are natural numbers and $\sigma_1(i,p), \sigma_2(i,p) \in \{-1, 1\}$.

For functions $Z_f(t)$ of the form (7), the following result plays a similar role to that of Theorem 2 relating the numbers of periodic orbits of each period and type with $\sigma_1(i,p)$ and $\sigma_2(i,p)$.

THEOREM 7. *Let $f : M \to M$ be a transversal map with Lefschetz zeta function of the form (7). For each odd $r \geq 1$ and $m = 0, 1, \ldots$, the following equalities hold:*

$$\sum_{k=0}^{m} EE_{2^k r} + EO_{2^m r} + \gamma(r,m) = OO_{2^m r} + \sum_{k=0}^{m} OE_{2^k r}, \tag{8}$$

where

$$\gamma(r,m) = \sum_{\substack{q|r \\ 2^m q \in P}} \sum_{i=1}^{N_{2^m q}} \sigma_2(i, 2^m q)\sigma_1(i, 2^m q)\beta_{r/q} + \sum_{j=0}^{m-1} \sum_{\substack{q|r \\ 2^j q \in P}} \sum_{i=1}^{N_{2^j q}} \sigma_2(i, 2^j q)\left(\beta_{2^{m-j} r/q} + \frac{\sigma_2(i, 2^j q) - 1}{2}\beta_{r/q}\right), \tag{9}$$

and we take the second summand equal zero when $m = 0$.

Consider a product of the form (7). A factor $\left(1 - \sigma_1(i,p)t^p\right)^{\sigma_2(i,p)}$ will be called *irreducible* if it remains after performing in correlative order the following reductions:

$$(1 + t^p)(1 + t^p)^{-1} = 1, \tag{R1}$$

$$(1 - t^p)(1 - t^p)^{-1} = 1, \tag{R2}$$

$$(1 + t^p)(1 - t^p) = 1 - t^{2p}, \tag{R3}$$

$$(1 + t^p)^{-1}(1 - t^p)^{-1} = (1 - t^{2p})^{-1}, \tag{R4}$$

$$(1 + t^p)^{-1}(1 - t^{2p}) = 1 - t^p, \tag{R5}$$

$$(1 + t^p)(1 - t^{2p})^{-1} = (1 - t^p)^{-1}, \tag{R6}$$

$$(1 + t^p)(1 + t^{2p}) = (1 - t^p)^{-1}(1 - t^{4p}), \tag{R7}$$

$$(1 + t^p)^{-1}(1 + t^{2p})^{-1} = (1 - t^p)(1 - t^{4p})^{-1}. \tag{R8}$$

THEOREM 8. *Let $f : M \to M$ be a transversal map of a compact manifold. Suppose that its Lefschetz zeta function $Z_f(t)$ is of the form (7) and that $(1 \pm t^n)^{\pm 1}$ is an ireducible factor.*

(a) *If n is odd then $n \in Per(f)$.*

(b) *If n is even then $\{\frac{n}{2}, n\} \cap Per(f) \neq \emptyset$.*

PROOF. (a)If we have an irreducible factor of the form $(1 - \sigma_1 t^n)^{\sigma_2}$ with n odd, then all the other possible irreducible factors associated to the same power of t are $(1 - \sigma_1 t^n)^{\sigma_2}$ or $(1 + \sigma_1 t^n)^{-\sigma_2}$ (see reduction rules (R1) to (R4)). So,

$$|\gamma(n,0)| = \left| \sum_{i=1}^{N_n} \sigma_2(i,n)\sigma_1(i,n) \right| \neq 0$$

and equation (8) for $m = 0$ ensures that $n \in \mathrm{Per}(f)$.

(b) Suppose $n = 2^k r$, with r odd and $k \geq 1$. Subtracting equations (8) for $m = k$ and $m = k+1$ we obtain

$$\begin{aligned}(10)\qquad &EE_{2^k r} + EO_{2^k r} - EO_{2^{k-1} r} + \gamma(r,k) - \gamma(r,k-1) = \\ &OO_{2^k r} + OE_{2^k r} - OO_{2^{k-1} r},\end{aligned}$$

where, using (9),

$$\begin{aligned}(11)\qquad \gamma(r,k) - \gamma(r,k-1) = &\sum_{i=1}^{N_{2^k r}} \sigma_2(i, 2^k r)\sigma_1(i, 2^k r) + \\ &\sum_{i=1}^{N_{2^{k-1} r}} \sigma_2(i, 2^{k-1} r)\left(\frac{1 - \sigma_1(i, 2^{k-1} r)}{2}\right).\end{aligned}$$

We shall prove the result for an irreducible factor of the form $1 - t^{2^k r}$ (in the other possible cases the proof is similar). For $1 - t^{2^k r}$ we have

$$\sigma_1(\cdot, 2^k r)\sigma_2(\cdot, 2^k r) = 1$$

and so the same equation will hold for every factor $(1 - \sigma_1(i, 2^k r)t^{2^k r})^{\sigma_2(i, 2^k r)}$ which persists after the reductions (R1–R4). On the other hand, the factors of the form $(1 - \sigma_1(i, 2^{k-1} r)t^{2^{k-1} r})^{\sigma_2(i, 2^{k-1} r)}$ which persist after the reductions verify

$$\sigma_2(i, 2^{k-1} r)\left(\frac{1 - \sigma_1(i, 2^{k-1} r)}{2}\right) \in \{0, 1\}.$$

Hence, from (11), $|\gamma(r,k) - \gamma(r,k-1)| > 0$ and using (10) we conclude that if $2^{k-1} r \notin \mathrm{Per}(f)$ then $2^k r \in \mathrm{Per}(f)$. □

From the preceeding theorem it follows that each irreducible factor of the Lefschetz zeta function forces one period. Using the properties of cyclotomic polynomials we shall derive upper bounds for the cardinal of the forced set of periods as well as for the largest forced period.

The n-th *cyclotomic polynomial*, $c_n(t)$, is by definition

$$c_n(t) = \frac{1 - t^n}{\prod_{d|n, d<n} c_d(t)}$$

for $n \in \mathbb{N} \setminus \{1\}$ and $c_1(t) = 1 - t$. The degree $\varphi(n)$ of $c_n(t)$ verifies

$$n = \sum_{d|n} \varphi(d)$$

and so $\varphi(n)$ is the Euler function, which may be computed through

$$\varphi(n) = n \prod_{\substack{p|n \\ p\ prime}} \left(1 - \frac{1}{p}\right).$$

The proof of the two following lemmas may be found in [**2**].

LEMMA 9. *Let $c_n(t)$ be of degree $\varphi(n) > 2$. Then $c_n(t)$ can be written as*

$$c_n(t) = \prod_{i=1}^{m}(1 - \sigma_1(i)t^{q_i})^{\sigma_2(i)}, \tag{12}$$

where $m \leq \varphi(n)/2$, $q_i \in \mathbb{N}$ and $\sigma_1(i), \sigma_2(i) \in \{-1, 1\}$ for $i = 1, ..., m$. Moreover, $q = \max_i q_i$ is smaller than or equal to $n/2$ if n is even or n if n is odd.

Given $d \in \mathbb{N}$, let $n_0(d)$ be defined by

$$n_0(d) = \left[d \prod_{i=1}^{k} \frac{p_i}{p_i - 1} \right] \tag{13}$$

where $[\cdot]$ denotes the integer part function and $k \in \mathbb{N}$ is the greatest number of consecutive primes p_i such that $p_1 = 2$ and $(p_1 - 1)...(p_k - 1) \leq d$.

LEMMA 10. *Given $d \in \mathbb{N}$, there exists $n_0 \in \mathbb{N}$ such that for $n > n_0$, $\varphi(n) > d$. Moreover, $n_0(d)$ is the best possible lower bound for n_0.*

Let $P(t)/Q(t)$ be a rational function. We define the *order* of $P(t)/ Q(t)$, denoted by order $(P(t)/Q(t))$, as the number

$$\left[\frac{\text{degree } P(t) + \text{degree } Q(t)}{2} \right] + 1.$$

PROPOSITION 11. *Let $f : M \to M$ be a transversal map of a compact manifold. Suppose that all the zeros and poles of $Z_f(t)$ are roots of unity. The number of different periods satisfying the hypothesis of Theorem 8 is smaller or equal than the order of $Z_f(t)$.*

PROOF. We know that $Z_f(t) = P(t)/Q(t)$ and that $P(t)$ and $Q(t)$ may be written as a product of cyclotomic polynomials. Let then $P(t) = P_1(t)P_2(t)$, $Q(t) = Q_1(t)Q_2(t)$, where $P_2(t)$ and $Q_2(t)$ contain every factor equal to $c_1(t)$, $c_2(t)$, $c_3(t)$ or $c_6(t)$. Now, by Theorem 8 and Lemma 9, the cardinal of the forced set of periods forced by $P_1(t)$ and $Q_1(t)$ is at most (degree $P_1(t)$ + degree $Q_1(t))/2$. But $P_2(t)$ or $Q_2(t)$ either force only fixed points or periods 1 and 3, in which case their degree is at least two. □

PROPOSITION 12. *Let $f : M \to M$ be as in Proposition 11 and let d be the maximum of the degrees of $P(t)$ and $Q(t)$, where $P(t)/Q(t) = Z_f(t)$. Then*

(a) *The largest odd period satisfying the hypothesis of Theorem 8 is at most $n_0(d)$.*
(b) *The largest even period satisfying the hypothesis of Theorem 8 is at most $2n_0(d)$ if $n_0(d)$ is odd and $2n_0(d) - 2$ if $n_0(d)$ is even.*

PROOF. Let $Z_f(t) = P(t)/Q(t)$ and let $c_n(t)$ be a cyclotomic polynomial which divides $P(t)$ or $Q(t)$. By Lemma 10, $n \leq n_0(d)$. By Lemma 9, the maximum power of t in the factorization (12) of $c_n(t)$ is is at most $n/2$ if n is even or n if n is odd. Therefore, before applying the reductions (R1) to (R8), the maximum power of t which appears in the factorization (7) of $Z_f(t)$ is $n_0(d)$ (respectively $n_0(d) - 1$) if n is odd (respectively n is even). The result follows taking into account the reduction rules and Theorem 8. □

REFERENCES

1. J. Casasayas, J. Llibre, A. Nunes, *Periodic Orbits for Transversal Maps*, preprint of the Centre de Recerca Matemàtica (1991).
2. J. Casasayas, J. Llibre, A. Nunes, *Periods and Lefschetz zeta functions*, to appear in Pacific J. of Math (1993).
3. J. Casasayas, J. Llibre, A. Nunes, *Algebraic properties of the Lefschetz zeta function, periodic points and topological entropy*, Publicacions Mat. UAB **36** (1992), 467–472.
4. J. M. Franks, *Homology and Dynamical Systems*, CBMS Regional Conf. Series, Vol. 49, Amer. Math. Soc., 1982.
5. J. Franks, *Some smooth maps with infinitely many hyperbolic periodic points*, Trans. Amer. Math. Soc. **226** (1977), 175-179.
6. D. Fried, *Periodic points and twisted coefficients*, in Lecture Notes in Maths. vol. 1007, Springer Verlag (1983), 175-179.
7. A. Manning, *Topological entropy and the first homology group*, Lecture Notes in Math. vol. 468, Springer Verlag (1975), 185-199.
8. M. Misiurewicz, F. Przytycki, *Topological Entropy and degree of smooth mappings*, Bulletin de l'Académie Polonaise des Sciences, Série des Sciences Math., Astr. et Phys. **XXV** (1977), 573-578.
9. T. Matsuoka, *The number of periodic points of smooth maps*, Ergod. Th. & Dynam. Sys. **9** (1989), 153-163.

DEPARTAMENT DE MATEMÀTICA APLICADA I ANÀLISI, UNIVERSITAT DE BARCELONA, GRAN VIA 585, 08071 BARCELONA, SPAIN

DEPARTAMENT DE MATEMÀTIQUES, UNIVERSITAT AUTÒNOMA DE BARCELONA, BELLATERRA, 08193 BARCELONA, SPAIN

CENTRE DE RECERCA MATEMÀTICA, UNIVERSITAT AUTÒNOMA DE BARCELONA, BELLATERRA, 08193 BARCELONA, SPAIN

Current address: Departamento de Física, Universidade de Lisboa, Campo Grande, Ed C1, Piso 4, 1700 Lisboa, Portugal

Contemporary Mathematics
Volume **152**, 1993

One-parameter Fixed Point Indices for Periodic Orbits [1, 2]

DONČO DIMOVSKI [3]

Abstract: In this paper, we give a short summary of the paper [2] where one-parameter fixed point indices were developed, we apply these indices to orbits for smooth flows, and we show how two of these indices are connected to Fuller's indices.

0. Introduction

One-parameter fixed point indices for a PL homotopy $F : X \times I \to X$, where X is a compact, connected PL n-dimensional manifold in $\mathbf{R}^n, n \geq 4$, were developed in [2]. The paper [2] is an improvement of the paper [3], where a careful and detailed geometrical exposition was given of the one-parameter Nielsen fixed point theory, bringing out the parallels between the classical and the one-parameter case. The classical Nielsen fixed point theory is concerned with the problem of when and how a map $f : X \to X$ can be deformed with a control, in order to remove some or all of the fixed point set $\text{Fix}(f) = \{x \mid x \in X, f(x) = x\}$. In [3] the one-parameter case was considered, i.e. the problem of when and how a homotopy $F : X \times I \to X$ can be deformed with control, in order to remove some or all of the fixed point set of F. A fixed point of F is a point $(x,t) \in X \times I$ such that $F(x,t) = x$. The set of all the fixed points of F is denoted by $\text{Fix}(F)$. In this setting, isolated circles of fixed points are the generic form of fixed points, as isolated individual fixed points are in the classical setting. In [2], two indices, $\text{ind}_1(F,V)$ and $\text{ind}_2(F,V)$ are defined for a family V of finitely many isolated circles of fixed points of F. The first index, $\text{ind}_1(F,V)$, is an element in the first homology group $H_1(E)$, where E is the space of paths in $X \times I \times X$ from the graph of F to the graph of P, and is a

[1] 1991 Mathematics Subject Classification. Primary 55M20, 58F22, 55M25.

[2] This paper is in final form and no version of it will be submitted for publication elsewhere

[3] Supported by a grant from the research council of Macedonia

slight generalization of the first obstruction discussed in [3], where $P : X \times I \to X$, $P(x,t) = x$, is the projection. It is mentioned in [3] that a solution to the one parameter fixed point problem in the transverse case can be found in [7], via an obstruction lying in the 1-dimensional framed bordism group of the function space E. This obstruction for a family of finitely many isolated transverse circles of fixed points V is zero if and only if $\mathrm{ind}_1(F,V) = 0$ and $\mathrm{ind}_2(F,V) = 0$. The second index, $\mathrm{ind}_2(F,V)$, is an element of the group with two elements $\mathbf{Z}_2$, and corresponds to the second obstruction given in [3], but the second obstruction in [3] can be defined only if we know that the first obstruction is zero, while $\mathrm{ind}_2(F,V)$ is defined independently of $\mathrm{ind}_1(F,V)$. The indices $\mathrm{ind}_1(F,V)$ and $\mathrm{ind}_2(F,V)$ are defined via another type of indices: $i_1(F,C)$ - which is a nonnegative integer, and $i_2(F,C)$ - which is an element in $\mathbf{Z}_2$, where C is an isolated circle of fixed points. As we will discuss in the next section, these indices $i_1(F,C)$ and $i_2(F,C)$ are defined as certain types of degrees $\deg_1$ and $\deg_2$ of the restriction of $P - F$ to the boundary of the neighborhood of the circle C, where $P - F$ is considered as a map into $\mathbf{R}^n$. The main improvements of the results from [3] proven in [2] are stated as Theorems 1.9, 1.10, 1.11 and 1.12 in this paper, i.e. it is proven that there is a compact neighborhood N of V and a homotopy from F to H rel $X \times I \backslash N$ such that $\mathrm{Fix}(H) = \mathrm{Fix}(F) \backslash V$ if and only if $\mathrm{ind}_1(F,V) = 0$ and $\mathrm{ind}_2(F,V) = 0$; moreover, it is shown how to modify F to join and create circles (transverse or not transverse) of fixed points with prescribed indices. In [2] we were concerned only with the fixed points of homotopies between selfmaps of X, where X is an n-manifold in $\mathbf{R}^n$. The general case of coincidences between two maps $F, G : M \to N$, where M is an $(m+1)$-dimensional and N is m-dimensional manifold, using the same techniques as in this paper, will be discussed in a subsequent paper. For this we need a geometric description of spin manifolds, which is going to appear in a joint paper with Ross Geoghegan. The general case of coincidences has also been considered in a paper by Jerzy Jezierski [8]. In [5], Ross Geoghegan and Andrew Nicas have developed an algebraic one-parameter Lefschetz-Nielsen fixed point theory and in a subsequent paper [6] they have applied it to dynamical systems, i.e. they use it to detect the Fuller homology class for periodic orbits and apply it to suspension semiflows.

The indices mentioned above can be defined for smooth homotopies in the same manner and applied to dynamical systems, i.e. flows, as follows ([6]). For a given smooth flow $\Phi : M \times \mathbf{R} \to M$, for any interval $[r,s]$ in $\mathbf{R}$, the restriction $F : M \times [r,s] \to M$ to Φ is a homotopy. If Γ is an isolated periodic orbit of Φ whose period is $k \in [r,s]$, we will see below that the circle $C = \{(x,k) \mid x \in \Gamma\}$ is an isolated circle of fixed points for F. Hence, we may assign to the orbit Γ the above-mentioned indices, i.e. we have $i_1(\Gamma)$, $\mathrm{ind}_1(\Gamma)$, and $i_2(\Gamma)$ which is equal to $\mathrm{ind}_2(\Gamma)$. We show that $i_1(\Gamma)$ is related to Fuller's index and $\mathrm{ind}_1(\Gamma)$ is related to Fuller's homology class [4], while $i_2(\Gamma)$ is a new type of index for orbits. In [2] we had the assumption $n \geq 4$, needed for the definition of i_2, but this assumption is not needed for the definition of i_1.

At the end of this introduction, I would like to thank Ross Geoghegan for numerous valuable conversation on these subjects, and for introducing me to Fuller's indices. Also I would like to thank Helga Schirmer for valuable discussions and for her hospitality and financial support during my visit to Carleton University.

1. One-parameter fixed point indices

In this section we give a brief summary of the results of [2].

For a positive integer m we denote by x the element $(x_1, x_2, \ldots, x_m)$ of $\mathbf{R}^m$ and by $\mid x \mid$ the length $\sqrt{x_1{}^2 + x_2{}^2 + \ldots + x_1{}^2}$. Let $D^m = \{x \mid x \in \mathbf{R}^m, \mid x \mid \leq 1\}$ and $S^m = \{x \mid x \in \mathbf{R}^{m+1}, \mid x \mid = 1\}$. We chose once and for all an orientation of $\mathbf{R}^m$, for all m, called the standard orientation, such that the standard orientation on $\mathbf{R}^{m+1} = \mathbf{R}^m \times \mathbf{R}$ is the product of the standard orientations on $\mathbf{R}^m$ and $\mathbf{R}$. We assume that D^m, and $I = [0,1]$ are oriented by the induced orientations from $\mathbf{R}^m$ and $\mathbf{R}$. Let S^m have the induced orientation from $\mathbf{R}^{m+1}$ and let $S^1 \times S^m$ and $S^1 \times D^m$ have the product orientations.

Definition 1.1. *Let $f : S^1 \times S^m \longrightarrow S^m$ be a given map. We define* $\deg_1(f) = \deg(\varphi)$, *where $\varphi : S^m \longrightarrow S^m$ is defined by $\varphi(x) = f(A, x)$ for a point $A \in S^1$, and* deg *is the usual degree of a map from S^m to S^m.*

It is easy to check that $\deg_1$ is well defined.

Next let $\mathcal{N}$ be the north pole $(0, 0, \ldots, 0, 1)$ of S^m, and let X be the factor space obtained from $S^1 \times S^m \cup D^2$ by the identification of $(x, \mathcal{N}) \in S^1 \times S^m$ with $x \in S^1 = \partial D^2$. Consider S^{m+1} as the boundary of $D^2 \times D^m$, i.e.

$$S^{m+1} = \partial(D^2 \times D^m) = (\partial D^2 \times D^m) \cup (D^2 \times \partial D^m) = (S^1 \times D^m) \cup (D^2 \times S^{m-1}).$$

Let $K : S^{m+1} \longrightarrow X$ be the map defined by:

$$K(x,y) = \begin{cases} (x, E(y)), & \text{for } (x,y) \in S^1 \times D^m; \\ x, & \text{for } (x,y) \in D^2 \times S^{m-1} \end{cases}$$

where $E : D^m \longrightarrow S^m$ is a map which sends ∂D^m to the north pole $\mathcal{N}$, and E restricted to $\mathrm{int}\mathrm{D}^m$ is a homeomorphism between $\mathrm{int}\mathrm{D}^m$ and $S^m \backslash \{\mathcal{N}\}$.

Definition 1.2. *Let $f : S^1 \times S^m \longrightarrow S^m$ be a given map. We define* $\deg_2(f)$ *to be the element $[\bar{f} \circ K]$ of $\pi_{m+1}(S^m) = \mathbf{Z}_2$ where $\bar{f} : X \longrightarrow S^m$ is defined by f on $S^1 \times S^m$ and by an extension on D^2 of f restricted to ∂D^2.*

The following facts are shown in [2].

Theorem 1.1. *(a) For $m \geq 3, \deg_2(f)$ is well defined.*

(b) Let $f, g : S^1 \times S^m \longrightarrow S^m$ be homotopic maps. Then, $\deg_1(f) = \deg_1(g)$, and for $m \geq 3, \deg_2(f) = \deg_2(g)$.

(c) Let $p : S^1 \times S^m \longrightarrow S^m$ be the projection, $p(z, y) = y$. Let $\varphi : S^m \longrightarrow S^m$ be a given map, and let $g : S^1 \times S^m \longrightarrow S^m$ be defined by $g = \varphi \circ p$. Then, $\deg_2(g) = 0$.

(d) Let $f : S^1 \times S^m \longrightarrow S^m$ be such that $f(S^1 \times \{\mathcal{N}\}) = Q$ for some $Q \in S^m$, and let $m \geq 3$. Let $h : S^1 \times S^m \longrightarrow S^m$ be the map defined by $h(z, y) = f(A, y)$ for $z \in S^1, y \in S^m$, where A is a point of S^1. Then, $\deg_2(f) = 0$ if and only if f is homotopic to the map h.

(e) Let $m \geq 3$ and $f, g : S^1 \times S^m \longrightarrow S^m$ be given. Then f is homotopic to g if and only if $\deg_1(f) = \deg_1(g)$ and $\deg_2(f) = \deg_2(g)$. ■

Next, we describe specific maps $h(m) : S^1 \times S^m \to S^m, m \geq 3$, with $\deg_2(h(m)) \neq 0$, obtained by suspensions, via the Hopf map from S^3 to S^2.

If X is a space, the suspension ΣX of X is the factor space $X \times D^1/\alpha$, where α is the equivalence whose classes are:

$$(x,t)^\alpha = \begin{cases} \{(x,t)\}, & \text{for } t \neq 1 \text{ and } t \neq -1; \\ \{(y,t) \mid y \in X\}, & \text{for } t = 1 \text{ and } t = -1. \end{cases}$$

For a given $f : S^1 \times S^m \to S^m$ we define a map $S(f) : S^1 \times S^{m+1} \to S^{m+1}$ by suspension of f on the second factor, where ΣS^m is identified with S^{m+1} as a subset of $D^{m+1} \times D^1$, i.e.

$$S(f)(x,(y,t)^\alpha) = (f(x,y),t)^\alpha.$$

Theorem 1.2. *Let $f : S^1 \times S^m \to S^m$ be a given map, $m \geq 3$. Then for $S(f) : S^1 \times S^{m+1} \to S^{m+1}$ defined as above,* $\deg_1(f) = \deg_1(S(f))$ *in* $\mathbf{Z}$, *and* $\deg_2(f) = \deg_2(S(f))$ *in* $\mathbf{Z}_2$. ∎

For $m \geq 2$, let $h(m) : S^1 \times S^m \to S^m$ be defined by:

$$h(m)(z,w,x) = (zw,x)$$

where $z \in S^1 \subseteq \mathbf{C}, (w,x) \in S^m \subseteq D^2 \times D^{m-1}, w \in D^2 \times \{0\} \subseteq \mathbf{C}, x \in D^{m-1}, \mathbf{C}$ is the set of complex numbers, and zw is the product of z and w as complex numbers. If we consider S^m as a subset of $D^2 \times D^{m-1}$, and if we identify ΣS^m with S^{m+1} as a subset of $D^2 \times D^{m-1} \times D^1$, from the definitions of $h(m)$ and $S(f)$, it follows that $S(h(m)) = h(m+1)$.

Theorem 1.3. *For* $m \geq 3, \deg_1(h(m)) = 1$ *in* $\mathbf{Z}$, *and* $\deg_2(h(m)) = 1$ *in* $\mathbf{Z}_2$. ∎

Theorem 1.4. *A map $f : S^1 \times S^m \to S^m$ has an extension $\bar{f} : S^1 \times D^{m+1} \to S^m$, if and only if* $\deg_1(f) = 0 = \deg_2(f)$. ∎

Let $SE : S^1 \times D^m \to \mathbf{R}^{m+1}$ be the embedding defined by:

$$SE(z,r,y) = ((2-r)z,y), \text{ for } z \in S^1, (r,y) \in D^m, r \in D^1.$$

We say that SE is the standard embedding. It follows from the definition of SE that SE is orientation preserving, where $S^1 \times D^m$ has the product orientation of the standard orientations, and the restriction of SE from $\mathcal{N} \times D^m$ to $\{0\} \times \mathbf{R}^m$ is orientation reversing. We will use the same notation SE for the image of the map SE. In other words,

$$SE = \{(z,x) \mid z \in \mathbf{R}^2, x \in \mathbf{R}^{m-1}, 1 \leq \mid z \mid \leq 3, \mid x \mid^2 \leq 1 - (\mid z \mid -2)^2\}.$$

Let $F : X \times I \to X$ be given, where $X \subseteq \mathbf{R}^m$ is an m-dimensional, compact, connected PL oriented manifold, such that $X \times I \subseteq \mathbf{R}^m \times \mathbf{R} = \mathbf{R}^{m+1}$, and let $m \geq 4$. We assume that X is oriented by the induced orientation from $\mathbf{R}^m$, and $X \times I$ has the product orientation. Let $SE : S^1 \times D^m \to \mathbf{R} \times \mathbf{R}^m$, be the standard embedding defined above. For each embedded oriented circle $C \subseteq X \times I$ and a regular neighborhood W of C in $X \times I$ we choose the isotopy class $\{\varphi\}$ of orientation preserving embeddings of pairs, $\varphi : (S^1 \times D^m, S^1 \times \{0\}) \to (W, C)$, by isotoping the standard embedding SE. Such a choice exists, because every two oriented embedded circles in $\mathbf{R}^{m+1}$ are isotopic, for $m \geq 3$.

Let $P : X \times I \to X$ be the projection, defined by $P(x,t) = x$, for every $x \in X$ and $t \in I$. A fixed point of the map F is a point $(x,t) \in X \times I$, such that $F(x,t) = x = P(x,t)$. The set of fixed points of F is denoted by Fix(F), i.e. $\mathrm{Fix}(F) = \{(x,t) \in X \times I \mid F(x,t) = x\}$. Let $C \subseteq X \times I$ be an isolated circle of fixed points, i.e. a circle of fixed points, which has a small enough neighborhood W, such that the only fixed points of F in W are the points of C. Since $X \subseteq \mathbf{R}^m$, the map $P - F : (W, C) \to (D^m_\epsilon, 0)$ is well defined, and we denote it by $\rho(F)$, where $D^m_\epsilon = \{x \in R^m \mid\mid x \mid\leq \epsilon\}$. Since C is an isolated circle of fixed points, we have $\rho(F)^{-1}(0) = C$. There are two orientations on C; we denote them by O_1 and O_2. Let φ be an embedding from the chosen isotopy class for such an oriented circle C and its neighborhood W. So we have a map $\zeta \circ \rho(F) \circ \varphi : S^1 \times D^m \to D^m$, where $\zeta : D^m_\epsilon \to D^m$ is the homeomorphism defined by multiplication by $1/\epsilon$. Let $\mu(F) : S^1 \times S^{m-1} \to S^{m-1}$ be the map $\mu(F) = \xi \circ \zeta \circ \rho(F) \circ \varphi'$ where $\xi : D^m \backslash \{0\} \to S^{m-1}$ is defined by $\xi(x) = \frac{1}{|x|} \cdot x$, and φ' is the restriction of φ to $S^1 \times S^{m-1}$.

Theorem 1.5. ([2]) *Let $\mu(F)$ and $\mu'(F)$ be defined as above, for the two orientations on C. Then, $\deg_1(\mu(F)) = -\deg_1(\mu'(F))$, and $\deg_2(\mu(F)) = \deg_2(\mu'(F))$.* ∎

Definition 1.3. *For a chosen orientation O on C, $\deg_1(\mu(F))$ will be denoted by $i_1(F, C, O)$. We say that an orientation on C is the natural orientation on C, if $\deg_1(\mu(F)) \geq 0$, and we say that $\deg_1(\mu(F))$ for this orientation is index 1 of F at C, denoted by $i_1(F, C)$.*

Remark. By the definition, $i_1(F, C) \geq 0$. In the case $\deg_1(\mu(F)) = 0$, both of the two orientations on C are natural, or using different words, C does not have a natural orientation.

The following notion is well defined by Theorem 1.5.

Definition 1.4. *Define index 2 of F at C, denoted by $i_2(F, C)$, to be $\deg_2(\mu(F))$.*

Theorem 1.6. ([2]) *Let X, F, C and W be as above. Then, $i_1(F, C) = 0 = i_2(F, C)$, if and only if F is homotopic to a map $G : X \times I \to X$ rel $X \times I \backslash W$, with $G(x,t) \neq x$ for each $(x,t) \in W$.* ∎

Definition 1.5. *Let A be an isolated fixed point of $F : X \times I \to X$. Let V be a small $(m+1)$-ball neighborhood of A in $X \times I$, such that $F(x) \neq P(x)$ for every $x \in V \backslash \{A\}$. Let varphi : $D^{m+1} \to V$ be a homeomorphism, and let $\mu(F) : S^m \to S^{m-1}$ be the map $\xi \circ \zeta \circ (P - F) \circ \varphi'$, where φ' is the restriction of φ to $S^m = \partial D^{m+1}$. Define index 2 of F at A, denoted by $i_2(F, A)$ to be the element $[\mu(F))] \in \pi_m(S^{m-1})$.*

Let C be an isolated circle of fixed points of $F : X \times I \to X$, B be an embedded disk in $\mathrm{int}(X \times I)$, with $\partial B = C$, and $H : (X \times I \times \{0\}) \cup (B \times I) \to X$ be a partial homotopy such that $H(x, 0) = F(x)$, for all $x \in X \times I$, $H(x, 1) = x$ for all $x \in B$, and $H(x,t) = x$ for all $x \in C$ and all $t \in I$.

Theorem 1.7. ([2]) *Let F, C, B and H be as above. Then there is a neighborhood N of B, and a map $G' : X \times I \to X$, homotopic to F rel $X \times I \backslash N$, such that $\mathrm{Fix}(G') = (\mathrm{Fix}(F) \backslash C) \cup \{A\}$, and $i_2(F, C) = i_2(G', A)$, where A is an isolated fixed point of G'. Moreover, if $i_2(F, C) = 0$, there is a neighborhood N of B, and a map $G : X \times I \to X$, homotopic to F rel $X \times I \backslash N$, such that $\mathrm{Fix}(G) = \mathrm{Fix}(F) \backslash C$.* ∎

Next, let $C \subseteq X \times I$ be an embedded circle on which F and P are ϵ-close, i.e. $N_{2\epsilon}(X)$ retracts to X in $\mathbf{R}^m$, and $d(F(x), P(x)) \leq \epsilon$ for each $x \in W$ where W is a small regular neighborhood of C in $X \times I$. Assume that for each $x \in \partial W$, $F(x) \neq P(x)$. Choose any orientation O for C, and let φ be an embedding from the chosen isotopy class of embeddings for (W, C). Let $f = P - F$, let g denote the restriction of $f \circ \varphi$ to $S^1 \times S^{m-1}$, and let $\deg_1(g) = k$, for $k \in \mathbf{Z}$.

Theorem 1.8. ([2]) *Let $X, F, C, W, \varphi, f, g, k$ be as above, let $m \geq 4$, and for $r \geq 1$, let $t_j, j = 1, \ldots, r$ be arbitrary numbers with $\sum t_i = \deg_1(g)$. Then F is homotopic to a map $G : X \times I \to X$, rel $X \times I \backslash W$ such that:*

(1) G has r isolated circles of fixed points $C_1, C_2, \ldots, C_r$ in W, all of them "parallel" to C;

(2) G does not have other fixed points in W;

(3) $\sum i_2(G, C_j) = \deg_2(g)$, where the sum is in $\pi_m(S^{m-1})$;

(4) If we orient C_j with an orientation O_j compatible with the orientation O on C (meaning that (C, O) and (C_j, O_j) determine the same element in the first homology group $H_1(W)$) then, for each $j = 1, 2, \ldots, r$, $i_1(G, C_j, O_j) = t_j$; and

(5) If for each j we hawe $t_j = 1$, then it is possible to make the circles C_j transverse, which means that the graph of G and the graph of the projection $P : X \times I \to X$ intersect transversely in $X \times I \times X$ at each C_j. ∎

Let $F : X \times I \to X$ be as above. Let A, B be two fixed points of F. It is said that A and B are in the same fixed point class if there is an arc $\alpha : [0,1] \to X \times I$ from A to B and a homotopy between $F \circ \alpha$ and $P \circ \alpha$ rel $\{0,1\}$. The relation of being in the same fixed point class is an equivalence. It follows directly from the definition, that if A, B belong to a circle of fixed points then they are in the same fixed point class. Two circles of fixed points are in the same fixed point class if and only if their points are in the same fixed point class.

Next we recall some definitions and facts from [3], where it was assumed that F is transverse to P, which implies that for each isolated circle of fixed points C, $i_1(F, C) = 1$.

Let E be the space of all (continuous, not necessarily PL) paths $\omega(t)$ in $X \times I \times X$ from the graph $\Gamma(F) = \{(x, t, F(x,t)) \mid (x,t) \in X \times I\}$ of F to the graph $\Gamma(P) = \{(x,t,x) \mid (x,t) \in X \times I\}$ of P, i.e. maps $\omega : [0,1] \to X \times I \times X$, such that $\omega(0) \in \Gamma(F)$ and $\omega(1) \in \Gamma(P)$. Let $C_1, \ldots, C_k$ be isolated circles in Fix$(F) \cap$ int$(X \times I)$, oriented by the natural orientations, and let $V = \cup C_j$. Then V determines a family of circles V' in E via the constant paths in E, i.e. each oriented isolated circle of fixed points $C : S^1 \to X \times I$ of F determines an oriented circle $C' : S^1 \to E$ defined by $C'(z)$=con$(C(z))$ where con$(C(z))$ is the constant path at $C(z) = (x, t_0)$. i.e. con$(C(z))(t) = (x, t_0, x)$ for each $t \in [0,1]$. The definitions of a fixed point class and E imply that two fixed points A and B are in a single fixed point class if and only if the corresponding points A' and B' in E are in a single path component of E. Since any two points A, B from a circle of fixed points are in a single fixed point class, it follows that a family V of circles of fixed points is in a single fixed point class if and only if there is a compact orientable surface S_0 and a map $\vartheta : S_0 \to E$, such that a part of ∂S_0 is mapped homeomorphically to V' . As it is shown in [3], such a surface S_0 and a map ϑ exist if and only if there is an embedded compact orientable surface S in $X \times I$ such that $V \subseteq \partial S$ and there is a partial homotopy $H : X \times I \times \{0\} \cup S \times I \to X$ satisfying: $H(x, 0) = F(x)$ for all $x \in X \times I$, $H(x, 1) = x$ for all $x \in S$, and $H(x, t) = x$ for all $x \in V$ and all t.

Theorem 1.9. ([2]) *Let $F : X \times I \to X$ be transverse to P with no fixed points in $\partial(X \times I)$ and transverse fixed point set as in* [3]. *Let $V = \cup C_j$ be a union of isolated circles of fixed points, such that V' lies in a path component of E. Then, there is a neighborhood, N, of V in $int(X \times I)$ containing no other fixed points of F, and a homotopy from F to H rel $X \times I \backslash N$, such that* $\text{Fix}(H) = \text{Fix}(F)\backslash V$, *if and only if: the geometric 1-cycle in E defined by V' with the natural orientations on the $C_j{}'s$ is $\mathbf{Z}$-homologous to zero; and* $\sum i_2(F, C_j) = 0$. ∎

In the above Theorem we used a family of isolated transverse circles of fixed points V and its corresponding family V' in the space of paths E. Next we generalize this notion slightly. Let V be a family of isolated circles of fixed points (not necessarily transverse) and isolated fixed points (which can not be transverse). For such a set V, let V' be the subset of E defined as above using constant paths. Then V' is a family of circles and points in E. A set V is in a single fixed point class if and only if V' is in a single path component of E. For an isolated, oriented circle of fixed points C, let $\{C'\}$ be the element from $H_1(E)$ determined by the geometric 1-cycle C'. For such sets V we define two indices.

Definition 1.6. *Let V be a family of isolated circles of fixed points $C_1, \ldots, C_k$ (not necessarily transverse) and isolated fixed points $A_1, \ldots, A_r$ in a single fixed point class. Let C_j be oriented by the natural orientation. We define the index 1 of V, denoted by $ind_1(F, V)$, to be the element $\sum i_1(F, C_j) \cdot \{C_j'\}$ in $H_1(E)$. We define the index 2 of V, denoted by $ind_2(F, V)$, to be the element $\sum i_2(F, C_j) + \sum i_2(F, A_i)$ in $\mathbf{Z}_2$.*

Theorem 1.9 is an improvement of the main theorems in [3] . Another improvement is the change of a subset in a single fixed point class to a single circle, which may be taken to be transverse.

Theorem 1.10. ([2]) *Let V be a family of isolated circles of fixed points and isolated fixed points of F in a single fixed point class. Then there is a neighborhood N of V missing other fixed points and a homotopy from F to G rel $X \times I \backslash N$, such that:*

(i) $\text{Fix}(G) = (\text{Fix}(F)\backslash V) \cup C$, *where C is an isolated circle of fixed points of G;*

(ii) $ind_1(G, C) = 0$ if and only if $ind_1(F, V) = 0$; and

(iii) $ind_2(G, C) = ind_2(F, V)$.

Moreover the circle C can be chosen to be transverse, which implies that $i_1(G, C) = 1$. ∎

Next we give another improvement of Theorem 1.9.

Theorem 1.11. ([2]) *Let $X \subseteq \mathbf{R}^m$ be an m-dimensional, compact, connected orientable manifold, $m \geq 4$, let $F : X \times I \to X$, and let $V \subseteq int(X \times I)$*

be a family of isolated circles of fixed points and isolated fixed points of F in a single fixed point class. Then, there exists a compact neighborhood N of V and a homotopy from F to G rel $X \times I \backslash N$ such that $\mathrm{Fix}(G) = \mathrm{Fix}(F) \backslash V$ *if and only if $ind_1(F, V) = 0$ in $H_1(E)$, and $ind_2(F, V) = 0$ in $\mathbf{Z}_2$.* ∎

Definition 1.7. *Let F have only isolated circles of fixed points and isolated fixed points. A fixed point class V is said to be inessential if $ind_1(F, V) = 0$ and $ind_2(F, V) = 0$, and essential otherwise. We denote by $N(F)$ the number of essential fixed point classes, and call it Nielsen number of F.*

Remark: In [5], Geoghegan and Nicas have defined a Nielsen number $N(F)$ which differs from the one mentioned above. Their Nielsen number does not use the ind_2. The map $F : I^n \times I \to I^n, n \geq 4$, defined in the second example (par 12. Examples from [3]) has $N(F) = 0$ as defined in [5] and has $N(F) = 1$ as defined above.

The following Theorem is about the number and types of fixed point classes whose proof follows from the above results. Let F, X be as above, and $m \geq 4$.

Theorem 1.12. *(1) F is homotopic to a map G_1 such that G_1 has exactly $N(F)$ fixed point classes.*

(2) F is homotopic to a map G_2 such that G_2 has exactly $N(F)$ isolated circles of fixed points.

(3) F is homotopic to a map G_3 such that G_3 has exactly $N(F)$ isolated transverse circles of fixed points.

(4) If F is homotopic to a map H such that H has only isolated circles of fixed points and isolated fixed points, then the number of fixed point classes of H is bigger than or equal to $N(F)$. ∎

2. Indices for periodic orbits

In the previous section we have worked in the PL category, but the methods used work as well as in the Diff category. In this section we will apply one-parameter fixed point indices to periodic orbits of flows. We will consider here only flows on differentiable manifolds M^n in $\mathbf{R}^n, n \geq 4$, because our indices were defined in this setting. The indices i_1 and ind_1 can be used in the general setting, and moreover for manifolds M^n when $n \geq 2$, which is also considered in [6].

Let f be a stationary velocity field defined over a manifold $M^n, n \geq 2$, and let $dx/dt = f(x)$ be an autonomous differential equation. (For example $M^n = \mathbf{R}^n$). This equation has, for each initial point x, a unique solution $y = \Phi(x, t)$ defined for t in an open interval $(-r, r)$ of $0 \in \mathbf{R}$. These solutions define a smooth flow $\Phi : M \times \mathbf{R} \to M$, which satisfies, where it is defined, the property $\Phi(\Phi(x, s), t) = \Phi(x, s + t)$. The solution points, (x, r) in $M \times \mathbf{R}$ of the equation $\Phi(x, r) = x$ for $r > 0$, called periodic points are of two types. The first type are the points (x, r), where $\Phi(x, t) = x$ for all t, i.e. x is a singularity of the field f. The second type are the points (x, r) such that $\Phi(x, r) = x$, $r > 0$, but there is $0 < s < r$, such that $\Phi(x, s) \neq x$. We will consider only the second type of periodic points, i.e. consider nonsingular fields. The property $\Phi(\Phi(x, s), t) = \Phi(x, s + t)$, implies that if $\Phi(x, r) = x$, then for each $0 \leq t \leq r$, $\Phi(\Phi(x, t), r) = \Phi(x, t)$. If (x, r) is a periodic point, let $q > 0$

be the least number such that $\Phi(x,q) = x$. Then $r = m \cdot q$, where m is an integer, and the set $C = \{(\Phi(x,t),r) \mid 0 \leq t \leq r\}$ is called a periodic orbit of period q and multiplicity m. The projection C'' of C in M is equal to $\{\Phi(x,t) \mid 0 \leq t \leq r\}$ and is also called a periodic orbit. Next, for any interval $[p,s]$ in $\mathbf{R}$, the restriction $F : M \times [p,s] \to M$ of Φ can be considered as a homotopy. If (x,r) is a periodic point and $r \in (p,s)$, then the periodic orbit $C = \{(\Phi(x,t),r) \mid 0 \leq t \leq r\}$ is in the fixed point set Fix(F). A periodic orbit C is called isolated, if it has a neighborhood in $M \times (0,\infty)$ which excludes all the other periodic points, and in this case C is an isolated circle of fixed points for F. Since a periodic orbit C is an isolated circle of fixed points for F, we have defined $i_1(F,C)$, $\text{ind}_1(F,C)$ and $i_2(F,C)$, and all the theorems from the previous section can be applied in this setting.

Next we will discuss the connection between our indices and Fuller's indices.

Let $C = \{(\Phi(x,t),r) \mid 0 \leq t \leq r\}$ be an isolated periodic orbit of period q and multiplicity m. Let C'' be its projection in M. Then C'' is equal to $\{\Phi(x,t) \mid 0 \leq t \leq q\}$. Since C'' is an integral curve, it is a smooth circle in M. Let $(x,r) \in C$, i.e. $x \in C''$, and let Σ be an $(n-1)$-dimensional submanifold of M, containing x and transversal to the field f at x. Each $y \in \Sigma$ near x serves as the initial point for a solution of the field f, which reintersects Σ at a point $\varphi(y)$, after a time $t(y)$ nearly equal to the period q of C. This map φ is a homeomorphism of a disk neighborhood B of x in Σ into Σ with x as an isolated fixed point (see[9]), and it is called a period map, or the "first return map". In other words, the first return map $\varphi : B \to \Sigma$ is defined by $\varphi(y) = \Phi(y,t(y))$. Since x is an isolated fixed point for φ, we may consider the "m-th return map", $\varphi^m : B \to \Sigma$, for sufficiently small B ([6]), and its classical index $i(\varphi^m,x)$ at its isolated fixed point x. In his paper [4], Fuller defines:

(1) The index i(C), which is the rational number $i(\varphi^m,x)/m$;

(2) The index cycle $\Lambda(C)$ which is the singular 1-cycle $i(C) \cdot S(C)$, where $S(C) : [0,r] \to M \times (0,\infty)$ is the 1-cycle defined by the path $S(C)(t) = (\Phi(x,t),r)$; and

(3) The integral homology class $\Lambda(C)$ in $H_1(M \times (0,\infty)) \equiv H_1(M)$, represented by the index cycle of C.

Remark: The integral homology class $\Lambda(C)$ is equal to the integral homology class represented by the 1-cycle $i(\varphi^m,x) \cdot \alpha$, where $\alpha : [0,q] \to M$ is defined by $\alpha(t) = \Phi(x,t)$, and moreover, $\Lambda(C) = i(\varphi^m,x) \cdot \{\alpha\}$ where $\{\alpha\}$ is the homology class represented by the 1-cycle α ([GN]).

Let E be the space of paths $\omega(t)$ from the graph of F to the graph of the projection P, as in the previous section, and let $h : E \to M$ be defined by $h(\omega) = w(0)$.

Theorem 2.1. *The integral homology class $\Lambda(C)$ is equal, up to a sign, to the image $h_*(ind_1(F,C))$ of $ind_1(F,C)$.*

Proof. From the definition of $\text{ind}_1(F,C) = i_1(F,C) \cdot \{C'\}$ it follows that

$$h_*(\text{ind}_1(F,C)) = i_1(F,C) \cdot h_*\{C'\} = i_1(F,C) \cdot \{C\} = i_1(F,C) \cdot \{\alpha\}.$$

Hence, we have to prove that $i_1(F,C) = i(\varphi^m,x)$. The index $i_1(F,C)$ is equal to the degree of the restriction of $P - F$ to $\partial(B \times [r-\epsilon, r+\epsilon])$, where P is the projection. Next, $\partial(B \times [r-\epsilon, r+e])$ can be written as $E_- \cup E_+$ where $E_- \cap E_+ = \partial(B'')$ and $B'' = \{(y, t(y)) \mid y \in B\}$. The restriction $F \mid: B'' \to B$ is equal to φ^m. Since B is an $(n-1)$-ball neighborhood of x in Σ, it follows that an n-ball neighborhood of x in M can be represented as $B \times [-u, u]$ with: $\{x\} \times [-u, u] \subseteq C''$; $B \times [-u, u] = E_- \cup E_+$ and $E_- \cap E_+ = B \times \{0\}$. Since the map F is defined via the flow Φ, it follows that $F(E_-) \subseteq E_-$, $F(E_+) \subseteq E_+$, $(P-F)(E_-) \subseteq E_-$ and $(P-F)(E_+) \subseteq E_+$, which implies that $\deg(P-F) = \deg(\mathrm{id}-\varphi^m)$ ([11]). Hence, $i(\varphi^m, x) = \pm i_1(F, C)$.∎

The above theorem shows the connection between Fuller's indices and i_1 and ind_1, while i_2 and ind_2 are new kinds of fixed point type indices for periodic orbits.

At the end we ask the following question.

Question: *Is it possible to define "useful" fixed point type indices for quasi-periodic orbits, limiting orbits and attractors?*

References

[1] Jiang, B.: Fixed Point Classes from a Differential viewpoint. In: Fixed Point Theory, Springer Lecture Notes in Mathematics, Vol. **886**, (1981), 163-170

[2] Dimovski, D.: One-parameter fixed point indices. Pac. J. Math. (to appear)

[3] Dimovski, D., Geoghegan, R.: One-parameter Fixed Point Theory. Forum Math. **2** (1990), 125-154

[4] Fuller F.B.: An index of fixed point type for periodic orbits. Amer. J. Math. **89**, (1967), 133-148

[5] Geoghegan, R., Nicas, A. : Parametrized Lefschetz-Nielsen fixed point theory and Hochschild homology traces. Amer. J. Math. (to appear)

[6] Geoghegan, R., Nicas, A. : Trace and torsion in the theory of flows. (preprint)

[7] Hatcher, A., Quinn, F.: Bordism invariants of intersections of submanifolds. Trans. Amer. Math. Soc. **200** (1974) 327-344

[8] Jezierski, J.: One codimensionalWecken type theorems. (to appear)

[9] Markus, L.: The behavior of the solutions of a differential system near a periodic solution. Ann. of Math., vol. **72** (1960),245-266

[10] Rourke, C., Sanderson,B.: Piecewise Linear Topology. Springer Verlag, New York, Heidelberg, Berlin 1972

[11] Spanier, E.F.: Algebraic Topology. McGraw-Hill, New York, 1966.

[12] Whitehead, G.: Elements of Homotopy Theory. Springer Verlag, New York, Heidelberg, Berlin 1978

Prirodno-matematički fakultet, University of Skopje,
PF 162, 91000 Skopje, Macedonia
e-mail: pmfdonco%nubsk@uni-lj.ac.mail.yu
donco@math.binghamton.edu

Contemporary Mathematics
Volume **152**, 1993

Dynamical zeta functions, Nielsen theory and Reidemeister torsion

ALEXANDER FEL'SHTYN AND RICHARD HILL

ABSTRACT. In this paper we continue to study the Reidemeister and Nielsen zeta functions. We prove functional equations and rationality of the Reidemeister zeta function of an eventually commutative endomorphism and of an eventually commutative continuous map of a compact polyhedron. We also prove functional equations and rationality of the Reidemeister zeta function of an endomorphism of any finite group and of a self-map of a polyhedron with finite fundamental group. We connect Reidemeister zeta function of a group endomorphism with the Lefschetz zeta function of the Pontryagin dual endomorphism, and as a consequence obtain a connection of Reidemeister zeta functions with the Reidemeister torsion. We also obtain arithmetical congruences for the Reidemeister and Nielsen numbers similar to those found by Dold for Lefschetz numbers. In the final section of the article we formulate some conjectures about Reidemeister numbers.

1. Introduction

We assume everywhere X to be a connected, compact polyhedron and $f : X \to X$ to be a continuous map. Let $p : \tilde{X} \to X$ be the universal covering of X and $\tilde{f} : \tilde{X} \to \tilde{X}$ a lifting of f, ie. $p \circ \tilde{f} = f \circ p$. Two liftings $\tilde{f}$ and $\tilde{f}'$ are called *conjugate* if there is a $\gamma \in \Gamma \cong \pi_1(X)$ such that $\tilde{f}' = \gamma \circ \tilde{f} \circ \gamma^{-1}$. The subset $p(Fix(\tilde{f})) \subset Fix(f)$ is called *the fixed point class of f determined by the lifting class* $[\tilde{f}]$. A fixed point class is called *essential* if its index is nonzero. The number of lifting classes of f (and hence the number of fixed point classes, empty or not) is called the *Reidemeister Number* of f, denoted by $R(f)$. It is a positive integer or infinity. The number of essential fixed point classes is called the *Nielsen number* of f, denoted by $N(f)$. The Nielsen number is always finite. $R(f)$ and $N(f)$ are homotopy type invariants. In the category of

1991 *Mathematics Subject Classification.* Primary 58F20.
The first author was supported by the Sonderforschungsbereich 170 in Göttingen.
The second author was supported by an SERC grant.
The detailed version of this paper will be submitted for publication elsewhere.

compact, connected polyhedra the Nielsen number of a map is equal to the least number of fixed points of maps with the same homotopy type as f.

Let G be a group and $\phi : G \to G$ an endomorphism. Two elements $\alpha, \alpha' \in G$ are said to be $\phi-$ *conjugate* iff there exists $\gamma \in G$ such that $\alpha' = \gamma.\alpha.\phi(\gamma)^{-1}$. The number of ϕ-conjugacy classes is called the *Reidemeister number* of ϕ, denoted by $R(\phi)$.

Taking the dynamical point of view, we consider the iterates of f and ϕ, and we may define several zeta functions connected with Nielsen fixed point theory (see [**4, 7, 8, 9, 21**]). We assume throughout this article that $R(f^n) < \infty$ and $R(\phi^n) < \infty$ for all $n > 0$. The Reidemeister zeta functions of f and ϕ and the Nielsen zeta function of f are defined as power series:

$$\begin{aligned} R_\phi(z) &:= \exp\left(\sum_{n=1}^{\infty} \frac{R(\phi^n)}{n} z^n\right), \\ R_f(z) &:= \exp\left(\sum_{n=1}^{\infty} \frac{R(f^n)}{n} z^n\right), \\ N_f(z) &:= \exp\left(\sum_{n=1}^{\infty} \frac{N(f^n)}{n} z^n\right). \end{aligned}$$

$R_f(z)$ and $N_f(z)$ are homotopy invariants. The function $N_f(z)$ has a positive radius of convergence which has a sharp estimate in terms of the topological entropy of the map f [**21**]. The above zeta functions are directly analogous to the Lefschetz zeta function

$$L_f(z) := \exp\left(\sum_{n=1}^{\infty} \frac{L(f^n)}{n} z^n\right),$$

where

$$L(f^n) := \sum_{k=0}^{\dim X} (-1)^k Tr\left[f^n_{*k} : H_k(X;\mathbb{Q}) \to H_k(X;\mathbb{Q})\right]$$

are the Lefschetz numbers of the iterates of f. The Lefschetz zeta function is a rational function of z and is given by the formula:

$$L_f(z) = \prod_{k=0}^{\dim X} \det\left(E - f_{*k}.z\right)^{(-1)^{k+1}},$$

where E is the identity matrix. A. Weil [**27**] introduced the function $L_f(z)$ in his study of the fixed points of the Frobenius map of an algebraic variety defined over a finite field. In the theory of discrete dynamical systems $L_f(z)$ was defined by Smale [**26**].

We begin the article by proving in §2 that $R_\phi(z)$ is a rational function with a functional equation in the case that ϕ is eventually commutative and G is finitely generated, and in the case that G is finite.

As a consequence we obtain in §3 rationality and a functional equation for $R_f(z)$ where either $\pi_1(X)$ is finite, or $f : X \to X$ is eventually commutative. The term eventually commutative will be explained later. In §4 we prove in special cases rationality of $N_f(z)$. As an application we calculate explicitly the Reidemeister and Nielsen zeta functions of continuous maps of lens spaces, nil manifolds and tori.

In his article **[3]**, Dold found a remarkable arithmetical property of the Lefschetz numbers for the iterations of a map f. He proved the following formula

$$\sum_{d|n} \mu(d).L(f^{n/d}) \equiv 0 \mod n$$

where n is any natural number and μ is the Möbius function. In the case that n is prime, this result was obtained by Zabreiko, Krasnosel'skii and Steinlein. In §5 we prove similar formulae:

$$\sum_{d|n} \mu(d).R(\phi^{n/d}) \equiv 0 \mod n,$$

$$\sum_{d|n} \mu(d).R(f^{n/d}) \equiv 0 \mod n$$

for eventually commutative endomorphisms ϕ and maps f, and for endomorphisms of finite groups and maps of spaces with finite fundamental groups. This result implies the corresponding congruences for the Nielsen numbers, which were proved by Heath, Piccinini, and You **[15]**.

In 1935 Reidemeister **[24]** classified upto PL equivalence the lens spaces S^3/Γ where Γ is a finite cyclic group of fixed point free orthogonal transformations. He used a certain new invariant which was quickly extended by Franz **[11]**, who used it to classify the generalized lens spaces S^{2n+1}/Γ. This invariant was a ratio of determinants concocted from a Γ-equivariant chain complex of S^{2n+1} and a nontrivial character $\rho : \Gamma \to U(1)$ of Γ. Such a ρ determines a flat bundle E over S^{2n+1}/Γ such that E has holonomy ρ. The new invariant is now called the *Reidemeister torsion*, or *R-torsion* of E. The results of Reidemeister and Franz were extended by de Rham **[22]** to spaces of constant curvature +1.

Later Milnor **[17]** identified the Reidemeister torsion with the Alexander polynomial, which plays a fundamental role in the theory of knots and links. After that Cheeger **[2]** and Müller **[19]** proved that the Reidemeister torsion coincides with the analytical torsion of Ray and Singer **[23]**.

Recently a connection between the Lefschetz type dynamical zeta functions and the Reidemeister torsion was established by D. Fried **[12, 13]**. The work of Milnor **[18]** was the first indication that such a connection exists. In §6 we obtain an expression for the Reidemeister torsion of the mapping tori of the dual map of a group endomorphism in terms of a Reidemeister zeta function of the endomorphism. This is obtained by expressing the Reidemeister zeta function in terms of the Lefschetz zeta function of the dual map.

2. The Reidemeister zeta function of a group endomorphism

PROBLEM. For which groups and endomorphisms is the Reidemeister zeta function a rational function? When does it have a functional equation? Is $R_\phi(z)$ an algebraic function?

2.1. A Convolution Product. When $R_\phi(z)$ is a rational function the infinite sequence $\{R(\phi^n)\}_{n=1}^\infty$ of Reidemeister numbers is determined by a finite set of complex numbers - the zeros and poles of $R_\phi(z)$.

LEMMA 1 ([**9**]). *$R_\phi(z)$ is a rational function if and only if there exists a finite set of complex numbers α_i and β_j such that $R(\phi^n) = \sum_j \beta_j^n - \sum_i \alpha_i^n$ for every $n > 0$.*

For two sequences (x_n) and (y_n) we may define the corresponding zeta functions:

$$X(z) := \exp\left(\sum_{n=1}^{\infty} \frac{x_n}{n} z^n\right),$$

$$Y(z) := \exp\left(\sum_{n=1}^{\infty} \frac{y_n}{n} z^n\right).$$

Alternately, given complex functions X and Y (defined in a neighbourhood of 0) we may define sequences

$$x_n := \frac{d^n}{dz} \log\left(X(z)\right)|_{z=0},$$

$$y_n := \frac{d^n}{dz} \log\left(Y(z)\right)|_{z=0}.$$

Taking the componentwise product of the two sequences gives another sequence, from which we obtain another complex function. We call this new function the *additive convolution* of X and Y, and we write it

$$(X * Y)(z) := \exp\left(\sum_{n=1}^{\infty} \frac{x_n \cdot y_n}{n} z^n\right).$$

It follows immediately from lemma 1 that if X and Y are rational functions then $X * Y$ is a rational function. In fact we may show using the same method the following

LEMMA 2 (CONVOLUTION OF RATIONAL FUNCTIONS). *Let*

$$X(z) = \prod_i (1 - \alpha_i z)^{m(i)}, \quad Y(z) = \prod_j (1 - \beta_j z)^{l(j)}$$

be rational functions in z. Then $X * Y$ is the following rational function

$$(X * Y)(z) = \prod_{i,j} (1 - \alpha_i \beta_j z)^{-m(i).l(j)}. \tag{1}$$

A consequence of this is the following

LEMMA 3 (FUNCTIONAL EQUATION OF A CONVOLUTION). *Let $X(z)$ and $Y(z)$ be rational functions satisfying the following functional equations*

$$X\left(\frac{1}{d_1.z}\right) = K_1 z^{-e_1} X(z)^{f_1}, \quad Y\left(\frac{1}{d_2.z}\right) = K_2 z^{-e_2} Y(z)^{f_2},$$

with $d_i \in \mathbb{C}$, $e_i \in \mathbb{Z}$,$K_i \in \mathbb{C}^\times$ and $f_i \in \{1,-1\}$. Suppose also that $X(0) = Y(0) = 1$. Then the rational function $X*Y$ has the following functional equation:

$$(X*Y)\left(\frac{1}{d_1 d_2 z}\right) = K_3 z^{-e_1 e_2}(X*Y)(z)^{f_1 f_2} \tag{2}$$

for some $K_3 \in \mathbb{C}^\times$.

2.2. Pontryagin Duality. Let G be a locally compact abelian topological group. We write $\hat{G}$ for the set of continuous homomorphisms from G to the circle $U(1) = \{z \in \mathbb{C} : |z| = 1\}$. This is a group with pointwise multiplication. We call $\hat{G}$ the *Pontryagin dual* of G. When we equip $\hat{G}$ with the compact-open topology it becomes a locally compact abelian topological group. The dual of the dual of G is canonically isomorphic to G.

A continuous endomorphism $f : G \to G$ gives rise to a continuous endomorphism $\hat{f} : \hat{G} \to \hat{G}$ defined by

$$\hat{f}(\chi) := \chi \circ f.$$

There is a 1-1 correspondence between the closed subgroups H of G and the quotient groups $\hat{G}/H^*$ of $\hat{G}$ for which H^* is closed in $\hat{G}$. This correspondence is given by the following:

$$H \leftrightarrow \hat{G}/H^*,$$
$$H^* := \{\chi \in \hat{G} \mid H \subset \ker \chi\}.$$

Under this correspondence, $\hat{G}/H^*$ is canonically isomorphic to the Pontryagin dual of H. If we identify G canonically with the dual of $\hat{G}$ then we have $H^{**} = H$.

If G is a finitely generated free abelian group then a homomorphism $\chi : G \to U(1)$ is completely determined by its values on a basis of G, and these values may be chosen arbitrarily. The dual of G is thus a torus whose dimension is equal to the rank of G.

If $G = \mathbb{Z}/n\mathbb{Z}$ then the elements of $\hat{G}$ are of the form

$$x \to e^{\frac{2\pi i y x}{n}}$$

with $y \in \{1, 2, \dots, n\}$. A cyclic group is therefore (uncanonically) isomorphic to itself.

The dual of $G_1 \oplus G_2$ is canonically isomorphic to $\hat{G}_1 \oplus \hat{G}_2$. From this we see that any finite, abelian group is (non-canonically) isomorphic to its own dual, and that the dual of any finitely generated discrete abelian group is the direct sum of a torus and a finite group.

Proofs of all these statements may be found, for example in [**25**]. We shall require the following statement:

PROPOSITION 1. *Let $\phi : G \to G$ be an endomorphism of an abelian group G. Then the* ker $\left[\hat{\phi} : \hat{G} \to \hat{G}\right]$ *is canonically isomorphic to the Pontryagin dual of Coker ϕ.*

PROOF. We construct the isomorphism explicitly. Let χ be in the dual of *Coker* $(\phi : G \to G)$. In that case χ is a homomorphism

$$\chi : G/Im\ (\phi) \longrightarrow U(1).$$

There is therefore an induced map

$$\overline{\chi} : G \longrightarrow U(1)$$

which is trivial on $Im\ (\phi)$. This means that $\overline{\chi} \circ \phi$ is trivial, or in other words $\hat{\phi}(\overline{\chi})$ is the identity element of $\hat{G}$. We therefore have $\overline{\chi} \in \ker(\hat{\phi})$.

If on the other hand we begin with $\overline{\chi} \in \ker(\hat{\phi})$, then it follows that χ is trivial on $Im\ \phi$, and so $\overline{\chi}$ induces a homomorphism

$$\chi : G/Im\ (\phi) \longrightarrow U(1)$$

and χ is then in the dual of *Coker* ϕ. The correspondence $\chi \leftrightarrow \overline{\chi}$ is clearly a bijection. □

2.3. Eventually commutative endomorphisms.

2.3.1. *Rationality of Reidemeister zeta functions of eventually commutative endomorphisms.* An endomorphism $\phi : G \to G$ is said to be eventually commutative if there exists a natural number n such that the subgroup $\phi^n(G)$ is commutative.

We are now ready to compare the Reidemeister zeta function of an endomorphism ϕ with the Reidemeister zeta function of $H_1(\phi) : H_1(G) \to H_1(G)$, where $H_1 = H_1^{Gp}$ is the first integral homology functor from groups to abelian groups.

THEOREM 1 ([**9**]). *If $\phi : G \to G$ is eventually commutative, then*

$$R_\phi(z) = R_{H_1(\phi)}(z) = \exp\left(\sum_{n=1}^{\infty} \frac{\#Coker\ (1 - H_1(\phi)^n)}{n} z^n\right).$$

This means that to find out about the Reidemeister zeta functions of eventually commutative endomorphisms, it is sufficient to study the zeta functions of endomorphisms of abelian groups. For the rest of this section G will be a finitely generated abelian group.

LEMMA 4 ([**9**]). *Let $\phi : \mathbb{Z}^k \to \mathbb{Z}^k$ be a group endomorphism. Then we have*

$$R_\phi(z) = \left(\prod_{i=0}^{k} \det(1 - \Lambda^i \phi.\sigma z)^{(-1)^{i+1}}\right)^{(-1)^r} \tag{3}$$

where $\sigma = (-1)^p$ *with* p *the number of* $\mu \in Spec\ \phi$ *such that* $\mu < -1$, *and* r *the number of real eigenvalues of* ϕ *whose absolute value is* > 1. Λ^i *denotes the exterior power.*

LEMMA 5. *Let* $\phi : G \to G$ *be an endomorphism of a finite abelian group* G. *Then we have*

$$R_\phi(z) = \prod_{[\gamma]} \frac{1}{1 - z^{\#[\gamma]}} \tag{4}$$

where the product is taken over the periodic orbits of ϕ *in* G.

We give two proofs of this lemma in this article. The first proof is given here and the second proof is a special case of the proofs of theorems 4 and 5

PROOF. Since G is abelian, we again have,

$$\begin{aligned} R(\phi^n) &= \#Coker\ (1-\phi^n) \\ &= \#G/\#Im\ (1-\phi^n) \\ &= \#G/\#(G/\ker(1-\phi^n)) \\ &= \#G/(\#G/\#\ker(1-\phi^n)) \\ &= \#\ker(1-\phi^n) \\ &= \#Fix\ (\phi^n) \end{aligned}$$

We shall call an element of G periodic if it is fixed by some iteration of ϕ. A periodic element γ is fixed by ϕ^n iff n is divisible by the cardinality the orbit of γ. We therefore have

$$\begin{aligned} R(\phi^n) &= \sum_{\substack{\gamma\ periodic \\ \#[\gamma] | n}} 1 \\ &= \sum_{\substack{[\gamma]\ such\ that, \\ \#[\gamma] | n}} \#[\gamma]. \end{aligned}$$

From this follows

$$
\begin{aligned}
R_\phi(z) &= \exp\left(\sum_{n=1}^{\infty} \frac{R(\phi^n)}{n} z^n\right) \\
&= \exp\left(\sum_{[\gamma]} \sum_{\substack{n=1 \\ \#[\gamma]|n}}^{\infty} \frac{\#[\gamma]}{n} z^n\right) \\
&= \prod_{[\gamma]} \exp\left(\sum_{n=1}^{\infty} \frac{\#[\gamma]}{\#[\gamma]n} z^{\#[\gamma]n}\right) \\
&= \prod_{[\gamma]} \exp\left(\sum_{n=1}^{\infty} \frac{1}{n} z^{\#[\gamma]n}\right) \\
&= \prod_{[\gamma]} \exp\left(-\log\left(1 - z^{\#[\gamma]}\right)\right) \\
&= \prod_{[\gamma]} \frac{1}{1 - z^{\#[\gamma]}}.
\end{aligned}
$$

□

For a finitely generared abelian group G we define the finite subgroup G^{finite} to be the subgroup of torsion elements of G. We denote the quotient $G^\infty := G/G^{finite}$. The group G^∞ is torsion free. Since the image of any torsion element by a homomorphism must be a torsion element, the function $\phi : G \to G$ induces maps

$$\phi^{finite} : G^{finite} \longrightarrow G^{finite}, \quad \phi^\infty : G^\infty \longrightarrow G^\infty.$$

THEOREM 2. *If G is a finitely generated abelian group and ϕ an endomorphism of G then $R_\phi(z)$ is a rational function and is equal to the following additive convolution:*

$$R_\phi(z) = R_\phi^\infty(z) * R_\phi^{finite}(z). \tag{5}$$

where $R_\phi^\infty(z)$ is the Reidemeister zeta function of the endomorphism $\phi^\infty : G^\infty \to G^\infty$, and $R_\phi^{finite}(z)$ is the Reidemeister zeta function of the endomorphism $\phi^{finite} : G^{finite} \to G^{finite}$. The functions $R_\phi^\infty(z)$ and $R_\phi^{finite}(z)$ are given by the formulae

$$R_\phi^\infty(z) = \left(\prod_{i=0}^{k} \det(1 - \Lambda^i \phi^\infty.\sigma z)^{(-1)^{i+1}}\right)^{(-1)^r}, \tag{6}$$

$$R_\phi^{finite} = \prod_{[\gamma]} \frac{1}{1 - z^{\#[\gamma]}}. \tag{7}$$

with the product in (8) being taken over all periodic ϕ-orbits of torsion elements $\gamma \in G$. Also, $\sigma = (-1)^p$ where p is the number of real eingevalues $\lambda \in Spec\ \phi^\infty$ such that $\lambda < -1$ and r is the number of real eingevalues $\lambda \in Spec\ \phi^\infty$ such that $|\lambda| > 1$.

PROOF. By proposition 1, the cokernel of $(1-\phi^n) : G \to G$ is the Pontrjagin dual of the kernel of the dual map $\widehat{(1-\phi^n)} : \hat{G} \to \hat{G}$. Since $Coker\ (1-\phi^n)$ is finite, we have

$$\#Coker\ (1-\phi^n) = \#\ker \widehat{(1-\phi^n)}.$$

The map $\widehat{1-\phi^n}$ is equal to $\hat{1} - \hat{\phi}^n$. Its kernel is thus the set of fixed points of the map $\hat{\phi}^n : \hat{G} \to \hat{G}$. We therefore have

$$R(\phi^n) = \#Fix\ \left(\hat{\phi}^n : \hat{G} \to \hat{G}\right). \tag{8}$$

The dual group of G^∞ is a torus whose dimension is the rank of G. This is canonically a closed subgroup of $\hat{G}$. We shall denote it $\hat{G}_0$. The quotient $\hat{G}/(\hat{G}_0)$ is canonically isomorphic to the dual of G^{finite}. It is therefore finite. From this we know that $\hat{G}$ is a union of finitely many disjoint tori. We shall call these tori $\hat{G}_0, \ldots, \hat{G}_r$.

We shall call a torus $\hat{G}_i$ periodic if there is an iteration $\hat{\phi}^s$ such that $\hat{\phi}^s(\hat{G}_i) \subset \hat{G}_i$. If this is the case, then the map $\hat{\phi}^s : \hat{G}_i \to \hat{G}_i$ is a translation of the map $\hat{\phi}^s : \hat{G}_0 \to \hat{G}_0$ and has the same number of fixed points as this map. If $\hat{\phi}^s(\hat{G}_i) \not\subset \hat{G}_i$ then $\hat{\phi}^s$ has no fixed points in $\hat{G}_i$.

From this we see

$$\#Fix\ \left(\hat{\phi}^n : \hat{G} \to \hat{G}\right) = \#Fix\ \left(\hat{\phi}^n : \hat{G}_0 \to \hat{G}_0\right) \times \#\{\hat{G}_i \mid \hat{\phi}^n(\hat{G}_i) \subset \hat{G}_i\}.$$

We now rephrase this

$$\begin{aligned} &\#Fix\ \left(\hat{\phi}^n : \hat{G} \to \hat{G}\right) \\ &= \#Fix\ \left(\widehat{\phi^\infty}^n : \hat{G}_0 \to \hat{G}_0\right) \times \#Fix\ \left(\widehat{\phi^{finite}}^n : \hat{G}/(\hat{G}_0) \to \hat{G}/(\hat{G}_0)\right). \end{aligned}$$

From this we have

$$R_\phi(z) = R_{(\phi^\infty)}(z) * R_{(\phi^{finite})}(z).$$

The rationality of $R_\phi(z)$ and the formulae for $R_\phi^\infty(z)$ and $R_\phi^{finite}(z)$ follow from the previous two lemmas and lemma 2. □

COROLLARY 1. *Let the assumptions of theorem 1 hold. Then the poles and zeros of the Reidemeister zeta function are complex numbers of the form $\zeta^a b$ where b is the reciprocal of an eigenvalue of one of the matrices*

$$\Lambda^i(\phi^\infty) : \Lambda^i(G^\infty) \longrightarrow \Lambda^i(G^\infty) \qquad 0 \le i \le rank\ G$$

and ζ^a is a ψth root of unity where ψ is the number of periodic torsion elements in G. The multiplicities of the roots or poles $\zeta^a b$ and $\zeta^{a'} b'$ are the same if $b = b'$ and $hcf(a, \psi) = hcf(a', \psi)$. Here hcf is the highest common factor.

2.3.2. *Functional equation for the Reidemeister zeta function of an eventually commutative endomorphism.*

LEMMA 6 (FUNCTIONAL EQUATION FOR THE TORSION FREE PART [**9**]). *Let $\phi : \mathbb{Z}^k \to \mathbb{Z}^k$ be an endomorphism. The Reidemeister zeta function $R_\phi(z)$ has the following functional equation:*

$$R_\phi\left(\frac{1}{dz}\right) = \epsilon_1 . R_\phi(z)^{(-1)^k} \tag{9}$$

where $d = \det \phi$ and ϵ_1 is a constant in $\mathbb{C}^\times$.

LEMMA 7 (FUNCTIONAL EQUATION FOR THE FINITE PART). *Let $\phi : G \to G$ be an endomorphism of a finite, abelian group G. The Reidemeister zeta function $R_\phi(z)$ has the following functional equation:*

$$R_\phi\left(\frac{1}{z}\right) = (-1)^p z^q R_\phi(z), \tag{10}$$

where q is the number of periodic elements of ϕ in G and p is the number of periodic orbits of ϕ in G.

The proof of this is a simple computation.

THEOREM 3 (FUNCTIONAL EQUATION). *Let $\phi : G \to G$ be an endomorphism of a finitely generated abelian group G. If G is finite, then the functional equation of R_ϕ is described in lemma 7. If G is infinite then R_ϕ has the following functional equation:*

$$R_\phi\left(\frac{1}{dz}\right) = \epsilon_2 . R_\phi(z)^{(-1)^{Rank\ G}} \tag{11}$$

where $d = \det(\phi^\infty : G^\infty \to G^\infty)$ and ϵ_2 is a constant in $\mathbb{C}^\times$.

PROOF. From theorem 2 we have $R_\phi(z) = R_\phi^\infty(z) * R_\phi^{finite}(z)$. In the previous two lemmas we have obtained functional equations for the functions $R_\phi^\infty(z)$ and $R_\phi^{finite}(z)$. Lemma 3 now gives the functional equation for $R_\phi(z)$. $\square$

2.3.3. *Connection of $R_\phi(z)$ with the Lefschetz zeta function of the dual map.*

THEOREM 4 (CONNECTION WITH LEFSCHETZ NUMBERS). *Let $\phi : G \to G$ be an endomorphism of a finitely generated abelian group. Then we have the following*

$$R(\phi^n) = \mid L(\hat{\phi}^n) \mid, \tag{12}$$

where $\hat{\phi}$ is the continuous endomorphism of $\hat{G}$ defined in §2.2 and $L(\hat{\phi}^n)$ is the Lefschetz number.

From this it follows:

$$R_\phi(z) = L_{\hat{\phi}}(\sigma z)^{(-1)^r}, \tag{13}$$

where r and σ are constants as described in theorem 2. If G is finite then this reduces to

$$R(\phi^n) = L(\hat{\phi}^n) \text{ and } R_\phi(z) = L_{\hat{\phi}}(z).$$

The proof is similar to that of Anosov [**1**] concerning continuous maps of nil-manifolds.

PROOF. We already know from formula (8) in the proof of theorem 2 that $R(\phi^n)$ is the number of fixed points of the map $\hat{\phi}^n$. If G is finite then $\hat{G}$ is a discrete finite set, so the number of fixed points is equal to the Lefschetz number. This finishes the proof in the case that G is finite. In general it is only necessary to check that the number of fixed points of $\hat{\phi}^n$ is equal to the absolute value of its Lefschetz number. We assume without loss of generality that $n = 1$. We are assuming that $R(\phi)$ is finite, so the fixed points of $\hat{\phi}$ form a discrete set. We therefore have

$$L(\hat{\phi}) = \sum_{x \in Fix\ \hat{\phi}} Index\ (\hat{\phi}, x).$$

Since ϕ is an endomorphism, the zero element is always fixed. Let x be any fixed point of ϕ. We then have a commutative diagram

$$\begin{array}{cccccc} g & \hat{G} & \xrightarrow{\hat{\phi}} & \hat{G} & g \\ \updownarrow & \updownarrow & & \updownarrow & \updownarrow \\ g+x & \hat{G} & \xrightarrow{\hat{\phi}} & \hat{G} & g+x \end{array}$$

in which the vertical functions are translations on $\hat{G}$ by x. Since the vertical maps map 0 to x, we deduce that

$$Index\ (\hat{\phi}, x) = Index\ (\hat{\phi}, 0)$$

and so all fixed points have the same index. It is now sufficient to show that $Index\ (\hat{\phi}, 0) = \pm 1$. This follows because the map on the torus

$$\hat{\phi} : \hat{G}_0 \to \hat{G}_0$$

lifts to a linear map of the universal cover, which is in this case the Lie algebra of $\hat{G}$. The index is then the sign of the determinant of the identity minus this linear map. This determinant cannot be zero, because $1-\hat{\phi}$ must have finite kernel by our assumption that the Reidemeister number of ϕ is finite (if $\det(1-\hat{\phi}) = 0$ then the kernel of $1-\hat{\phi}$ is a positive dimensional subgroup of $\hat{G}$, and therefore infinite). □

2.4. Endomorphisms of finite groups. In this section we consider finite non-abelian groups. We shall write the group law multiplicatively. We generalize our results on endomorphisms of finite abelian groups to endomorphisms of finite non-abelian groups. We shall write $\{g\}$ for the ϕ-conjugacy class of an element $g \in G$. We shall write $< g >$ for the ordinary conjugacy class of g in G. We continue to write $[g]$ for the ϕ-orbit of $g \in G$, and we also write now $[< g >]$ for the ϕ-orbit of the ordinary conjugacy class of $g \in G$. We first note that if ϕ is an endomorphism of a group G then ϕ maps conjugate elements to conjugate elements. It therefore induces an endomorphism of the set of conjugacy classes of G. If G is abelian then a conjugacy class consists of a single element. The following is thus an extension of lemma 5:

THEOREM 5. *Let G be a finite group and let $\phi : G \to G$ be an endomorphism. Then $R(\phi)$ is the number of ordinary conjugacy classes $< x >$ in G such that $< \phi(x) > = < x >$.*

PROOF. From the definition of the Reidemeister number we have,

$$R(\phi) = \sum_{\{g\}} 1$$

where $\{g\}$ runs through the set of ϕ-conjugacy classes in G. This gives us immediately

$$\begin{aligned} R(\phi) &= \sum_{\{g\}} \sum_{x \in \{g\}} \frac{1}{\#\{g\}} \\ &= \sum_{\{g\}} \sum_{x \in \{g\}} \frac{1}{\#\{x\}} \\ &= \sum_{x \in G} \frac{1}{\#\{x\}} \end{aligned}$$

We now calculate for any $x \in G$ the order of $\{x\}$. The class $\{x\}$ is the orbit of x under the G-action

$$(g, x) \longmapsto gx\phi(g)^{-1}.$$

We verifty that this is actually a G-action:

$$\begin{aligned}(id, x) &\longmapsto id.x.\phi(id)^{-1} \\ &= x, \\ (g_1g_2, x) &\longmapsto g_1g_2.x.\phi(g_1g_2)^{-1} \\ &= g_1g_2.x.(\phi(g_1)\phi(g_2))^{-1} \\ &= g_1g_2.x.\phi(g_2)^{-1}\phi(g_1)^{-1} \\ &= g_1(g_2.x.\phi(g_2)^{-1})\phi(g_1)^{-1}.\end{aligned}$$

We therefore have from the orbit-stabilizer theorem,

$$\#\{x\} = \frac{\#G}{\#\{g \in G \mid gx\phi(g)^{-1} = x\}}.$$

The condition $gx\phi(g)^{-1} = x$ is equivalent to

$$x^{-1}gx\phi(g)^{-1} = 1 \quad \Leftrightarrow \quad x^{-1}gx = \phi(g)$$

We therefore have

$$R(\phi) = \frac{1}{\#G}\sum_{x\in G} \#\{g \in G \mid x^{-1}gx = \phi(g)\}.$$

Changing the summation over x to summation over g, we have:

$$R(\phi) = \frac{1}{\#G}\sum_{g\in G} \#\{x \in G \mid x^{-1}gx = \phi(g)\}.$$

If $< \phi(g) > \neq < g >$ then there are no elements x such that $x^{-1}gx = \phi(g)$. We therefore have:

$$R(\phi) = \frac{1}{\#G}\sum_{\substack{g\in G \text{ such that} \\ <\phi(g)>=<g>}} \#\{x \in G \mid x^{-1}gx = \phi(g)\}.$$

The elements x such that $x^{-1}gx = \phi(g)$ form a coset of the subgroup satisfying $x^{-1}gx = g$. This subgroup is the centralizer of g in G which we write $C(g)$. With this notation we have,

$$\begin{aligned}R(\phi) &= \frac{1}{\#G}\sum_{\substack{g\in G \text{ such that} \\ <\phi(g)>=<g>}} \#C(g) \\ &= \frac{1}{\#G}\sum_{\substack{<g>\subset G \text{ such that} \\ <\phi(g)>=<g>}} \# < g > .\#C(g).\end{aligned}$$

The last identity follows because $C(h^{-1}gh) = h^{-1}C(g)h$.

From the orbit stabilizer theorem, we know that $\# < g > .\#C(g) = \#G$. We therefore have

$$R(\phi) = \#\{< g > \subset G \mid < \phi(g) > = < g >\}.$$

□

From this theorem we have immediately,

THEOREM 6. *Let ϕ be an endomorphism of a finite group G. Then $R_\phi(z)$ is a rational function with a functional equation. In particular we have,*

$$R_\phi(z) = \prod_{[<g>]} \frac{1}{1 - z^{\#[<g>]}},$$

$$R_\phi\left(\frac{1}{z}\right) = (-1)^a z^b R_\phi(z).$$

The product here is over all periodic ϕ-orbits of ordinary conjugacy classes of elements of G. The number $\#[< g >]$ is the number of conjugacy classes in the ϕ-orbit of the conjugacy class $< g >$. In the functional equation the numbers a and b are respectively the number of periodic ϕ-orbits of conjugacy classes of elements of G and the number of periodic conjugacy classes of elements of G. A conjugacy class $< g >$ is called periodic if for some $n > 0$, $< \phi^n(g) >=< g >$

PROOF. From the previous theorem we know that $R(\phi^n)$ is the number of conjugacy classes $< g >\subset G$ such that $\phi^n(< g >) \subset< g >$. We can rewrite this

$$R(\phi^n) = \sum_{\substack{[< g >] \text{ such that} \\ \#[< g >] \mid n}} \#[< g >].$$

From this we have,

$$R_\phi(z) = \prod_{[<g>]} \exp\left(\sum_{\substack{n=1 \text{ such that} \\ \#[< g >] \mid n}}^{\infty} \frac{\#[< g >]}{n} z^n \right).$$

The first formula now follows by using the power series expansion for $\log(1 - z)$. The functional equation follows from the previous theorem in exactly the same way as lemma 7 follows from lemma 5. □

COROLLARY 2. *Suppose that ϕ_1 and ϕ_2 are two endomorphisms of a finite group G with*

$$\forall g \in G,\ \phi_1(g) = h\phi_2(g)h^{-1}$$

for some fixed element $h \in G$. Then $R_{\phi_1}(z) = R_{\phi_2}(z)$.

COROLLARY 3. *Let ϕ be an inner automorphism. Then*

$$R_\phi(z) = \frac{1}{(1 - z)^b}$$

where b is the number of conjugacy classes in the group. In particular, all but finitely many of the symmetric and alternating groups have the property that any automorphism is an inner automorphism, and so this corollary applies.

REMARK 1. If we think of the set of conjugacy classes in G as a discrete set then the Reidermeister number of ϕ is equal to the Lefschetz number of the induced map on the conjugacy classes.

2.5. The Reidemeister zeta function and group extensions. Suppose we are given a commutative diagram

$$\begin{array}{ccc} G & \xrightarrow{\phi} & G \\ \downarrow p & & \downarrow p \\ \overline{G} & \xrightarrow{\overline{\phi}} & \overline{G} \end{array} \tag{14}$$

of groups and homomorphisms. In addition let the sequence

$$0 \to H \to G \to \overline{G} \to 0 \tag{15}$$

be exact. Then ϕ restricts to an endomorphism $\phi\mid_H : H \to H$.

DEFINITION 1. The short exact sequence (16) of groups is said to have a *normal splitting* if there is a section $\sigma : \overline{G} \to G$ of p such that $Im\ \sigma = \sigma(G)$ is a normal subgroup of G. An endomorphism $\phi : G \to G$ is said to *preserve* this normal splitting if ϕ induces a morphism of (16) with $\phi(\sigma(\overline{G})) \subset \sigma(\overline{G})$.

In this section we study the relation between the Reidemeister zeta functions $R_\phi(z)$, $R_{\overline{\phi}}(z)$ and $R_{\phi|_H}(z)$.

THEOREM 7. *Let the sequence (14) have a normal splitting which is preserved by $\phi : G \to G$. Then we have*

$$R_\phi(z) = R_{\overline{\phi}}(z) * R_{\phi|_H}(z).$$

In particular, if $R_{\overline{\phi}}(z)$ and $R_{\phi|_H}(z)$ are rational functions then so is $R_\phi(z)$. If $R_{\overline{\phi}}(z)$ and $R_{\phi|_H}(z)$ are rational functions with functional equations as described in lemma 3 then so is $R_\phi(z)$.

PROOF. From the assumptions of the theorem it follows that for every $n > 0$

$$R(\phi^n) = R(\overline{\phi}^n).R(\phi^n\mid_H) \quad (\textit{see } [\mathbf{14}]).$$

□

3. The Reidemeister zeta function of a continuous map.

Let $f : X \to X$ be given, and let a specific lifting $\tilde{f} : \tilde{X} \to \tilde{X}$ be chosen as reference. Let Γ be the group of covering translations of $\tilde{X}$ over X. Then every lifting of f can be written uniquely as $\gamma \circ \tilde{f}$, with $\gamma \in \Gamma$. So elements of Γ serve as coordinates of liftings with respect to the reference $\tilde{f}$. Now for every $\gamma \in \Gamma$ the composition $\tilde{f} \circ \gamma$ is a lifting of f so there is a unique $\gamma' \in \Gamma$ such that $\gamma' \circ \tilde{f} = \tilde{f} \circ \gamma$. This correspondence $\gamma \to \gamma'$ is determined by the reference $\tilde{f}$, and is obviously a homomorphism.

DEFINITION 2. The endomorphism $\tilde{f}_* : \Gamma \to \Gamma$ determined by the lifting $\tilde{f}$ of f is defined by

$$\tilde{f}_*(\gamma) \circ \tilde{f} = \tilde{f} \circ \gamma.$$

LEMMA 8. *Lifting classes of f are in 1-1 correspondence with $\tilde{f}_*$-conjugacy classes in π, the lifting class $[\gamma \circ \tilde{f}]$ corresponding to the $\tilde{f}_*$-conjugacy class of γ. We therefore have $R(f) = R(\tilde{f}_*)$.*

We shall say that the fixed point class $p(Fix(\gamma \circ \tilde{f}))$, which is labeled with the lifting class $[\gamma \circ \tilde{f}]$, *corresponds* to the $\tilde{f}_*$-conjugacy class of γ. Thus $\tilde{f}_*$-conjugacy classes in π serve as coordinates for fixed point classes of f, once a reference lifting $\tilde{f}$ is chosen. Using lemma 8 we may apply the theorems of §2 to the Reidemeister zeta functions of continuous maps.

THEOREM 8. *Let X be a polyhedron with finite fundamental group $\pi_1(X)$ and let $f : X \to X$ be a continuous map. Then $R_f(z)$ is a rational function with a functional equation:*

$$R_f(z) = R_{\tilde{f}_*}(z) = \prod_{[<g>]} \frac{1}{1 - z^{\#[<g>]}},$$

$$R_f\left(\frac{1}{z}\right) = (-1)^a z^b R_f(z).$$

The product in the first formula is over all periodic $\tilde{f}_*$-orbits of ordinary conjugacy classes of elements of $\pi_1(X)$. The number $\#[< g >]$ is the number of conjugacy classes in the f_*-orbit of $< g >$. In the functional equation the numbers a and b are respectively the number of periodic $\tilde{f}_*$ orbits of cojugacy classes of elements of $\pi_1(X)$, and the number of periodic conjugacy classes of elements of $\pi_1(X)$.

DEFINITION 3. A map $f : X \to X$ is said to be *eventually commutative* if there exists an natural number n such that $f_*^n(\pi_1(X, x_0))$ $(\subset \pi_1(X, f^n(x_0)))$ is commutative.

It is easily seen that f is eventually commutative iff $\tilde{f}_*$ is eventually commutative (see [**16**]).

Theorems 2 and 3 yield

THEOREM 9. *Let $f : X \to X$ be eventually commutative. Then $R_f(z)$ is a rational function and is given by:*

$$R_f(z) = R_{\tilde{f}_*}(z) = R_{f_{1*}}(z) = R^{\infty}_{f_{1*}}(z) * R^{finite}_{f_{1*}}(z),$$

where $R^{\infty}_{f_{1*}}(z)$ is the Reidemeister zeta function of the endomorphism $f^{\infty}_{1*} : H_1(X, \mathbb{Z})^{\infty} \to H_1(X, \mathbb{Z})^{\infty}$ and $R^{finite}_{f_{1*}}(z)$ is the Reidemeister zeta function of the endomorphism $f^{finite}_{1*} : H_1(X, \mathbb{Z})^{finite} \to H_1(X, \mathbb{Z})^{finite}$. The functions $R^{\infty}_{f_{1*}}(z)$ and $R^{finite}_{f_{1*}}(z)$ are given by the formulae:

$$R^{\infty}_{f_{1*}}(z) = \left(\prod_{i=0}^{k} \det\left(1 - \Lambda^i f^{\infty}_{1*} \sigma z\right)^{(-1)^{i+1}}\right)^{(-1)^r}$$

$$R^{finite}_{f_{1*}}(z) = \prod_{[h]} \frac{1}{1 - z^{\#[h]}}$$

With the product over $[h]$ being taken over all periodic f_{1*} orbits of torsion elements $h \in H_1(X, \mathbb{Z})$, and with $\sigma = (-1)^p$ where p is the number of $\mu \in Spec\ f^{\infty}_{1*}$ such that $\mu < -1$. The number r is the number of real eigenvalues of f^{∞}_{1*} whose absolute value is > 1.

THEOREM 10 (FUNCTIONAL EQUATION). *Let $f : X \to X$ be eventually commutative. If $H_1(X; \mathbb{Z})$ is finite, then $R_f(z)$ has the following functional equation:*

$$R_f\left(\frac{1}{z}\right) = (-1)^p z^q R_f(z),$$

where p is the number of periodic orbits of f_{1*} in $H_1(X; \mathbb{Z})$ and q is the number of periodic elements of f_{1*} in $H_1(X; \mathbb{Z})$.

If $H_1(X; \mathbb{Z})$ is infinite then $R_f(z)$ has the following functional equation:

$$R_f\left(\frac{1}{dz}\right) = \epsilon_2 . R_f(z)^{(-1)^{Rank\ H_1(X;\mathbb{Z})}},$$

where $d = \det(f^{\infty}_{1*})$ and $\epsilon_2 \in \mathbb{C}^{\times}$ is a constant.

3.1. The Reidemeister zeta function and Serre bundles. Let $p : E \to B$ be a Serre bundle in which E, B and every fibre are connected, compact polyhedra and $F_b = p^{-1}(b)$ is a fibre over $b \in B$. A Serre bundle $p : E \to B$ is said to be *(homotopically) orientable* if for any two paths w, w' in B with the same endpoints $w(0) = w'(0)$ and $w(1) = w'(1)$, the fibre translations $\tau_w, \tau_{w'} : F_{w(0)} \to F_{w(1)}$ are homotopic. A map $f : E \to E$ is called a *fibre map* if there is an induced map $\bar{f} : B \to B$ such that $p \circ f = \bar{f} \circ p$. Let $p : E \to B$ be an orientable Serre bundle and let $f : E \to E$ be a fibre map. Then for any two fixed points b, b' of $\bar{f} : B \to B$ the maps $f_b = f\,|_{F_b}$ and $f_{b'} = f\,|_{F_{b'}}$ have the same homotopy type; hence they have the same Reidemeister numbers $R(f_b) = R(f_{b'})$ [**16**].

The following theorem describes the relation between the Reidemeister zeta functions $R_f(z)$, $R_{\bar{f}}(z)$ and $R_{f_b}(z)$ for a fibre map $f : E \to E$ of an orientable Serre bundle $p : E \to B$.

THEOREM 11. *Suppose that $f : E \to E$ admits a Fadell splitting in the sense that for some e in $Fix f$ and $b = p(e)$ the following conditions are satisfied:*

(i) *the sequence*

$$0 \longrightarrow \pi_1(F_b, e) \xrightarrow{i_*} \pi_1(E, e) \xrightarrow{p_*} \pi_1(B, e) \longrightarrow 0$$

is exact,

(ii) *p_* admits a right inverse (section) σ such that $Im\ \sigma$ is a normal subgroup of $\pi_1(E, e)$ and $f_*(Im\ \sigma) \subset Im\ \sigma$.*

We then have

$$R_f(z) = R_{\bar{f}}(z) * R_{f_b}(z).$$

If $R_{\bar{f}}(z)$ and $R_{f_b}(z)$ are rational functions then so is $R_f(z)$. If $R_{\bar{f}}(z)$ and $R_{f_b}(z)$ are rational functions with functional equations as described in lemma 3 then so is $R_f(z)$.

PROOF. This follows from theorem 7. □

4. The Nielsen zeta function

4.1. The Jiang subgroup and the Nielsen zeta function. From the homotopy invariance theorem (see [**16**]) it follows that if a homotopy $\{h_t\} : f \cong g : X \to X$ lifts to a homotopy $\{\tilde{h}_t\} : \tilde{f} \cong \tilde{g} : \tilde{X} \to \tilde{X}$, then we have $Index(f, p(Fix\ \tilde{f})) = Index(g, p(Fix\ \tilde{g}))$. Suppose $\{h_t\}$ is a cyclic homotopy $\{h_t\} : f \cong f$; then this lifts to a homotopy from a given lifting $\tilde{f}$ to another lifting $\tilde{f}' = \alpha \circ \tilde{f}$, and we have

$$Index(f, p(Fix\ \tilde{f})) = Index(f, p(Fix\ \alpha \circ \tilde{f})).$$

In other words, a cyclic homotopy induces a permutation of lifting classes (and hence of fixed point classes); those in the same orbit of this permutation have the same index. This idea is applied to the computation of $N_f(z)$.

DEFINITION 4. The *trace subgroup of cyclic homotopies* (the *Jiang subgroup*) $I(\tilde{f}) \subset \pi$ is defined by

$$I(\tilde{f}) = \left\{ \alpha \in \pi \;\middle|\; \begin{array}{l} \text{there exists a cyclic homotopy} \\ \{h_t\} : f \cong f \text{which lifts to} \\ \{\tilde{h}_t\} : \tilde{f} \cong \alpha \circ \tilde{f} \end{array} \right\}$$

(see [**16**]).

From theorem 9 and the results of Jiang [**16**] we have:

THEOREM 12. *Suppose that there is a natural number m such that $\tilde{f}_*^m(\pi) \subset I(\tilde{f}^m)$ and $L(f^n) \neq 0$ for every $n > 0$. Then $N_f(z) = R_f(z)$ is rational and is given by:*

$$N_f(z) = R_f(z) =$$
$$= \left(\left(\prod_{i=0}^{k} \det\left(1 - \Lambda^i f_{1*}^{\infty} \sigma z\right)^{(-1)^{i+1}} \right)^{(-1)^r} \right) * \left(\prod_{[h]} \frac{1}{1 - z^{\#[h]}} \right), \tag{16}$$

This function has a functional equation as described in theorem 10

COROLLARY 4. *Let the assumptions of theorem 12 hold. Then the poles and zeros of the Nielsen zeta function are complex numbers of the form $\zeta^a b$ where b is the reciprocal of an eigenvalue of one of the matrices*

$$\Lambda^i(f_{1*}^{\infty}) : \Lambda^i(H_1(X;\mathbb{Z})^{\infty}) \longrightarrow \Lambda^i(H_1(X;\mathbb{Z})^{\infty}) \qquad 0 \leq i \leq rank\ G$$

and ζ^a is a ψth root of unity where ψ is the number of periodic torsion elements in $H_1(X;\mathbb{Z})$. The multiplicities of the roots or poles $\zeta^a b$ and $\zeta^{a'} b'$ are the same if $b = b'$ and $hcf(a,\psi) = hcf(a',\psi)$. Again $hcf(a,\psi)$ is the highest common factor of a and ψ.

COROLLARY 5. *Let $I(id_{\tilde{X}}) = \pi$ and $L(f^n) \neq 0$ for all $n > 0$. Then formula (16) is valid.*

COROLLARY 6. *Suppose that X is aspherical, f is eventually commutative and $L(f^n) \neq 0$ for all $n > 0$. Then formula (16) is valid.*

EXAMPLE 1. Let $f : T^n \to T^n$ be a hyperbolic endomorphism. Then $N_f(z) = R_f(z)$ and this is a rational function given by:

$$N_f(z) = \left(\prod_{i=0}^{k} \det\left(1 - \Lambda^i f_{1*} \sigma z\right)^{(-1)^{i+1}} \right)^{(-1)^r} = L_f(\sigma z)^{(-1)^r}$$

where σ and r are as in theorem 9.

4.2. Polyhedra with finite fundamental group. For a compact polyhedron X with finite fundamental group, $\pi_1(X)$, the universal cover $\tilde{X}$ is compact, so we may explore the relation between $L(\tilde{f}^n)$ and $Index\ (p(Fix\ \tilde{f}^n))$.

From the results of Boju Jiang [**16**] and theorem 6 of this article, one deduces immediately:

THEOREM 13. *Let X be a connected, compact polyhedron with finite fundamental group π. Suppose that the action of π on the rational homology of the universal cover $\tilde{X}$ is trival, i.e. for every covering translation $\alpha \in \pi$,*

$\alpha_* = id : H_*(\tilde{X},\mathbb{Q}) \to H_*(\tilde{X},\mathbb{Q})$. Let $L(f^n) \neq 0$ for all $n > 0$. Then N_f is a rational function given by

$$N_f(z) = R_f(z) = \prod_{[<h>]} \frac{1}{1 - z^{\#[<h>]}}, \tag{17}$$

where the product is taken over all periodic $\tilde{f}_*$-orbits of ordinary conjugacy classes in the finite group $\pi_1(X)$. This function has a functional equation as described in theorem 6 or theorem 8.

LEMMA 9. *Let X be a polyhedron with finite fundamental group π and let $p : \tilde{X} \to X$ be its universal covering. Then the action of π on the rational homology of $\tilde{X}$ is trivial iff $H_*(\tilde{X};\mathbb{Q}) \cong H_*(X;\mathbb{Q})$.*

COROLLARY 7. *Let $\tilde{X}$ be a compact 1-connected polyhedron which is a rational homology n-sphere, where n is odd. Let π be a finite group acting freely on $\tilde{X}$ and let $X = \tilde{X}/\pi$. Then theorem 13 applies.*

COROLLARY 8. *If X is a closed 3-manifold with finite π, then theorem 13 applies.*

COROLLARY 9. *Let $X = L(m, q_1, \dots, q_r)$ be a generalized lens space and $f : X \to X$ a continuous map with $f_{1*}(1) = d$ where $\mid d \mid \neq 1$. The Nielsen and Reidemeister zeta functions are then rational and are given by the formula:*

$$N_f(z) = R_f(z) = \prod_{[h]} \frac{1}{1 - z^{\#[h]}} = \prod_{t=1}^{\varphi_d(m)} (1 - e^{2\pi i t/\varphi_d(m)} z)^{-a(t)}.$$

where $[h]$ runs over the periodic f_{1*}-orbits of elements of $H_1(X;\mathbb{Z})$. The numbers $a(t)$ are natural numbers given by the formula

$$a(t) = \sum_{\substack{s \mid m \text{ such that} \\ \varphi_d(m) \mid t\varphi_d(s)}} \frac{\varphi(s)}{\varphi_d(s)},$$

where φ is the Euler totient function and $\varphi_d(s)$ is the order of the multiplicative subgroup of $(\mathbb{Z}/s\mathbb{Z})^\times$ generated by d.

5. Congruences for Reidemeister and Nielsen numbers

Let $\mu(d)$, $d \in \mathbb{N}$ be the Moebius function, i.e.

$$\mu(d) = \begin{cases} 1 & if\ d = 1, \\ (-1)^k & if\ d\ is\ a\ product\ of\ k\ distinct\ primes, \\ 0 & if\ d\ is\ not\ square-free. \end{cases}$$

We define the numbers $S(d)$, $d \in \mathbb{N}$, by

$$S(d) = \sum_{d_1 | d} \mu(d_1) R\left(\phi^{d/d_1}\right).$$

One then has

THEOREM 14 (INFINITE PRODUCT FORMULA [**9**]).

$$R_\phi(z) = \prod_{d=1}^{\infty} \sqrt[d]{(1-z^d)^{-S(d)}}. \tag{18}$$

In theorems 2 and 11 we prove that the Reidemeister zeta function of an eventually commutative map or endomorphism is rational. These results together with theorem 14 and the congruences of Dold for the Lefschetz numbers give motivation for the following:

THEOREM 15 (CONGRUENCES FOR THE REIDEMEISTER NUMBERS). *Let $\phi : G \to G$ be an endomorphism of the group G. If ϕ is eventually commutative or if G is finite, then one has for all natural numbers n,*

$$\sum_{d|n} \mu(d) R(\phi^{n/d}) \equiv 0 \mod n.$$

For finite groups, these congruences follow from those of Dold for Lefschetz numbers since we have identified in theorem 4 and remark 1 the Reidemeister numbers with the Lefschetz numbers of induced maps. We give here a simple direct proof of the congruences.

PROOF. We first consider the case that ϕ is eventually commutative. Using theorem 1 we may assume that G is abelian, and one has $R(\phi^n) = \#Coker\ (1 - \phi^n)$.

From proposition 1 and our assumption that $R(\phi^n)$ is finite, we have $R(\phi^n) = \#\ker(\hat{1} - \hat{\phi}^n)$. The kernel of $\hat{1} - \hat{\phi}^n$ is the set of fixed points of $\hat{\phi}^n$. We therefore have,

$$R(\phi^n) = \#Fix\ \left[\hat{\phi}^n : \hat{G} \to \hat{G}\right].$$

Let P_n denote the number of periodic points of $\hat{\phi}$ of least period n. One sees immediately that

$$R(\phi^n) = \#Fix\ \left[\hat{\phi}^n\right] = \sum_{d|n} P_d.$$

Applying Möbius' inversion formula, we have,

$$P_n = \sum_{d|n} \mu(d) R(\phi^{n/d}).$$

On the other hand, we know that P_n is always divisible be n, because P_n is exactly n times the number of $\hat{\phi}$-orbits in $\hat{G}$ of length n.

In the case that G is finite, we know from theorem 5 that $R(\phi^n)$ is the number conjugacy classes in G which are fixed by ϕ^n. The proof then follows as in the previous case. □

COROLLARY 10. *Let $f : X \to X$ be a self-map. If f is eventually commutative or if $\pi_1(X)$ is finite, then one has for any n,*

$$\sum_{d|n} \mu(d).R(f^{n/d}) \equiv 0 \mod n.$$

PROOF. $R(f^n) = R(\tilde{f}_*^n) = R(f_{1*}^n)$. □

COROLLARY 11. *Suppose that $f : X \to X$ is a self-map with either $\tilde{f}_*(\pi) \subset I(\tilde{f})$, or $\pi_1(X)$ finite and $H_*(\tilde{X};\mathbb{Q}) \cong H_*(X;\mathbb{Q})$. If for every d dividing a certain natural number n we have $L(f^{n/d}) \neq 0$, then one has for that particular n,*

$$\sum_{d|n} \mu(d).N(f^{n/d}) \equiv 0 \mod n.$$

6. Connection with Reidemeister Torsion

6.1. Preliminaries. Like the Euler characteristic, the Reidemeister torsion is algebraically defined. Roughly speaking, the Euler characteristic is a graded version of the dimension, extending the dimension from a single vector space to a complex of vector spaces. In a similar way, the Reidemeister torsion is a graded version of the absolute value of the determinant of an isomorphism of vector spaces. Let $d^i : C^i \to C^{i+1}$ be a cochain complex C^* of finite dimensional vector spaces over $\mathbb{C}$ with $C^i = 0$ for $i < 0$ and large i. If the cohomology $H^i = 0$ for all i we say that C^* is *acyclic*. If one is given positive densities Δ_i on C^i then the Reidemeister torsion $\tau(C^*, \Delta_i) \in (0, \infty)$ for acyclic C^* is defined as follows:

DEFINITION 5. Consider a chain contraction $\delta^i : C^i \to C^{i-1}$, ie. a linear map such that $d \circ \delta + \delta \circ d = id$. Then $d + \delta$ determines a map $(d+\delta)_+ : C^+ := \oplus C^{2i} \to C^- := \oplus C^{2i+1}$ and a map $(d+\delta)_- : C^- \to C^+$. Since the map $(d+\delta)^2 = id + \delta^2$ is unipotent, $(d+\delta)_+$ must be an isomorphism. One defines $\tau(C^*, \Delta_i) :=\mid \det(d+\delta)_+ \mid$ (see **[13]**).

Reidemeister torsion is defined in the following geometric setting. Suppose K is a finite complex and E is a flat, finite dimensional, complex vector bundle with base K. By choosing a flat density on E we obtain a preferred density Δ_i on the cochain complex $C^i(K, E)$ with coefficients in E. One defines the R-torsion of (K, E) to be $\tau(K; E) = \tau(C^*(K; E), \Delta_i) \in (0, \infty)$.

6.2. The Reidemeister zeta Function and the Reidemeister Torsion of the Mapping Tori. Let $f : X \to X$ be a homeomorphism of a compact polyhedron X. Let $T_f := (X \times I)/(x, 0) \sim (f(x), 1)$ be the mapping tori of f. We shall consider the bundle $p : T_f \to S^1$ over the circle S^1. We assume here that E is a flat, complex vector bundle with finite dimensional fibre and base S^1. We form its pullback p^*E over T_f. Note that the vector spaces $H^i(p^{-1}(b), c)$

with $b \in S^1$ form a flat vector bundle over S^1, which we denote H^iF. The integral lattice in $H^i(p^{-1}(b), \mathbb{R})$ determines a flat density by the condition that the covolume of the lattice is 1. We suppose that the bundle $E \otimes H^iF$ is acyclic for all i. Under these conditions D. Fried [**13**] has shown that the bundle p^*E is acyclic, and we have

$$\tau(T_f; p^*E) = \prod_i \tau(S^1; E \otimes H^iF)^{(-1)^i}. \tag{19}$$

Let g be the prefered generator of the group $\pi_1(S^1)$ and let $A = \rho(g)$ where $\rho : \pi_1(S^1) \to GL(V)$. Then the holonomy around g of the bundle $E \otimes H^iF$ is $A \otimes f_i^*$. Since $\tau(E) = \mid \det(I - A) \mid$ it follows from (19) that

$$\tau(T_f; p^*E) = \prod_i \mid \det(I - A \otimes f_i^*) \mid^{(-1)^i}. \tag{20}$$

We now consider the special case in which E is one-dimensional, so A is just a complex scalar λ of modulus one. Then in terms of the rational function $L_f(z)$ we have [**13**]:

$$\tau(T_f; p^*E) = \prod_i \mid \det(I - \lambda . f_i^*) \mid^{(-1)^i} = \mid L_f(\lambda) \mid^{-1} \tag{21}$$

From this formula and theorem 4 we have

THEOREM 16. *Let $\phi : G \to G$ be an automorphism of a finitely generated abelian group G. If G is infinite then one has*

$$\tau\left(T_{\hat{\phi}}; p^*E\right) = \mid L_{\hat{\phi}}(\lambda) \mid^{-1} = \mid R_{\phi}(\sigma\lambda) \mid^{(-1)^{r+1}},$$

and if G is finite one has

$$\tau\left(T_{\hat{\phi}}; p^*E\right) = \mid L_{\hat{\phi}}(\lambda) \mid^{-1} = \mid R_{\phi}(\lambda) \mid^{-1},$$

where λ is the holonomy of the one-dimensional flat complex bundle E over S^1, r and σ are the constants descrebed in theorem 2.

THEOREM 17. *Let $f : X \to X$ be a homeomorphism of a compact polyhedron X. If $\pi_1(X)$ is abelian then one has when $\pi_1(X)$ is infinite,*

$$\tau\left(T_{\widehat{(f_{1*})}}; p^*E\right) = \left| L_{\widehat{(f_{1*})}}(\lambda) \right|^{-1} = \left| R_f(\sigma\lambda) \right|^{(-1)^{r+1}},$$

and when $\pi_1(X)$ is finite,

$$\tau\left(T_{\widehat{(f_{1*})}}; p^*E\right) = \left| L_{\widehat{(f_{1*})}}(\lambda) \right|^{-1} = \left| R_f(\lambda) \right|^{-1},$$

where r and σ are the constants described in theorem 9.

THEOREM 18. *Let $f : X \to X$ be a hyperbolic automorphism of a torus or a nilmanifold X. Then*

$$\tau(T_f; p^*E) = \mid L_f(\lambda) \mid^{-1} = \mid N_f(\sigma.\lambda) \mid^{(-1)^{r+1}} = \mid R_f(\sigma.\lambda) \mid^{(-1)^{r+1}} \quad (22)$$

where $\sigma = (-1)^p$, p is the number of real eigenvalues of f_{*1} in the region $(-\infty, -1)$ and r is the number of real eigenvalues of f_{*1} whose absolute value is greater that 1.

PROOF. From [**9**] it follows that $N_f(z) = R_f(z) = (L_f(\sigma.z))^{(-1)^r}$. The theorem then follows formula (21). □

7. Concluding remarks, problems

7.1. Reidemeister and Nielsen zeta functions modulo a normal subgroup. In the theory of (ordinary) fixed point classes, we work on the universal covering space. The group of covering transformations plays a key role. It is not surprising that this theory can be generalized to work on all regular covering spaces. Let K be a normal subgroup of the fundamental group $\pi_1(X)$. Consider the regular covering $p_K : \tilde{X}/K \to X$ corresponding to K. A map $\tilde{f}_K : \tilde{X}/K \to \tilde{X}/K$ is called a lifting of $f : X \to X$ if $p_K \circ \tilde{f}_K = f \circ p_K$. We know from the theory of covering spaces that such liftings exist if and only if $f_*(K) \subset K$. If K is a fully invariant subgroup of $\pi_1(X)$ (in the sense that every endomorphism sends K into K) such as, for example the commutator subgroup of $\pi_1(X)$, then there is a lifting of any continuous map. We may define the mod K-Reidemeister and Nielsen zeta functions (see [**6, 10**]) and develop a similar theory for them by simply replacing $\tilde{X}$ and $\pi_1(X)$ by $\tilde{X}/K$ and $\pi_1(X)/K$ in every definition, every theorem and every proof, since everything was done in terms of liftings and covering translations.

7.2. Automorphisms of non-abelian groups. In our study of the Reidemeister zeta function of a group endomorphism we restrict our attention to eventually commutative endomorphisms and endomorphisms of finite groups. The situation for an endomorphism of an infinite non-abelian group is still unclear, although we do have a conjecture in this direction.

We have not found any group endomorphism whose zeta function is not rational. For this question it is sufficient to consider only isomorphisms. [1]

We shall write $IRR\ G$ for the set of isomorphism classes of irreducible unitary representations of G. If $\phi : G \to G$ is an endomorphism, then there is an induced map $IRR\ \phi : IRR\ G \to IRR\ G$ given by $IRR\ \phi(\rho) := \rho \circ \phi$. With these definitions, IRR is a functor. If ϕ is an inner automorphism then $IRR\ \phi$ is

[1] It now seems that certain endomorphisms of the semi-direct product of $\mathbb{Z}/2\mathbb{Z}$ and $\mathbb{Z}^n$ for $n > 1$ have irrational zeta functions. Here $\mathbb{Z}/2\mathbb{Z}$ operates on $\mathbb{Z}^n$ by sending every element to its inverse. On the other hans, we can now show that an endomorphism of a direct sum of a finite group with $\mathbb{Z}^n$ always has a rational zeta function with a functional equation.

the identity map. If G is abelian then all its irreducible representations are 1-dimensional, and so $IRR\ G$ and $IRR\ \phi$ coincide with $\hat{G}$ and $\hat{\phi}$. We are lead to the following:

QUESTION Is $R(\phi)$ equal to the number of fixed points of the map $IRR\ \phi : IRR\ G \to IRR\ G$?

and equivalently in the topological situation:

QUESTION Let $f : X \to X$ be a homeomorphism of a compact polyhedron. Is $R(f)$ equal to the number of isomorphism classes of complex irreducible flat vector bundles which are preserved by the map induced by f?

In this article we have obtained (in a slightly different notation) a 'yes' in the case that our maps are eventually commutative (this case reduces to formula (8) and for this we do not use the fact that G is finitely generated) and in the case that the group is finite (this is essentially the same as theorem 5).

WE WOULD LIKE TO THANK THE SONDERFORSCHUNGSBEREICH 170 IN GÖTTINGEN AND THE MATHEMATICS INSTITUTE OF THE UNIVERSITY OF GÖTTINGEN FOR THEIR KIND HOSPITALITY, SUPPORT AND THE USE OF THEIR FACILITIES. RICHARD HILL WOULD LIKE TO THANK THE SERC FOR THEIR SUPPORT DURING THE LAST THREE YEARS. WE ARE PARTICULARLY GRATEFUL TO H. S. HOLDGRÜN, D. NOTBOHM, A. PANCHISHKIN AND S. J. PATTERSON FOR VALUABLE CONVERSATIONS AND COMMENTS.

REFERENCES

1. D. V. Anosov The Nielsen numbers of maps of nil-manifolds, Russian Mathem. Surveys, 40:4 (1985), 149-150.
2. J. Cheeger, Analytic torsion and the heat equation, Annals of Math., 109 (1979), 259-322.
3. A. Dold, Fixed point indices of iterated maps, Inventiones Math. 74 (1983), 419-435.
4. A. L. Fel'shtyn, New zeta function in dynamics, in Tenth Internat. Conf. on Nonlinear Oscillations, Varna, Abstracts of Papers, Bulgar. Acad. Sci., 1984, 208
5. -, A new zeta function in Nielsen theory and the universal product formula for dynamic zeta functions, Functsional Anal. i Prilozhen 21 (2) (1987), 90-91 (in Russian); English transl.: Functional Anal. Appl. 21 (1987), 168-170.
6. -, Zeta functions in Nielsen theory, Functsional Anal. i Prilozhen 22 (1) (1988), 87-88 (in Russian); English transl.: Functional Anal. Appl. 22 (1988), 76-77.
7. -, New zeta functions for dynamical systems and Nielsen fixed point theory, in : Lecture Notes in Math. 1346, Springer, 1988, 33-35.
8. -, Dynamical zeta-functions and the Nielsen theory, in : Baku Internat. Topological Conf., Abstracts of papers, Akad. Nauk SSSR, 1988, 311.
9. -, The Reidemeister zeta function and the computation of the Nielsen zeta function, Colloquium Mathematicum 62 (1) (1991), 153-166.
10. -, The Reidemeister, Nielsen zeta functions and the Reidemeister torsion in dynamical systems theory, Mathematica Göttingensis. Heft 47 (1991) 32p.
11. W. Franz, Über die Torsion einer Überdeckung, J. Reine Angew. Math., 173 (1935), 245-254.
12. D. Fried, Homological identities for closed orbits, Invent. Math., 71 (1983), 219-246.
13. D. Fried, Lefschetz formula for flows, The Lefschetz centennial conference, Contemp. Math., 58 (1987), 19-69.
14. P. R. Heath, Product formulae for Nielsen numbers of fibre maps, Pacific J. Math. 117 (2) (1985), 267-289.
15. P. R. Heath, R. Piccinini, C. You, Nielsen-type numbers for periodic points I, Lecture Notes in Math. Vol. 1411 (1988) p.86-88.

16. B. Jiang, Nielsen Fixed Point Theory, Contemp. Math. 14, Birkhäuser, 1983.
17. J. Milnor, A duality theorem for Reidemeister torsion, Ann. of Math., 76 (1962), 137-147.
18. J. Milnor, Infinite cyclic covers, Proc. Conf. "Topology of Manifolds" in Michigan 1967, 115-133.
19. W. Müller, Analytic torsion and R-torsion of Riemannian manifolds, Adv. in Math. 28 (1978), 233-305.
20. J. Nielsen, Untersuchung zur Topologie des geschlossenen zweiseitigen Fläche, Acta Math. 50 (1927), 189-358.
21. V. B. Pilyugina and A. L. Fel'shtyn, The Nielsen zeta function, Funktsional. Anal. i Prilozhen. 19 (4) (1985), 61-67 (in Russian); English transl.: Functional Anal. Appl. 19 (1985), 300-305.
22. G. de Rham, Complexes a automorphismes et homeomorphie differentiable, Ann. Inst. Fourier, 2 (1950), 51-67.
23. D. Ray and I. Singer, R-torsion and the Laplacian an Riemannian manifolds, Adv. in Math. 7 (1971), 145-210.
24. K. Reidemeister, Automorphismen von Homotopiekettenringen, Math. Ann. 112 (1936), 586-593.
25. W. Rudin : Fourier Analysis on Groups, Interscience tracts in pure and applied mathematics number 12, 1962.
26. S. Smale, Differentiable dynamical systems, Bull. Amer. Math. Soc. 73 (1967), 747-817.
27. A. Weil, Numbers of solutions of equations in finite fields, ibid. 55 (1949), 497-508.

Department of Mathematics, St. Petersburg Institute of Technology, Moskowsky Prospekt 26, St. Petersburg, 198013 Russia.

Current address: Sonderforschungsbereich 170, Mathematisches Institüt der Georg August Universität, Bunsenstraße 3-5, 3400 Göttingen, Germany.

E-mail address: felshtyn@cfgauss.uni-math.gwdg.de

Sonderforschungsbereich 170, Mathematisches Institüt der Georg August Universität, Bunsenstrasse 3-5, 3400 Göttingen, Germany.

E-mail address: hill@cfgauss.uni-math.gwdg.de

Contemporary Mathematics
Volume **152**, 1993

Cycles for Disk Homeomorphisms and Thick Trees

JOHN FRANKS AND MICHAŁ MISIUREWICZ

ABSTRACT. We investigate the problem of coexistence of cycles (i.e. periodic orbits) of orientation preserving homeomorphisms of the two dimensional disk. The study of many dynamical properties of such homeomorphisms can be reduced to an investigation of piecewise monotone maps of certain trees into themselves. The trees are closely related to the "train tracks" associated to pseudo-Anosov homeomorphisms in the work of Thurston. A homeomorphism and a cycle of that homeomorphism (considered up to conjugacy and isotopy relative to the cycle) determine an equivalence class called a *pattern*. We address such questions as when one pattern forces another, when a pattern forces the existence of cycles of given period, and what one can say about a lower bound for the topological entropy forced by a given pattern.

1. Introduction

In this article we investigate the problem of coexistence of cycles (i.e. periodic orbits) of orientation preserving homeomorphisms of the two dimensional disk. A successful technique in the study of many problems of dynamical systems is to reduce the problem to the study of a simpler system on a space of lower dimension. The underlying idea of this investigation is to reduce the problem from a two dimensional one to a one dimensional one. We do this by trying to associate to each disk homeomorphism with a distinguished cycle P a piecewise monotone map on a tree which carries much of the dynamical information of the original two dimensional homeomorphism.

Most of the paper is devoted to a description of an algorithm which start with a disk homeomorphism with a distinguished cycle and either produces a tree map (together with some extra structure) which carries a great deal of

1991 *Mathematics Subject Classification*. Primary 58F20; Secondary 54H20, 57M99, 58F03.

This paper is in final form and no version of it will be submitted for publication elsewhere

dynamical information about the original homeomorphism, or produces a way of reducing the investigation of the original homeomorphism to two simpler ones to which the algorithm can be re-applied.

Of course the correspondence of the homeomorphims to tree maps is not in general one-to-one, and, in fact, there may be finitely many possibilities for the tree map. The result depends only on the isotopy class of the the original homeomorphism relative to the distinguished cycle P and hence the resulting tree map contains information about only those dynamic properties which are invariant under such isotopies. Of particular interest, for our purposes, are the fact that one can read off the minimal entropy from the tree map and from the tree map (with extra structure) it is sometimes possible to deduce information about which "cycles up to isotopy" or *patterns* force the existence of others. This question of which patterns force which others (see below for careful definitions) is one of the central problems of the dynamics homeomorphisms of surfaces and is the goal toward which this paper is directed.

Underlying much of what we do is the work of Thurston on the classification of surface homeomorphisms [**T**]. The existence of a one dimensional map of a "train track" associated with a disk homeomorphism follows from the work of Thurston. His train track contains essentially the same information as our tree map and is only technically different (train tracks are never trees, for example). Our goal, however, was to provide a practical procedure for starting with a homeomorphism and constructing the tree map. Our methods are close to techniques of Bestvina and Handel [**BH1**] In fact our reduction methods are almost the same as the ones of a new paper [**BH2**] of Bestvina and Handel. However, their paper was not available to us during the writing of ours.

The outline of this paper is as follows. In Section 2 we define the concept of "thick tree" and thick tree map which formalizes the extra structure we wish to include with the one dimensional tree map. In Sections 3 through 9 we describe the algorithmic procedure for either reducing to simpler disk homeomorphisms or producing a Markov tree map with very nice properties. We also give three examples illustrating how the algorithm works. In Section 10 we show that if the Markov tree map so obtained is irreducible and not periodic then the original disk homeomorphism is isotopic (relative to the cycle P) to a pseudo-Anosov homeomorphism (Theorem 10.1). Moreover, this pseudo-Anosov homeomorphism has a Markov partition equivalent to the partition of the tree map into edges. This is used to establish relationships between the periodic points and entropy of the tree map and those of the pseudo-Anosov homemorphism (10.2-5), which in turn is related to the original disk homeomorphism via the work of Thurston [**T**].

In Section 11 we show that while the tree map we obtain is not completely canonical, there are only finitely many possibilities which can be obtained from one another by a finite procedure. In Section 12 we consider the topological entropy of Markov tree maps and derive some estimates which are of interest in light of the results of previous sections.

Finally Sections 13 and 14 investigate the forcing relation in patterns. In particular we address the question of what patterns are "primary", i.e. force no other patterns of the same period as themselves. This is a problem which is well understood, for example, for maps of the interval. While we are unable to solve it completely, it is possible to obtain considerable information concerning such patterns (see Theorem 14.4).

We proceed now to some formal definitions. Let $\mathbb{D} \subset \mathbb{R}^2$ will be the unit closed disk. Let $\mathfrak{A}$ be the class of all orientation preserving homeomorphisms $F : \mathbb{D} \to \mathbb{D}$. We follow the general scheme of [**M**] (see also [**ALM**). Let $\mathcal{P}$ be the class of all pairs (P, F), where $F \in \mathfrak{A}$ and P is a cycle of F, disjoint from $\partial\mathbb{D}$ (the boundary of $\mathbb{D}$). We shall say that the pairs (P, F) and (Q, G) from $\mathcal{P}$ are *equivalent* (and write $(P, F) \sim (Q, G)$) if there exists an orientation preserving homeomorphism $H : \mathbb{D} \to \mathbb{D}$ such that $H(P) = Q$ and $H \circ F \circ H^{-1}$ and G are *isotopic rel.* Q (that is, there exists an isotopy $(K_t)_{t\in[0,1]}$ with $K_0 = H \circ F \circ H^{-1}$, $K_1 = G$ and $K_t(x) = G(x)$ for all $t \in [0, 1]$, $x \in Q$). The equivalence classes of the relation $\sim$ will be called *patterns*. The equivalence class to which a pair (P, F) belongs will be denoted $[(P, F)]$. Patterns are called "braid types" by some authors. However, we prefer the term "pattern" as they are direct generalizations of one-dimensional patterns (see [ALM]) and we want to stress the similarities of the one and two dimensional theories. Moreover, unlike many authors (for instance [**BGN, B1, B2, GLM, H1, LM, Ma**]), we do not use the theory of braids.

We shall say that a homeomorphism $F \in \mathfrak{A}$ *exhibits* a pattern A if there is a cycle P of F such that $[(P, F)] = A$. A pattern A *forces* a pattern B if every $F \in \mathfrak{A}$ which exhibits A exhibits also B.

Clearly, if $(P, F) \sim (Q, G)$ then the periods (which we define as the minimal periods) of P and Q are the same. Therefore we can speak about the *period* of a pattern. We shall say that a pattern A is *primary* if it does not force any other pattern of the same period. We define also the *entropy* of a pattern A (denoted $h(A)$) as the infimum of topological entropies of all homeomorphisms $F \in \mathfrak{A}$ exhibiting A.

The basic property of the forcing relation among patterns is that it is an ordering relation (see [**B3**]).

We can change slightly the classes of maps and isotopies under consideration, without changing the set of patterns. Namely, we can replace the class $\mathfrak{A}$ by the class $\mathfrak{B}$ of all orientation preserving homeomorphisms of $\mathbb{D}$ onto a subset of $\mathbb{D}$, and consider in the definition of the equivalence relation $\sim$ the isotopies $(K_t)_{t\in[0,1]}$ in the sense that each K_t is a diffeomorphism not necessarily of $\mathbb{D}$ onto itself, but of $\mathbb{D}$ onto its image under K_t. Then, we can replace the class $\mathfrak{B}$ by the class $\mathfrak{C}$ of all homeomorphisms of $\mathbb{D}$ onto a subset of the interior of $\mathbb{D}$. Notice that in such a case any cycle has to be contained in the interior of the image of $\mathbb{D}$. The notion of a pattern and the notions of forcing, entropy and primarity are independent on whether we consider the class of maps $\mathfrak{A}$, $\mathfrak{B}$, or $\mathfrak{C}$ (except that the forcing relation may be slightly different because of peripheral cycles; see Sections 10 and 11). Indeed, we can add a "collar" to $\mathbb{D}$ and extend our homeomorphism to it in order to switch from one class to another; to get the homeomorphism defined on $\mathbb{D}$ each time, we simply "rescale" the disk. Clearly, all this can be done in such a way that that no new nonwandering points in the interior of the disk are created, so no new patterns appear and the entropy stays the same. Therefore, in the rest of the paper we shall work in the class which suits our needs the best, and the results proved for one class will be true for the others. Although we will work with the classes $\mathfrak{A}$ and $\mathfrak{C}$, the class $\mathfrak{B}$ is the largest one (it contains both $\mathfrak{A}$ and $\mathfrak{C}$) and the reader may prefer to think of the results stated for $\mathfrak{B}$ as of the most general ones.

To describe the structure of various patterns, we need some more definitions. We shall say that a pattern A is *twist* if it is exhibited by some rotation of the disk. Clearly, the entropy of a twist pattern is zero.

Suppose that $F \in \mathfrak{B}$ and P is a cycle of period n. Then (P, F) is equivalent to (P_0, F_0) for a $F_0 \in \mathfrak{A}$ with the property that F_0^n is the identity on a neighborhood of P_0 and that F_0 is the identity on the boundary of $\mathbb{D}$. Then we can choose disks $D_1, D_2, \dots, D_n \subset \mathbb{D}$, disjoint from $\partial\mathbb{D}$, such that $F_0(D_i) = D_{i+1}$ for $i = 1, 2, \dots, n-1$ and $F_0(D_n) = D_1$ and with P_0 contained in the interiors of the disks D_i. Let B denote the pattern $[(P, F)] = [(P_0, F_0)]$. Suppose that A is some pattern of period m exhibited by G and G is the identity on the boundary of $\mathbb{D}$. Then there is a cycle Q of period m of $G : \mathbb{D} \to \mathbb{D}$ with $A = [(Q, G)]$. We choose an orientation preserving homeomorphism $H : D_1 \to \mathbb{D}$ and consider the homeomorphism $H \circ G \circ H^{-1}$ defined from D_1 to itself. We extend this to a homeomorphism of $\mathbb{D}$ by letting it be the identity outside of D_1 and we denote the resulting homeomorphism $\widetilde{G}$. We can now define an *A-extension* of B. Let $\widetilde{F} = \widetilde{G} \circ F_0$ and let $R = \bigcup_{i=0}^{n-1} \widetilde{F}^i(H^{-1}(Q))$. Then R is a cycle of period nm for $\widetilde{F}$ and we say that $C = [(R, \widetilde{F})]$ is an A-extension of B.

Notice that A and B do not determine C uniquely. When we build a homeomorphism exhibiting an A-extension of B we choose an F_0 exhibiting B with the property that F_0 is the identity on $\bigcup_{i=1}^{n} D_i$. In doing this we have some choice. Namely, we can surround D_1 by a "collar" on which we can put any Dehn twist. Equivalently, when we choose G exhibiting A which is the identity on $\partial\mathbb{D}$, we can put a collar on boundary of $\mathbb{D}$ and put any Dehn twist on it. One can avoid this nonuniqueness by replacing $\mathfrak{A}$ by the set of homeomorphisms which are the identity on $\partial\mathbb{D}$ and requiring that the isotopy $(K_t)_{t\in[0,1]}$ is the identity on $\partial\mathbb{D}$ for each t.

The construction of the extensions is associative, in the sense that any (A-extension of B)-extension of C is an A-extension of a (B-extension of C) and conversely. This is because any A-extension of B is built as a composition of two homeomorphisms ($\widetilde{G} \circ F_0$ above). Hence an extension of an extension will be constructed as a composition of three homeomorphisms and associativity follows from the associativity of composition.

We would like to thank Z. Nitecki for reading a preliminary version of this paper and making many valuable suggestions which we have adopted.

2. Thick tree structures

We have to describe the objects with which we will work. They are not very complicated, unfortunately the rigorous description requires many definitions.

By a *tree* we mean a connected finite one-dimensional CW-complex which does not contain any subset homeomorphic to a circle. We shall refer to the *vertices* and *edges* of a tree. From each vertex of a tree there emerges a certain number of edges; this number will be called the *valence* of a vertex. When we look at a tree from the topological point of view then we do not see the vertices of valence 2; nevertheless we will admit such vertices. A vertex of valence 1 will be called an *end* of a tree. If e is an edge of a tree T then it contains two vertices; call them v_1 and v_2. We shall call the set $e \setminus \{v_1, v_2\}$ an *open edge* and denote it $\mathrm{Int}(e)$. Notice that when we regard e as a subset of the topological space T then $\mathrm{Int}(e)$ does not necessarily coincide with the interior of e (we shall denote this topological interior by $\mathrm{int}(e)$; similarly we shall denote the closure of a set by $\mathrm{cl}(\cdot)$).

Let T be a tree with the set of vertices V, P a finite set in the interior of $\mathbb{D}$, and R a finite subset of V containing all the ends of T. We shall say that a map $\pi : \mathbb{D} \to T$ is a *thick tree structure of* $(\mathbb{D}, P)$ *over* (T, R) if:

(i) π is continuous and onto,
(ii) π maps P bijectively onto R,
(iii) for every vertex v of T, the set $\pi^{-1}(v)$ is homeomorphic to a closed disk,
(iv) if $x \in P$ then x is an interior point of $\pi^{-1}(\pi(x))$,
(v) for every edge e of T there is a homeomorphism H_e of $\mathrm{Int}(e) \times [0,1]$ onto some subset of $\mathbb{D}$ such that $\pi \circ H_e$ is the projection to the first coordinate (that is, $\pi \circ H_e(x,t) = x$ for $x \in \mathrm{Int}(e)$ and $t \in [0,1]$),
(vi) if e_1 and e_2 are distinct edges of T then the closures of $\pi^{-1}(\mathrm{Int}(e_1))$ and $\pi^{-1}(\mathrm{Int}(e_2))$ are disjoint.

The way of thinking about the thick tree structures can be illustrated by the following figure (Figure 1). In particular, it is good to draw not the disk $\mathbb{D}$, but a set homeomorphic to it having the shape indicating the thick tree structure.

A tree T with the thick tree structure π (if we want to be more precise, we should mention also P and R) will be called a *thick tree*. To stress its connection with T, we shall denote it $\widehat{T}$. We shall also say that the thick tree $\widehat{T}$ and the tree T are *associated* to each other.

Every edge of a tree is homeomorphic to an interval. Therefore we can speak about *subintervals* of an edge. If f maps a subinterval I of some edge into a tree T then we shall say that it is *monotone* on I if $f(I)$ is homeomorphic to an interval and the map between intervals obtained by composing $f|_I$ with homeomorphisms from both sides, is monotone (not necessarily strictly). Clearly, these notions do not depend on the choice of homeomorphisms involved.

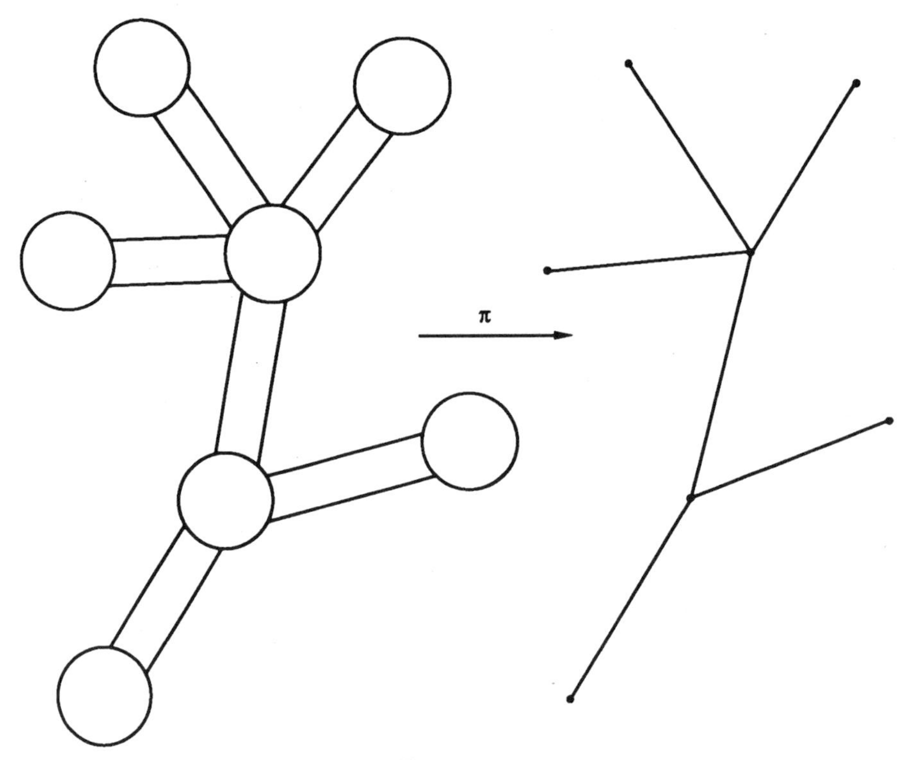

FIGURE 1

We shall say that a map $f : T \to T$ is *Markov* if it is continuous and every edge of T can be divided into subintervals which are mapped by f in a monotone way onto some edges of T. Notice that if f is Markov then $f(V) \subset V$. If f fails to be Markov only because some edge is mapped to a vertex (or several edges are mapped to vertices), we will call it *weakly Markov.*

If T is a tree and v its vertex then we can speak about *edge germs at* v. They are infinitesimal pieces of edges containing v. If $f : T \to T$ is a Markov map then it induces a map from the set of edge germs at v to the set of edge germs at $f(v)$. We shall refer to this map also as f.

We can introduce a *cyclic ordering* of edge germs at v. If a graph is embedded into $\mathbb{D}$ then the natural orientation of $\mathbb{D}$ induces a cycling ordering of the edge germs at v. If we have a homeomorphism h of trees then any cyclic ordering of edge germs at v is carried by h to a cyclic ordering of edge germs at $h(v)$. Similarly, if we have a thick tree structure π of $(\mathbb{D}, P)$ over (T, R) then there is a cyclic ordering of edge germs at any vertex v of T, induced by the natural orientation of $\mathbb{D}$ and π. We shall refer to this cyclic ordering as the natural cyclic ordering.

Suppose we have some thick tree structure π of $(\mathbb{D}, P)$ over (T, R) and an embedding $\widehat{f} : T \to \mathbb{D}$ such that:

(i) $\widehat{f}(\pi(P)) = P$,
(ii) at any vertex of T, the cyclic ordering of edge germs carried by $\widehat{f}$ from the natural cyclic ordering is the same as the cyclic ordering induced by the natural orientation of $\mathbb{D}$ at the same vertex,
(iii) the map $\pi \circ \widehat{f}$ is Markov.

This is the situation which we shall consider most of the time. We shall say then that $\widehat{f}$ is a *thick tree map* of $\widehat{T}$; the maps $\widehat{f}$ and $f = \pi \circ \widehat{f}$ will be called *strictly associated* to each other.

This last notion is too restrictive. We shall say that two Markov maps $f, g : T \to T$ are *equivalent* if for every edge e of T there exists an orientation preserving homeomorphism $h : e \to e$ and a partition of e into intervals such that for every interval I from this partition f and $g \circ h$ coincide on the endpoints of I and map I in a monotone way onto the same edge of T. Now we say that $\widehat{f}$ and f are *associated* to each other if there exists g equivalent to f which is strictly associated to $\widehat{f}$.

We would like to have some connection between disk homeomorphisms and thick tree maps. For this end we define a *core* of a thick tree $\widehat{T}$ as an embedding $c : T \to \operatorname{int}(\mathbb{D})$ such that:

(i) $c(\pi(x)) = x$ for every $x \in P$,
(ii) $c(v) \subset \operatorname{int}(\pi^{-1}(v))$ for every $v \in V$,
(iii) $\pi \circ c$ maps every edge of T in a monotone way onto itself,
(iv) whenever $c(T)$ intersects the boundary of $\pi^{-1}(v)$ for any vertex v of T, this intersection is transversal.

Clearly, each thick tree has a core. Now we shall say that a homeomorphism $F \in \mathfrak{C}$ for which P is a cycle and a thick tree map $\widehat{f}$ of $\widehat{T}$ are *associated* to each other if there exists a homeomorphism $G \in \mathfrak{C}$ isotopic to F rel. P and a core c of $\widehat{T}$ such that $G \circ c = \widehat{f}$. Notice that for $G \circ c$ conditions (i) and (ii) of the definition of a thick tree map are automatically fulfilled.

Our aim is to investigate patterns via thick tree maps. In the definition of the equivalence of cycles there were two steps. The first one was to conjugate everything by an orientation preserving homeomorphism of $\mathbb{D}$ and the second step was to deform the resulting system by an isotopy rel. the cycle. The first step – conjugating everything by a homeomorphism – can be performed later and this will not change anything. It can be performed after making an isotopy or even on the level of the trees; we can also transport a thick tree structure by it. Therefore we can forget about this step, keeping in mind only that we can replace our tree maps by conjugate ones (which is anyhow obvious).

Thus, if we are given a pattern A then we will proceed as follows. We start with any $F \in \mathfrak{C}$ and its cycle P such that $[(P, F)] = A$. Then we consider the class of all thick tree maps associated to F. In this class we try to find some special element. It will either give us an extension structure of A or will be in some sense "simple". In the first case, if A is a B-extension of C, we continue working with B and C. In the second case a (thick) tree map will give us important information about A.

The procedure described above is basically nothing else but application of Thurston theory on classification of surface homeomorphisms. If we remove P from the disk then we can deform F by a homotopy to a form where it is either reducible (this corresponds to the case of an extension) or is periodic (which corresponds to a twist pattern) or is pseudo-Anosov. In the last two cases, all the information about minimal entropy and other cycles which are forced is in the homeomorphism we get. Therefore, if we apply our procedure to the trees, we will also obtain all this information, that is the information about the entropy of A and about what patterns are forced by A.

In fact, we want not only to prove the existence of this "simple" form of a (thick) tree map, but we also want to provide an algorithm for finding it.

Suppose we are given a map $F \in \mathfrak{C}$ having a cycle P. Then we start by choosing a tree $T \subset \mathrm{int}(\mathbb{D})$ such that $P \subset V$ (where V is the set of the vertices of T) and every endpoint of T belongs to P. This is clearly possible. We set $R = P$. Next we take a small closed neighborhood K of T, homeomorphic to a disk, with a thick tree structure over (T, R) on it, with id_T as a core. This step is also clear. We can also assume that the boundary of K is a topological circle. We can take another topological circle $S \subset \mathrm{int}(K)$ surrounding T, which bounds some region L. Then there will be a homeomorphism between $K \setminus \mathrm{int}(L)$ and $\mathbb{D} \setminus \mathrm{int}(L)$, which is the identity on the boundary of L. We extend this homeomorphism to the whole of K by making it the identity on L and then we can conjugate by it to transport the thick tree structure from K to $\mathbb{D}$. In this way we get a thick tree structure π of $(\mathbb{D}, P)$ over (T, R) with a core id_T.

Now we want to get a map associated with F which has all the properties of a thick tree map except it is only weakly Markov instead of Markov. We have to modify F by an isotopy rel. P to a homeomorphism G such that $\pi \circ G|_T$ is weakly Markov and take $G|_T$ as the thick tree map. By choosing appropriate isotopies we can move $G|_T$ in any way we want as long as it stays a homeomorphism onto its image, G stays the same on P and $G(T)$ does not touch $\partial\mathbb{D}$.

Here is a sketch of one way to get the desired G. By a small perturbation we may assume that any piece of one edge, which is mapped to another (or the same) edge by $\pi \circ F$, is mapped in a piecewise monotone way. Given the thick tree structure it is clear one can construct a flow φ_t on $\mathbb{D}$ which has a sink at each vertex of T and saddle in each edge whose unstable manifold approximately coincides with the edge. Moreover we can choose the saddle point $p_e \in e$ so that it is not the image of a turning point of any of the piecewise monotone restrictions of $\pi \circ F$ mentioned above. We also can construct the flow so that the stable manifold of p_e is sufficiently close to $\pi^{-1}(p_e)$ that no point of this stable manifold is mapped by π to the image of a turning point. Consider now the isotopy $G_t = \varphi_t \circ F$. For large t the flow φ_t will push all turning points on the F image of an edge into $\pi^{-1}(V)$ which is a neighborhood of all the sinks. Thus for large t the homeomorphism $\pi \circ G_t$ is weakly Markov on T.

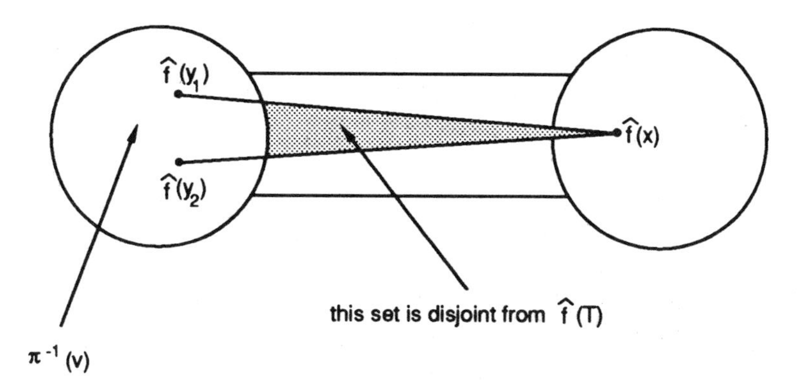

FIGURE 2

We shall describe next steps in the consecutive sections.

3. Tightening

Assume that we have a map $F \in \mathfrak{C}$ with a cycle P, a thick tree $\widehat{T}$ with a core c and a thick tree map $\widehat{f} = F \circ c$. We shall use our usual notations: π for the thick tree structure, T for the tree associated to $\widehat{T}$, R for $\pi(P)$, V for the set of vertices of T, f for the Markov tree map associated to $\widehat{f}$. If we want to replace $\widehat{f}$ by some simpler thick tree map associated to F, our first question is: can we do it without changing $\widehat{T}$, or at least by changing $\widehat{T}$ only in some simple way?

Suppose that we have the following situation. There is a subinterval I of an edge of T which is a union of two subintervals $I = I_1 \cup I_2$ with $I_1 \cap I_2 = \{x\}$ for some point x. Each of the intervals I_1 and I_2 is mapped by f in a monotone way onto the same edge of T. For $i = 1, 2$, let y_i be the endpoint of I_i different than x. Notice that $v = f(y_1) = f(y_2)$ is a vertex of T. Assume further that the component of $\mathbb{D} \setminus (\widehat{f}(I) \cup \pi^{-1}(v))$ which does not contain $\partial\mathbb{D}$ is disjoint from $\widehat{f}(T)$ (see Figure 2). In this situation we shall say that $\widehat{f}$ has an *edge loop*.

A *vertex loop* is similar. It looks as follows. There is a vertex $x \in V \setminus R$ and subintervals $I_1, I_2, \dots, I_s$ of *all* edges emerging from x, one subinterval per edge. Each of them has x as one endpoint and is mapped by f in a monotone way onto the same edge of T. For $i = 1, 2, \dots, s$, let y_i be the endpoint of I_i different than x. Notice that $v = f(y_1) = f(y_2) = \dots = f(y_s)$ is a vertex of T. Finally, we assume that every component of $\mathbb{D} \setminus (\widehat{f}(\bigcup I_i) \cup \pi^{-1}(v))$ which does not contain points of $\partial\mathbb{D}$ is disjoint from $\widehat{f}(T)$.

If $\widehat{f}$ has a loop then we can remove this loop by modifying F by an isotopy rel. P to a $G \in \mathfrak{C}$ such that the only difference between $\widehat{f}$ and $\widehat{g} = G \circ c$ is that all subintervals I_i are mapped by $\widehat{g}$ into $\pi^{-1}(v)$. Now we want to get a thick tree map from a map $\widehat{g}$ which has all the properties of a thick tree map except that it is only weakly Markov instead of Markov. This can be done by collapsing every edge which is mapped to a vertex to a point. This gives us a new tree T'. The modification of the thick tree structure is obvious: we just compose π with the collapsing map $T \to T'$. It is also obvious that we can modify the core c only on the parts mapped to π^{-1} of the collapsing edges, to get a core c' of the new thick tree $\widehat{T}'$. The new map $g' : T' \to T'$ induced by g is associated to $\widehat{g}' = G \circ c'$. Now $\widehat{g}'$ is either a thick tree map or it has all the necessary properties except it is only weakly Markov. In the latter case, we repeat the construction, and we do it until we get a thick tree map. This has to happen eventually, because each time the number of edges of the tree decreases.

The construction described above, leading from a weakly Markov map which is otherwise a thick tree map to a true thick tree map, will be also used later. We shall call it *collapsing.* In fact, we sometimes have to use collapsing right after the first step described in the previous section.

Thus, if $\widehat{f}$ has a loop then we can find a thick tree map $\widehat{g}$, also associated to F, such that the total number of the subintervals of edges which are mapped to entire edges, is smaller for g than for f. We shall call this operation *tightening.*

Each time we describe some modification of a thick tree map, we will try to determine what it does to the entropy of the associated tree map (here by *entropy* we mean *topological entropy*). Since this map is Markov, there is a standard way to compute its entropy (cf. e.g. [**ALM**]). Let $f : T \to T$ be a Markov map. Then we build a generalized oriented graph $\mathcal{G}(f)$ (the *Markov graph* of f) by taking the edges of T as vertices of $\mathcal{G}(f)$ and putting m arrows from an edge e to an edge d if there are subintervals $I_1, \dots, I_m$ of e with pairwise disjoint interiors such that $f(I_i) = d$ for all i. The corresponding transition matrix $M(f)$ has rows and columns indexed by the vertices of $\mathcal{G}(f)$ and the e, d entry of $M(f)$ is m if there are m arrows from e to d. The corresponding subshift of finite type $\Sigma(f)$ has arrows of $\mathcal{G}(f)$ as states and there is a transition from a to b if a terminates at the same vertex of $\mathcal{G}(f)$ as b originates. Then we define the entropy of $\mathcal{G}(f)$ as the entropy of $\Sigma(f)$. It is the same as the entropy of f and is equal to the logarithm of the spectral radius of $M(f)$.

We shall call a Markov tree map *transitive* if for every pair of edges e and d of T there is $n \geqslant 1$ such that the e, d entry of $M(f)^n$ is positive. This is the same as to say that for every vertices e and d of $\mathcal{G}(f)$ there is a path from e to d. Notice that transitivity of f is not the same as topological transitivity. In fact, if f is transitive then there are two possibilities. The first one is that there is only one loop in $\mathcal{G}(f)$; then the edges of T are just permuted by f and $h(f) = 0$. The other possibility is that there are at least two loops in $\mathcal{G}(f)$. Then it is easy to see that the entropy of $\mathcal{G}(f)$ (and therefore of f) is positive and there is g equivalent to f which is topologically transitive.

If two Markov maps are equivalent, then they have the same Markov graphs. Therefore they have the same entropy and if one of them is transitive, so is the other one.

The operation of tightening on the level of Markov graphs amounts to removing some arrows, and perhaps removing some vertices at which no longer any arrow originates. Therefore on the level of subshifts of finite type it amounts to restricting the shift to a closed invariant subspace. Thus, during this operation the entropy cannot increase. Moreover, if f is transitive, so is $\Sigma(f)$, so the entropy actually has to decrease (see e.g. [**S**]).

We shall call a thick tree map *tight* if it has no loop. Thus, we see that if we start with a thick tree map $\widehat{f}$ associated to F, which has a loop then after finite number of tightenings (we can perform only finitely many of them since each time the total number of the subintervals of the edges which are mapped to the whole edges, decreases) we get a tight tree map $\widehat{g}$ also associated to F, and $h(g) \leqslant h(f)$, where f and g are Markov tree maps associated to $\widehat{f}$ and $\widehat{g}$ respectively. Moreover, if f is transitive then $h(g) < h(f)$.

4. Modifications of trees

The simplest modifications of a tree T we will make are adding or removing one or more vertices of valence 2. From the topological point of view, no change will occur. In particular, if $f : T \to T$ is a continuous map, the entropy of f will not change. However, when we work with thick trees, it is important which points of T are vertices. Therefore we have to describe the necessary modifications of a thick tree structure. We can think about the edges and a vertex involved in the following way. The open edge without an additional vertex on it in the thick tree can be represented as a rectangle $X = (0,6) \times [0,1]$ with the map $\pi : X \to (0,6)$ given by $\pi(x,y) = x$. The same edge with a vertex on it can be represented as the same rectangle X with the map $\pi' : X \to (0,6)$ given by $\pi'(x,y) = 3x/2$ for $x \in (0,2)$, $\pi'(x,y) = 3$ for $x \in [2,4]$ and $\pi'(x,y) = 3(x-2)/2$ for $x \in (4,6)$. This corresponds to an open edge $(0,6)$ and two open edges $(0,3)$ and $(3,6)$ with a vertex at 3, respectively. Thus, adding a vertex is switching from π to π', and removing a vertex is the converse operation. Since every situation can be reduced to this model (by transporting via homeomorphisms), this is a sufficient description. A core of the thick tree changes in a simple way. If we remove a vertex, there is no problem at all. If we add a vertex, we just have to place the image of the new vertex in the interior of its inverse image under π', but this is also clearly possible.

Suppose we have a thick tree map $\widehat{f}$ of $\widehat{T}$. The only problem which can occur when we add or remove vertices of valence 2 is whether after this operation the associated tree map stays Markov. This produces the following restrictions. We can remove a vertex of valence 2 only if it is not an image (under f) of any other vertex. Moreover, we can remove a whole cycle of vertices of valence 2 if none of them is an image of any vertex not belonging to this cycle. Conversely, we can add a vertex at x if $f(x)$ is a vertex. Moreover, we can add vertices at all points of an arbitrary cycle. With those restrictions, the thick tree map remains a thick tree map.

We noticed already that the entropy does not change during adding or removing valence 2 vertices. We have also to find out what happens to transitivity. If

f was transitive and we removed some vertices then in the Markov graph $\mathcal{G}(f)$ some vertices will be glued together. However, no arrows will be erased (at most they will be glued together), so all old paths will survive. Therefore the map will remain transitive. If f was transitive and we added some vertices then we have to distinguish two cases. The first case is when f maps some edge to more than one edge. In this case, as we already noticed, there exists g equivalent to f which is topologically transitive. Since adding vertices does not change anything from the topological point of view, g remains topologically transitive, and therefore transitive. Thus, f is also transitive. The second case is when f just permutes the edges. Then adding vertices can destroy transitivity. However, in this case there will be no need to add vertices.

Another modification of a tree, already more complicated, is *gluing* two edges together (called *Stalling's folding operation* in [**BH1**]). We can make this modification if these edges emerge from the same vertex and are neighbors in the cyclic order at this vertex. Let these edges be e_1 and e_2, their common vertex v and let the endpoints of e_i different than v be w_i for $i = 1, 2$. We can assume that in the cyclic ordering v_1 goes just before v_2. Then we can consider the following representation of e_1, e_2, v, w_1 and w_2 (as in the description of the adding/removing of a vertex, we can get this form by transporting everything via some homeomorphisms). Those edges and vertices in the thick tree can be represented as a rectangle $X = [0, 10] \times [0, 1]$ with the map $\pi : X \to [0, 4]$ given by $\pi(x, y) = 0$ for $x \in [0, 2]$, $\pi(x, y) = x - 2$ for $x \in (2, 4)$, $\pi(x, y) = 2$ for $x \in [4, 6]$, $\pi(x, y) = x - 4$ for $x \in (6, 8)$ and $\pi(x, y) = 4$ for $x \in [8, 10]$ (see Figure 3). This corresponds to the vertices w_1 at 0, v at 2 and w_2 at 4 and open edges $\mathrm{Int}(e_1) = (0, 2)$ and $\mathrm{Int}(e_2) = (2, 4)$. The segment $[4, 6] \times \{0\}$ of the boundary of $\pi^{-1}(v)$ is contained in the boundary of the disk, whereas from the segment $[4, 6] \times \{1\}$ some other thick edges can emerge. After the gluing, we get an edge e from e_1 and e_2 and a vertex w from the vertices w_1 and w_2. The thick tree structure can be altered in the following way. We define a new mapping π', which is the same as π outside X and $\pi'(X) = [0, 2]$ (we can assume that $e = [0, 2]$). If $z \in X$ then we set $\pi'(z) = 0$ if $z \in [4, 6] \times [4/5, 1]$ and $\pi'(z) = 2$ if $z \in [0, 2] \times [0, 1] \cup [8, 10] \times [0, 1] \cup [0, 10] \times [0, 2/5]$. For the rest of z's we draw the segment joining $(5, 1)$ with z, measure which part of this segment lies outside $z \in [4, 6] \times [4/5, 1]$ $(= (\pi')^{-1}(v))$, call this number $\xi(z)$ (we get $\xi(z) \in (0, 2/3)$) and set $\pi'(z) = 3\xi(z)$. All this is correct if we assume that there is no point of R in $[4, 6] \times [2/5, 4/5]$ (which we clearly can assume) and at most one of the points w_1, w_2 belongs to R. The latter condition is a real restriction, however we shall see later that whenever we have to make gluing, it will be satisfied.

For gluings, we will discuss the problems connected with the modifications of a thick tree map and further restrictions resulting from it, as well as the behavior of the entropy, in Section 5.

We can also perform the reverse operation: *split* e into e_1 and e_2 (which results also in splitting w into w_1 and w_2). The model for it is the same as for gluing,

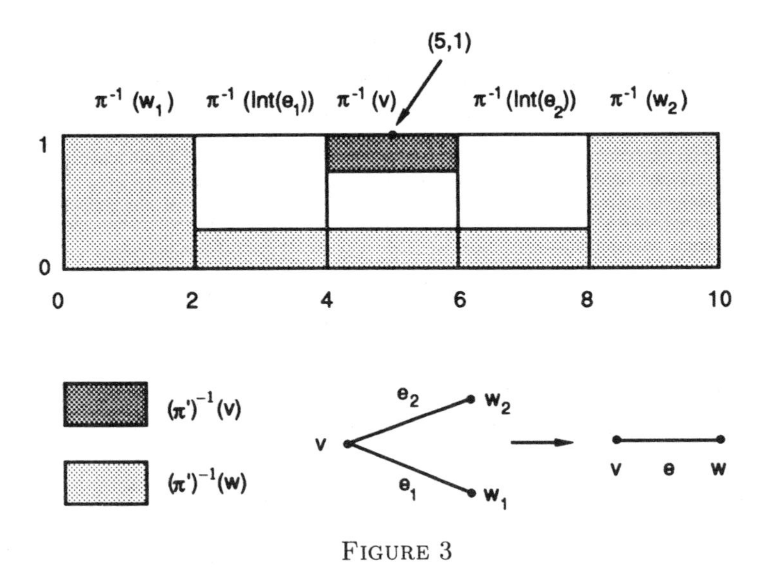

FIGURE 3

but we go backwards, that is replace π' by π. We discuss further splittings in Section 8.

In both cases, gluing and splitting, the modification of the core is more or less obvious; we can make it locally in a small neighborhood of the edges involved.

5. Gluing reductions

As always, we assume that we have a thick tree map $\widehat{f}$ of a thick tree $\widehat{T}$, associated to a homeomorphism $F \in \mathfrak{C}$ with a cycle P. The tree associated to $\widehat{T}$ is T, it has the set of vertices V and f is the Markov tree map associated to $\widehat{f}$. Moreover, π is the thick tree structure we consider and $R = \pi(P)$.

Let e_1 and e_2 be edges emerging from a vertex v of T and assume that they are neighbors in the cyclic order at this vertex. We want to know when we can glue those edges together and modify $\widehat{f}$ in such a way that it stays associated to F. An obvious condition which allows us easily to define a new Markov map on the modified tree is that both e_1 and e_2 are mapped in a monotone way onto the same edge of T. This condition is by no means necessary. Nevertheless, in order to simplify the descriptions of the operations we will perform, we shall restrict our attention to it. If we really want to glue together some longer edges, we can divide them to shorter pieces by adding valence 2 vertices and then glue them step by step. Notice that if the endpoints of e_i different than v are w_i then under our condition $f(w_1) = f(w_2)$, so at least one of the points w_1 and w_2 does not belong to R, so there are no troubles with gluing e_1 with e_2.

When we glue together e_1 with e_2 and we get a new tree T' from T then there is a natural projection $p : T \to T'$. Suppose that both e_1 and e_2 are mapped in a monotone way onto the same edge of T. Then we can define a new Markov map $f' : T' \to T'$ in a natural way. Namely, if $x \in T' \setminus e$, where $e = p(e_1) = p(e_2)$, then $f'(x) = p(f(x))$. If $x \in e$ then $f'(x) = p((p|_{e_1})^{-1}(x))$. Modifications to $\widehat{f}$ by composing with an isotopy rel. P (so that it stays associated to F) required to get a thick tree map $\widehat{f'}$ associated to f', are similar.

To see what happens to the entropy of a Markov map f when we perform gluing, consider the Markov graph $\mathcal{G}(f)$. If the edges e_1 and e_2 of T were mapped by f in a monotone way onto the same edge of T, then from each of the vertices e_1 and e_2 of $\mathcal{G}(f)$ one arrow originates, and those arrows terminate at the same vertex of $\mathcal{G}(f)$. Gluing e_1 and e_2 together amounts to replacing them in $\mathcal{G}(f)$ by one vertex e. All arrows terminating at e_1 or e_2 terminate now at e; the arrows originating at e_1 and e_2 are replaced by one arrow originating at e and terminating at the same vertex as before (or at e if they terminated before at one of e_1, e_2). The entropy of $\mathcal{G}(f)$ can be computed by taking the limit of $(1/n)\log$ of the number of paths of length n in $\mathcal{G}(f)$. Our modification of the graph $\mathcal{G}(f)$ does not change the number of paths of length n, except that it cuts the number of paths originating at e_1 and e_2 (they are replaced by the paths originating at e). Since $\lim_{n\to\infty}(1/n)\log(a_n/2) = \lim_{n\to\infty}(1/n)\log a_n$, the entropy does not change.

Another thing to check is transitivity. Suppose that before gluing f was transitive. This means that for every pair of vertices d and c in $\mathcal{G}(f)$ there was a path from d to c. It is clear that this path remains intact after gluing (only except passing through e_1 or e_2 it will pass through e). Therefore our Markov map remains transitive after gluing.

Since we shall usually divide edges by adding valence 2 vertices first, we shall consider now edge germs rather than full edges.

We shall say that edge germs e and d at a vertex $v \in V$ are *gluable in n steps* if their images under f^n are equal. If they are gluable in 1 step, we shall simply say that they are *gluable.* If there exists n such that they are gluable in n steps then we shall call them *eventually gluable.*

Notice that all three kinds of gluability are equivalence relations. Moreover, since f preserves (weakly) the cyclic ordering of the germs, the equivalence classes are "intervals" in this cyclic ordering. Clearly, if e and d are gluable in n steps and $k \geqslant n$ then they are also gluable in k steps. Moreover, if n is sufficiently large then eventual gluability and gluability in n steps coincide (of course how large n has to be, depends on f).

Suppose that e and d are two distinct edge germs at $v \in V$ which are eventually gluable. Then there is n such that $f^n(e) \neq f^n(d)$ but $f^{n+1}(e) = f^{n+1}(d)$. Look at $\widehat{f}(f^n(e))$ and $\widehat{f}(f^n(d))$ and at the points $x(e)$ and $x(d)$ respectively, where they cross the boundary X of $\pi^{-1}(f^{n+1}(v))$, entering $\pi^{-1}(f^{n+1}(\mathrm{Int}(e)))$ (see Figure 4). (Although e and d are infinitesimal pieces of edges, at this moment we should think about them as having some small length.) Suppose that when we go along the common boundary of $\pi^{-1}(f^{n+1}(v))$ and $\pi^{-1}(f^{n+1}(\mathrm{Int}(e)))$ following the natural orientation of X then we meet first $x(d)$ and then $x(e)$. In this case we shall say that *there is a left turn from e to d.* If we meet first $x(e)$ and then

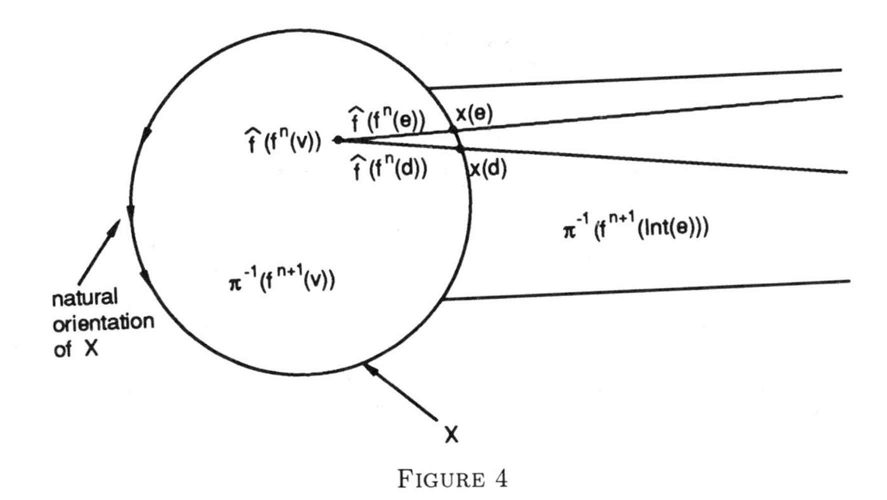

FIGURE 4

$x(d)$ then we shall say that *there is a right turn from e to d*. The meaning of these expressions is as follows. If we go along $\widehat{f}(f^n(e))$ towards $\widehat{f}(f^n(v))$, reach $\widehat{f}(f^n(v))$ and then turn to $\widehat{f}(f^n(d))$; then we have to turn to the left or to the right depending on whether there is a left or a right turn from e to d.

Our aim is to perform several gluings (some of them preceded by adding valence 2 vertices), get to a situation when the resulting map is not tight, and tighten it. Then, if the map from which we started was transitive, we lower the entropy. However, for this we have first to detect where to make the gluings. This is essentially connected with existence of loops for iterates of our map (we assume that the map itself is tight). (When we speak about the map, sometimes we mean the thick tree map and sometimes the associated Markov map of the tree). Thus, we will have two kinds of *gluing reduction possibilities* (GRP in short).

A *GRP of edge type at a vertex $v \in V$* occurs in the following situation. There are eventually gluable edge germs e_1 and e_2 at v, which are neighbors in the cyclic order at v and there is an edge d of T and its two subintervals I_1 and I_2 with disjoint interiors and a common endpoint x such that f maps I_i onto e_i in a monotone way for $i = 1, 2$ (here we use the same names for the edge germs and the whole edges). Moreover, if $v \in R$ then we have an additional requirement (see Figure 5). Let y_1 and y_2 be the points where $\widehat{f}(I_1)$ and $\widehat{f}(I_2)$ respectively cross the boundary between $\pi^{-1}(v)$ and $\pi^{-1}(\mathrm{Int}(e_i)))$. Then the image under $\widehat{f}$ of the subinterval with the endpoints $\widehat{f}^{-1}(y_1)$ and $\widehat{f}^{-1}(y_2)$ divides $\pi^{-1}(v)$ into two parts: X_1 and X_2. The first of them shares with $\pi^{-1}(v)$ the part of the boundary from y_1 to y_2 (in the natural orientation); the second one the part of the boundary from y_2 to y_1. We require that the only point z of $P \cap \pi^{-1}(v)$ is in X_1 if there is a left turn from e_1 to e_2 and in X_2 if there is a right turn from e_1 to e_2. These conditions assure us that when after n steps (iterates) the germs e_1 and e_2 will be glued together, the images of I_1 and I_2 (or their subintervals)

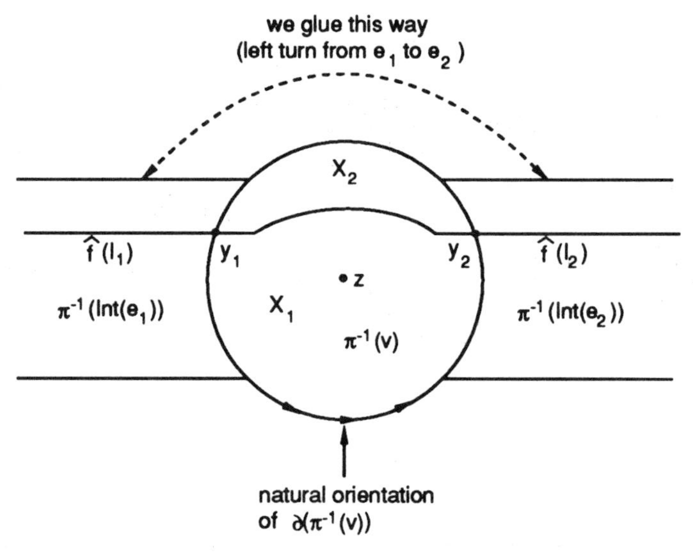

FIGURE 5

will form a loop.

A *GRP of vertex type at a vertex* $v \in V$ occurs when there is a vertex $v \in V \setminus R$ with all edge germs at v eventually gluable. This condition gives us a vertex loop for some iterate of the map. Of course it is always edges which can be glued in a GRP. The modifier "edge type" or "vertex type" refers to the kind of loop that can be removed.

Assume now that our map is transitive. When we find a GRP then we introduce new valence 2 vertices if necessary, make one or more gluings and we get a loop, so we can tighten the map decreasing the entropy. However, we have to be careful *where* we add new vertices. We shall illustrate our point on an example, which also shows why it is better to make gluings one step at a time.

Suppose that $P = \{x, y, z\}$ with $F(x) = y$, $F(y) = z$ and $F(z) = x$ and T is an interval $[\pi(x), \pi(z)]$ with and an additional vertex $\pi(y)$ inside T. Denote the edges of T by $e_1 = [\pi(x), \pi(y)]$ and $d = [\pi(y), \pi(z)]$. Suppose also that the images of T under the core c and $\widehat{f}$ look like Figure 6 (we assume that π is more or less projection onto the horizontal line, with a necessary subsequent collapsing of intervals which are to be mapped to the vertices).

The edge germs of e_1 and d are gluable at y, with the right turn from e_1 to d. When we "bend" the interval at y, making a right turn when we go from $c(e_1)$ to $c(d)$ then we see that $\widehat{f}(d)$ is "inside" this bend. Therefore there is a GRP of edge type there. There is a point $\widetilde{t} \in (\pi(y), \pi(z))$ such that $f(t) = \pi(y)$. Thus, we can add a vertex at $\widetilde{t}$ and get a new tree T' with the edges $e_1 = [\pi(x), \pi(y)]$, $e_2 = [\pi(y), \widetilde{t}]$ and $d' = [\widetilde{t}, \pi(z)]$. The images of T under the new core c' and $\widehat{f}'$ are shown in Figure 7.

Now the images of e_1 and e_2 under f are equal, so we can be tempted to glue

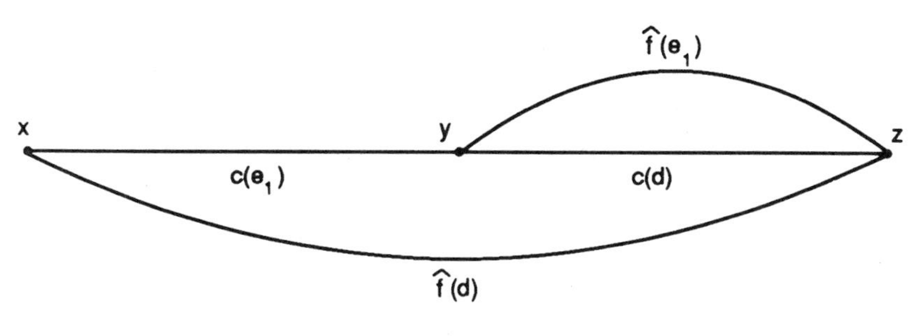

FIGURE 6

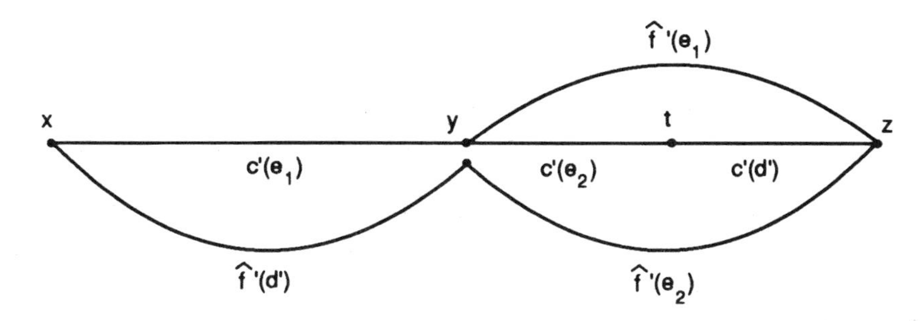

FIGURE 7

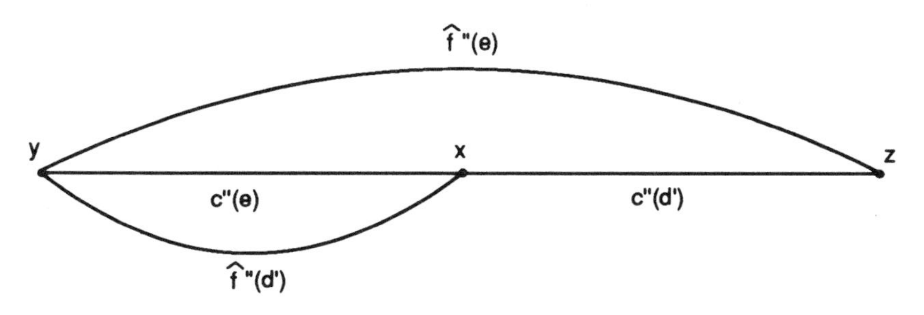

FIGURE 8

together e_1 and e_2. However, when we do it, we get the situation illustrated in Figure 8, which is basically Figure 6 turned by 180 degrees (so shouldn't we call it Figure 9?). The loop, which was supposed to appear, disappeared in a miraculous way. In fact, this happened because *we glued too much.* Therefore before we start actual gluing, we have to introduce new valence 2 vertices, which in turn may require introducing new ones, etc. Thus it seems that a much better idea is to choose new vertices which we want to add more carefully.

In a general case, under the assumption of transitivity, let us think about gluing as a continuous process. We can assume that the edges of T are straight segments. By replacing f by an equivalent map, we can assume that each edge

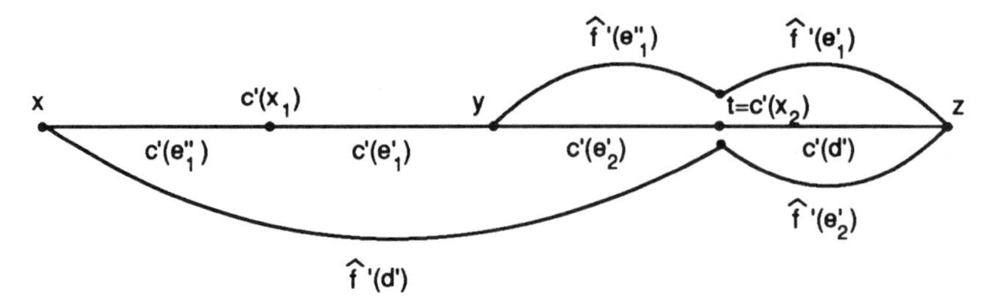

FIGURE 9

can be partitioned into subintervals which are mapped onto edges in a *linear* way. Now, our assumption on the transitivity gives us 2 possibilities. One is that the map just permutes the edges, but then clearly no gluing is possible (or necessary). The other possibility is that sooner or later the image of every edge is larger than just one edge. Consequently, when we come back to the same edge, the corresponding iterate of the map is expanding. In fact, we can even assign the lengths to the edges in such a way that f itself is piecewise expanding (we can even make this expansion uniform with the constant equal to the exponential of the entropy of f; then the lengths of the edges will be given by the components of the appropriate eigenvector of the matrix $M(f)$).

Suppose we have a GRP of edge type. When we think about gluing as a continuous process, we glue together longer and longer pieces of our edges e_1 and e_2. If their germs are gluable in n steps, then we look at their images under f^k for $k = 0, 1, \dots, n$. That is, if we imagine gluing e_1 with e_2 up to points $x_1 \in e_1$ and $x_2 \in e_2$ then we trace the points $f^k(x_1)$ and $f^k(x_2)$ as the points x_1 and x_2 move outward from the common vertex v of e_1 and e_2. We continue until either one (or more) of those points meets a vertex or two (or more) of them collide with each other (of course, since always $f^n(x_1) = f^n(x_2)$, we do not treat this as a collision). When this happens, we stop and add new vertices (if they are not vertices already) at those points. If we do it with the example from Figure 6, instead of Figure 7 we will get a situation illustrated on Figure 9 (this time, $\widetilde{t} = f(x_2)$ is a fixed point).

In the example, we now glue e_1' with e_2', get a loop and after tightening get a permutation of 3 edges of a triod. In the general case, the situation is more complicated. We can assume that the germs of e_1 and e_2 are not gluable in $n-1$ steps. Look at the images d_1 and d_2 of those germs under f^{n-1}. They are gluable, so we introduce a new valence 2 vertex on one of d_1, d_2 if necessary and we get two edges which can be glued. We glue them and then e_1 and e_2 become gluable in $n-1$ steps (or, if $n = 1$, we get a loop already). Then we simply repeat the whole procedure (more formally, we use induction). The only thing that can go wrong is that our gluing influences the images of I_1 and I_2 (see the definition of the GRP of edge type) which were supposed to create a loop (if $n = 1$) or to give a GRP after gluing (if $n > 1$). This is what happened in our example when we made the gluing too long. However, when we think again about the gluing as a continuous process, this means that the first image of the intervals which we are gluing, approaches v along one of e_1, e_2, and gets there during gluing. This implies that some collision of this image with one of x_1, x_2 occurs before we finish gluing, and contradicts our choice of the vertices we added. Therefore, with the construction described everything works well.

If there is a GRP of vertex type, we do a similar construction. We replace two edges and two points on them by s of them (all the edges at the vertex where GRP occurs). Then we make gluings again step by step, each time reducing by 1 the sum of numbers of steps necessary to perform all gluings of neighboring pairs of edges at this vertex. This time the preliminary step with adding vertices when the collision of images of x_i's occur, saves us from a situation when after some gluing new edges become attached to our vertex.

Thus if f is transitive and we find a GRP, we can follow the described procedure, and after a finite number of adding valence 2 vertices and gluings edges, we get a loop, which we destroy by tightening. During this last step, the entropy drops down, so the resulting map g has strictly smaller entropy than f. We shall call this procedure a *gluing reduction*.

6. Cleaning reductions

Suppose we start with a transitive f and try to get rid of all GRP's by making gluing reductions. It may happen that we succeed – then we are happy. It may happen that at some moment we get a map which is not transitive – we shall discuss this situation later. However, can we be sure that that one of the above situations will occur after finitely many gluing reductions? After each reduction the entropy decreases. However, we add more and more new valence 2 vertices. Therefore the Markov maps can be more and more complicated, so the loss of the entropy at each step can be smaller and smaller. Thus, it is a good idea after each step to get rid of the valence 2 vertices which are not in R.

There is no problem with those valence 2 vertices from $V \setminus R$ which are not images under any iterate of f of any vertex of valence larger than 2. We can just remove them all, as described in the previous section. We shall now describe the operation of *dragging* which prepares other valence 2 vertices from $V \setminus R$ for the removal.

Let $v \in V \setminus R$ be a vertex of T of valence 2. Then there are two edges, e and d, emerging from v. Let w and u be the endpoints of e and d respectively, different from v. If v is the image of some vertices, we can "drag" these images to w or u without enlarging the entropy of the map (see [**BH1, ACG**]). More precisely, we can compose f with the map which either collapses the whole e into the point w and maps d in a monotone way onto $e \cup d$ or collapses the whole d into the point u and maps e in a monotone way onto $e \cup d$. The modifications of the associated thick tree map $\widehat{f}$ are obvious; it will clearly remain associated to F. The only problem is that it may happen that the new map is only weakly Markov instead of Markov. Then we perform the collapsing operation described in Section 3 (we shall call it *interval collapsing*, to distinguish it from another collapsing operation which we will introduce soon). In such a way we get a new tree T', a new map $f' : T' \to T'$, associated to a thick tree map $\widehat{f'}$ of a thick tree $\widehat{T}'$, associated in turn to the map F (the same as $\widehat{f}$). The tree T' has no more vertices than T, and if it has the same number of vertices as T then it has a vertex of valence 2 which is not an image of any vertex. This vertex can be removed and this effectively reduces the number of vertices of the tree. Moreover, the entropy of the resulting Markov map does not exceed the entropy of the original map f.

Thus, by successive draggings we can get rid of all valence 2 vertices which are not elements of R, and do not enlarge the entropy, while staying in the class of the thick tree maps associated to F. The practical problem emerging is how to recognize whether we should drag the images to w or to u. This is also described in [**BH1**] (Lemma 1.7) and [**ACG**] (Lemma 6). In fact, the situation we have here is the one described in Lemma 1.7 of [**BH1**]; Lemma 6 of [**ACG** is dual, that is the matrix appearing there can be thought about as $M(f)$ transposed. We have to take a right eigenvector $(a_e)_{e \in E}$ of $M(f)$ (where E is the set of edges of T), corresponding to the largest eigenvalue of $M(F)$, that is to $exp(h(f))$. If $a_e < a_d$ then we drag to u and the entropy decreases. If $a_d < a_e$ then we drag to w and the entropy decreases. If $a_e = a_d$ then we can drag to u or to w and in both cases the entropy stays the same.

When applying this procedure in practice, determining which one of the numbers a_e and a_d is smaller may be not so simple. Fortunately, from our experience it seems that really dragging is not necessary. The reason why we describe it is that we cannot prove that it is not necessary.

Now suppose that f is not transitive. Then there is a proper subset X of T, which is a union of some edges of T, which is f-invariant, that is $f(X) \subset X$. What we do in such a situation, depends on X. If no connected component of X contains more than one element of R, then we can collapse each connected component of X to a point, in the same way as we produced interval collapsing. We shall call it *component collapsing*. Clearly, this collapsing also cannot increase the entropy.

Thus, we can define a *cleaning reduction* as getting rid of valence 2 vertices which are not in R by successive draggings and then applying all possible component collapsings. This reduction does not increase the entropy of the map. The resulting tree has no valence 2 vertices which are not in R. The resulting map has the property that there is no proper invariant subset of the tree which is a union of edges, and such that each of its components contains at most one element of R. Notice that any invariant subset of T which intersects R has to contain it, since the orbit of every element of R is equal to the whole R.

Now we can resolve our problem mentioned at the beginning of this section.

Suppose we start with a transitive map, perform a gluing reduction, perform a cleaning reduction, get a transitive map, and repeat this procedure as long as possible. Do we necessarily stop? The answer is positive. To see this, notice first that the trees with which we deal have no valence 2 vertices which are not in R. Moreover, all ends of those trees belong to R.

We shall prove a lemma, which will be also useful later. We need some notation. Let V_1 be the set of all vertices of T of valence 1, V_2 the set of all vertices of T of valence 2, and V_3 the set of all vertices of T of valence 3 and more. Let t_i denote the number of elements of V_i $(i = 1, 2, 3)$, s the number of edges of T, n the number of elements of R and r the number of elements of $V \setminus R$. Finally, let $\alpha(v)$ denote the valence of a vertex v minus 1.

LEMMA 6.1. *If* $V_1 \cup V_2 \subset R$ *then* $r \leqslant n-2$ *and* $s \leqslant n + t_1 - 3$. *Moreover,* $\sum_{v\in R} \alpha(v) \leqslant n-2$.

PROOF. We have

$$\sum_{v\in V_1} \alpha(v) = 0, \tag{1}$$

$$\sum_{v\in V_2} \alpha(v) = t_2, \tag{2}$$

$$\sum_{v\in V_3} \alpha(v) \geqslant 2t_3. \tag{3}$$

By induction (add edges one by one starting from the interval) we prove easily that

$$s = t_1 + t_2 + t_3 - 1 \tag{4}$$

and

$$s = 1 + \sum_{v\in V} \alpha(v). \tag{5}$$

Since $V_1 \cup V_2 \subset R$, we have

$$t_1 + t_2 \leqslant n. \tag{6}$$

From (1), (2), (3) and (5) we get $s \geqslant 1 + t_2 + 2t_3$; from this and (4) we get

$$t_3 \leqslant t_1 - 2. \tag{7}$$

Since $n + r = t_1 + t_2 + t_3$, by (6) we get $r \leqslant t_3$. From this, (7) and (6) we get $r \leqslant n-2$. This proves the first statement of the lemma.

From (4) and (7) we get $s \leqslant 2t_1 + t_2 - 3$, so by (6) we get $s \leqslant n + t_1 - 3$. This completes the proof of the first part of the lemma.

From (4), (5) and since $n + r = t_1 + t_2 + t_3$, we get

$$\sum_{v \in R} \alpha(v) = n + r - 2 - \sum_{v \in V \setminus R} \alpha(v).$$

For every $v \in V \setminus R$ we have $\alpha(v) > 1$, so $\sum_{v \in V \setminus R} \alpha(v) \geqslant r$. Therefore $\sum_{v \in R} \alpha(v) \leqslant n - 2$. $\square$

Since $s < n + t_1 \leqslant 2n$, once R is specified, the number of edges of T is bounded, so the size of the matrix $M(f)$ is bounded.

LEMMA 6.2. *If T has m edges and f is transitive then $h(f) \geqslant (1/m) \log q$, where q is the largest entry of $M(f)$.*

PROOF. There are vertices a and b of the graph $\mathcal{G}(f)$ with q arrows from a to b. By transitivity, there is a path from b to a in $\mathcal{G}(f)$. Choose the shortest such path and denote its length by l (if $a = b$ then set $l = 0$). Clearly, $l \leqslant m - 1$. Thus we get q distinct paths of length $l + 1$ from a to itself. This means that for every k there are at least q^k paths in $\mathcal{G}(f)$ of length $k(l + 1)$. Consequently, $h(f) \geqslant (1/(l + 1)) \log q \geqslant (1/m) \log q$. $\square$

Thus, once R is specified and some bound on the entropy of f is set, then for transitive f's there are only finitely many choices for the matrix $M(f)$, that is, a finite choice of the values of $h(f)$. Therefore, if after each step the entropy decreases, there are only finitely many steps to make.

We should stress that our process can terminate in two ways. The first possibility is that we get a transitive map without a GRP. We shall deal with this case later. The second possibility is that we get a map which is not transitive. We shall consider this case in the next section.

7. Big reductions

Denote by $\mathcal{E}$ the class of all subsets of T which are unions of some edges of T. Clearly, if $X \in \mathcal{E}$ then $f(X) \in \mathcal{E}$. For most of this section we assume that f is not transitive but there is no invariant set $X \neq T$ which belongs to $\mathcal{E}$ and such that every component of X contains at most one element of R. Then there exists an invariant set $X \neq T$ which belongs to $\mathcal{E}$ and contains R and which has at least one component containing more than one element of R.

Set $Y = \bigcap_{n=0}^{\infty} f^n(X)$. All the sets $f^n(X)$ belong to $\mathcal{E}$ and the sequence $(f^n(X))$ is descending, so the set Y also belongs to $\mathcal{E}$. Moreover, since $f(R) = R$, we have $R \subset Y$. It follows that Y contains all edges of T adjacent to the ends of T. Clearly, $f(Y) = Y$.

Let Λ be the set of all connected components of Y. For every $Z \in \Lambda$ there is $\xi(Z) \in \Lambda$ such that $f(Z) \subset \xi(Z)$. This defines a map $\xi : \Lambda \to \Lambda$.

LEMMA 7.1. (a) *The map ξ is a cyclic permutation.*

(b) *Each element of Λ contains the same number of elements of R; this number is at least 2.*

(c) *The number of elements of Λ is at least 2.*

PROOF. Since $f(Y) = Y$, ξ maps Λ *onto* Λ. Since Λ is finite, this proves that ξ is a permutation.

If $Z \in \Lambda$ then the set $W = \bigcup_{n=0}^{\infty} f^n(Z)$ belongs to $\mathcal{E}$ and $f(W) \subset W$. By our assumptions, W cannot be disjoint from R. Therefore there is $k \geqslant 0$ such that $\xi^k(Z)$ contains a point of R. Since ξ is a permutation, it follows that $\xi^{k+j}(Z) = Z$ for some $j \geqslant 0$ and thus Z also contains a point of R. Together with the facts that ξ is a permutation and that the orbit of any element of R equals to the whole R, this proves (a).

Since ξ is a cyclic permutation, for every $Z, U \in \Lambda$ there exists $j \geqslant 0$ such that $\xi^j(Z) = U$. Since f^j is one-to-one on R, it follows that the number of elements of R contained in U is larger than or equal to the number of elements of R contained in Z. Therefore each element of Λ contains the same number of elements of R. By our assumptions, this number cannot be less than 2 This proves (b).

To prove (c), assume that Λ has only one element. Then this element has to contain all ends of T (because they belong to R). Since it is connected, it has to be equal to T, contrary to the assumption that $X \neq T$. $\square$

Denote by k the number of elements of Λ. We have now k pairwise disjoint sets $\pi^{-1}(Z)$ where $Z \in \Lambda$, each of them is homeomorphic to $\mathbb{D}$, their union contains P and F permutes them cyclically. This description matches exactly the description of an extension from Section 1, if we take into account that now the map F belongs to $\mathfrak{C}$ rather than to $\mathfrak{A}$. Therefore in this situation the pattern A which we investigate is a B-extension of C for certain patterns B and C. Moreover, the patterns B and C are non-trivial, that is cycles belonging to them have period larger than 1.

The description of an extension from Section 1 gives us also immediately the way to obtain thick tree maps corresponding to the patterns B and C. For C, the corresponding thick tree map can be obtained by collapsing each component of Λ to a point. This collapsing differs from the component collapsings described in Section 6 only by the fact that now there are at least 2 elements of R in each collapsing component, so the set R changes. Then, if necessary, we perform interval collapsings.

For B, we choose an element Z of Λ and an orientation preserving homeomorphism $H : \mathbb{D} \to \pi^{-1}(Z)$. Now we can treat Z as a new tree and $\pi \circ H$ as a thick tree structure over it. The old core and the k-th iterate if the old thick tree map (in the obvious sense) restricted to Z give a new core and a new thick tree map of Z, respectively. The only thing that can go wrong is that perhaps Z has ends which are not elements of R (or we should say rather, of the new cycle $R \cap Z$, but this makes no difference). We solve this problem immediately by collapsing the "unnecessary" branches of Z and then, if necessary, performing interval collapsings.

The procedure described above reduces a thick tree map representing a pattern A, which is a B-extension of C, to two thick tree maps, representing B and C

respectively. We shall call it a *big reduction*. After performing it, we try to reduce further both B and C.

There is another case when we can make a big reduction. This happens if there are connected subsets $Z_1, Z_2, \ldots, Z_k$ of T and a vertex v such that $Z_i \cap Z_j = \{v\}$ for every $i \neq j$ and $f(Z_i) = Z_{i+1}$ for $i = 1, 2, \ldots, k-1$ and $f(Z_k) = Z_1$. Moreover we assume that at least one of the sets Z_i consists of more than one edge (otherwise the pattern we consider is just a twist pattern and we cannot make any more non-trivial reductions). Then we start in a slightly different way as before, namely blow up v to a whole "star" (k-od), so that Z_i's become disjoint. On the new edges we define the map just as a permutation, agreeing with the map on Z_i's. It is obvious how to modify a thick tree and the map on it. Then Lemma 7.1 holds and we continue as before.

The possibility of making a big reduction in the second way is strongly connected with transitivity not only of f, but of all its iterates.

LEMMA 7.2. *If f is transitive and we cannot make a big reduction then either f is (up to an equivalence) rotation of a star (k-od) or all iterates of f are also transitive.*

PROOF. Suppose that f is transitive but some iterate of f is not transitive. Then the set E of all edges of T can be decomposed into the union of pairwise disjoint subsets E_i, $i = 0, 1, \ldots, n-1$ such that the image of any $e \in E_i$ is a union of elements of E_{i+1} (here the addition is mod. n). We may assume that the sets E_i are minimal possible (that is, n is maximal possible; then clearly $n \geqslant 2$). Then each of the sets $X_i = \bigcup_{e \in E_i} e$ is connected. Each pair of distinct sets X_i and X_{i+j} can intersect at at most one point and this point is a vertex. Let v be a fixed point of f, which must exist since T is a tree. Then $v \in X_i$ for some i and hence $v = f^j(v) \in X_{i+j}$ for all j. Therefore all X_i's intersect at the single vertex v. Thus, we can make a big reduction, unless T is a star and f permutes its edges. Since f preserves the ordering of edge germs, f is a rotation of T (up to equivalence). $\square$

As for gluing reductions, we will say that there is no BRP for f (or for $\widehat{f}$) if there is no possibility to make a big reduction.

8. Splitting reductions

In our efforts to find the simplest thick tree maps associated to a given $F \in \mathfrak{C}$, we came to a moment when we obtained a transitive map without GRP or BRP and the tree has no vertices of valence 2 in $V \setminus R$. It turns out that we can do better. Namely, we can get a map for which there are no gluable pairs of edge germs at the vertices of $V \setminus R$. The natural way to try to do this is to glue together whatever is gluable. Although this gave positive results in all examples we considered, we are not sure that this procedure will always stop after a finite number of gluings. Therefore we will apply splittings instead. It may seem strange at first that in such a way we do not obtain new gluable pairs of edge

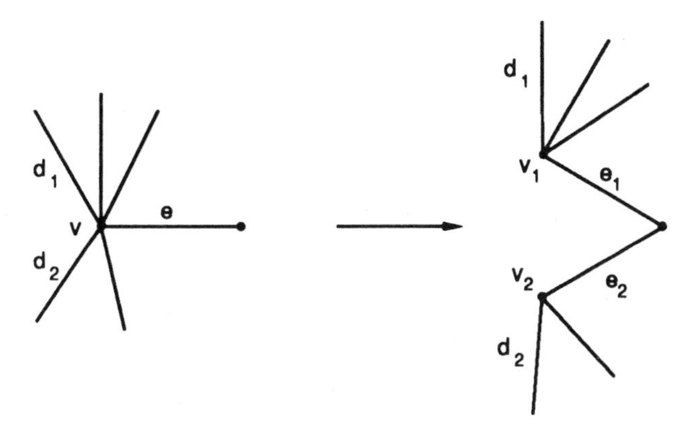

FIGURE 10

germs, but on the contrary, remove them. However, one has to remember that our aim is not to get rid of *all* gluable pairs, which is not possible, but only of those which are at the vertices not belonging to R.

Splitting has been described in Section 4, as the operation reverse to gluing. We should add to that description that if we split an edge e into two edges e_1 and e_2 then we also split one of the endpoints v of e into two vertices v_1 and v_2. If $v \in V \setminus R$ then then among the edge germs at v there are two neighbors in the cyclic order, say d_1 and d_2, different from e and such that after the splitting d_i becomes an edge germ at v_i for $i = 1, 2$. In such a situation we shall say that we *start the splitting at v between d_1 and d_2* (see Figure 10). If $v \in R$ then the situation is slightly more complicated, but we shall avoid such splittings.

Suppose that we have a thick tree map $\widehat{f}$ of $\widehat{T}$ associated to $F \in \mathfrak{C}$ and we want to split an edge e of T starting at v between d_1 and d_2. We have to find conditions under which $\widehat{f}$ can be modified to a map of a new thick tree so that it remains associated to F. Clearly, the only problem is what to do with the images under f which pass through v. Then an obvious condition which allows us to modify $\widehat{f}$ in an easy way is that the component Y of the set $\pi^{-1}(v) \setminus F(c(T))$ containing the part X of the boundary of $\pi^{-1}(v)$ between $\mathrm{cl}(\pi^{-1}(\mathrm{Int}(d_1)))$ and $\mathrm{cl}(\pi^{-1}(\mathrm{Int}(d_2)))$, intersects the closure of $\pi^{-1}(\mathrm{Int}(e))$ (see Figure 11). In less formal way, this condition says that if we enter $\pi^{-1}(v)$ between $\pi^{-1}(\mathrm{Int}(d_1))$ and $\pi^{-1}(\mathrm{Int}(d_2))$ then we can reach $\pi^{-1}(\mathrm{Int}(e))$ without crossing the image under F of the core of $\widehat{T}$. We shall call it a *splitting possibility of order 1 between d_1 and d_2 towards e.*

In the above situation, if we draw a curve in the thick tree, starting at some point of X, passing through Y, then entering $\pi^{-1}(\mathrm{Int}(e))$, continuing till it meets $\pi^{-1}(u)$ (where $u \neq v$ is the other endpoint of e), in such a way that it does not intersect $F(c(T))$, then we can think about our splitting as being performed *along this curve.*

If in the description of the splitting possibility of order 1 we replace in the

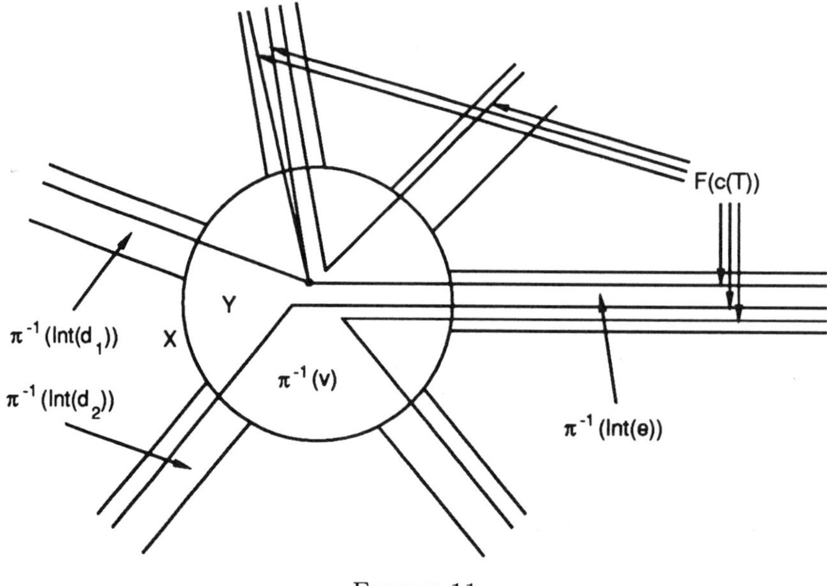

FIGURE 11

definition of Y the set $\pi^{-1}(v) \setminus F(c(T))$ by the set $\pi^{-1}(v) \setminus F^n(c(T))$, then we get a *splitting possibility of order n between d_1 and d_2 towards e.* We shall use this notion later.

Since splitting is the reverse operation to gluing and gluing does not change the entropy, also splitting does not change the entropy. To see that splitting does not destroy transitivity, look at what changes in the Markov graph of f. First notice that under our assumptions Markov tree maps are always onto, whether transitive or not. This follows immediately from the assumptions that all ends of T belong to R and that f maps R onto R. Now, when we split an edge e into edges e_1 and e_2 then all paths in $\mathcal{G}(f)$ passing through e before the splitting, pass through e_1 or e_2 after the splitting. Moreover, since the map stays onto, for both $i = 1$ and $i = 2$ there is an arrow terminating at e_i. Clearly, there is also an arrow originating at e_i. Thus, the only thing that can go wrong is that e_1 (or e_2) is mapped only to itself. However, this would mean that before the splitting e was mapped only to itself. Since f was transitive, the tree could consist of one edge only, and in this case splitting was not possible.

Now, we can apply the above result to the iterates of f and in view of Lemma 7.2 we see that if there were no BRP's before the splitting, there will be none after.

The next problem we have to consider is when there were no GRP's before splitting, can it happen that they appear after splitting? Fortunately, this cannot happen. Eventual gluability of pairs of germs does not change during splitting, except that we obtain a new pair of germs, say e_1 and e_2 at u. However, there cannot be a GRP due to gluability of e_1 and e_2 at u, since then there would be a loop before the splitting. Another reason for a new GRP could be the change of the image. It could happen a priori that all the edge germs at v_1 or v_2 (the new vertices) are eventually gluable. However, since the tree map is onto, so there is something mapped to those germs. The image of this "something" (before the splitting) is separated in the thick tree from the image of anything mapped to the rest of the germs at v; the separation line is the same line along which we make the splitting. Thus, the conditions for GRP are satisfied already for f before the splitting. This proves that indeed, if there were no GRP's before splitting, then there are no GRP's after splitting.

Notice also that if we remove those valence 2 vertices from $V \setminus R$ which either are not images of any vertices or form a cycle and none of them is an image of any vertex not from this cycle, then clearly if there were no GRP's before this operation then there will be no GRP's after. Clearly, removing valence 2 vertices does not affect also the property of no BRP's.

To have an adequate language for dealing with numbers of gluable edge germs we introduce the notion of an *eventual valence* of a vertex v. We define it as the maximal number of edge germs at v such that no pair among them is eventually gluable. We shall denote by $\beta(v)$ the eventual valence of v minus 1. Notice that for every $v \in R$ we have $\beta(v) = 0$. Moreover, if there is no GRP then $\beta(v) > 0$ for every $v \in V \setminus R$ (otherwise we would have GRP of vertex type).

LEMMA 8.1. *For every $v \in V$ we have*

(a) $\beta(v) \leqslant \beta(f(v))$,

(b) $\beta(v) \geqslant \sum_{u \in f^{-1}(v) \cap V} \beta(u) - 1$.

PROOF. (a) follows from the fact that the image of not eventually gluable pair of edge germs is not eventually gluable. To prove (b), denote by $v_1, v_2, \ldots, v_s$ the elements of $f^{-1}(v) \cap V$. For every i choose pairwise not eventually gluable edge germs $e_i^{(0)}, e_i^{(1)}, \ldots, e_i^{(\beta(v_i))}$ at v_i. We may assume that this is their cyclic order. We may assume also that $v_1, v_2, \ldots, v_r$ are those vertices among $v_1, v_2, \ldots, v_s$ for which $\beta(v_i) > 0$ and that the cyclic order of the images of all those germs is $f(e_1^{(0)}), f(e_1^{(1)}), \ldots, f(e_1^{(\beta(v_1))}), f(e_2^{(0)}), f(e_2^{(1)}), \ldots, f(e_2^{(\beta(v_2))}), \ldots,$ $f(e_r^{(0)}), f(e_r^{(1)}), \ldots, f(e_r^{(\beta(v_r))})$. In fact, some of the above germs can be already glued together, but this applies only to the pairs of the form $f(e_i^{(\beta(v_i))})$ with $f(e_{i+1}^{(0)})$ (and $f(e_r^{(\beta(v_r))})$ with $f(e_1^{(0)})$). Moreover, only those pairs can be eventually gluable. This leaves at least $\sum_{i=1}^{r} \beta(v_i)$ pairwise not eventually gluable edge germs at v. Therefore $\beta(v) \geqslant \sum_{i=1}^{r} \beta(v_i) - 1 = \sum_{i=1}^{s} \beta(v_i) - 1$. □

From Lemma 8.1(a) we get immediately the following corollary.

COROLLARY 8.2. *Eventual valences at all points of a cycle of vertices are the same.*

LEMMA 8.3. *If there are no GRP's then eventual valence at any vertex which is not periodic, is at most 2.*

PROOF. Let a vertex v_1 be not periodic. Then there is $m \geqslant 0$ such that the vertex $v_2 = f^m(v_1)$ is not periodic but the vertex $v_3 = f^{m+1}(v_1)$ is periodic. The vertex v_3 is the image of some vertex v_4 from the same cycle (there is nothing wrong if $v_4 = v_3$). By Lemma 8.1(a), $\beta(v_2) \geqslant \beta(v_1)$. By Lemma 8.1(b), $\beta(v_3) \geqslant \beta(v_2) + \beta(v_4) - 1$. By Corollary 8.2, $\beta(v_4) = \beta(v_3)$. Therefore

$$\beta(v_3) \geqslant \beta(v_2) + \beta(v_3) - 1 \geqslant \beta(v_1) + \beta(v_3) - 1,$$

that is $\beta(v_1) \leqslant 1$. □

Our aim is to get a map for which valence and eventual valence at every vertex from $V \setminus R$ is the same. Lemma 8.3 shows that for such maps all vertices which are not periodic have valence 2. Therefore they can be just removed one by one, so that each time we remove a vertex which is not an image of any vertex. From what we said it follows that if we start with a transitive map without GRP's or BRP's, perform several splittings in such a way that the valence and the eventual valence of every vertex not from R becomes the same and then remove valence 2 vertices which are not in R then we arrive to a map which has the same entropy and additionally some very nice properties. Those properties are: transitivity, no GRP's or BRP's, every vertex is periodic, every vertex of valence 2 belongs to R and for any vertex $v \in V \setminus R$ the set of all edge germs at v is mapped by f bijectively onto the set of all edge germs at $f(v)$. Since at that moment our reductions basically stop, we shall call the Markov maps with those properties (or more precisely, the associated thick tree maps) *irreducible*. The only ingredient still missing in the procedure described above is how to perform splittings in such a way that the valence and the eventual valence of every vertex not from R becomes the same. In order to do it, we will need several lemmas.

LEMMA 8.4. *Let v be a vertex of T periodic of period m. Then there is $j \geqslant 1$ such that if d_1 and d_2 are edge germs at v and the germs $f^{mj}(d_1)$ and $f^{mj}(d_2)$ are distinct then $f^{mj}(d_1)$ and $f^{mj}(d_2)$ are not eventually gluable at v.*

PROOF. Look at the images of the set of all germs at v under $f^m, f^{2m}, f^{3m}, \dots$ We get a descending sequence of finite sets, so this sequence has to stabilize. Therefore for some j the set of the images of all edge germs at v under f^{mj} is just permuted by f^m. Thus, any pair of distinct elements of this set cannot be eventually gluable. □

LEMMA 8.5. *Let e_1 and e_2 be eventually gluable edge germs at a vertex v of T. Then there is $k \geqslant 1$ such that whenever u is a vertex such that $f^k(u) = v$ then the images under f^k of the edge germs at u miss at least one of e_1, e_2.*

PROOF. If v is not periodic then there is $k \geqslant 1$ such that $f^{-k}(v)$ contains no vertex and then there is nothing to prove.

Assume that v is periodic of period m By Lemma 8.4, there is $j \geqslant 1$ such that the images under f^{mj} of the edge germs at v miss at least one of e_1, e_2. There is also $i \geqslant 0$ such that the image under f^i of every vertex is periodic. Set $k = mj + i$. Suppose that u is a vertex such that $f^k(u) = v$ and d_1, d_2 are edge germs at u such that $f^k(d_1) = e_1$ and $f^k(d_2) = e_2$. Then $f^i(u)$ belongs to the orbit of v and $v = f^k(u) = f^{mj+i}(u) = f^{mj}(f^i(u))$, so we must have $f^i(u) = v$. Set $d_3 = f^i(d_1)$ and $d_4 = f^i(d_2)$. Then d_3 and d_4 are edge germs at v and $f^{mj}(d_3) = e_1$, $f^{mj}(d_4) = e_2$, a contradiction. □

LEMMA 8.6. *Assume that there are no GRP's for f. Let e_1 and e_2 be eventually gluable edges at a vertex $v \notin R$, which are neighbors in the cyclic order at v. Then there is $k \geqslant 1$ such that there is a splitting possibility of order k between e_1 and e_2 towards some edge e_3 which is not eventually gluable with either of e_1, e_2.*

PROOF. Let k be the constant from Lemma 8.5. Look at the component Y of the set $\pi^{-1}(v) \setminus F^k(c(T))$ containing the part X of the boundary of $\pi^{-1}(v)$ between $\mathrm{cl}(\pi^{-1}(\mathrm{Int}(d_1)))$ and $\mathrm{cl}(\pi^{-1}(\mathrm{Int}(d_2)))$. It cannot contain any other part of the boundary of $\mathbb{D}$, since $F^k(c(T))$ is connected and f is onto. It cannot contain more than one component of the common boundary of $\pi^{-1}(\mathrm{Int}(e_1))$ and $\pi^{-1}(v)$ since otherwise there would be a loop for f^k (the same holds for e_2 instead of e_1). Suppose that it does not contain any point of the common boundary of $\pi^{-1}(\mathrm{Int}(e_3))$ and $\pi^{-1}(v)$ for any edge $e_3 \neq e_1, e_2$ at v. Then the boundary of Y contains a piece of $F^k(c(T))$ which touches both $\pi^{-1}(\mathrm{Int}(e_1))$ and $\pi^{-1}(\mathrm{Int}(e_2))$. If this piece is contained in $F^k(c(d))$ for some edge d then there is a loop for some iterate of f (k + the number of iterates necessary to glue e_1 with e_2), a contradiction. If not, then this piece contains $F^k(c(u))$ for some vertex u and this contradicts our choice of k. In any case we get a contradiction. Therefore, by the definition there is a splitting possibility of order k between e_1 and e_2 towards some $e_3 \neq e_1, e_2$.

Suppose that e_3 is gluable with e_1 or e_2 in r steps. Then there is a loop for f^{k+r}, a contradiction. Thus, e_3 is not eventually gluable with any of e_1, e_2. □

We shall call a curve $\gamma : [a, b] \to \mathbb{D}$ a *splitting path of order n* if:

(i) $\gamma(a) \in \partial\mathbb{D}$,
(ii) $\pi(\gamma(a)) \in V \setminus R$,
(iii) $\pi(\gamma(b)) \in R$,
(iv) $\gamma([a, b]) \cap F^n(\mathbb{D}) = \emptyset$,
(v) $[a, b]$ can be divided into finitely many intervals, each of them mapped by $\pi \circ \gamma$ in a monotone way onto a different edge of T,
(vi) for any $t \in (a, b)$ and an edge e if $\gamma(t) \in \pi^{-1}(\mathrm{Int}(e))$ then the component of the set $\pi^{-1}(\mathrm{Int}(e)) \setminus F^n(c(T))$ to which $\gamma(t)$ belongs is disjoint from $\partial\mathbb{D}$.

We assume in (iv) disjointness of $\gamma([a, b])$ from $F^n(\mathbb{D})$, rather than from $F^n(c(T))$ because then automatically it is disjoint from $F^m(\mathbb{D})$ for all $m \geqslant n$; otherwise this would make no difference.

Notice that if we have a splitting path of order 1 then we can perform splitting reductions along γ starting from $\gamma(a)$ and continuing until we reach $\gamma(b)$.

For a curve $\gamma : [a, b] \to \mathbb{D}$ we shall refer to $\gamma(a)$ and $\gamma(b)$ as the *beginning* and the *end* of γ respectively.

LEMMA 8.7. *Assume that there are no GRP's for f. If there exists a pair of eventually gluable edge germs at some vertex from $V \setminus R$ then there exists a splitting path of order n for some $n \geqslant 1$ (perhaps after replacing F by another homeomorphism $G \in \mathcal{C}$ isotopic to F rel P).*

PROOF. We build up our curve γ gradually. If there exists a pair of eventually gluable edge germs at some vertex from $V \setminus R$ then by Lemma 8.6 we can start between those edges. When we come with our curve to the boundary of $\pi^{-1}(u)$ for some vertex $u \notin R$ and just before that we traversed $\pi^{-1}(\mathrm{Int}(d))$ for some edge d, and the curve up to this moment is disjoint from $F^m(\mathbb{D})$ for some $m \geqslant 1$ then we look at the component of $\pi^{-1}(u) \setminus F^m(\mathbb{D})$ containing the end of this curve. By the same reasons as in the proof of Lemma 8.6, either we can continue our curve along $\pi^{-1}(e)$ for some edge $e \neq d$ or we are "blocked" by a component of $\pi^{-1}(u) \cap F^m(\mathbb{D})$, containing $\pi^{-1}(f^m(v))$ for some vertex v. Moreover, the edge germs at v whose m-th images block our path, are neighbors in the cyclic order and their images under f^m are equal to the germ of d at u. Therefore they are eventually gluable and we can use Lemma 8.6. This lemma gives us a possibility to continue the curve; this time we avoid the set $F^{m+k}(\mathbb{D})$, where k is the integer from Lemma 8.6. We do not continue "back" through $\pi^{-1}(d)$, since then there would be a loop for f^{m+k}.

In such a way we can construct our curve without "going back", so after some time we reach $\pi^{-1}(w)$ for some $w \in R$ and then we stop. If we are careful at each step of the construction to have the appropriate interval mapped by $\pi \circ \gamma$ onto the corresponding edge in a monotone way, then clearly the curve we get is a splitting path of order n for some n. The obstructions we can meet can be only local and can be easily corrected by moving F by an isotopy rel P (this requires "straightening" the images of $\mathbb{D}$ under the iterates of F; in fact one global modification is also possible: think for instance about making G an Axiom A diffeomorphism with points $c(V)$ attracting and the images under c of the edges along the leaves of the unstable foliation). $\square$

LEMMA 8.8. *Assume that there are no GRP's for f and that there exists a pair of eventually gluable edge germs at some vertex from $V \setminus R$. Then it is possible to make finite number of splittings such that the sum of valences of points of R will increase.*

PROOF. By Lemma 8.7, there exists a splitting path $\gamma : [a, b] \to \mathbb{D}$ of order n for some $n \geqslant 1$. By composing γ with an isotopy $(K_t)_{t\in[0,1]}$ rel P such that for every $s \in [a, b]$ and $t \in [0, 1]$ we have $K_t(\gamma(s)) \notin F^n(\mathbb{D})$ and $\pi(K_t(\gamma(s))) = \pi(\gamma(s))$ (we also modify F as at the end of the previous proof if necessary), we can get another splitting path of order n, $\gamma' = K_1 \circ \gamma$. In such a case we shall call γ and γ' equivalent. Now, among all splitting paths equivalent to γ we choose one which is a splitting path of order m for minimal possible m (this is easy to do in practice). We may assume that this is actually the path γ. If $m = 1$ then we can perform splittings along γ. Only the last one will involve an element of R and it will increase the valence at this vertex. Therefore in this case we are done. Assume now that $m > 1$. There is a point $z \in [a, b]$ such that $\gamma([z, b]) \subset F^{m-1}(\mathbb{D})$ and $\gamma(z) \in \partial F^{m-1}(\mathbb{D})$. Again by replacing γ by another equivalent splitting path of order m if necessary, we can assume that this z is as large as possible (again this step can be easily performed in practice). We have then in particular $\gamma(z) \in \pi^{-1}(v)$ for some vertex v from $V \setminus R$.

The path $\delta' : [z, b] \to \mathbb{D}$ defined by $\delta'(t) = F^{1-m}(\gamma(t))$ satisfies (perhaps after "straightening" inside π^{-1} of the interiors of the edges, so that we get monotonicity after composition with π) all the conditions for the splitting path of order 1, except that it ends in a middle of π^{-1} of some edge. Then we extend it further to the nearest π^{-1} of a vertex, and we get a path $\delta : [z, w] \to \mathbb{D}$ which either is a splitting path of order 1 or fails to be it only because $\pi(\delta(w))$ belongs to $V \setminus R$ instead of R. In both cases we can make splittings along δ and we see that one of the following cases occurs (notice that $\pi(\delta(x)) \notin R$ for any $x < w$, since $f(R) = R$).

Case 1. $\pi(\delta(w)) \in R$. Then δ was actually a splitting path of order 1 and as in the case of $m = 1$ we are done.

Case 2. $\pi(\delta(w)) \notin R$ and γ does not intersect δ. Then γ becomes a splitting path of order smaller than n in the new thick tree obtained after performing splittings along δ.

Case 3. $\pi(\delta(w)) \notin R$ and γ intersects δ. By replacing γ and δ by equivalent paths if necessary, we can make this intersection as close to the end of γ as possible (if we remove it in such a way, we get Case 2). Then this intersection happens in $\pi^{-1}(v)$ of some vertex v. Then we replace γ by a shorter curve γ' starting at $\pi^{-1}(v)$ in the new thick tree obtained after performing splittings along δ. Then γ' is a splitting path of order smaller than n in the new thick tree.

In Cases 2 and 3 we continue in the same way, each time making the order of the splitting path smaller. Therefore after a finite number of steps we have to get either $m = 1$ or Case 1, which completes the proof. □

The operation described in Lemma 8.8, that is making consecutive splittings leading to enlarging the sum of valences at the elements of R, will be called a *splitting reduction.* Since the number of ends of T is bounded by the period of R, the valence of any vertex cannot be larger than the period of R. Therefore the sum of valences of elements of R cannot be larger than the square of the period of R. Thus, if we start from a map without GRP's and perform consecutive splitting reductions, after finitely many of them we come to a situation when valence and eventual valence of every vertex not in R are the same. This completes the description of obtaining an irreducible thick tree map associated to a given map $F \in \mathfrak{C}$.

9. The algorithm

To summarize the whole procedure leading from any thick tree map associated

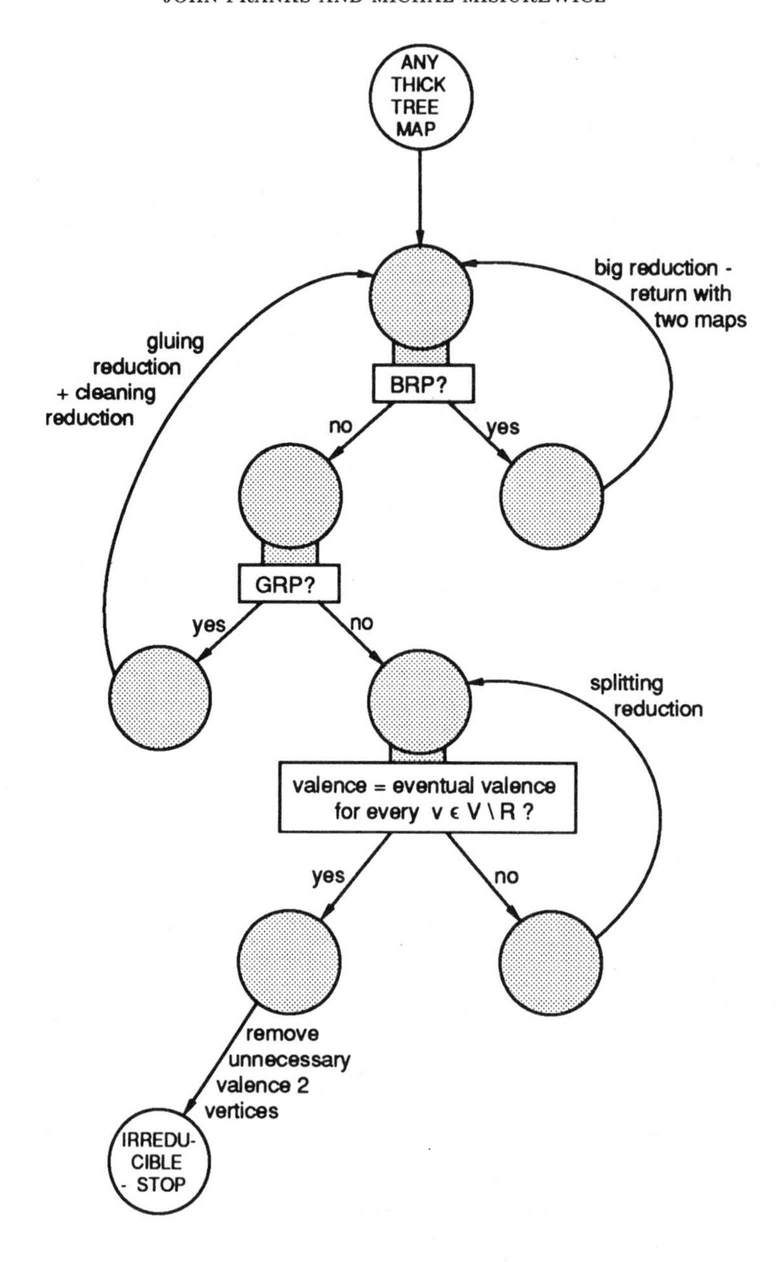

FIGURE 12

to $F \in \mathfrak{C}$ to an irreducible one (or several irreducible ones, in case when we perform big reductions) we present the following flowchart (see Figure 12).

We would like to make a comment, which originates from our experience rather than theoretical reasoning. Namely, if one is given a thick tree map and the aim is to get an irreducible one out of it, it is good to try to do it without draggings or splittings. The procedure presented in the preceding sections is not the unique one. Usually one can simplify the map by performing mainly gluings, occasionally adding or removing valence 2 vertices, collapsing something and performing big reductions (this helps of course, but is not so frequent). When it is obvious that you can make a big reduction or collapsing or remove some valence 2 vertices – do it. Otherwise try gluing. When you glue to make a gluing reduction – check how far you can glue (remember the example from Section 5!). If all this does not work – try the formal algorithm.

Another minor remark is that sometimes making a collapsing of a star is not worthwhile, if in a moment we would have to blow up this point to a star to make a big reduction.

Remember also that an irreducible form of a thick tree map associated to $F \in \mathfrak{C}$ is usually not unique (even up to equivalence!). Other ones can be obtained by gluings (we shall show this in Section 11). If some additional properties are desirable, try to get them.

Now we give some examples how to use the algorithm. To represent a thick tree and a thick tree map, we make a simple drawing containing all the necessary information. We draw the tree T, replacing the points of the cycle R by circles with capital letters denoting the vertices (see Figures 13, 14, 15). Each vertex is mapped to the one denoted by the next letter, and the last vertex is mapped to the one denoted by A (that is, if there are vertices A, B, C, D then A is mapped to B to C to D to A). The edges are denoted by natural numbers. The images (under the thick tree map) of the edges are drawn as dashed lines and denoted by the corresponding numbers with a prime (thus the image of the edge 3 will be denoted $3'$). Notice that they are drawn correctly only up to an isotopy rel. P, so one has to figure out through which thick edges they pass. For instance, in Figure 13a, the image $2'$ passes through the thick edges 4, 4, 2 and 3. We could try to draw a picture indicating this more clearly, but in general such drawings would be less transparent. When we speak of edges and vertices, we will omit the words "edge", "vertex" and leave only numbers and letters. Thus, "the germs of 1 and 2 at A" means "the germs of edges 1 and 2 at the vertex A".

Let us start with the situation presented in Figure 13a.

We can glue all of 1 with a part of 2. A common vertex of 1, 2 and 4 becomes a valence 2 vertex, so we delete it and rename the edge that contained it 1. A loop in the image of 3 appears, so we tighten the map. Thus, we perform a gluing reduction and we get the situation as in Figure 13b.

Now we can glue parts of 2 and 3, up to a fixed point on 3. A new edge appears, and we call it 5. Again a loop in the image of 3 appears, and we tighten a map. At this moment we get already a standard model for our pattern (see Figure 13c).

Now let us take another pattern, with a thick tree map as in Figure 14a.

We see a GRP at C. If we glue the germs of 4 and 5 at D then (since there is a left turn from 4 to 5) we can glue the germs of 2 and 3 at C and this will create a loop in the image of 5. The first step to perform this gluing reduction

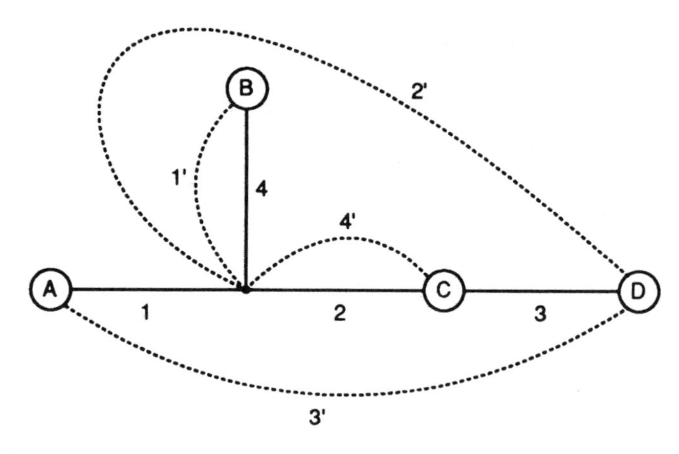

FIGURE 13a

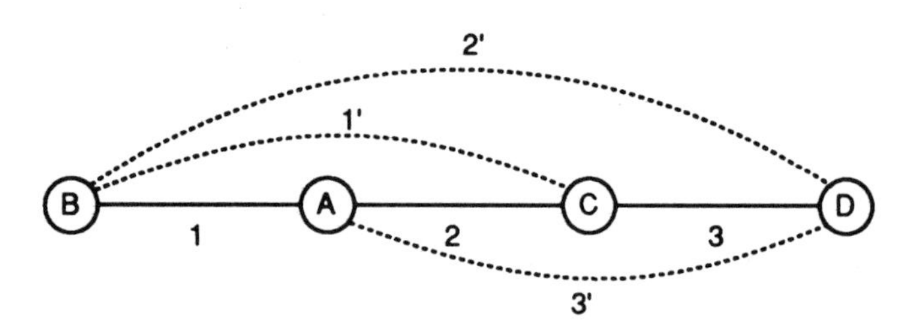

FIGURE 13b

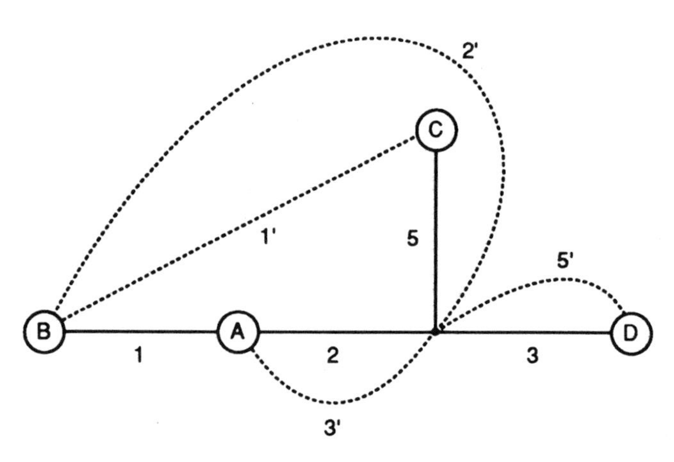

FIGURE 13c

is to glue parts of 4 and 5. A new vertex is created, and a new edge which we call 6 (see Figure 14b).

Now we glue parts of 2, 3 and 4. This creates a new vertex and a new edge,

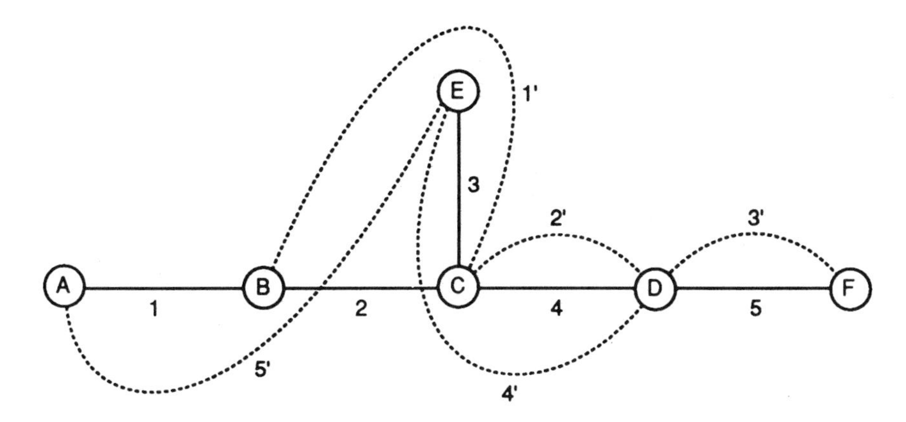

FIGURE 14a

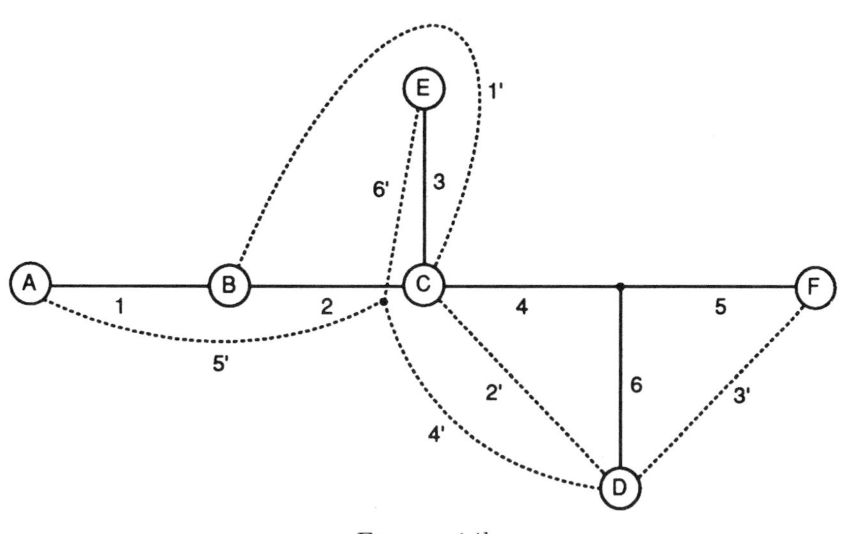

FIGURE 14b

which we call 7. In the image of 4, 5, and 6 a vertex loop is created, so we tighten the map. The result is shown in Figure 14c.

Now, 4 is mapped only to itself, so we can contract it to a point (see Figure 14d).

We glue together 2 and a part of 1. There is a right turn from 1 to 2, so B goes "up". Therefore an edge loop in the image of 5 is created. We tighten the map (see Figure 14e).

At this moment there is a BRP. The edges can be grouped: 2 with 3, 7 with 5, and 6 with 1. Each group is mapped to the next. Thus we see that our pattern is an extension of a twist pattern of period 3 by the twist pattern of period 2.

The third example is the twist by 3/5 on 5 points followed by the full Dehn

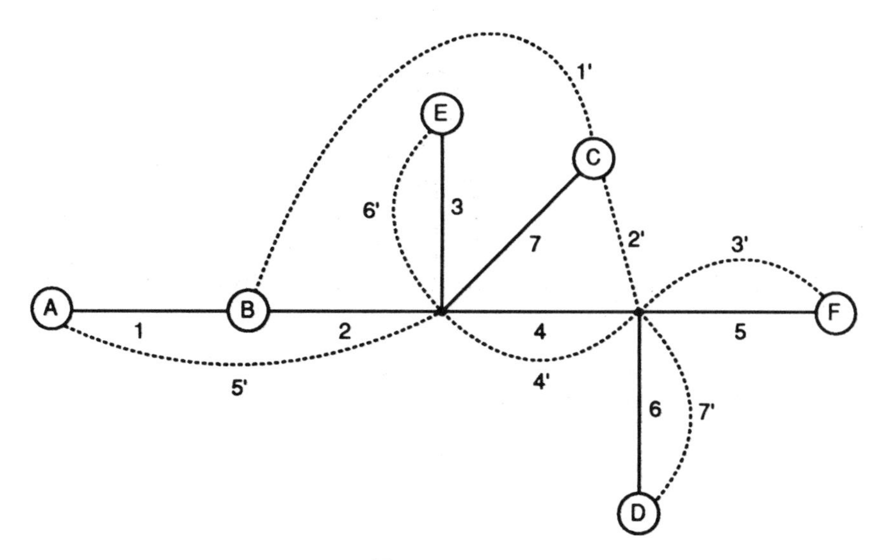

FIGURE 14c

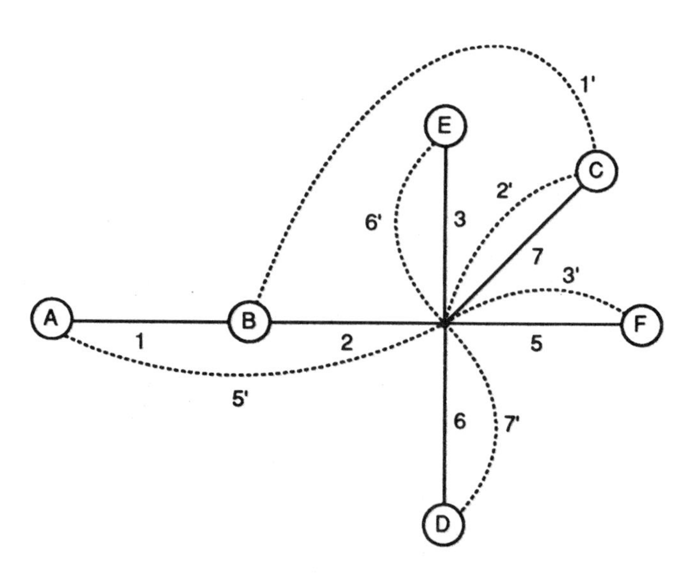

FIGURE 14d

twist on two neighboring points (A and D), see Figure 15a.

One can easily check that there are no GRP's. We can follow the algorithm and start performing splittings. We split 1, starting at the central point between 2 and 3. Part of 1 and the whole of 2 are separated after this by a vertex of valence 2, which we remove and rename the whole edge 2 (see Figure 15b).

Now we split 3, starting at the central point between 4 and 5. Part of 3 and all of 4 become separated by a vertex of valence 2, which we remove and rename the whole edge 4. This is already a standard model for our pattern (see Figure

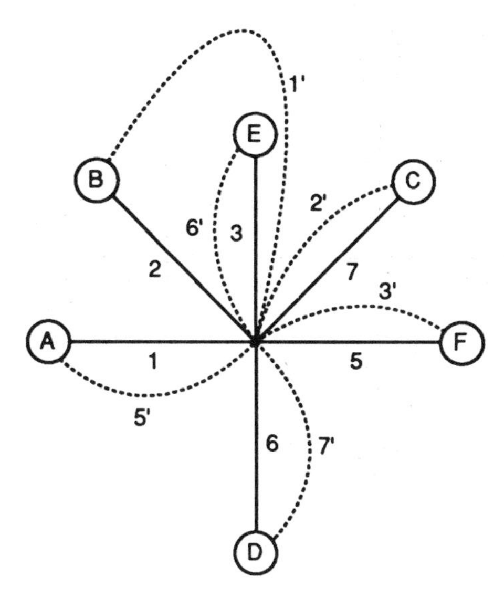

FIGURE 14e

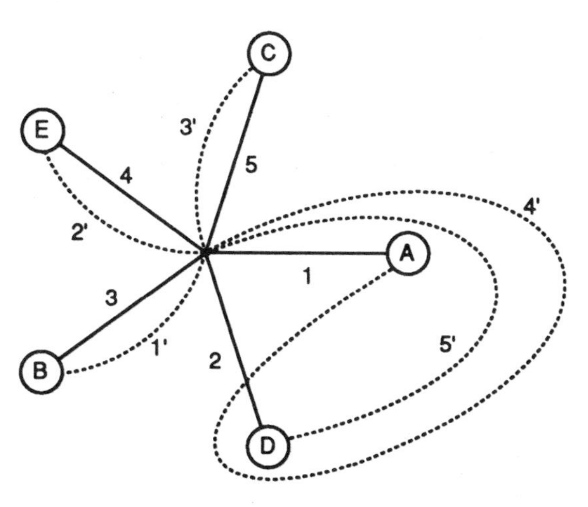

FIGURE 15a

15c).

Instead of splittings, we can try gluings, as we suggested at the beginning of this section. Starting from the situation as in Figure 15a, we can glue together 5 and a part of 4 (see Figure 15d).

Now we can glue 3 and a part of 2. We get again a standard model (see Figure 15e).

The two standard models we got are definitely different. However, as we mentioned earlier, we can get to each of them from the other one by gluings. If we start from Figure 15e, we can glue 4 with a part of 5 (see Figure 15f).

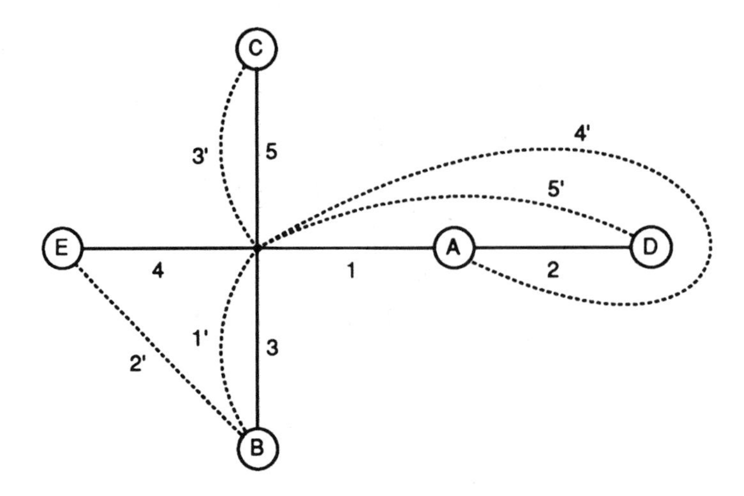

FIGURE 15b

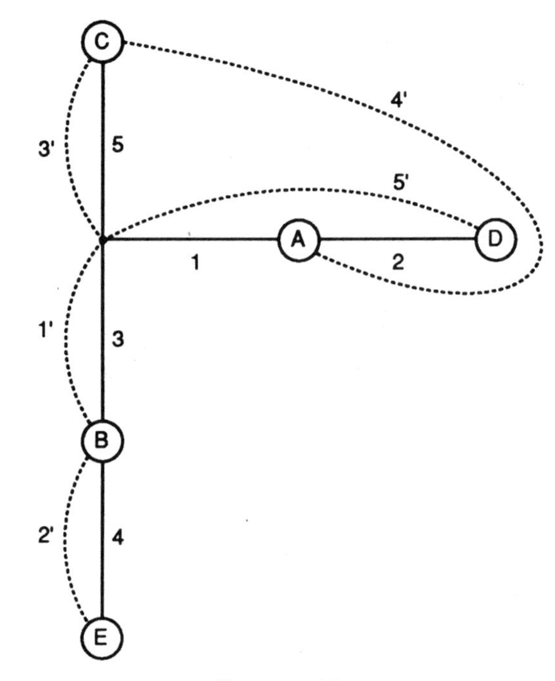

FIGURE 15c

Now we can glue 2 with a part of 3 (see Fig 15g).

Now we may notice that the situations presented in Figures 15g and 15c differ only by a cyclic permutation of the marked vertices. By performing gluings inverse to the first two splittings (modulo the permutation of vertices) we get back to our initial model (Figure 15a) with the marked vertices permutted cyclically. This means that we get a "cycle" of of six gluings which performed consecutively

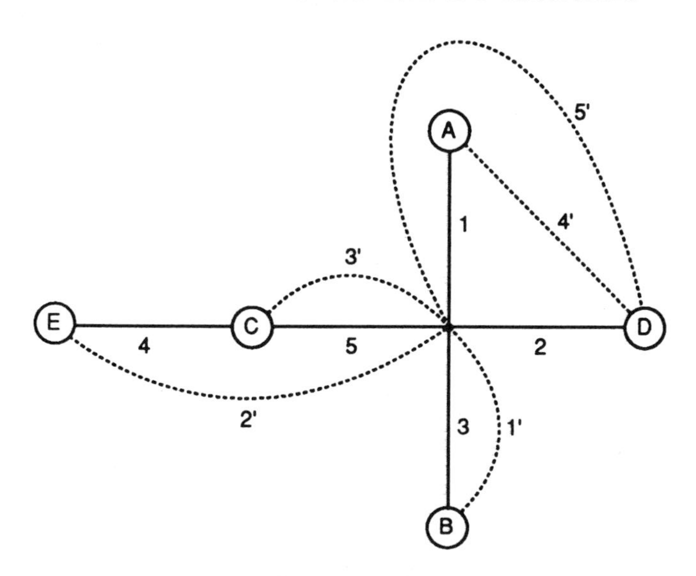

FIGURE 15d

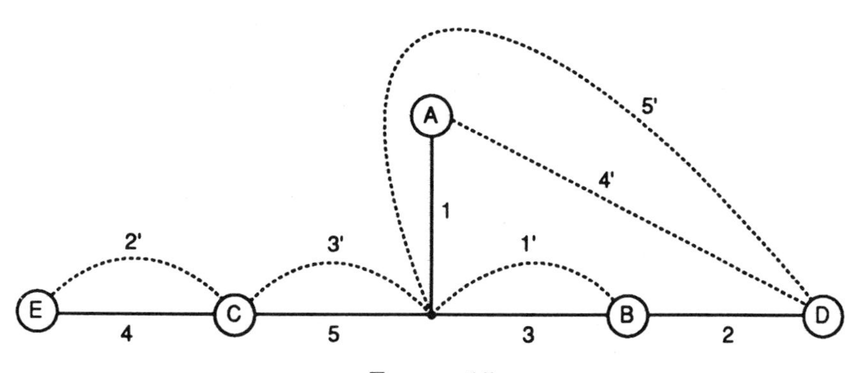

FIGURE 15e

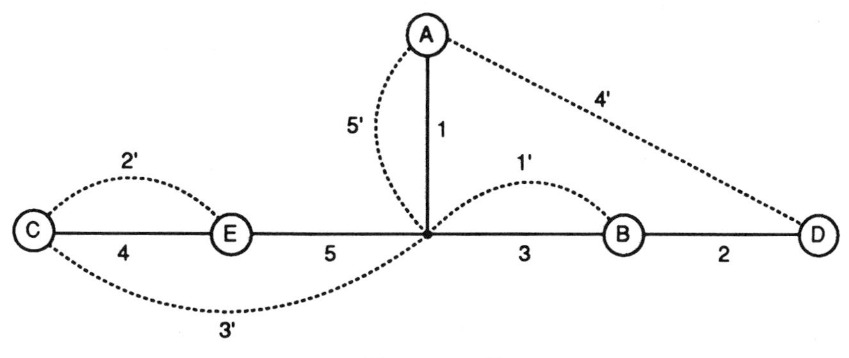

FIGURE 15f

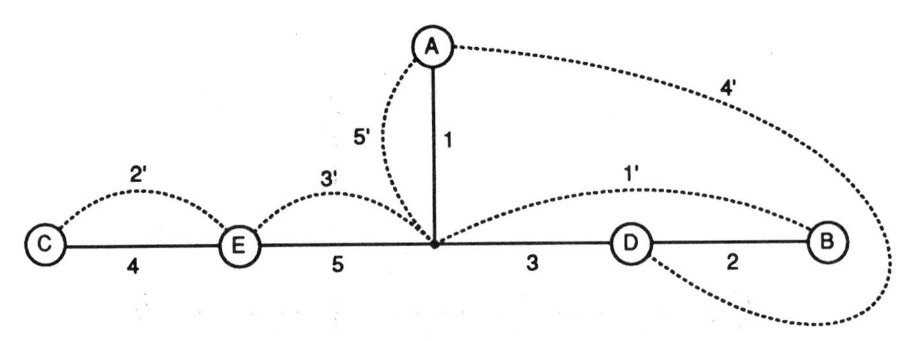

FIGURE 15g

give us the original model with cyclically permuted marked vertices. Continuing this procedure we get a loop of 30 gluings such that we go back exactly to the initial model. Along this loop we encounter both our standard models.

In fact, the cyclic permutation on the marked vertices above is simply F^3 restricted to P. In general, if we take a standard model of any pattern and just permute vertices by some iterate of F then we get also a standard model for the same pattern. This is so, because the new model is obtain from the previous one by a conjugacy via an iterate of F. Since F commutes with all its iterates, the homeomorphism obtained from F by applying a conjugacy via a power of F, is also F.

Therefore we may think about two standard models obtained from each other in the above way as equivalent. Then getting the models from Figures 15g and 15c from each other up to equivalence requires only a short loop of gluings: Figure 15a to 15d to 15e to 15f to 15g (equivalent to 15c) to 15b to 15a.

10. Pseudo-Anosov homeomorphisms

In this section we show how, given a cycle P of a homeomorphism F in the class $\mathfrak{C}$ whose associated tree map is irreducible and not periodic, to construct a pseudo-Anosov homeomorphism isotopic to F rel. P. By results of earlier sections we can and will assume that the associated Markov tree map has a particularly nice form. In particular we will assume in this section that the associated Markov tree map $f : T \to T$ is piecewise expanding, that it permutes all the vertices of T, and that it is a bijection on edge germs at each vertex which is not in the cycle $R = \pi(P)$. That we can do it, follows from the definition of irreducibility.

THEOREM 10.1. *Suppose $F \in \mathfrak{C}$ has the cycle P with pattern A and is associated to a thick tree map $\widehat{f}$. Suppose $\widehat{f}$ is associated to a Markov tree map $f : T \to T$ which is irreducible and not periodic. Then there is a homeomorphism $G : \mathbb{D} \to \mathbb{D}$ which has a cycle P with pattern A such that $G : \mathbb{D} \setminus P \to \mathbb{D} \setminus P$ is pseudo-Anosov and has a Markov partition equivalent to the Markov partition for f given by edges of the tree T.*

PROOF. Let M be the non-negative integer transition matrix corresponding to the Markov partition of T by edges and let $l = (l_1, \ldots, l_n)$, and $r = (r_1, \ldots, r_n)$ be the left and right eigenvectors respectively of M corresponding to its Perron-Frobenius eigenvalue β. We assume that the edges of T are constructed to have lengths so that f uniformly expands the length of each edge by a factor β. To achieve this make the ith edge have length equal to r_i. Let R_i be a Euclidean rectangle of size r_i by l_i, i.e. let $R_i = [0, r_i] \times [0, l_i]$. We will glue these rectangles together along parts of their boundaries to form a surface homeomorphic to the disk and then define the desired homeomorphism on this disk. The map will be constructed so that its stable and unstable foliations will restrict to the foliations of R_i by line segments parallel to the boundary edges.

The first step in this process is to choose an orientation for each edge of T and create a disk by replacing the ith edge by R_i. To do this first consider a vertex v of T which is not an element of the cycle $R = \pi(P)$. Renumbering the edges, if necessary, we can assume that $e_1, e_2, \ldots e_k$ are the edges incident to v written in an order consistent with the cyclic order they have at v in the thick tree. We will identify ends of all the rectangles R_i whose corresponding edge e_i is incident to v. If e_i is oriented away from v we use the end $0 \times [0, l_i]$ and otherwise use the end $r_i \times [0, l_i]$. (See Figure 16). For this to to fit as shown in Figure 16 we must show that there are positive numbers a_i and b_i such that $l_i = a_i + b_i$, for $1 \leqslant i \leqslant k$, $b_i = a_{i+1}$ for $1 \leqslant i < k$, and $b_k = a_1$. We show this as follows.

Let n be a positive integer chosen so that f^n fixes all vertices of T and fixes the germs of edges incident to all vertices (that is, those germs which are fixed by some iterate). Let $A = M^n$, let $g = f^n$, and let $\widehat{g}$ be a thick tree map associated to F^n and g. Remember that $\widehat{g}$ maps T into $\mathbb{D}$. Then A_{ji} is the number of times that that edge e_j covers edge e_i under the tree map g. Looking at the thick tree map $\widehat{g}$ we observe that if E_i is the "thickened edge" e_i, i.e. $\pi^{-1}(\mathrm{Int}(e_i))$, in the notation of Section 1, then A_{ji} is the number of times $\widehat{g}(e_j)$ passes through E_i, where $c : T \to \mathbb{D}$ is the core map described in Section 1. We single out $\widehat{g}(e_i) \subset E_i$ for special attention and count the number of times $\widehat{g}(e_j)$ passes through E_i on each side of the germ of $\widehat{g}(e_i)$. Thus let A^+_{ji} denote the number of times $\widehat{g}(e_j)$ passes through E_i to the "left" of the germ of $\widehat{g}(e_i)$ (as viewed from v looking outward) and let A^-_{ji} denote the number of times $\widehat{g}(e_j)$ passes through E_i to the "right" of the germ of $\widehat{g}(e_i)$.

Note then that if $i \neq j$, we have $A_{ji} = A^+_{ji} + A^-_{ji}$, and $A_{ii} = A^+_{ii} + A^-_{ii} + 1$ since in this case $\widehat{g}(e_i)$ itself contributes 1 to A_{ii}. A key observation from the geometry of the thick tree map (see Figure 17) is that

$$A^+_{ji} = A^-_{j(i+1)}, \tag{8}$$

and

$$A^+_{jk} = A^-_{j1}. \tag{9}$$

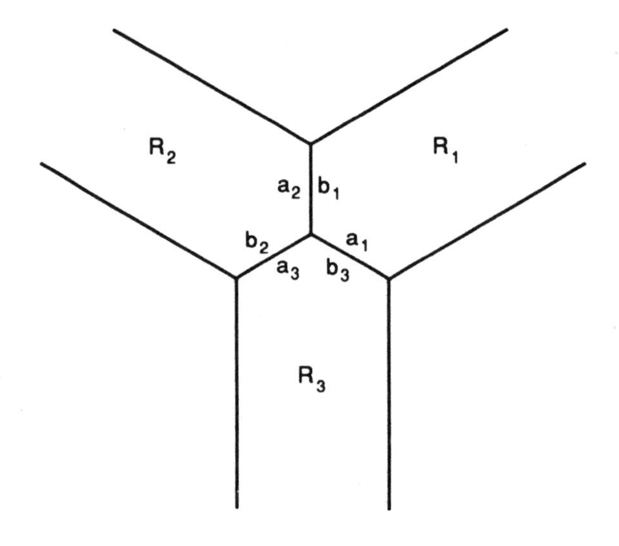

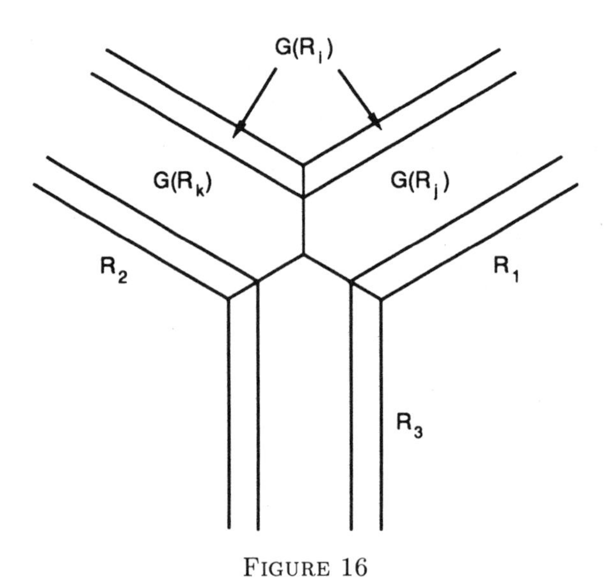

FIGURE 16

Since l is a left eigenvector of A, we have

$$\lambda l_i = \sum_j l_j A_{ji} = \sum_j l_j A^+_{ji} + \sum_j l_j A^-_{ji} + l_i, \tag{10}$$

where $\lambda = \beta^n$ is the Perron-Frobenius eigenvalue of $A = M^n$.

Motivated by the geometry of the thick tree we hope to construct, we note

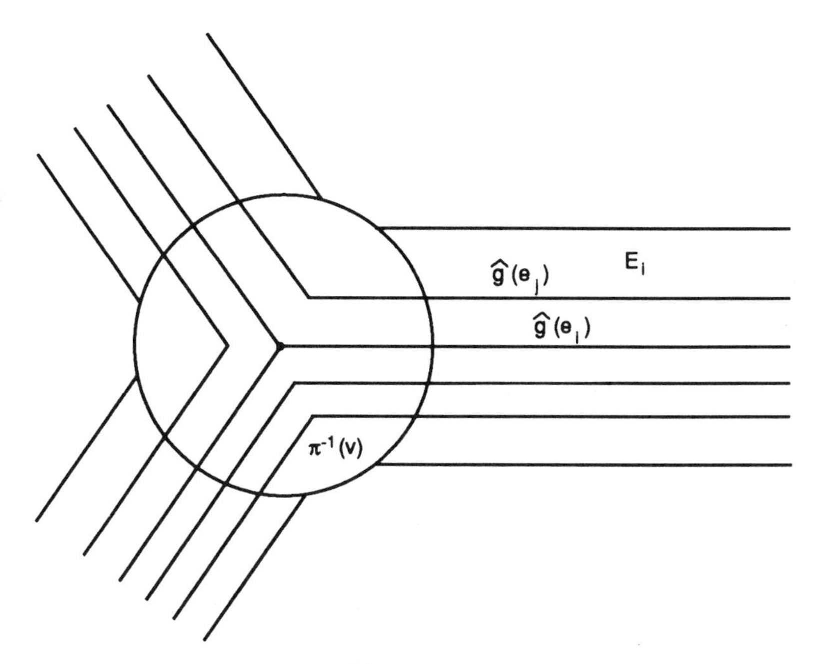

FIGURE 17

that the numbers a_i and b_i we seek should satisfy

$$a_i = \frac{1}{\lambda}\left(\sum_j l_j A^-_{ji} + a_i\right),$$

and the analogous equation for b_i. Thus we consider the equations

$$\lambda a_i = \sum_j l_j A^-_{ji} + a_i, \tag{11}$$

and

$$\lambda b_i = \sum_j l_j A^+_{ji} + b_i. \tag{12}$$

Hence we let

$$a_i = \frac{1}{\lambda - 1}\sum_j l_j A^-_{ji},$$

and

$$b_i = \frac{1}{\lambda - 1}\sum_j l_j A^+_{ji}.$$

It now follows from (8) and (9) above that $b_i = a_{i+1}$ for $1 \leqslant i < k$, and $b_k = a_1$. And from (10) it follows that $l_i = a_i + b_i$.

If v is an element of the cycle R, and also an end of the tree T then there is a single edge e_i incident to it, which we assume oriented away from v. The edge

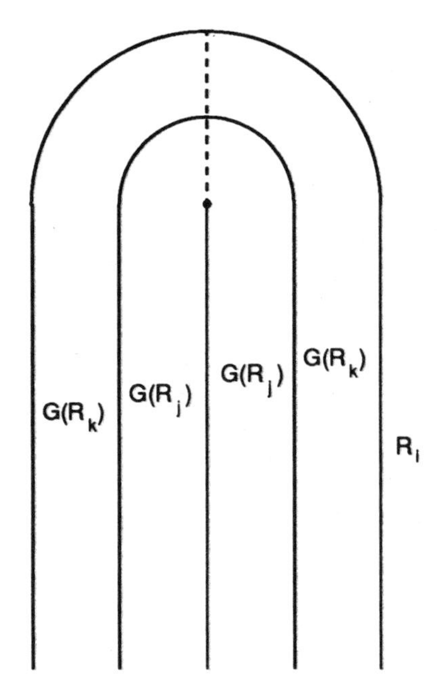

FIGURE 18

$0 \times [0, l_i]$ of R_i is given identifications by folding it at its center (see Figure 18). More precisely for $t \in [0, l_i/2]$ the point $(0, t) \in R_i$ is identified with $(0, l_i - t)$.

Finally suppose $v \in R$ is an interior vertex of T, n is the period of R, and $g = f^n$. Then g maps the germs of all edges incident to v onto the germ of a single edge e_m. In fact, referring again to the geometry of $\widehat{g}$ we can assume that the edges incident to v are numbered so that in the list $e_1, \dots, e_m, \dots, e_k$ they are in their cyclic order around v and that for $1 \leqslant i < m$, the germ of $\widehat{g}(e_i)$ is to the right of the germ of $\widehat{g}(e_m)$ while for $m < i \leqslant k$, the germ of $\widehat{g}(e_i)$ is to the left of the germ of $\widehat{g}(e_m)$ (as viewed from v looking outward; see Figure 19).

We want to make identifications in a fashion similar to what we did in the case that v is an end of T. The construction is quite similar. Let

$$l_v = \sum_{i=1}^{k} l_i$$

and identify the end $0 \times [0, l_i]$ of R_i with the subinterval of $[0, l_v]$ given by $[s_i, s_i + l_i]$ where

$$s_i = \sum_{j=1}^{i-1} l_j.$$

That is, we line up the ends of the R_i in the correct order on the interval $[0, l_v]$. We then fold this interval, identifying t with $l_v - t$ for each $t \in [0, l_v/2]$. It is important to show that the point $l_v/2 \in [0, l_v/2]$ lies in the subinterval

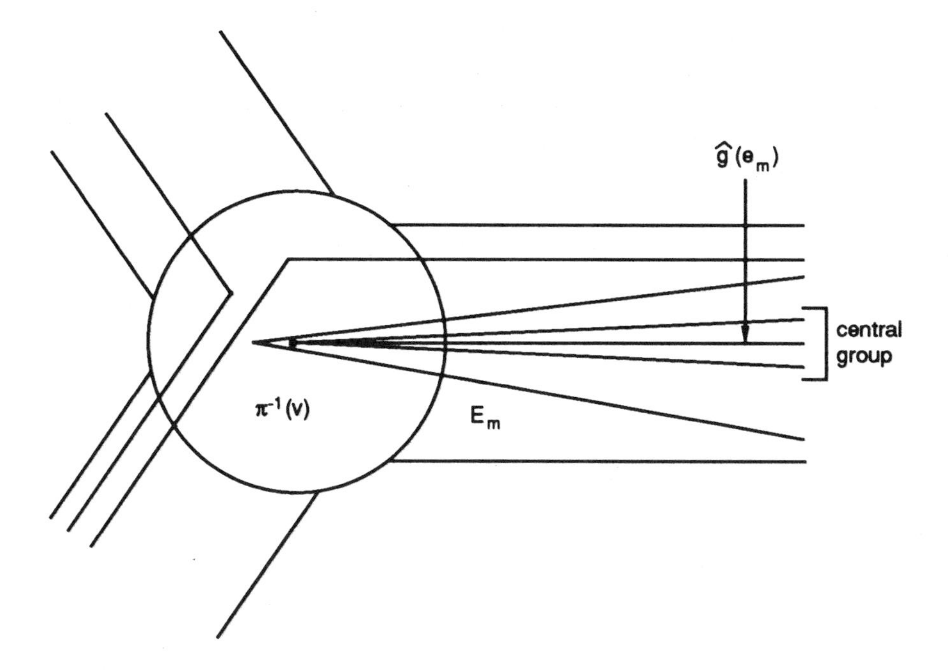

FIGURE 19

$[s_m, s_m + l_m]$ identified with the end of R_m. That is, we must show $s_m < l_v/2 < s_m + l_m$.

This requires a computation similar to the one above. Again let $A = M^n$ and let λ be the Perron-Frobenius eigenvalue of A and suppose that s is the total number of edges in T. Then we have

$$\lambda l_i = \sum_{j=1}^{s} l_j A_{ji}.$$

We focus on the "thickened edge" E_m and count the images of $\widehat{g}(e_j)$ which pass through it. The k segments of the arcs $\widehat{g}(e_j)$, $1 \leqslant j \leqslant k$ (one segment per arc) which end at v form an adjacent "central group" in E_m. For each i let A^-_{im} denote the number of times $\widehat{g}(e_i)$ passes through E_m to the right of this group (as viewed from v looking outward). Likewise let A^+_{im} denote the number of times $\widehat{g}(e_i)$ passes through E_m to the left of this group. If $i \leqslant k$ we don't count the time that $\widehat{g}(e_i)$ passes through E_m as part of the central group in either A^+_{im} or A^-_{im}. Thus we have $A_{im} = A^+_{im} + A^-_{im} + 1$, for $1 \leqslant i \leqslant k$ and $A_{im} = A^+_{im} + A^-_{im}$ for $i > k$. We observe that

$$\sum_{j=1}^{s} \sum_{i<m} l_j A_{ji} + \sum_{j=1}^{s} l_j A^-_{jm} = \sum_{j=1}^{s} \sum_{i>m} l_j A_{ji} + \sum_{j=1}^{s} l_j A^+_{jm}. \tag{13}$$

This can be seen by observing that, since f is irreducible, each of the arcs which make up $\widehat{g}(e_j) \cap \pi^{-1}(v)$ has one end in one of the E_i with $i < m$ or has one end in E_m which continues to the right of the central group (again "right" as viewed from v looking outward). Thus if we multiply the number of such arcs by l_j, and sum over j, we have the left hand side of equation (13). But the other end of each of these arcs lies in one of the E_i with $i > m$ or lies in E_m and continues to the left of the central group. Thus multiplying the number of such arcs by l_j and summing over j also gives the right hand side of equation (13).

Note that the fact that $A_{im} = A^+_{im} + A^-_{im} + 1$ implies that the sum of all terms on the left of (13) plus all terms on the right, plus

$$\sum_{j=1}^{k} l_j$$

equals

$$\sum_{i=1}^{k} \sum_{j=1}^{s} l_j A_{ji} = \sum_{i=1}^{k} \lambda l_i = \lambda l_v.$$

From this it follows that if K is the quantity on either side of equation (13), $2K < \lambda l_v$, or $K/\lambda < l_v/2$. But

$$s_m = \sum_{i=1}^{m-1} l_i = \frac{1}{\lambda} \sum_{j=1}^{s} \sum_{i<m} l_j A_{ji} < \frac{1}{\lambda} K < \frac{l_v}{2}.$$

A similar argument shows

$$\sum_{i=m+1}^{k} l_i < \frac{l_v}{2}$$

and hence that $s_m + l_m > l_v/2$.

Let D denote the union of the R_i with the identifications described above. There is a natural map $\tilde{\pi} : D \to T$ obtained by projecting each R_i onto say $[0, r_i] \times 0$ and identifying this line segment with the edge e_i in a way which preserves lengths and orientations. The topological space D is a thickened version of T and $\tilde{\pi} : D \to T$ is a homotopy equivalence, so D is topologically a two dimensional disk.

The foliations of the rectangles R_i by lines parallel to the sides, induce two foliations with singularities on D which will be the stable and unstable foliations of the pseudo-Anosov map we construct. The singular points of the foliations are in one-to-one correspondence with the vertices of T. Those which correspond to a vertex not in the cycle R will be prong singularities of pseudo-Anosov type (see Figure 16). Their coordinates in a rectangle R_i containing them will be $(0, a_i)$ or (r_i, a_i). The singular points corresponding to points in the cycle R will be "one-prong" singularities as found in "generalized pseudo-Anosov" maps [**FLP**]. We will eventually puncture at these points. If the corresponding vertex in T is an endpoint then this singular point will be the midpoint of the end of a rectangle containing it. If the corresponding vertex v is in R but not an endpoint then the singular point will be the midpoint of the line segment $[0, l_v]$ constructed above.

We want now to construct a pseudo-Anosov map $H : D \to D$ which satisfies $\widetilde{\pi} \circ H = f \circ \widetilde{\pi}$. We do this by defining it on each R_i and checking compatibility with the identifications. Intuitively we take each R_i, stretch it by a factor of β in its first coordinate, shrink it by a factor of $1/\beta$ in the second coordinate and place it in D in accordance with the geometry of the picture for $\widehat{f}$. Thus $H(R_i)$ will be a rectangle of dimensions βr_i by l_i/β. It will cross R_j completely M_{ij} times. Note the fact that $\beta r_i = \sum_j M_{ij} r_j$ is compatible with this. Also the fact that

$$l_j = \frac{1}{\beta} \sum_j l_i M_{ij} \tag{14}$$

says that each R_j can be exactly filled by the parts of the $H(R_i)$ which cover it.

To define H we observe that the semi-conjugacy equation $\widetilde{\pi} \circ H = f \circ \widetilde{\pi}$ tells us one coordinate of the map. That is, if $x \in R_i$ and $f(\widetilde{\pi}(x)) \in e_j$ then we want $H(x) \in R_j$ and in fact $H(x)$ should be in the line segment $a \times [0, l_j]$ where a is chosen so that $\widetilde{\pi}(a \times [0, l_j]) = f(\widetilde{\pi}(x))$. To determine the second coordinate we note that from the geometry of the thick tree map $\widehat{f}$ we can read off the order in which we want the pieces of the various $H(R_i)$ to cover R_j. Since we also know the width of each of these pieces, l_i/β, the second coordinate is determined. At this point we don't yet have a well defined map on each of the rectangles R_i, but it is well defined for each $x \in R_i$ such that $f(\widetilde{\pi}(x))$ is not a vertex of T. If $f(\widetilde{\pi}(x))$ is a vertex and the point $\widetilde{\pi}(x)$ corresponds to an allowable transition from e_i to e_j then $H(x)$ is a well defined point in R_j

We need to check that when $H(x)$ is multiply defined, its different values (in different R_j's) are all identified when we make the identifications used to construct D. If $x \in (0, r_i) \times [0, l_i]$, i.e. if $\widetilde{\pi}(x)$ is not a vertex of T, but $f(\widetilde{\pi}(x)) = \widetilde{\pi}(H(x))$ is a vertex then our definition of H would give $H(x)$ in the ends of two of the R_j's, however parts of these two ends are identified, and in particular the two points corresponding to $H(x)$ are identified in D (consider Figures 16 and 17). It is also clear that H is continuous at x.

The next step is to show that H permutes the singular points of D. Suppose $v \in R$ and $f(v) = w$. If w_0 and v_0 are the corresponding singular points, then v_0 is the midpoint of a line segment which is either a rectangle end $[0, l_m]$ or the segment $[0, l_v]$ constructed above which is a union of ends of rectangles. The same is true for w_0. By the definition of H the line segment containing v_0 is mapped to the interior of the line segment containing w_0. The construction also is such that the image of the first segment must be centered in the second. To see these two facts observe that w is an endpoint so that the point of P corresponding to it must have the images of edges going around it on both sides (see Figures 18 and 19). This means that J_1, the image under H of the line segment containing v_0, is situated in J_2, the line segment containing w_0 in such a way that the H image of any rectangle which intersects J_2 must intersect it on both sides of J_1. Thus the midpoint of the interval containing v_0 must be mapped to the midpoint of J_2. That is, v_0 must be mapped to w_0. From this it is clear that for any points in $\widetilde{\pi}^{-1}(v)$, (i.e. points on the folded interval $[0, l_m]$ or the folded $[0, l_v]$), the definition of H is compatible with the identifications and H is continuous at these points.

Suppose now that $v \notin R$ is a vertex of T and $f(v) = w$. If v_0 and w_0 are the corresponding singularities we need to see that $H(v_0)$ is well defined and equal to w_0. To see this observe that if $v_0 \in R_i$ then equations (11) and (12) above imply that v_0 is a fixed point of H^n (restricted to one end of R_i). But if the germ of e_i is mapped by f to the germ of e_j then the end of R_i containing v_0 is mapped to the end of R_j containing w_0. Since w_0 is also a fixed point of H^n restricted to this end, it must be the case that $H(v_0) = w_0$. From this and equations (11) and (12), plus the fact that $a_{i+1} = b_i$ it follows that for any point in $\widetilde{\pi}^{-1}(v)$ the definition of H is compatible with the identifications, and that H is continuous.

At this point we have constructed a continuous map $H : D \to D$ which is essentially pseudo-Anosov in that it has invariant foliations which are uniformly expanded and contracted. It is not quite a homeomorphism, however, since parts of the boundary of D are mapped to the interior of D. More precisely there are points of the boundary of a rectangle R_i which are mapped by H to interior points of some other rectangle. The image of such a point will also be the image of another point in the boundary of a different rectangle. Thus H is one-to-one on the interior of D and that part of the boundary whose image also lies in the boundary of D. But there are intervals on the boundary of D whose image is in the interior and the map is two-to-one on the union of such intervals.

We observe however, that the combinatorics of this construction guarantees that $H : D \setminus S \to D \setminus S$, where S is the set of singularities is homotopy equivalent to $F : \mathbb{D} \setminus P \to \mathbb{D} \setminus P$.

In order to get a homeomorphism we will identify points on the boundary of D forming a two-sphere on which H is well defined and a homeomorphism. The boundary of D consists of a finite set of arcs meeting in "cusps" (see Figure 20). Each of these arcs is a union of expanding edges of the rectangles R_i constructed above. We orient the boundary of D and let C_j, $1 \leqslant j \leqslant r$ denote the boundary arcs ordered in a fashion compatible with the orientation.

Notice that the image $H(C_i)$ consists of one of the arcs C_j together with two arcs in the interior of D which end at the endpoints of C_j. The correspondence $i \to j$ when $C_j \subset H(C_i)$ determines a permutation p of these intervals. It follows that there is a $k > 0$ such that $C_i \subset \mathrm{Int}(H^k(C_i))$ for all i. Thus, since H is uniformly expanding on C_i, there is a unique fixed point $z_i \in C_i$ of H^k. Clearly H permutes $\{z_i\}$ in the same fashion it permutes the C_i's, i.e. $H(z_i) = z_{p(i)}$. Let α_i and ω_i denote the endpoints of C_i and let a_i and b_i be the lengths of the subintervals from α_i to z_i and z_i to ω_i respectively. Thus $\omega_i = \alpha_{i+1}$ and $\omega_r = \alpha_1$.

Next define c_i to be the the length of the subinterval of $H(C_i)$ with endpoints $H(\alpha_i)$ and $\alpha_{p(i)}$, i.e. the length of the beginning segment of $H(C_i)$ which sticks

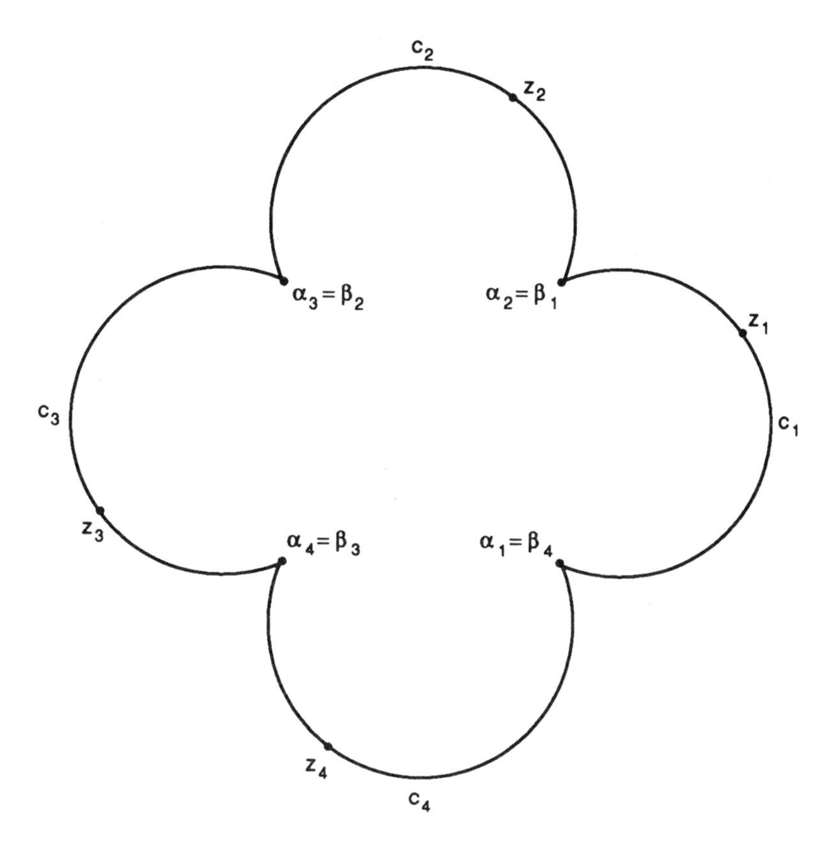

FIGURE 20

into the interior of D. Likewise define d_i to the the length of the subinterval of $H(C_i)$ with endpoints $H(\omega_i)$ and $\omega_{p(i)}$. Clearly $d_i = c_{i+1}$ and $d_r = c_1$.

If β is the Perron-Frobenius eigenvalue of M, i.e. the expansion factor for H, then

$$\beta a_i = a_{p(i)} + c_i,$$

and

$$\beta b_i = b_{p(i)} + d_i.$$

Considering these as two systems of equations, one with unknowns a_i and one with unknowns b_i, we observe that they have unique solutions. From the relation between the c_i's and the d_i's together with the fact that $p(i+1) = p(i) + 1$, mod r, it follows that $b_i = a_{i+1}$ and $b_r = a_1$.

Hence the segment of C_i from z_i to ω_i can be isometrically identified with the segment of C_{i+1} from α_{i+1} to z_{i+1}. Doing this for all i we obtain a quotient space homeomorphic to S^2. The map H is compatible with these identifications and hence induces a map on S^2. This map is easily seen to be a homeomorphism. All of the points $\{z_i\}$ are identified to a single fixed point in S^2 which we denote z. It is of pseudo-Anosov or "generalized pseudo-Anosov" type with r prongs.

If we blow up the fixed point z of $H : S^2 \to S^2$ we obtain the desired homeomorphism $G : \mathbb{D} \to \mathbb{D}$. The rectangles $\{R_i\}$ form the required Markov partition. □

We next wish to investigate the correspondence between periodic points of the pseudo-Anosov homeomorphism G and the corresponding tree map $f : T \to T$. It is nearly true that there is a one-to-one correspondence between the periodic points of f and those of G. The only place this fails is for a finite set of periodic points of G on the boundary of $\mathbb{D}$. We shall say that a cycle of any homeomorphism of $\mathbb{D}$ is *peripheral* if it lies entirely in the boundary of $\mathbb{D}$. While we defined the concept of pattern only for cycles in the interior of $\mathbb{D}$, it is clear one could consider peripheral patterns also, but they are not particularly interesting since they are all twist (i.e. exhibited by rotations of $\mathbb{D}$) and any two exhibited by the same homeomorphism must be equivalent.

We can see how peripheral cycles arise in the construction from the proof of Theorem 10.1. Suppose E_i is the "thickened edge" e_i, (i.e. $\pi^{-1}(\mathrm{Int}(e_i))$), and for some m there is a component of $F^m(E_i) \cap E_i$ which is entirely to one side of E_i. That is, we suppose that one component of the intersection of the boundary of the disk with E_i is not separated from $F^m(E_i) \cap E_i$ in E_i by any component of $F^m(E_j) \cap E_i$ for any $j \neq i$. If m is minimal with this property, it is clear that there is a periodic point of period m in the component of $F^m(E_i) \cap E_i$ which is next to the boundary. Following through the construction of G is easy to see that this gives rise to a point of period m for G on the boundary of $\mathbb{D}$. Viewed as a periodic point of the circle map $G|_{\partial\mathbb{D}}$ the orbit of this point is repelling, but in $\mathbb{D}$ it is an endpoint of a stable manifold segment. Of course the thick tree periodic point corresponds to a periodic point for the corresponding Markov tree map $f : T \to T$, namely its image under $\widetilde{\pi}$. Hence we will also refer to this periodic point for f as peripheral. Note that this only makes sense because f comes from a thick tree; it would not make sense to speak of a peripheral periodic orbit for an arbitrary Markov tree map.

From the construction of $G : \mathbb{D} \to \mathbb{D}$ in the proof of Theorem 10.1, it is easy to see that all periodic points of $G|_{\partial\mathbb{D}}$ which are repelling are the endpoints of stable manifold segments in $\mathbb{D}$ and hence correspond to peripheral cycles of $f : T \to T$. However, each two adjacent such points in $\partial\mathbb{D}$ are separated by a attracting periodic point of $G|_{\partial\mathbb{D}}$, also of period m. These points arise from the blowing up construction and are the endpoints of "one-sided" unstable manifolds for G. There are no points of $f : T \to T$ corresponding to them.

COROLLARY 10.2. *Given A, f and G as in Theorem 10.1, the period of any peripheral cycle for G will be less than or equal to $n-2$, where n is the period of the pattern A. Moreover, all peripheral cycles for f and G have the same period, say m, and the index of any point of a peripheral cycle in T for the map f^m is -1.*

PROOF. From the proof of Theorem 10.1, it is clear that there is exactly one repelling peripheral periodic point in each of the arcs of the boundary of D whose endpoints are the cusps described above (see Figure 20). Hence the period of any such point is bounded above by the number of cusps. But from the construction of D in Theorem 10.1 it follows that the number of cusps is equal to $\sum_{v\in R}\alpha(v)$ where R is the cycle in the tree T corresponding to the pattern A and $\alpha(v)$ denotes the valence of the vertex v minus 1. From Lemma 6.1 we know that $\sum_{v\in R}\alpha(v)\leqslant n-2$, so any repelling peripheral cycle has period $\leqslant n-2$. Since all peripheral cycles for G are cycles of a single circle homeomorphism they all have the same period, so this bound is valid for all of them.

Recall from the proof of Theorem 10.1 that there is a semiconjugacy $\widetilde{\pi}: D\to T$ from $H: D\to D$ to f which collapses segments of stable manifolds in D to points. Each repelling peripheral periodic point of H is the endpoint of a stable manifold segment and hence corresponds to a peripheral periodic point for f of the same period, say m, as those of H. This semiconjugacy must take a repelling peripheral fixed point for H (it is repelling in ∂D; in D it is "half a saddle") to an expanding fixed point of f^m. Also the local unstable manifold must have its orientation preserved by f^m at these points since the same is true for the local unstable manifolds of the points in the repelling peripheral cycles for H. Thus the peripheral periodic points for f all have index -1. □

For non-peripheral cycles there is a very nice correspondence between periodic points of G and those of the tree map $f: T\to T$.

COROLLARY 10.3. *Given A, f and G as in Theorem 10.1, there is a bijection between the non-peripheral periodic points of $f: T\to T$ and the non-peripheral periodic points of $G:\mathbb{D}\to\mathbb{D}$ which preserves the periods and index (as fixed points of the same iterate of G and f). Moreover, $h(A)=h(f)$.*

PROOF. The semiconjugacy $\widetilde{\pi}: D\to T$ determines a bijection between the non-peripheral periodic points of $f: T\to T$ and those of $H: D\to D$. This is because for each periodic point $z\in T$, of period n, the interval $\widetilde{\pi}^{-1}(z)$ is mapped to itself and contracted by the map H^n. It follows that $\widetilde{\pi}^{-1}(z)$ contains a unique point of period n for H. Moreover, it is in the interior of the interval $\widetilde{\pi}^{-1}(z)$ and hence in the interior of D if and only if z is non-peripheral. It is also easy to see that $G:\mathbb{D}\to\mathbb{D}$ has the same interior periodic points as $H: D\to D$. These points are precisely the same and have the same period.

A fixed point of G^n has index $1-k$ if its unstable manifold has k prongs whose germs are preserved by G^n. It has index 1 if these germs are rotated. Likewise a fixed point of f^n has index $1-k$ if it has valence k in T and its edge germs are preserved by f^n. It has index 1 if these germs are rotated. Since the semiconjugacy takes the unstable manifold germs to edge germs it follows that corresponding points have the same index.

Since any pseudo-Anosov homeomorphism has minimal entropy in its isotopy class, we have $h(G)=h(A)$. Since the Markov partitions for G and f are equivalent, we get $h(G)=h(f)$. Thus, $h(A)=h(f)$. □

PROPOSITION 10.4. *Let A, P, f and G be as in Theorem 10.1. If $P_0\neq P$ is* *a cycle of G and has pattern B, then $h(B) < h(A)$.*

PROOF. Let $P_0 \neq P$ be a cycle of G with pattern B. Let $R = \pi(P)$ and R_0 be the cycles for $f : T \to T$ corresponding to P and P_0 respectively. Let T_0 be the minimal subtree of T containing R_0. We define a retraction $q : T \to T_0$ by letting $q(x)$ be the closest point in T_0 to x, and define $f_0 : T_0 \to T_0$ by $f_0 = q \circ f$. The map f_0 is nearly Markov. It fails in that some subintervals will be mapped to single points.

If $\mathcal{P}$ is the partition of T by edges let $\mathcal{P}_1 = \mathcal{P} \vee f^{-1}(\mathcal{P})$ and let M be the transition matrix for f for this Markov partition. Form a matrix M_0 from M by changing M_{ij} to 0 if either i or j corresponds to a partition element disjoint from the interior of T_0. Then M_0 can be thought of as a transition matrix for f_0. Each element of M is greater than or equal the corresponding entry of M_0 and some are strictly greater. Indeed some entries of M will be positive with the corresponding entry of M_0 being 0. From this it follows that $h(f) > h(f_0)$ (see for instance [**S**]).

The homeomorphism G with the cycle P_0 is associated to a thick tree map $\widehat{f_0}$, which is in turn associated to a tree map obtained from f_0 by collapsing all intervals whose image is a point (and making further cleaning, if necessary). We can proceed with the construction of the previous sections and since no step increases the entropy and by Corollary 10.3 and Theorem 12.2 (it will be proved later, but its proof does not use the proposition we are proving now), we get $h(B) \leqslant h(f_0)$. Therefore $h(B) = h(f_0) < h(f) = h(A)$. □

COROLLARY 10.5. *If in the situation described in Theorem 10.1, the pattern A is primary then $G : \mathbb{D} \to \mathbb{D}$ has only one cycle of period n, where n is the period of A.*

11. Standard models for patterns

A pattern $A = [(P, F)]$ will be called *pseudo-Anosov* if F is isotopic rel P to a pseudo-Anosov homeomorphism G. For such a pattern we can construct a thick tree map $\widehat{f}$ and an associated Markov tree map $f : T \to T$ which is irreducible and not periodic. From the previous section it follows that the following pieces of information are sufficient to construct G, and therefore to reconstruct A:

(i) the tree T, up to a homeomorphism, with a given cyclic ordering of edges at each vertex,
(ii) the set R,
(iii) the Markov map f, up to equivalence,
(iv) the ordering of the images in the thick tree, whenever those images coincide in the tree, at the vertices.

We shall call the above collection (i)-(iv) a *standard model for A*. It is also clear that the same definition works for a twist pattern A.

If A is a B-extension of C and we have standard models for both B and C then we can construct a standard model for A in a rather obvious way, replacing the points of the cycle from the model for C by the tree from the model for B. When defining the map on a new tree, we have to be careful and use the information we ignored when making the corresponding big reduction (but which can be easily extracted from there). This information tells us how many times (and in which direction) we have to go around a small (glued in) tree with an image of an edge of a tree from the model of C. This is in fact the information on the Dehn twists which occur in the reductions in the Thurston's theory. We leave the details of this construction to the reader. In fact, such standard models have been constructed by other authors (for instance by Toby Hall [**H1**]).

When dealing with a pseudo-Anosov pattern, it is convenient to consider the generalized pseudo-Anosov homeomorphism $H : S^2 \to S^2$, constructed at the end of the proof of Theorem 10.1. It has a fixed point z, of pseudo-Anosov type with r prongs (generalized pseudo-Anosov if $r = 1$), with pieces of unstable prongs corresponding to the boundary of the disk. Each element of P is of generalized pseudo-Anosov type (with 1 prong). Moreover, to each vertex v of T which does not belong to R there corresponds a periodic point of H of pseudo-Anosov type with the number of prongs equal to the valence of v. Its period is the same as the period of v. Notice that since the valence of v is larger than 2, there is really a singularity at this point. From the construction of H it follows that on the rest of S^2 the homeomorphism H is smooth, so there are no more singularities. Thus, this correspondence is one-to-one. It follows also that the Markov partition constructed in the proof of Theorem 10.1 consists of the sets into which S^2 is divided by the following curves:

(i) pieces of all the unstable prongs of z; each of them starts at z and ends up at an interior point of one of the curves listed in (ii),

(ii) pieces of the stable prongs of all the points of P; each of them starts at a point of P and ends up at an interior point of one of the curves listed in (i),

(iii) pieces of all the stable prongs of all the other singular points of H; each of them starts at such a point and ends up at an interior point of one of the curves listed in (i).

The interiors of all the curves listed above are pairwise disjoint.

Using the above model, we can prove the following theorem.

THEOREM 11.1. *For a pseudo-Anosov pattern A, there are finitely many standard models for A. In each of them, the number and periods of periodic vertices which do not belong to the representative of A, are the same. Each of them can be obtained from each other by finitely many gluings.*

PROOF. Let $H : S^2 \to S^2$ be the pseudo-Anosov homeomorphism constructed for the pattern A. It is unique up to a conjugacy (see e.g. [**FLP**]), so every standard model can be obtained by taking the partition of S^2 by curves as above, and extracting the necessary data from H with this partition. From the proof of Theorem 10.1 and the considerations preceding this theorem, it is obvious how to do it. Since there is a one-to-one correspondence between the singularities of H (excluding z) and the vertices of T, the number of choices for T is finite. Since the entropies of H and the Markov map $f : T \to T$ are equal, by Lemma 6.2 it follows that the number of choices for f (up to equivalence) is finite. Thus, the number of standard models for A is finite.

Choose two such standard models. For both of them, the periodic vertices which do not belong to the representative of A are in one-to-one correspondence with the same singularities of H, and this correspondence preserves the periods. This proves the second assertion of the theorem.

When we choose the curves described in (i), then there is at most one way of choosing the curves described in (ii) and (iii). Therefore each standard model for A is determined by a system of curves described in (i). Notice that if we replace some set of such curves by their images under H^n with $n \in \mathbb{Z}$ then we get the same standard model. Therefore, if we have two standard models for A and the corresponding sets of curves described in (i), we may assume that every curve for the first model is longer than or equal to the corresponding curve for the second model. Hence, to get the second model from the first one, we have to shorten those curves. Since they correspond to the pieces of the boundary of the disk, the effect will be the same as if we glued together pieces of the boundary of the disk. It is not difficult to see that on the level of thick tree maps, each such operation amounts to a finite number of gluings (although we cannot expect that the tree map will be irreducible after *each* gluing, it will be after *all* of them). This proves the last assertion of the theorem. □

REMARK 11.2. If we replace in the last assertion of Theorem 11.1 "gluings" by "splittings" then it remains true (just make the reverse operations).

The following fundamental theorem follows immediately from Section 4 of [**H2**] (also see [**AF**]).

THEOREM 11.3. *Let P be a cycle of a homeomorphism $F : \mathbb{D} \to \mathbb{D}$ such that $F : \mathbb{D} \setminus P \to \mathbb{D} \setminus P$ is pseudo-Anosov. Then the pattern $[(P, F)]$ forces all patterns exhibited by F.*

Thus, to determine patterns forced by a given pseudo-Anosov pattern A, we can construct a standard model for this pattern, and then take non-peripheral cycles of the corresponding tree map. For each such cycle, we fix it as our new cycle R and proceed with the whole algorithm to get a corresponding standard model. It will correspond to a pattern forced by A, and all patterns forced by A can be found in such a way.

If we start with a pattern which is not pseudo-Anosov, we have to add two ingredients to the whole procedure. The first one is the obvious remark that a twist pattern forces only itself and the pattern of period 1, and the second one is Theorem 12.1 (see the next section). When using Theorem 12.1, one has to figure out which D-extension of C is forced by a given B-extension of C, that is which Dehn twist should be applied before gluing back a small disk into the large one. However, in concrete situations this is not difficult; this information can be also easily extracted from the thick tree map.

Another problem is connected with the fact that sometimes we consider homeomorphisms of the disk into its interior, and then peripheral cycles are no more peripheral. One can show that a repelling peripheral cycle of the homeomorphism G constructed in Theorem 10.1 is unremovable in the sense of Hall [**H2**], and therefore for the homeomorphisms of the disk into its interior Theorem 13.3 applies also to them.

12. Topological entropy

In this section we investigate topological entropy of patterns. Since here we do not introduce any other kind of entropy, we shall call topological entropy simply entropy. The first observation we should make is that from the definitions of entropy and forcing of patterns it follows immediately that if a pattern A forces a pattern B then $h(A) \geqslant h(B)$.

When we are given a pattern and we want to see what its entropy is and what other patterns it forces, we have to check first whether it is a non-trivial extension. If it is, then we reduce our problem to the one requiring investigation of simpler patterns. If our aim is to determine which patterns are forced by A, this method works because of the following theorem, derived easily from the results of [**H1**].

THEOREM 12.1. *If a pattern A is a B-extension of C then A forces all patterns forced by C, some D-extension of C for every pattern D forced by B, and no other patterns. Moreover, it does not force any B-extension of C other than A.*

PROOF. Let A be a B-extension of C. By the results of Section 5.2 (in particular Proposition 5.8) of [**H1**], there exists a homeomorphism F exhibiting A, with the small disks permuted as in the definition of the extension, such that it exhibits only those patterns which are forced by A. (In fact, the Thurston's theory describes how to construct it.) Let Q be a cycle of F with pattern E. Since the union of the small disks is invariant, Q lies either entirely in this union or entirely outside it. If it lies outside, then E is forced by C, since after collapsing small disks to points we get the corresponding "minimal" model for C (this again follows from the construction based on the Thurston's theory, as described in [**H1**]). If it lies inside the union of the small disks, we take one of them and consider F^n on it, where n is the period of C. Again, this gives a "minimal" model for B, so in this case E is a D-extension of C for some D forced by B.

Conversely, every pattern forced by A has to be exhibited by the map after collapsing the small disks, so it will have a representative in F outside the union of the small disks. If D is forced by B then F^n on a small disk will exhibit D, so some D-extension of C will have a representative in F inside the union of the small disks. This proves the first part of the theorem.

Now suppose that A forces some D which is a B-extension of C different than A. It has to differ from A by some Dehn twist on the "collar" of a small disk. Since forcing is a partial ordering, D cannot be forced by A, and this means that whenever we take a model for A with the small disks permuted as in the definition of the extension and "minimal" in the outside part (the one obtained after collapsing the small disks), there has to be representative of D inside those small disks. Looking at F^n on a small disk, we conclude that whenever there is a representative of B for some G, there must be also another one, differing from the first one by some fixed Dehn twist on the "collar" of the whole disk (we may assume G fixed on the boundary of the disk). However, we can repeat this argument for the new representative and iterate this procedure. Therefore whenever a homeomorphism exhibits B, it has to have infinitely many representatives of B. However, for the "minimal" model as discussed above this is clearly not true, so we obtain a contradiction. This completes the proof. $\square$

In the case of entropy, we have to characterize the entropy of a B-extension of C in terms of B and C.

THEOREM 12.2. *If a pattern A is a B-extension of C and C has period n then*

$$h(A) = \max(h(C), (1/n)h(B)).$$

PROOF. By Theorem 12.1, A forces C, so $h(A) \geqslant h(C)$. Clearly, if F exhibits A then F^n exhibits B, so $h(F) \geqslant (1/n)h(B)$. Therefore $h(A) \geqslant (1/n)h(B)$. On the other hand, we can easily build a homeomorphism F exhibiting A with structure the same as in the definition of extension. We can do this in such a way that the entropy of the homeomorphism obtained after collapsing the small disks to points is as close to $h(C)$ as we want and the entropy of F^n on any small disk is as close to $h(B)$ as we want (the entropy of F^n on a small disk does not depend on which small disk we take; on all n of them the maps F^n are conjugate). The set of nonwandering points of F is the union of two closed invariant sets: one in the small disks and the other outside small disks (including their boundaries). Therefore the entropy of F is the maximum of two numbers: the entropy of F on $\mathbb{D}$ minus the interiors of small disks and the entropy of F on the union of small disks. The first number is equal to the entropy of the map which we get after collapsing each small disk to a point (the map on the boundaries of small disks will not contribute to the entropy; in fact we can assume that it is just rotation). The second number is equal to $1/n$ of the entropy of F^n on any of the small disks. This shows that $h(A) \leqslant \max(h(C), (1/n)h(B))$. $\square$

Thus, to say something about the entropy of the patterns, it is enough to consider only those which are not non-trivial extensions. By Corollary 10.3, their entropy can be computed via the corresponding Markov tree maps. Moreover, we saw that those maps can be chosen irreducible (more precisely, the associated

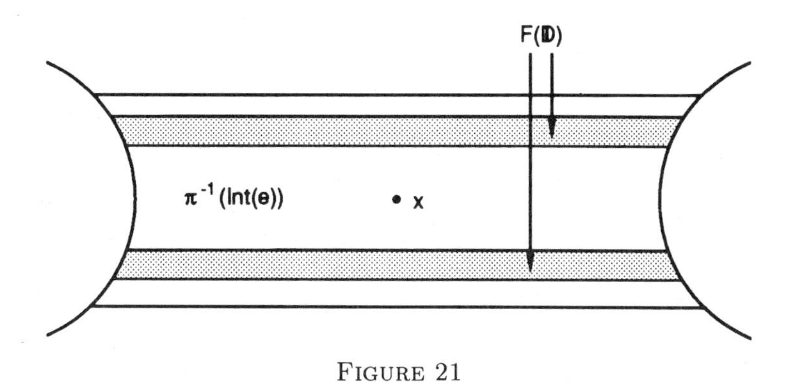

FIGURE 21

thick tree maps are irreducible). To make some estimates better, we shall prove one more property of the irreducible maps.

LEMMA 12.3. *If $\widehat{f}$ is irreducible and all vertices from R have valence* 1 *then f permutes the edges of T.*

PROOF. Suppose that $\widehat{f}$ is irreducible and all vertices from R have valence 1, but there is an edge of T whose image is larger than just one edge (or perhaps it is one edge covered twice). Then there is an edge e which is an image of 2 or more edges (or is covered twice by the image of one edge). Therefore there is a point $x \in \pi^{-1}(\text{Int}(e))$ which cannot be joined with $\partial\mathbb{D}$ in $\pi^{-1}(\text{Int}(e))$ by a curve not intersecting $F(\mathbb{D})$ (see Figure 21).

Since $\mathbb{D} \setminus F(\mathbb{D})$ is arcwise connected and contains $\partial\mathbb{D}$, the point x can be joined by a curve with $\partial\mathbb{D}$ in $\mathbb{D} \setminus F(\mathbb{D})$. As in the case of the curves along which we made splittings in Section 8, we can assume that this curve can be divided into finitely many pieces and on each of them the composition with π is monotone. Since $\widehat{f}$ is irreducible and all vertices from R have valence 1, the edge germs at each vertex are mapped bijectively to the edge germs at its image. Hence, when we move along this curve crossing $\pi^{-1}(v)$ for some vertex v, we enter $\pi^{-1}(\text{Int}(d))$ for a new edge d again in a component of $\pi^{-1}(\text{Int}(d)) \setminus F(\mathbb{D})$ disjoint from $\partial\mathbb{D}$ (see Figure 22 for $v \in V \setminus R$ and Figure 23 for $v \in R$). Therefore we cannot reach $\partial\mathbb{D}$ when moving along our curve, a contradiction. □

In the rest of this section we will work only with Markov tree maps and will not use the structure of the associated thick tree maps. Therefore we need some notation for the class of all Markov tree maps associated to irreducible thick tree maps. We shall use for this class the symbol $\mathcal{M}$. The tree on which the elements of $\mathcal{M}$ act, will be always denoted by T. The basic properties of a map $f \in \mathcal{M}$ are: it is Markov, transitive, every vertex is periodic, all vertices of valence 1 and 2 belong to the same cycle R, at every vertex v not in R the set of edge germs is mapped bijectively onto the set of edge germs at $f(v)$ and the cyclic ordering of the edge germs is preserved. As was explained in Section 3, since f is transitive, there are two possibilities. Either f permutes the edges of T and

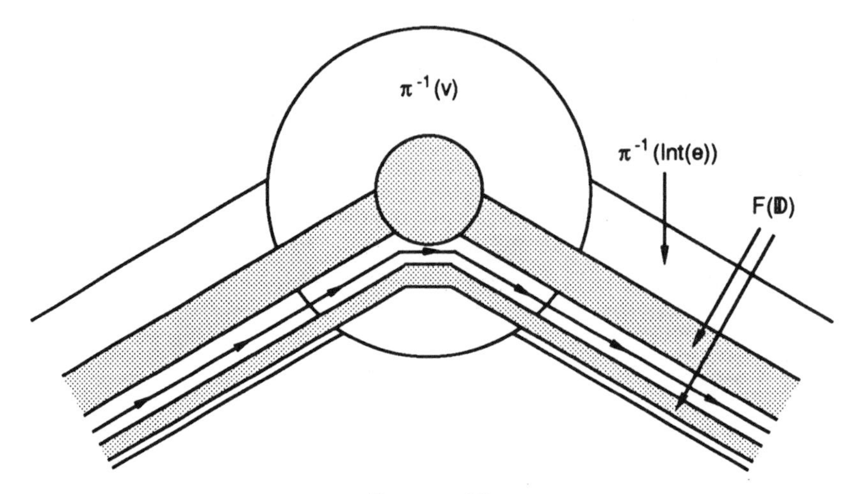

FIGURE 22

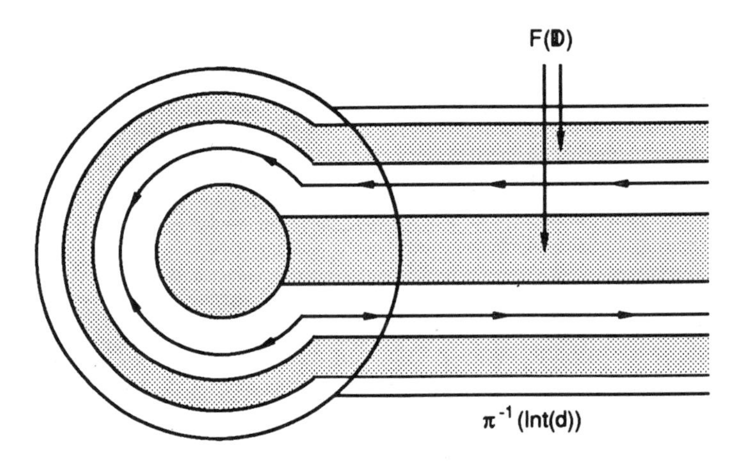

FIGURE 23

then $h(f) = 0$, we shall denote the set of those elements of $\mathcal{M}$ by $\mathcal{M}_0$; or there is an edge mapped to more than one edge (or at least to the same edge covering it twice) and then $h(f) > 0$, we shall denote the set of those elements of $\mathcal{M}$ by $\mathcal{M}_+$. Now we shall try to get some more information about the elements of $\mathcal{M}_0$ and $\mathcal{M}_+$.

LEMMA 12.4. *If $f \in \mathcal{M}_0$ then T is an n-od (a star) and f (up to equivalence) is a rotation.*

PROOF. Assume that $f \in \mathcal{M}_0$. Since the image of every edge is an edge, the set of the edge germs at any vertex $v \in R$ is embedded into the set of edge germs at $f(v)$. Since there is a vertex of R which is an end of T, this means that the valence of any element of R cannot be larger than 1. Thus, all elements of R are ends of T. Consequently, the image of any edge adjacent to an end of T is an edge adjacent to an end of T. Since f is transitive, we see that every edge of T is adjacent to an end of T. This means that T is an n-od (called sometimes a star). Since f preserves the cyclic ordering of germs, it has to be (up to equivalence) a rotation of T. $\square$

Clearly, if T is a star and f its rotation then the corresponding pattern (that is, the pattern $[(P, F)]$) is a twist pattern. Therefore from Theorem 12.2 and Lemma 12.4 we get a complete characterization of zero entropy patterns. This characterization has been known already, see [**GST**] and [**LM**].

THEOREM 12.5. *A pattern A has entropy zero if and only if there exists twist patterns $A_1, \dots, A_n$ such that A is an A_n-extension of an A_{n-1}-extension of ... of A_1 (if $n = 1$ then simply $A = A_1$).*

By Lemma 7.2, if $f \in \mathcal{M}_+$ then all iterates of f are transitive. Now we will try to obtain entropy estimates for maps from $\mathcal{M}_+$. If $x, y \in T$ then we shall use the notation $\langle x, y\rangle$ for the smallest connected subset of T containing both x and y. In the following lemmas we assume that $f \in \mathcal{M}_+$.

LEMMA 12.6. *Assume that $a \in T$, $f(a) = b$ and a is the only preimage of b under f. Then any connected component of $T \setminus \{a\}$ is mapped by f into a connected component of $T \setminus \{b\}$.*

PROOF. If C is a connected component of $T \setminus \{a\}$ and $f(C)$ intersects more than one connected component of $T \setminus \{b\}$ then, since $f(C)$ is connected, it contains b. Then there is a point in $C \subset T \setminus \{a\}$ whose image is b, a contradiction. $\square$

LEMMA 12.7. *Let z be a fixed point of f, X the set of all ends of T, and n the number of elements of X. Then every point of $T \setminus (\{z\} \cup \bigcup_{i=0}^n f^i(X))$ has at least two preimages under f^n.*

PROOF. Suppose that there is a point $a \in T \setminus (\{z\} \cup \bigcup_{i=0}^n f^i(X))$ with only one preimage under f^n. Since f is onto, a has only one preimage under f^i for $i = 0, 1, \dots, n$. Denote this preimage a_i. Then $a_i \in \langle z, x_i\rangle$ for some $x_i \in X$. Since X has only n elements, for some k, j with $0 \leqslant k < j \leqslant n$ we have $x_k = x_j$. Then $g = f^{j-k}$ maps a_j to a_k and a_j is the only preimage of a_k under g. Moreover, both a_j and a_k belong to $\langle z, x\rangle$, where $x = x_k = x_j \in X$.

Let C_1, C_2 be the components of $T \setminus \{a_j\}$ and D_1, D_2 the components of $T \setminus \{a_k\}$. We can number them in such a way that $z \in C_1 \cap D_1$. By Lemma 12.6 and since $f(z) = z$, we have $g(C_1) \subset D_1$. Since $a \notin \bigcup_{i=0}^n f^i(X \cup \{z\})$, the points a_k and a_j do not belong to $X \cup \{z\}$, so all C_1, C_2, D_1, D_2 are nonempty. We have either $D_1 \subset C_1$ or $C_1 \subset D_1$. If $D_1 \subset C_1$ then, since $g(C_1) \subset D_1$, we get $g(D_1) \subset D_1$, contrary to transitivity of g. Therefore $C_1 \subset D_1$. Then $D_2 \subset C_2$. By Lemma 12.6, either $g(C_2) \subset D_1$ or $g(C_2) \subset D_2$. If $g(C_2) \subset D_1$ then, since $g(C_1) \subset D_1$, we get $g(T)$ contained in the closure of D_1, a contradiction since g

is onto. Therefore $g(C_2) \subset D_2 \subset C_2$, again contrary to transitivity of g. Thus, in all cases we get a contradiction. □

THEOREM 12.8. *If $f \in \mathcal{M}_+$ and T has n ends then $h(f) \geqslant (1/n)\log 2$.*

PROOF. By replacing f with an equivalent map if necessary, we can assume that f is piecewise expanding with constant $\beta = \exp(h(f))$. Since f has a fixed point, we can use Lemma 12.7. If $h(f) < (1/n)\log 2$ then $\beta^n < 2$ and therefore there is at least a whole interval of points which have only one preimage under f^n, a contradiction. □

Theorems 12.2 and 12.8, together with the obvious fact that the twist patterns have entropy zero, give us a method of estimating the entropy of a pattern. Of course, if we come to the moment when we have to use Theorem 12.8 then we should find an irreducible thick tree map corresponding to A such that the tree has minimal possible number of ends (remember that this irreducible thick tree map is not unique and even the tree is not unique). On the other hand, if we want to get a general result then we can use Lemma 12.3 which tells us that if $f \in \mathcal{M}_+$ then the number of the ends of T is at most the period of R minus 1. Therefore we obtain the following result.

THEOREM 12.9. *If A is a pattern of period k which is not twist and is not an extension then $h(A) \geqslant (1/(k-1))\log 2$.*

13. The forcing relation

The aim of this section is to use the tools developed in previous sections for investigation of the forcing relation among patterns for our classes of maps.

As usual, we work with a map $F \in \mathfrak{C}$ having a cycle P with $[(P, F)] = A$ and associated to a thick tree map $\widehat{f}$ which is in turn associated to a tree map $f : T \to T$. Our strategy is to assume that $f \in \mathcal{M}_+$ and see what cycles f has to have. This method gives results on the forcing relation because of Theorem 11.3.

We have to use the Lefschetz fixed point theory. For this it is useful to have fixed points of a given iterate of a map isolated. Therefore we will work with the class $\mathcal{M}_{pe}$ of those maps from $\mathcal{M}_+$ which are piecewise expanding. Since every element of $\mathcal{M}_+$ is equivalent to one of those maps, this does not lead to any loss of generality.

Let $f \in \mathcal{M}_{pe}$, let q be a positive integer smaller than the period of R and let z be a fixed point of f^q. Then we can define the *index* of z in the usual way (see e.g. [**D**]). Since q is smaller than the period of R, the point z does not belong to R. In particular, it is not an end of T. If z is a vertex of T then f^q either keeps the edge germs at z fixed or rotates them. If z is not a vertex of T then we can add a vertex at z to get a new tree T'. Then we can also speak about the edge germs at z. It is clear that in this case also f^q either keeps those germs fixed or rotates them. It is easy to check (taking into account that the map is piecewise expanding) that the index of z is negative if f^q keeps the germs fixed and it is 1 if f^q rotates the germs.

The main technical tool which we shall use in this section is the following lemma. As for cycles, when we speak about a *period* of an edge germ we mean its *least period.*

LEMMA 13.1. *Let $f \in \mathcal{M}_{pe}$, let q be a positive integer and let Q be a nonempty invariant set disjoint from R such that every $z \in Q$ is a fixed point of f^q of negative index and each edge germ at z is periodic for f of period q. Let S be the closure of one of the connected components of $T \setminus Q$ and let m be the number of edges of S. Then for every $i \geqslant m$, the map f has a cycle of period iq. Moreover, if $q = 1$ then we can require additionally that those cycles are contained entirely in S.*

PROOF. Let T' be the tree obtained from T by adding vertices at all the points of Q (which are not yet vertices of T). For $z \in Q \cap S$ denote by $e(z)$ the edge of S adjacent to z. We claim that there is $y \in Q$ such that in the Markov graph of $f^q : T' \to T'$ there are two different loops b_1 and b_2 passing through $e(y)$ exactly once, passing only through edges of S and through each of them at most once, and having lengths respectively 1 and $k \leqslant m$. We proceed with the proof of this claim.

The points of Q are not ends of T, so S is a proper subset of T'. By Lemma 7.2, f^q is transitive (as a map of T, but it is obvious that also as a map of T').

Let $z \in Q \cap S$. Since the index of z is negative, the germ of $e(z)$ at z is mapped by f^q to itself. Hence, there is an arrow (in the Markov graph of f^q) from $e(z)$ to itself. In the simple case when there is no arrow from $e(z)$ to any other edge of S there must be two arrows from $e(z)$ to itself (independently of whether the other endpoint of $e(z)$ belongs to Q or not; remember that f^q is piecewise expanding), and we are done. Therefore in the rest of the proof we shall assume that for every $y \in Q \cap S$ there is an arrow from $e(y)$ to some other edge of S. This means in particular that there are at least two edges in S, so the other endpoint of $e(z)$ does not belong to Q.

The sets $X_i = f^{qi}(e(z) \setminus Q)$, $i = 0, 1, 2, \ldots$, form an ascending sequence. By transitivity, as long as they are contained in $S \setminus Q$, this sequence does not stabilize. Therefore there is $k(z)$ such that $X_{k(z)-1}$ is contained in $S \setminus Q$ but $X_{k(z)}$ is not. Then there is an edge $d \subset X_{k(z)-1}$ such that $f^q(d) \supset e(\xi(z))$ for some $\xi(z) \in Q \cap S$. This gives us a path from $e(z)$ to $e(\xi(z))$ passing only through edges of S and such that the first transition is from $e(z)$ to a different edge of S.

The map $\xi : Q \cap S \to Q \cap S$ has a cycle. Concatenating the paths found above along this cycle we get a loop from $e(y)$ (for some $y \in Q \cap S$) to itself, passing only through the edges of S and passing through some edge different from $e(y)$. The shortest of such loops is the loop b_2 we are looking for. It passes through any edge of T' at most once (since it is the shortest one) and it passes only through edges of S, so its length is not larger than m. The loop b_1 can be obtained by taking the arrow which goes from $e(y)$ to itself. This proves the claim.

Our strategy now is to find cycles of f of period iq which correspond to loops in the Markov graph for f^q having length $i \geqslant m$ which are formed by concatenating b_1 and b_2. This is much easier if $i > k$ (recall that $i \geqslant m \geqslant k$) so we consider that case first.

Clearly, we can choose b_1 in such a way that it corresponds to the minimal subinterval I of $e(y)$ with y as one of the endpoints, such that $f^q(I) = e(y)$. By minimality, for each $j = 0, 1, \dots, q-1$ the set $f^j(I)$ is contained in some edge of T' and f is one-to-one on $f^j(I)$. We claim that if $0 < j < q$ then I and $f^j(I)$ have disjoint interiors. Since the edge germs at y are periodic of period q, the only possible situation in which this could be false is when $f^j(y)$ is the other endpoint of $e(y)$. However, then the endpoint of $f^j(I)$ different from $f^j(y)$ is an interior point of I. Its image under f^{q-j} (and therefore also its image under f^q) is a vertex of T'. Thus, the image under f^q of an interior point of I is a vertex of T'. This contradicts minimality of I. This proves our claim that if $0 < j < q$ then I and $f^j(I)$ have disjoint interiors.

Now, both loops in the Markov graph of f^q give rise to q times longer loops in the Markov graph of f. This is so because f is Markov. Call those loops a_1 (the loop of length q, corresponding to the interval I and its images) and a_2 (the loop of length kq). Fix a positive integer p and consider the loop a of length $(p+k)q$ obtained by going around a_1 p times and around a_2 once. By a standard argument (see e.g. [**BGMY, ALM**]) there exists a point $x \in I$ such that $f^{(p+k)q}(x) = x$ and $f^i(x)$ belongs to the i-th vertex of the loop, when we start counting from 0 (remember that a vertex of a Markov graph is an edge of a tree).

We claim that x is an interior point of I. Indeed, if $x = y$ then $f^{pq}(x) = x$, but $f^{pq}(x)$ belongs to a subinterval of $e(y)$ which corresponds to b_2, and this subinterval does not contain y (b_2 is different from b_1 and passes through $e(y)$ only once). If x is the other endpoint of I then it is an interior point of $e(y)$, but its image under f^q (and therefore under all higher iterates of f) is a vertex of T'. Thus, in this case x cannot be a periodic point. This proves the claim that x is an interior point of I.

We next want to show that if $0 \leqslant i < (p+k)q$ then $f^i(x) \in \mathrm{Int}(I)$ if and only if $i = jq$ with $0 \leqslant j < p$. Clearly, $f^i(x) \in \mathrm{Int}(I)$ if for $i = jq$ with $0 \leqslant j < p$. Since b_2 is different from b_1 and passes through $e(y)$ only once, the point $f^i(x)$ does not belong to $\mathrm{Int}(I)$ for $i = jq$ with $p \leqslant j < p+k$. Since the interiors of I and its images under f^j, $j = 1, 2, \dots q-1$ were disjoint, if $|i - i'| < q$ then it cannot happen that both $f^i(x)$ and $f^{i'}(x)$ belong to $\mathrm{Int}(I)$.

We have now taken care of all values of i except those not divisible by q with $pq < i < (p+k-1)q$. In this case, we argue as follows. If for some i like this $f^i(x) \in \mathrm{Int}(I)$ then some part of the loop a_2 looks as follows. There is some d_1 on a place with number divisible by q, then j places further (with $0 < j < q$) there is $e(y)$ (this is the place corresponding to $f^i(x) \in \mathrm{Int}(I)$), q places further there is again $e(y)$ (since $f^q(\mathrm{Int}(I)) = \mathrm{Int}(e(y))$), and $q - j$ places further there is some d_2. This means that there is a path from d_1 to $e(y)$ of length j and a path from $e(y)$ to d_2 of length $q-j$. Therefore there is a path of length q from d_1 to d_2. Consequently, in the Markov graph of f^q there is an arrow from d_1 to d_2. Clearly, we may assume that b_2 is the shortest loop with the desired properties. In this loop we used a path of length 2 to get from d_1 to d_2. However, we can get from d_1 to d_2 in one step, which contradicts the minimality of b_2. Thus we have shown that if $0 \leqslant i < (p+k)q$ then $f^i(x) \in \mathrm{Int}(I)$ if and only if $i = jq$ with $0 \leqslant j < p$.

Now we can prove that the period of x is $(p+k)q$. Let us write an infinite 0-1 sequence with 1 on i-th place if and only if $f^i(x) \in \mathrm{Int}(I)$. If the period of x is l then this sequence is periodic of period which is a divisor of l. However, our sequence consists of the blocks of length q of two types: $1, 0, \dots, 0$ and $0, 0, \dots, 0$. It has p blocks of the first type, then k blocks of the second type, and then everything starts to repeat periodically. Therefore its period is $(p+k)q$. Thus, the period of x is also $(p+k)q$.

This proves the main assertion of the lemma for the periods larger than kq. If $k < m$ we are finished. Therefore all that remains is to consider the case when $k = m$ and show that there is a cycle of f of period kq. The obvious candidate for such a cycle is the one corresponding to the loop a_2. As before, we can choose a point x of this cycle and a minimal interval J such that $x \in J \subset e(y)$, the images under f^i of J are contained in the consecutive elements of a_2 and $f^{kq}(J) = e(y)$, $f^{kq}(x) = x$. Moreover, we know that the interiors of I and J are disjoint. In particular, $x \neq y$. We have to show that the period of x is kq. First we show that we can assume that x is not a vertex of T'.

Let v be the endpoint of $e(y)$ different from y. Since $x \neq y$, it remains to show that $x \neq v$. As before, we can assume that b_2 is of minimal length. Therefore when we look at the consecutive images under the iterates of f^q of the edge $e(y)$, each time we get only one edge more, until we go out of S. If $f^q(v) \notin S \setminus \{y\}$ then $k = 1$ and $f^q(v) \neq v$, so $x \neq v$. If $f^q(v) \in S$ and $f^q(v) \neq v$ then we take the point w of $e(y)$ closest to v such that $f^q(w) = v$. Then the image under f^q of the interval with the endpoints w and v is an edge of S. The image of some point of this edge under $f^{(k-1)q}$ does not belong to $S \setminus \{y\}$. Therefore there is a point u between w and v such that $f^{kq}(u) \notin S \setminus \{y\}$. On the other hand, $f^{kq}(w) = f^{(k-1)q}(v)$ is a vertex of S different from y. Therefore there is a point x' in the interior of the interval with the endpoints w and u with $f^{kq}(x') = x'$. The point $f^q(x')$ belongs to $S \setminus e(y)$. Thus, the loop in the Markov graph of f^q corresponding to the orbit of x' starts at $e(y)$, then goes to some edge of S different from $e(y)$ and has length m. If it passes through some edge not contained in S then b_2 could be chosen shorter than m, a contradiction. Otherwise, we can assume that this is b_2, and then $x = x'$. However, $x' \neq v$, so then $x \neq v$. The third case to consider is $f^q(v) = v$. Then the first arrow in b_2 can be chosen in at least two ways. One of these choices gives $x = v$, but then the second choice does not. In this case, of course, we make the second choice.

Thus, we have proven that we can assume that x is not a vertex of T', so we make this assumption. Since the set of vertices of T' is invariant, we see then that no point of the cycle of f to which x belongs, is a vertex of T'. Now we want to prove that the period of x is kq or that there is another periodic point of that period. Since b_2 passes through each edge of T' at most once and x is not a vertex of T', the period of x for f^q is k. Therefore if the period of x for f is not kq then it is not divisible by q.

Suppose first that the only edge of the form $e(z)$, $z \in Q$ contained in S is $e(y)$. Consequently, none of the points $f^i(v)$ for $i = 1, 2, \ldots, q-1$ belongs to S. The points $f^i(x)$, $i = 1, 2, \ldots, q-1$ belong to the interiors of the edges adjacent to the points $f^i(v)$, and those interiors are disjoint from S. Therefore the points $f^i(x)$ do not belong to S. If the period of x for f is not divisible by q then there is an i not divisible by q such that $f^i(x) = x$. Then there is $j < q$ such that $i+j$ is divisible by q and then we have $f^j(x) = f^{i+j}(x) \in S$, a contradiction. It can also happen that $S = e(y) = e(z)$ for some $z \neq y$. In this case there are clearly no problems unless $f^{q/2}(y) = z$ and $f^{q/2}(z) = y$ and $f^{q/2}$ maps the edge germ of $e(y)$ at y to the edge germ of $e(z)$ at z and vice versa. However, then in the Markov graph of $f^{q/2}$ there are at least 3 arrows from $e(y)$ to itself. Only one of the loops of length 2 formed by those arrows can correspond to y, only one to z and only one to a period $q/2$ cycle. The remaining two loops have to correspond to period q cycles.

Suppose now that there is more than one edge of the form $e(z)$, $z \in Q$ in S. By the minimality of b_2, if we start from any $e(z) \subset S$ and take its consecutive images under the iterates of f^q then each time we get only one edge more, until we get all edges (here we intersect the image each time with S). If $e(z_1), e(z_2) \subset S$ and the smallest connected set containing both of them contains some other edge, then one has to pass through this edge on the way from $e(z_1)$ to $e(z_2)$, as well as on the way from $e(z_2)$ to $e(z_1)$, a contradiction. Therefore all the edges of the form $e(z)$ have a common vertex $u \notin Q$. If $f^q(u) \notin S \setminus Q$ then b_2 is not minimal, a contradiction. If $f^q(u) \in S \setminus Q$ but $f^q(u) \neq u$ then there are arrows in the Markov graph of f^q from all of the edges of the form $e(z)$ to the same edge d. Then there is a path from d to one of the $e(z)$ not passing through other ones, and this contradicts the minimality of b_2. Therefore we have $f^q(u) = u$. The edge germs at u have some period l for f^q. We can form a loop $a_2' \neq a_1$ of length l through those edges, and by the minimality of a_2 we get $l = k = m$. Thus, S is an k-od with u as the central point. Moreover, the edge germs at u are permuted cyclically by f^q. Let t be the period of u for f. Then $q = tr$ for some integer r. Since the edge germs at u are permuted cyclically by $f^q = (f^t)^r$ and there are k of them, r and k are coprime.

Look at the Markov graph of f^t. Since f^t is transitive, there is an edge germ d of S, whose image under f^t starts at u, goes to the other end of another edge d', goes outside S, comes back and leaves S along yet another edge. Thus, there are at least two arrows from d to d'. One of these arrows corresponds to the first passage of $f^t(d)$ (and therefore to the edge germ of d at u being mapped to the edge germ of d' at u), the second one to the second passage. The arrows corresponding to the permutation of the edge germs at u form a loop of length k in the Markov graph of f^t. Go around this loop r times and replace one arrow from d to d' by the second arrow constructed above. The loop in the Markov graph of f^t we obtain clearly has period kr. The corresponding loop in the Markov graph of f passes through the edges of S exactly every t times (since the arrows corresponding to f^t mapping an edge germ at u to an edge germ at u give rise to paths through edge germs at the images of u, which are outside u), except perhaps between d and d' that is on one block of length t. Since our loop consists of $kr > 1$ blocks of length t and at most one block is exceptional in the above sense, the period of this loop has to be divisible by t. Since its period for f^t is kr, its period for f is $krt = kq$. Clearly, the corresponding periodic point cannot either be u or belong to Q. Therefore it has also period kq and we are done. This completes the proof of the main statement of the lemma.

If $q = 1$ then we use only loops which pass only through the edges of S, so the cycles we get are contained in S. □

Each continuous map from a tree into itself has a fixed point. Therefore we can divide $\mathcal{M}_{pe}$ into two classes: $\mathcal{M}_{pe}^1$ consisting of maps with only one fixed point and $\mathcal{M}_{pe}^2$ consisting of maps with more fixed points. The elements of $\mathcal{M}_{pe}^2$ are easier to study.

Proposition 13.2. *Any map of $\mathcal{M}_{pe}^2$ has cycles of all periods larger than or equal to $n-2$, where n is the period of R. Moreover, it has a cycle of period n different from R.*

Proof. By the Lefschetz fixed point formula (see e.g. [**D**]) the sum of the indices of the fixed points of f is equal to the Euler characteristic of T, that is to 1. The indices of the fixed points of f are either 1 or negative. If all of them are 1 then the sum of the indices is equal to the number of the fixed points. Therefore then f has only one fixed point. Since $f \in \mathcal{M}_{pe}^2$, we get that it has a fixed point z of negative index.

Let s be the number of edges of T. By Lemma 6.1 we have $s \leqslant n + t_1 - 3$, where t_1 is the number of ends of T. In view of Lemma 12.3, $t_1 \leqslant n - 1$. Therefore $s \leqslant 2n - 4$. Let S be the component of $T \setminus \{z\}$ with the minimal number of edges. This number is not larger than $s/2$, so it is not larger than $n-2$. Since the index of z is negative, f fixes the edge germs at z. Thus we can use Lemma 13.1 with $Q = \{z\}$ and $q = 1$. Consequently, f has cycles of all periods larger than or equal to $n-2$. Moreover, the cycle of period n which we get is contained in S, so it cannot be equal to R. □

Now we start to investigate the maps of $\mathcal{M}_{pe}^1$.

Lemma 13.3. *If $f \in \mathcal{M}_{pe}^1$ then there is an integer $q > 1$ dividing the period of R such that either f has a cycle of period q different from R and such that* *the index of every point of this cycle for f^q is negative, or the edge germs at the fixed point z of f are periodic (under f) of period q.*

PROOF. By the construction of Section 10, if $f \in \mathcal{M}^1_{pe}$ then there is a generalized pseudo-Anosov homeomorphism $G : \mathbb{D} \to \mathbb{D}$ equivalent to F having only one fixed point z. This fixed point cannot be on the boundary of $\mathbb{D}$ since, if it were, the homeomorphism of S^2 obtained by collapsing this boundary to point would be a generalized pseudo-Anosov homeomorphism with only one fixed point. This is impossible since the total index for this homeomorphism would be 2 and the index of any fixed point of a generalized pseudo-Anosov homeomorphism is less than or equal to 1.

Thus z is an interior point of $\mathbb{D}$ and we can blow up z to obtain a homeomorphism $\widetilde{G}$ of an annulus X without fixed points in the interior. Since $f \in \mathcal{M}^1_{pe}$, the pattern of P is not twist and therefore the cycle P of $\widetilde{G}$ is not topologically monotone. By [**F**], if the lifting of $\widetilde{G}|_{\mathrm{Int}(X)}$ to its universal covering space has a cycle then it has a fixed point. Then $\widetilde{G}|_{\mathrm{Int}(X)}$ itself has a fixed point, a contradiction. This shows that the rotation number of P is not an integer. Write this rotation number as p/q, where p and q are coprime. Then by [**B3**], $\widetilde{G}$ has a topologically monotone cycle Q of period q, which is not the same as P since it is monotone.

Clearly, all the points of Q have the same index for $\widetilde{G}^q$. Suppose that this index is not negative. Note that this means that it is not possible for Q to lie on the boundary of the annulus, since any fixed points of $\widetilde{G}^q$ on the boundary must be saddles with no flip in their stable or unstable manifolds and hence must have negative index. (The value of the index of these points should be counted as $-1/2$ as can be seen by considering the doubled annulus obtained by reflecting the original annulus through its boundary. However, for us it suffices to know this value is negative.)

The generalized pseudo-Anosov homeomorphism G^n has no fixed points of index zero for any $n > 0$ except the points of the cycle P (this is a general property of pseudo-Anosov homeomorphisms; see [FLP] or [T]). It follows that for $\widetilde{G}^q$ the index of every point of Q is positive.

Look at all the fixed points of $\widetilde{G}^q$ of rotation number p/q for $\widetilde{G}$. They form a fixed point class of $\widetilde{G}^q$. Since by an isotopy we can pass from $\widetilde{G}^q$ to a rotation without any fixed points, the index of this fixed point class is 0. This index is equal to the sum of indices of the points from our fixed point class. Thus, since the indices of the points of Q are positive, there must be another cycle of $\widetilde{G}$ of rotation number p/q such that its elements are fixed points of $\widetilde{G}^q$ with negative index. Since p and q are coprime, the period of this cycle is q.

Consequently, $\widetilde{G}$ has a cycle with rotation number p/q such that its elements are fixed points of $\widetilde{G}^q$ with negative index. Now there are two possibilities. If this cycle does not lie on the component Y of the boundary of the annulus obtained from blowing up z, then G has a cycle of period q, different from P. The points of this cycle have negative index for G^q. Hence, if this cycle is non-peripheral, by Corollary 10.3 there is a corresponding periodic cycle of f which has the same period and whose points have the same negative index for f^q. If the cycle of G is peripheral but not in Y then by Corollary 10.2 there is a peripheral cycle for f of period q and its points have index -1 for f^q.

If the cycle we are considering of $\widetilde{G}$ lies on Y, it corresponds to the fixed point z of f. In this case the homeomorphism $\widetilde{G}$ restricted to Y has rotation number p/q. Since the edge germs at z correspond to the unstable prongs of periodic points on Y, and the numbers p and q are coprime, we get that the edge germs at z are periodic of period q. In both cases, since p/q is not an integer, $q > 1$. Since p/q is a rotation number of P, q divides the period of P (which is the same as the period of R). □

PROPOSITION 13.4. *Any map of $\mathcal{M}^1_{pe}$ has cycles of all periods ni, $i \geq 2$, where n is the period of R.*

PROOF. Let $f \in \mathcal{M}^1_{pe}$. By Lemma 13.3, there is an integer $q > 1$ dividing n such that either f has a cycle of period q different than R and such that the index of every point of this cycle for f^q is negative, or the edge germs at the fixed point z of f are periodic (under f) of period q. In both cases the assumptions of Lemma 13.1 are satisfied (in the second case with $Q = \{z\}$). The number of components of $T \setminus Q$ is at least q. As in the proof of of Proposition 13.2, from Lemmas 6.1 and 12.3 we obtain that the number of edges of T is not larger than $2n - 4$. Thus, the number of edges of T' (the tree obtained from T by adding vertices at the points of Q) is not larger than $q + 2n - 4$. Therefore there is a component of $T \setminus Q$ with the number of edges $m \leqslant (q + 2n - 4)/q < 1 + 2n/q$. If $i \geqslant 2$ then $in = kq$ for some k and we get $k \geqslant 2n/q$. We obtain $k + 1 \geqslant 1 + 2n/q > m$, and since both k and m are integers, $k \geqslant m$. By Lemma 13.1, f has a cycle of period $kq = in$. □

14. Primary patterns

In this section we shall use the results of the previous sections to try to identify primary patterns. Recall that a pattern is called primary if it does not force any other pattern of the same period.

First we need some terminology. As in Section 11, a pattern $A = [(P, F)]$ is pseudo-Anosov if F is isotopic rel P to a pseudo-Anosov homeomorphism G. It is well known (see e.g. [**FLP**]) that this pseudo-Anosov homeomorphism is unique up to conjugacy. If G has only one fixed point then we shall say that A is *single-fixed-point pseudo-Anosov*; if it has more fixed points then we shall say that A is *multi-fixed-point pseudo-Anosov.* In other words, single-fixed-point (respectively multi-fixed-point) pseudo-Anosov patterns are those patterns which have corresponding tree maps from $\mathcal{M}^1_{pe}$ (respectively $\mathcal{M}^2_{pe}$).

By our construction (this follows also from the Thurston's theory; see e.g. [**FLP**], [**T**]), each pattern is either twist or pseudo-Anosov or is an extension.

Therefore each pattern A admits a *canonical representation* $(A_1, \ldots, A_m)$. Namely, there exist non-trivial (that is, of period greater than 1) patterns A_i, $i = 1, \ldots, m$, each of them twist or pseudo-Anosov, such that A is an A_m-extension of an A_{m-1}-extension of ... of A_1 (if the period of A is 1 then we take $m = 0$). In fact, for a given pattern the canonical representation is unique. Indeed, assume that A is a B-extension of C and a B'-extension of C', where C and C' are non-trivial twist or pseudo-Anosov patterns. By Theorem 12.1, since A forces C' and C' is twist or pseudo-Anosov, C forces C'. Similarly, C' forces C. By the antisymmetry of the forcing relation, $C = C'$. Now, again by Theorem 12.1, since A forces itself, we get that B forces B' and B' forces B, so $B = B'$. Using this argument repeatedly we see that the patterns $A_1, A_2, \ldots$ in a canonical representation of a given pattern A are determined uniquely.

Now we start to analyze various patterns to see whether they are primary or not.

LEMMA 14.1. *Every twist pattern is primary.*

PROOF. This follows immediately from their definition. □

LEMMA 14.2. *No multi-fixed-point pseudo-Anosov pattern is primary.*

PROOF. If A is a multi-fixed-point pseudo-Anosov pattern of period n then by Proposition 13.2 a corresponding irreducible tree map has at least 2 cycles of period n. By Corollary 10.3, the corresponding pseudo-Anosov map has has also at least 2 cycles of period n. Therefore, by Corollary 10.5, A is not primary. □

LEMMA 14.3. *Let B and C be non-trivial patterns and let A be a B-extension of C. Then:*

(a) *if B is not primary then A is not primary,*
(b) *if C is twist and B is primary then A is primary,*
(c) *if C is pseudo-Anosov then A is not primary.*

PROOF. Let k and n be the periods of B and C respectively. Then the period of A is kn. If B is not primary then it forces some pattern $D \neq B$ of period k. By Theorem 12.1, A forces some D-extension of C (call it E). By the uniqueness of the extension decomposition (Proposition 5.2 of [**H1**]), $E \neq A$. Since the period of E is kn, the same as for A, the pattern A is not primary. This proves (a).

Now we prove (b). Assume that C is twist and A is primary. Suppose that A forces some $E \neq A$ of period kn. Since C is twist, it does not force E. Therefore by Theorem 12.1, E is a D-extension of C for some pattern D forced by B. The period of D is $kn/n = k$. Since B is primary, we get $D = B$. Thus, by Theorem 12.1 we get a contradiction. This proves (b).

Finally, we prove (c). Assume that C is pseudo-Anosov. By Propositions 13.2 and 13.4 a tree map from $\mathcal{M}_{pe}$ corresponding to C has a cycle of period kn. By Corollary 10.3, the corresponding pseudo-Anosov map has a cycle of period kn, and by Theorem 11.3, C forces some pattern E of period kn. By Theorem 12.1,

A forces E. Suppose that $E = A$. Then A forces C and C forces A. Since forcing is an ordering relation, it follows that $C = A$, which is impossible (they have different periods). Thus, A is not primary. This proves (c). □

Now we can prove the main result of this section.

THEOREM 14.4. *Let* $(A_1, \ldots, A_m)$ *be the canonical representation of a pattern* A. *Then* A *is primary if and only if* $A_1, \ldots, A_{m-1}$ *are twist and* A_m *is either twist or primary single-fixed-point pseudo-Anosov.*

PROOF. We shall use induction on m. If $m = 0$ then A is trivial, and clearly it is primary. On the other hand, there are no A_i's, so all conditions for them are satisfied in void.

Let $m = 1$. Then $A = A_m$. By Lemmas 14.1 and 14.2, A is primary if and only if it (and therefore A_m) is either twist or primary single-fixed-point pseudo-Anosov.

Assume now $m > 1$ and that the assertion of the theorem holds for all smaller m's. Then A is an A_m-extension of C where C has the canonical representation $(A_1, \ldots, A_{m-1})$. Both B and C are non-trivial. Assume that A is primary. By Lemma 14.3(a), A_m is primary, so by Lemma 14.2 it is either twist or primary single-fixed-point pseudo-Anosov. If for some $i < m$, A_i is not twist then it is pseudo-Anosov. By associativity of extensions, A is a B'-extension of C' for some B' with canonical form $(A_i, \ldots, A_m)$ and C' with canonical form $(A_1, \ldots, A_{i-1})$. In turn, B' is a B''-extension of A_i for some B'' with canonical form $(A_{i+1}, \ldots, A_m)$. Since $i < m$, the pattern B'' is non-trivial. Therefore by Lemma 14.3(c), B' is not primary. If $i = 1$ then $B' = A$, so A is not primary, a contradiction. If $i > 1$ then C' is non-trivial, so by Lemma 14.3(a), A is not primary, also a contradiction. This proves that if A is primary then all $A_i, \ldots, A_{m-1}$ are twist.

Assume now that $A_1, \ldots, A_{m-1}$ are twist and A_m is either twist or primary single-fixed-point pseudo-Anosov. The pattern A is a B'''-extension of A_1, where B''' has canonical form $(A_2, \ldots, A_m)$. By the induction hypothesis, B''' is primary, so by Lemma 14.3(b), A is also primary. This completes the proof. □

Notice that in view of Lemma 13.3 there is no single-fixed-point pseudo-Anosov pattern of a prime period. Indeed, if a pattern A of prime period n corresponds to a tree map $f \in \mathcal{M}^1_{pe}$ then the integer q from Lemma 13.3 has to be equal to n. By Lemmas 12.3 and 12.4, the number of ends of the tree is smaller than n, so the number of edge germs at the fixed point of f is also smaller than n. Therefore by Lemma 13.3, f has a cycle of period n different than R, so A is not primary.

In fact, we do not know any single-fixed-point pseudo-Anosov primary pattern. Therefore we make the following conjecture.

CONJECTURE. *Let* $(A_1, \ldots, A_m)$ *be the canonical representation of a pattern* A. *Then* A *is primary if and only if* $A_1, \ldots, A_m$ *are twist. That is, a pattern is primary if and only if it has entropy zero.*

We finish this section by the following result, analogous to one for interval maps (see e.g. [**ALM**]).

THEOREM 14.5. *A pattern A of period n is primary if and only if there exists a homeomorphism $F \in \mathfrak{A}$ which has only one cycle of period n and this cycle has pattern A.*

PROOF. If such a homeomorphism exists then A is primary by the definition. Assume now that A is primary. Let $(A_1, \ldots, A_m)$ be the canonical form of A. By Theorem 14.4, the patterns $A_1, \ldots, A_{m-1}$ are twist, and a_m is either twist or primary single-fixed-point pseudo-Anosov. We can build F step by step using the definition of an extension, and on each step we can avoid producing unnecessary cycles. With twist patterns this is obvious. In the last step, if A_m is primary single-fixed-point pseudo-Anosov then we use Corollary 10.5. □

REFERENCES

[ALM] Ll. Alsedà, J. Llibre and M. Misiurewicz, *Combinatorial dynamics and entropy in dimension one*, World Scientific, Singapore (to appear).

[ACG] J. Ashley, E. Coven and W. Geller, *Entropy of semipatterns or, how to connect the dots to minimize entropy*, Proc. Amer. Math. Soc. (to appear).

[AF] D. Asimov and J. Franks, *Unremovable closed orbits*, Geometric dynamics, Lecture Notes in Math., vol. 1007, Springer, Berlin, Heidelberg, New York, 1983, pp. 22–29, and correction (preprint, 1989).

[BGN] D. Benardete, M. Gutierrez and Z. Nitecki, *Braids and the Nielsen-Thurston classification*, preprint.

[BH1] M. Bestvina and M. Handel, *Train tracks and automorphisms of free groups*, preprint.

[BH2] M. Bestvina and M. Handel, *Train tracks for surface homeomorphisms*, preprint.

[BGMY] L. Block, J. Guckenheimer, M. Misiurewicz and L.-S. Young, *Periodic points and topological entropy of one dimensional maps*, Global theory of dynamical systems, Lecture Notes in Math., vol. 819, Springer, Berlin, Heidelberg, New York, 1980, pp. 18–34.

[B2] P. Boyland, *An analog of Šarkovskiĭ's theorem for twist maps*, Cont. Math. **81** (1988), 219–233.

[B1] ———, *Braid types and a topological method of proving positive entropy*, preprint.

[B3] ———, *Rotation sets and monotone periodic orbits for annulus homeomorphisms*, preprint.

[D] A. Dold, *Fixed point index and fixed point theorem for Euclidean neighborhood retracts*, Topology **4** (1965), 1–8.

[FLP] A. Fathi, F. Laudenbach and V. Poenaru, *Travaux de Thurston sur les surfaces*, Asterisque **66-67** (1979).

[F] J. Franks, *Recurrence and fixed points of surface homeomorphisms*, Ergod. Th. and Dynam. Sys. **8*** (1988), 99–107.

[GST] J. M. Gambaudo, S. van Strien and C. Tresser, *The periodic orbit structure of orientation preserving diffeomorphisms on D^2 with topological entropy zero*, Ann. Inst. H. Poincaré (Phys. théor.) **49** (1989), 335–356.

[GLM] J. Guashi, J. Llibre, R. S. MacKay, *A classification of braid types for periodic orbits of diffeomorphisms of surfaces of genus one with topological entropy zero*, Publ. Matem. UAB (to appear).

[H1] T. Hall, *The bifurcations creating horseshoes*, preprint (1990).

[H2] ———, *Unremovable periodic orbits of homeomorphisms*, preprint (1991).

[LM] J. Llibre and R. MacKay, *A classification of braid types for diffeomorphisms of surfaces of genus zero with topological entropy zero*, J. London Math. Soc. **42** (1990), 562–576.

[Ma] T. Matsuoka, *Braids of periodic points and a two-dimensional analogue of Šarkovskiĭ's ordering*, Dynamical systems and nonlinear oscillations, Kyoto, 1985.

[M] M. Misiurewicz, *Formalism for studying periodic orbits of one dimensional maps*, European Conference on Iteration Theory (ECIT 87),, World Scientific, Singapore, 1989, pp. 1–7.

[S] E. Seneta, *Non-negative matrices: an introduction to theory and applications*, Second ed., Springer, Berlin, Heidelberg, New York, 1981.

[T] W. Thurston, *On the geometry and dynamics of diffeomorphisms of surfaces*, Bull. Amer. Math. Soc. **19** (1988), 417–431.

Department of Mathematics, Northwestern University, Evanston, IL 60208

E-mail address: john@math.nwu.edu

Department of Mathematics, Northwestern University, Evanston, IL 60208 and Department of Mathematics, Princeton University, Princeton, NJ 08544

Current address: Department of Mathematical Sciences, Indiana University – Purdue University at Indianapolis, Indianapolis, IN 46202

E-mail address: mmisiure@indyvax.iupui.edu

Contemporary Mathematics
Volume **152**, 1993

Lefschetz Trace Formulae, Zeta Functions and Torsion in Dynamics

ROSS GEOGHEGAN AND ANDREW NICAS

ABSTRACT. We outline our one-parameter Lefschetz-Nielsen fixed point theory and its applications to flows on compact manifolds. These applications include a Lefschetz trace theorem for a finite portion of a flow which can detect such features of a flow as the tranverse Euler class and the Fuller index of closed orbits. New K_1 and torsion invariants of a suspension flow are described; this involves a theory of "non-commutative" zeta functions.

Introduction

The Lefschetz trace formula is one of the most useful theorems for proving the existence of a fixed point of a continuous transformation $f: M \to M$ of a closed manifold M. It equates a geometric invariant, the intersection number of the graph of f and the graph of the identity map, with an algebraic invariant, $L(f)$, the alternating sum of the traces of the action of f on homology. In particular, the non-vanishing of $L(f)$ implies that the fixed point set of f, $\mathrm{Fix}(f)$, is non-empty.

The classical Nielsen theory of a map $f: M \to M$ is a refinement of Lefschetz's fixed point theory which takes place in the universal cover, $\tilde{M}$. The invariant $L(f)$ is replaced by the *Reidemeister trace*, $R(f)$, which is the alternating sum of $\mathbb{Z}(\pi_1(M))$-traces on cellular chains, reduced modulo a semiconjugacy relation in $\pi_1(M)$, and the intersection number is replaced by a sum of terms of the form (integer) · (Nielsen fixed point class). The (Nielsen) number of non-zero "components" is a lower bound for the number of fixed points of f.

1991 *Mathematics Subject Classification*. Primary 19J10, 55M20, 58F20; Secondary 58F25.
The first author was supported in part by NSF Grant #DMS-9005508.
The second author was supported in part by NSERC Grant #OGP0038057.
The detailed version of this paper will be submitted for publication elsewhere.

One–parameter fixed point theory, which we introduced in [**GN**$_1$], does something analogous for homotopies $F : M \times [0,1] \to M$. In the simplest case, in which F_0 and F_1 are fixed point free and M has dimension greater than one, the generic fixed point set is a finite union of disjoint circles. The geometric intersection invariant is then an element of $H_1(M)$. The refinement of this geometric invariant which is analogous to classical Nielsen theory involves a sum of 1–dimensional homology classes in suitable covering spaces, one for each Nielsen class of circles. The corresponding algebraic invariants are traces, $L(F)$ and $R(F)$, analogous to $L(f)$ and $R(f)$. These are the new features of [**GN**$_1$]. They are described in §2, below. As in classical fixed point theory, the algebraic traces "detect" the circles of fixed points.

It is natural to apply this theory to a non-singular flow $\Phi : M \times \mathbb{R} \to M$ or, perhaps, to the restriction of Φ to some compact portion, $M \times [a,b]$ of the flow. The (non-transverse) circles of fixed points are then closed orbits, and we find that our new traces detect features of these closed orbits which were previously familiar in dynamics: specifically the Fuller index and the transverse Euler class. This is described in §3 and §4.

But the algebraic setting for our traces turns out to be useful for a much more far reaching study of Φ, where we consider the restriction of Φ to the *non-compact* $M \times (0,\infty)$. Here, the classical motivation is the dynamical zeta functions of [**M**] and [**F**], and Fried's torsion invariants [**F**]. Our algebraic setting makes it possible to construct new K_1 and torsion invariants of a (suspension) flow, which have the advantage of being "non-commutative" in the sense of being associated with the universal cover, $\tilde{M}$, rather than "commutative" as in [**F**], i.e. associated with the universal *abelian* cover. From our invariants it is possible to recover the previous zeta functions and torsions.

In the first half of this paper we outline our one-parameter fixed point theory, and in the second half we apply it to dynamics. Details and proofs can be found in [**GN**$_1$] and [**GN**$_2$].

§1. Background Algebra

Let R be a ring and $\mathcal{M}$ an R–R bimodule (i.e. a left and right R–module satisfying $(r_1 m)r_2 = r_1(mr_2)$ for all $m \in \mathcal{M}$, and $r_1, r_2 \in R$), The *Hochschild chain complex* $\{C_*(R,\mathcal{M}), d\}$ consists of $C_n(R,\mathcal{M}) = R^{\otimes n} \otimes \mathcal{M}$ where $R^{\otimes n}$ is the tensor product of n copies of R and the boundary operator is given by:

$$\begin{aligned} d(r_1 \otimes \ldots \otimes r_n \otimes m) &= r_2 \otimes \ldots \otimes r_n \otimes m r_1 \\ &+ \sum_{i=1}^{n-1} (-1)^i r_1 \otimes \ldots \otimes r_i r_{i+1} \otimes \ldots \otimes r_n \otimes m \\ &+ (-1)^n r_1 \otimes \ldots \otimes r_{n-1} \otimes r_n m. \end{aligned}$$

The tensor products are taken over the integers. The n–th homology of this complex is the n–th *Hochschild homology of R with coefficient bimodule $\mathcal{M}$*. It is denoted by $HH_n(R, \mathcal{M})$. If $\mathcal{M} = R$ with the standard R–R bimodule structure then we write $HH_n(R)$ for $HH_n(R, \mathcal{M})$. A useful reference for this material is [**I**].

We will be concerned mainly with HH_1 and HH_0 which are computed from

$$\begin{array}{cccccc} \ldots \longrightarrow R \otimes R \otimes \mathcal{M} & \xrightarrow{d} & R \otimes \mathcal{M} & \xrightarrow{d} & \mathcal{M} \\ r_1 \otimes r_2 \otimes m & \mapsto & r_2 \otimes mr_1 - r_1 r_2 \otimes m + r_1 \otimes r_2 m & & \\ & & r \otimes m & \mapsto & mr - rm \end{array}$$

If R is a ring and $\phi : R \to R$ is a ring endomorphism, we define R^ϕ to be the R–R bimodule whose underlying abelian group is R and whose bimodule structure is given by: $r \cdot m = rm$ and $m \cdot r = m\phi(r)$ for $r \in R$ and $m \in R^\phi$. As applied below, $R = \mathbb{Z}G$, the integral group ring of a group G and $\phi : \mathbb{Z}G \to \mathbb{Z}G$ is induced from a group homomorphism $\phi : G \to G$.

Elements g_1 and g_2 of a group G are *semiconjugate* if and only if there exists $g \in G$ such that $g_1 = g g_2 \phi(g^{-1})$. We write $C(g)$ for the semiconjugacy class containing g and G_ϕ for the set of semiconjugacy classes. The partition of G into the union of its semiconjugacy classes induces a direct sum decomposition of $HH_*(\mathbb{Z}G, (\mathbb{Z}G)^\phi)$ as follows: each generating chain $c = g_1 \otimes \cdots \otimes g_n \otimes m$ can be written in *canonical form* as $g_1 \otimes \cdots \otimes g_n \otimes g_n^{-1} \cdots g_1^{-1} g$ where we think of $g \equiv g_1 \cdots g_n m \in G$ as "marking" a semiconjugacy class. All the generating chains occurring in the boundary $d(c)$ are easily seen to have markers in $C(g)$ when put into canonical form. For $C \in G_\phi$ let $C_*(\mathbb{Z}G, (\mathbb{Z}G)^\phi)_C$ be the subgroup of $C_*(\mathbb{Z}G, (\mathbb{Z}G)^\phi)$ generated by those generating chains whose markers lie in C. The decomposition $(\mathbb{Z}G)^\phi \cong \bigoplus_{C \in G_\phi} \mathbb{Z}C$ as a direct sum of abelian groups determines a decomposition of chain complexes $C_*(\mathbb{Z}G, (\mathbb{Z}G)^\phi) \cong \bigoplus_{C \in G_\phi} C_*(\mathbb{Z}G, (\mathbb{Z}G)^\phi)_C$ yielding a natural isomorphism $HH_*(\mathbb{Z}G, (\mathbb{Z}G)^\phi) \cong \bigoplus_{C \in G_\phi} HH_*(\mathbb{Z}G, (\mathbb{Z}G)^\phi)_C$ where the summand $HH_*(\mathbb{Z}G, (\mathbb{Z}G)^\phi)_C$ corresponds to the homology classes of Hochschild cycles marked by the elements of C. We call this summand the *C-component*. Given any $\mathbb{Z}G$–$\mathbb{Z}G$ bimodule N let $\overline{N}$ be the left $\mathbb{Z}G$ module whose underlying abelian group is N and whose left module structure is given by $gm = g \cdot m \cdot g^{-1}$. There is a natural isomorphism $HH_*(\mathbb{Z}G, N) \cong H_*(G, \overline{N})$ which is induced from an isomorphism of the Hochschild complex to the bar complex for computing group homology, see [**I, theorem 1.d**]. The decomposition $\overline{(\mathbb{Z}G)^\phi} \cong \bigoplus_{C \in G_\phi} \mathbb{Z}C$ is a direct sum of left $\mathbb{Z}G$ modules, inducing a direct sum decomposition $H_*(G, \overline{(\mathbb{Z}G)^\phi}) \cong \bigoplus_{C \in G_\phi} H_*(G, \mathbb{Z}C)$. Choosing representatives $g_C \in C$ we have an isomorphism of left $\mathbb{Z}G$ modules $\mathbb{Z}C \cong \mathbb{Z}(G/Z(g_C))$ where $Z(h) = \{g \in G \mid h = gh\phi(g^{-1})\}$ denotes the *semicentralizer* of $h \in G$. Since $H_*(G, \mathbb{Z}(G/Z(g_C)))$ is naturally isomorphic to $H_*(Z(g_C))$, we obtain a natural isomorphism $HH_*(\mathbb{Z}G, (\mathbb{Z}G)^\phi) \cong \bigoplus_{C \in G_\phi} H_*(Z(g_C))$; furthermore, $HH_*(\mathbb{Z}G, (\mathbb{Z}G)^\phi)_C$ corresponds to the summand $H_*(Z(g_C))$ under this identification. In particular $HH_0(\mathbb{Z}G, (\mathbb{Z}G)^\phi) \cong \mathbb{Z}G_\phi$, the free abelian group generated by the semiconjugacy classes, and $HH_1(\mathbb{Z}G,(\mathbb{Z}G)^\phi) \cong \bigoplus_{C\in G_\phi} H_1(Z(g_C))$, the direct sum of the abelianizations of the semicentralizers.

Next, we review traces in Hochschild homology. Let R be a ring and $\mathcal{M}$ a R–R bimodule. If B is a square matrix over $\mathcal{M}$, its *trace*, denoted by trace(B), is the element $\sum_i B_{ii} \in \mathcal{M}$. This element can be viewed as a Hochschild 0–cycle and thus defines an element of $HH_0(R,\mathcal{M})$ which we will denote by $T_0(B)$.

If A is a $n \times m$ matrix over R and B is a $m \times n$ matrix over $\mathcal{M}$, define $A \otimes B$ to be the $n \times n$ matrix with entries in the abelian group $R \otimes \mathcal{M}$ given by $(A \otimes B)_{ij} = \sum_k A_{ik} \otimes B_{kj}$. The *trace* of $A \otimes B$ is then $\sum_{jk} A_{jk} \otimes B_{kj} \in R \otimes \mathcal{M}$. We interpret this trace as a Hochschild 1–chain. Clearly, it is a cycle if and only if trace$(AB) =$ trace(BA). in which case we denote its homology class by $T_1(A \otimes B) \in HH_1(R,\mathcal{M})$.

We recall the definition of K_1 of a ring. Let $GL(n,R)$ denote the general linear group consisting of all $n \times n$ invertible matrices over R, and let $GL(R)$ be the direct limit of the sequence $GL(1,R) \subset GL(2,R) \subset \cdots$. A matrix in $GL(R)$ is called *elementary* if coincides with the identity except for a single off–diagonal entry. The subgroup $E(R) \subset GL(R)$ generated by the elementary matrices is precisely the commutator subgroup of $GL(R)$ and the abelian quotient group $GL(R)/E(R)$ is, by definition, $K_1(R)$. The *Dennis trace* homomorphism DT: $K_1(R) \longrightarrow HH_1(R)$ is given as follows: let $\alpha \in K_1(R)$ be represented by an invertible $n \times n$ matrix A, then $\mathrm{DT}(\alpha) = T_1(A \otimes A^{-1})$.

Returning to the situation of a group G with an endomorphism ϕ, let W be a given subset of G_ϕ. Define

$$HH_*(\mathbb{Z}G,(\mathbb{Z}G)^\phi; W) \equiv \bigoplus_{C\in G_\phi - W} HH_*(\mathbb{Z}G,(\mathbb{Z}G)^\phi)_C$$

regarded as a subgroup of $HH_*(\mathbb{Z}G,(\mathbb{Z}G)^\phi)$. It may happen in topological applications that trace$(A \otimes B) \in C_1(\mathbb{Z}G,(\mathbb{Z}G)^\phi)$ is not a cycle, but that for an appropriate geometrically defined $W \subset G_\phi$ its C–component $[\mathrm{trace}(A \otimes B)]_C \in C_1(\mathbb{Z}G,(\mathbb{Z}G)^\phi)_C$ is a cycle for all $C \notin W$. Then we write $T_1(A \otimes B; W)$ for the element of $HH_1(\mathbb{Z}G,(\mathbb{Z}G)^\phi; W) \subset HH_1(\mathbb{Z}G,(\mathbb{Z}G)^\phi)$ whose C–component is represented by $[\mathrm{trace}(A \otimes B)]_C$ for each $C \in G_\phi - W$.

§2. One-parameter fixed point theory

Let X be a finite connected oriented CW complex and let $I = [a,b]$ with the usual oriented CW structure. Let $F : X \times I \to X$ be a cellular homotopy, where $X \times I$ has the product CW structure and its cells are given the product orientation. We introduce algebraic invariants which will detect $\mathrm{Fix}(F) \equiv \{(x,t) \in X \times [a,b] \mid F(x,t) = x\}$.

We pick a vertex $(v,a) \in X \times I$ as our basepoint and choose a basepath τ from v to $F(v,a)$. *We identify $\pi_1(X \times I,(v,a))$ with $\pi_1(X,v) \equiv G$ via the isomorphism induced by the projection $p : X \times I \to X$.* In particular, we write $\phi : G \to G$ for the homomorphism

$$\pi_1(X \times I, (v, a)) \xrightarrow{F_\#} \pi_1(X, F(v, a)) \xrightarrow{c_{[\tau^{-1}]}} \pi_1(X, v).$$

Let $\tilde{\tau}$ be the lift of the basepath τ to the universal cover which starts at the basepoint, $\tilde{v} \in \tilde{X}$, and let $\tilde{F}$ be the unique lift of F mapping $(\tilde{v}, a)$ to $\tilde{\tau}(1)$. Choose an oriented lift, $\tilde{e}$, for each cell e. Regard $C_k(\tilde{X})$ as a right $\mathbb{Z}G$ module. $\tilde{F}$ induces a chain homotopy $\tilde{D}_k : C_k(\tilde{X}) \to C_{k+1}(\tilde{X})$. More precisely, $\tilde{D}_k(\tilde{e}) = (-1)^{k+1}\tilde{F}_k(\tilde{e} \times I)$, and $\tilde{D}_k(\tilde{e}g) = \tilde{D}_k(\tilde{e})\phi(g)$. The boundary $\tilde{\partial}_k : C_k(\tilde{X}) \to C_{k-1}(\tilde{X})$ satisfies $\tilde{\partial}_k(\tilde{e}g) = \tilde{\partial}_k(\tilde{e})g$. Define endomorphisms of $\oplus_k C_k(\tilde{X})$ by $\tilde{D}_* = \oplus_k(-1)^{k+1}\tilde{D}_k$, $\tilde{\partial}_* = \oplus_k \tilde{\partial}_k$, $\tilde{F}_{a\ *} = \oplus_k(-1)^k \tilde{F}_{a\ k}$, and $\tilde{F}_{b\ *} = \oplus_k(-1)^k \tilde{F}_{b\ k}$. We reuse the symbols $\tilde{D}_*$, $\tilde{\partial}_*$, $\tilde{F}_{a\ *}$, and $\tilde{F}_{b\ *}$ for the matrices of the corresponding endomorphisms with respect to the finite $\mathbb{Z}G$–bases provided by $\{\tilde{e}\}$. The chain homotopy relation yields the $\mathbb{Z}G$–matrix equation:

$$\tilde{D}_*\phi(\tilde{\partial}_*) - \tilde{\partial}_*\tilde{D}_* = \tilde{F}_{a\ *} - \tilde{F}_{b\ *}.$$

The minus sign appearing on the left arises from our convention concerning the alternation of signs. Note that the entry of the matrix $\tilde{D}$ corresponding to an n–cell $\tilde{e}_1$ and an $(n+1)$–cell $\tilde{e}_2$ is the coefficient of $\tilde{e}_2$ in the $(n+1)$–chain $\tilde{F}_*(\tilde{e}_1 \times I)$.

Let $\omega = F(v \times I)$. Then $\tau\omega$ is a path from v to $F_1(v)$. Note that it is this path which must be used to determine the lift $\tilde{F}_b$ of F_b, and so even if $F_a = F_b$ it is possible that $\tilde{F}_a \neq \tilde{F}_b$.

We want to define the *one–parameter trace* of F, denoted by $R(F)$, lying in

$$HH_1(\mathbb{Z}G, (\mathbb{Z}G)^\phi; G_\phi(\partial F)) \equiv \bigoplus_{C \in G_\phi - G_\phi(\partial F)} HH_1(\mathbb{Z}G, (\mathbb{Z}G)^\phi)_C$$
$$\cong \bigoplus_{C \in G_\phi - G_\phi(\partial F)} H_1(Z(g_C)).$$

In the language of §1, consider trace$(\tilde{\partial}_* \otimes \tilde{D}_*) \in \mathbb{Z}G \otimes (\mathbb{Z}G)^\phi$. This is a Hochschild 1–chain whose boundary is

$$\text{trace}(\tilde{D}_*\phi(\tilde{\partial}_*) - \tilde{\partial}_*\tilde{D}_*) = \text{trace}(\tilde{F}_{a\ *} - \tilde{F}_{b\ *}).$$

The latter might not be zero, so trace$(\tilde{\partial}_* \otimes \tilde{D}_*)$ might not be a cycle; but in the important special case in which F_a and F_b have no fixed points then it can be shown that it is indeed a cycle. In this special case:

$$R(F) = T_1(\tilde{\partial}_* \otimes \tilde{D}_*) \in HH_1(\mathbb{Z}G, (\mathbb{Z}G)^\phi).$$

For the general case (where F_a or F_b has a fixed point) let $G_\phi(\partial F)$ be the subset $\{C_1, \ldots, C_k\}$ of G_ϕ consisting of Nielsen semiconjugacy classes associated to fixed points of F_a or F_b. Then we define

$$R(F) = T_1(\tilde{\partial}_* \otimes \tilde{D}_*; G_\phi(\partial F)) \in HH_1(\mathbb{Z}G, (\mathbb{Z}G)^\phi; G_\phi(\partial F)) \subset HH_1(\mathbb{Z}G, (\mathbb{Z}G)^\phi).$$

[One should think of $R(F)$ as being analogous to the classical Reidemeister trace of a map $f : X \to X$, $R(f) \equiv T_0(\tilde{f}^*) \in HH_0(\mathbb{Z}G, (\mathbb{Z}G)^\phi) \cong \mathbb{Z}G_\phi$.]

The C–component of $R(F)$, denoted by $\iota(F, C) \in HH_1(\mathbb{Z}G, (\mathbb{Z}G)^\phi)_C \cong H_1(Z(g_C))$, is the *fixed point index* of F corresponding to $C \in G_\phi$; it can be viewed as an element of $H_1(\tilde{X}/Z(g_C))$. The number of nonzero fixed point indices is the *one–parameter Nielsen number*, $N(F)$, of F.

The augmentation $\epsilon : \mathbb{Z}G \to \mathbb{Z}$ can be viewed as a morphism $\epsilon : (\mathbb{Z}G)^\phi \to \mathbb{Z}$ of $\mathbb{Z}G$–$\mathbb{Z}G$ bimodules where $\mathbb{Z}$ is given the trivial bimodule structure, or as a morphism $\epsilon : \overline{(\mathbb{Z}G)^\phi} \to \overline{\mathbb{Z}}$ of left $\mathbb{Z}G$ modules where $\overline{\mathbb{Z}}$ is the trivial module. There is a commutative diagram:

$$\begin{array}{ccc} HH_*(\mathbb{Z}G, (\mathbb{Z}G)^\phi) & \xrightarrow{\epsilon} & HH_*(\mathbb{Z}G, \mathbb{Z}) \\ \cong \downarrow & & \cong \downarrow \\ H_*(G, \overline{(\mathbb{Z}G)^\phi}) & \xrightarrow{\epsilon} & H_*(G, \overline{\mathbb{Z}}) \end{array}$$

giving a natural homomorphism $HH_*(\mathbb{Z}G, (\mathbb{Z}G)^\phi) \to H_*(G)$. The *one–parameter Lefschetz class*, $L(F)$, is defined to be the image of the Hochschild homology class $R(F)$ in $H_1(G)$ under this homomorphism. Note that

$$L(F) = \sum_{C \in G_\phi - G_\phi(\partial F)} j_C(\iota(F, C))$$

where $j_C : H_1(Z(g_C)) \to H_1(G)$ is induced by the inclusion $Z(g_C) \subset G$. This $L(F)$ is analogous to the Lefschetz number, $L(f) \in H_0(G) \cong \mathbb{Z}$, of a map $f: X \to X$.

REMARK. In the above formulation, $L(F)$ arose from the trivial representation of G in the sense that the augmentation is the extension of the trivial representation to $\mathbb{Z}G \to \mathbb{Z}$. More generally, let S be a ring (typically, S would be the algebra of $n \times n$ complex matrices) and let $\rho : G \to S^*$ be a representation of G in the group of units of S such that $\rho\phi = \rho$. One can define the *ρ–twisted one–parameter Lefschetz class*, $L(F, \rho) \in HH_1(\mathbb{Z}G, S) \cong H_1(G, \bar{S})$, where S has the $\mathbb{Z}G$–$\mathbb{Z}G$ bimodule structure determined by ρ, to be the image of $R(F)$ under the induced homomorphism $\rho_* : HH_*(\mathbb{Z}G, (\mathbb{Z}G)^\phi) \to HH_*(\mathbb{Z}G, S)$.

Two fixed points (x, t) and (y, t') of F are *in the same fixed point class* if and only if for some path μ from (x, t) to (y, t'), the loop $(p \circ \mu)(F \circ \mu)^{-1}$ is homotopically trivial. This defines an equivalence relation on $\mathrm{Fix}(F)$. Just as in the classical case, there is an injective function Ψ from the set of fixed point classes of F to the set, G_ϕ, of semiconjugacy classes: the class containing (x, t) is mapped to the semiconjugacy class, C, containing $g_C \equiv [(p \circ \mu)(F \circ \mu)^{-1}\tau^{-1}]$, where μ is any path from the basepoint (v, a) to (x, t). It is easy to check that Ψ is well–defined, that F has only finitely many fixed point classes, and that fixed points in the same path component of $\mathrm{Fix}(F)$ are in the same fixed point class. Indeed, with the same notation, let ω be a loop in $\mathrm{Fix}\, F \subset X \times [a, b]$ based at (x, t), and let $h = [\mu\omega\mu^{-1}] \in G$.

PROPOSITION 2.1. *The element h lies in the semicentralizer $Z(g_C)$.*

The principal general theorems in [$\mathbf{GN}_1$] are:

THEOREM 2.2 (ONE–PARAMETER LEFSCHETZ FIXED POINT THEOREM). *Suppose $L(F) \neq 0$. Then every map homotopic to F rel $X \times \{a,b\}$ has a fixed point not in the same fixed point class as any fixed point in $X \times \{a,b\}$. In particular, if F_a and F_b are fixed point free, every map homotopic to F rel $X \times \{a,b\}$ has a fixed point.*

THEOREM 2.3 (ONE–PARAMETER NIELSEN–WECKEN FIXED POINT THEOREM).
Every map homotopic to F rel $X \times \{a,b\}$ has at least $N(F)$ fixed point classes other than the fixed point classes which meet $X \times \{a,b\}$. In particular, if F_a and F_b are fixed point free, the fixed point set of every map homotopic to F rel $X \times \{a,b\}$ has at least $N(F)$ path components.

THEOREM 2.4 (INVARIANCE).
(a) Let $F, H : X \times I \to X$ be cellular; if $F \simeq H$ rel $X \times \{a,b\}$ then $R(F) = R(H)$.
(b) If $h : X \to X'$ is a cellular simple homotopy equivalence, and if $F : X \times I \to X$ and $F' : X' \times I \to X'$ are cellular homotopies such that $hF \simeq F'(h \times 1)$ rel $X \times \{a,b\}$, then $h_{\dagger}R(F) = R(F')$.

When X is a compact polyhedron, it is possible to extend the definition of $R(F)$ and $L(F)$ to arbitrary continuous homotopies $F : X \times I \to X$ by means of simplicial approximation as follows. Let K be a triangulation of X and let J be the standard triangulation of I (with two vertices and one 1–simplex). Consider a simplicial approximation $E \colon Q \to K$ to F where Q is subdivision of $K \times J$ ($K \times J$ denotes a product triangulation obtained without adding more vertices). Since $|E| \colon |K| \times I \to |K|$ is cellular (for the cell structure determined by K), we may form $R(|E|)$ and $L(|E|)$.

THEOREM 2.5. *There exists a sufficiently fine triangulation, K, of X so that $R(F)$ and $L(F)$ are well-defined by $R(|E|)$ and $L(|E|)$ respectively, i.e. if K' is any subdivision of K and $E' : Q' \to K'$ is a simplicial approximation to F then $R(|E|) = R(|E'|)$ and $L(|E|) = L(|E'|)$. Furthermore, if $F' : X \times I \to X$ is a continuous homotopy such that $F \simeq F'$ rel $X \times \{a,b\}$ then $R(F) = R(F')$ and $L(F) = L(F')$.*

We now turn to the manifold case. Let M be a PL (respectively smooth) n–manifold. Every map $M \times I \to M$ is homotopic to a PL (respectively smooth) map F whose graph is transverse to the graph of the projection, and is such that $\mathrm{Fix}(F) \cap M \times \{t\}$ is finite for each $t \in I$. It can be arranged that the graph of F then consists of circles and arcs which, for all but finitely many values of t, cross $M \times \{t\}$ transversely, and which miss $(\partial M) \times I$ entirely. If (x,t) lies in such a circle (or arc) crossing $M \times \{t\}$ transversely, orient the circle (or arc) in the direction of positive time if $\iota(F_t, x)$ is positive, and in the other direction if it is negative.

We are only interested in the circles, indeed in circles not in the same fixed point class as any arc. Let V be such a circle, let $(x,t) \in V$, let μ be a path in $M \times [a,b]$ from (v,a) to (x,t), and let ω be the loop based at (x,t) obtained by traversing V once in the direction of its orientation. With notation as above, Proposition 2.1 implies that the corresponding element, h, of G lies in $Z(g_C)$. In this way, we associate with V an element of $H_1(Z(g_C)) \cong (Z(g_C))_{\mathrm{ab}}$. If there are two circles V_1 and V_2 in the same fixed point class, one reaches the same semicentralizer $Z(g_C)$ from both circles provided the path used for $(x_1,t_1) \in V_1$ is μ, and the path used for $(x_2,t_2) \in V_2$ is $\mu\nu$, where $p(\nu)F(\nu)^{-1}$ is homotopically trivial. One treats any (finite) number of circles similarly. Thus, for each fixed point class, B, of F which does not meet $M \times \{a,b\}$, we have defined an element $\iota(F,B) \in H_1(Z(g_C))$, where $g_C \equiv [p(\mu)F(\mu)^{-1}\tau^{-1}]$ represents the semiconjugacy class C, and $\Psi(B) = C$. In fact, one can always reduce to the case in which only one circle occurs in each fixed point class; see [**Di**$_1$] or [**Di**$_2$].

Let $B_1, \dots, B_k$ be the fixed point classes of F which do not meet $M \times \{a,b\}$. Then, for $j = 1, \dots, k$ we have $\iota(F,B_j) \in H_1(Z(g_{C_j}))$ where $C_j = \Psi(B_j)$.

DEFINITION 2.6. The *transverse intersection invariant* of F is:

$$\Theta(F) = \sum_{j=1}^{k} \iota(F,B_j) \in \bigoplus_{C \in G_\phi} H_1(Z(g_C)) \cong HH_1(\mathbb{Z}G, (\mathbb{Z}G)^\phi).$$

The main geometric theorem of [**GN**$_1$] can be paraphrased as:

THEOREM 2.7. *Let M be an oriented smooth or PL compact n-manifold. Suppose the smooth or PL map $F : M \times I \to M$ has the above transversality properties. Then $\Theta(F) = -R(F)$. Also, $\sum_{j=1}^{k} j_C(\iota(F,B_j)) = -L(F)$, where j_C is induced by the inclusion $Z(g_C) \subset G$.*

REMARK. Proofs in [**GN**$_1$] are given for the PL case, but they go through with the obvious changes in the smooth case.

The extent to which $\Theta(F)$ (or $R(F)$) is the total obstruction to removing the relevant circles of fixed points by a homotopy of F rel $M \times \{a,b\}$ is discussed in [**GN**$_1$], [**DG**], [**Di**$_1$] and [**Di**$_2$].

We close with some remarks on homological computations.

Recall that when f is cellular, the Lefschetz number, $L(f)$, can be computed from the induced homomorphism $f_* : C_*(X) \to C_*(X)$ on cellular chains: $L(f) = \sum_i (-1)^i \operatorname{trace}(f_i)$. There is a similar method for computing the one-parameter Lefschetz class $L(F)$, where $F : X \times I \to X$ is cellular, $I = [a,b]$. Let $A_1 : G \to H_1(G) \equiv G_{\mathrm{ab}}$ be the abelianization homomorphism; its natural extension to a ring homomorphism $\mathbb{Z}G \to \mathbb{Z}G_{\mathrm{ab}}$ will also be denoted by A_1. Let $A_2 : \mathbb{Z}G_{\mathrm{ab}} \to G_{\mathrm{ab}}$ be the natural homomorphism and let $A : ZG \to G_{\mathrm{ab}}$ be the composite $A = A_2A_1$. Let X be as above. Let $\bar{X}$ be the universal *abelian* cover of X (i.e. the covering space of X corresponding to $\ker A_1$). Let $\bar{\partial}_*$ be the boundary operator of the cellular chain complex $C_*(\bar{X})$ and let D_* be the chain homotopy determined by F.

PROPOSITION 2.8. *Suppose F has no fixed points in $X \times \{a,b\}$. Then*

$$L(F) = A(\operatorname{trace}(\tilde{\partial}_* D_*)) = A_2(\operatorname{trace}(\bar{\partial}_* D_*)) \in G_{\mathrm{ab}}.$$

REMARK. In the special case of a "cyclic" homotopy, i.e. a self-homotopy $F : f \simeq f$, $L(F)$ can be computed (over a field of coefficients) at the level of homology, see [**GN**$_3$]. This formula involves the cohomology algebra of X and the homomorphism $H_*(X \times S^1) \to H_*(X)$ determined by the cyclic homotopy F; it can be viewed as a one-parameter analog of the Hopf trace formula which asserts, in the classical case, that a chain level trace is equal to a homology trace.

§3. Detecting the Fuller homology classes via traces

Let M^n be a closed connected oriented C^∞-manifold. Let a C^∞ vector field $\mathcal{X}$ be given on M, and let $\Phi : M \times \mathbb{R} \to M$ be the flow obtained by integrating $\mathcal{X}$. The point $(x,t) \in M \times (0,\infty)$ is a *periodic point* if $\Phi(x,t) = x$. There are two kinds of periodic points:

(i) if $\Phi(\{x\} \times \mathbb{R}) = \{x\}$ then every point of $\{x\} \times (0,\infty)$ is called a *stationary point*;

(ii) if there is a least positive number q such that $\Phi(x,q) = x$, then the set $\{(\Phi(x,t),q) \mid 0 \le t \le q\}$ is called a *periodic orbit of multiplicity* 1. More generally, let $k = mq$, where m is a positive integer, and call the set $\{(\Phi(x,t),k) \mid 0 \le t \le k\}$ a *periodic orbit of period k and multiplicity m.*

Thus, a periodic point $(x,t) \in M \times (0,\infty)$ is either stationary or lies on a periodic orbit.

A *periodic set*, P, is a subset of $M \times (0,\infty)$ which is the union of periodic orbits. This set is *isolated* if it has a compact neighborhood in $M \times (0,\infty)$ (called an *isolating neighborhood*) containing no other periodic points.

In his fundamental paper [**Fu**], Fuller assigns an integral homology class $\Lambda(P) \in H_1(M \times (0,\infty)) \equiv H_1(M)$ to each compact periodic set P. When P is an isolated periodic orbit of period k write $P \equiv \{(\Phi(x,t),k) \mid 0 \le t \le k\}$. Then $\Lambda(P)$ is the homology class represented by the *index cycle*, i.e. the singular 1-cycle $\iota(f^m,x) \cdot \omega$. Here, $(x,k) \in P$; the *primitive loop*, ω, is the 1-cycle $[0,q] \to M$ given by $t \mapsto \Phi(x,t)$; an open $(n-1)$-disk, D, containing x and transverse to P is chosen by the "method of sections" together with a "first return map", f, taking a neighborhood of x in D back into D, where f has the form $f(y) = \Phi(y,t(y))$; and $\iota(f^m,x) \in \mathbb{Z}$ is the classical fixed point index of "the m-th return map", f^m, at its isolated fixed point $x \in D$. When P is the union of finitely many isolated periodic orbits P_i, $\Lambda(\cup P_i)$ is defined to be $\sum_i \Lambda(P_i)$.

When P is an arbitrary compact isolated periodic set for which V is an isolating neighborhood, one applies [**Fu, Lemma 3.1**] to perturb the vector field $\mathcal{X}$ to a new vector field $\mathcal{X}'$ whose corresponding periodic set P' (for which V is again isolating) is the union of finitely many isolated periodic orbits; and then one defines $\Lambda(P)$ to be $\Lambda(P')$. Fuller proves that $\Lambda(P)$ is well-defined and is invariant under appropriate deformation of the vector field $\mathcal{X}$ ("continuation"); see [**Fu, Theorem 1**].

This $\Lambda(P)$ is the *Fuller homology class* of P. If P is a single orbit of multiplicity m, we may rewrite our representative cycle as $(\iota(f^m, x)/m) \cdot m\omega$. The rational number $\iota(f^m, x)/m$ is called the *Fuller index of* P. There is a consistent homotopically invariant way of extending this $\mathbb{Q}$–valued Fuller index to arbitrary compact isolated periodic sets so that it is additive on finite unions of periodic orbits; see [**Fu, Theorem 2**].

We can use the one–parameter Lefschetz class of §2 to detect periodic orbits as follows. Assume Φ has no stationary points. Let $0 < \epsilon < U < \infty$ be such that Φ has no periodic orbit of period ϵ or U. Let P be the union of all the periodic orbits whose periods lie in (ϵ, U).

PROPOSITION 3.1. *Let $F : M \times [\epsilon, U] \to M$ be the restriction of Φ to $M \times [\epsilon, U]$. Then $\Lambda(P) = -L(F)$.*

The partition of the fixed points of Φ into fixed point classes was essentially explained in §2, the only difference here being that the domain, $M \times \mathbb{R}$, is not compact, so that there might be infinitely many fixed point classes. However, the restriction $F = \Phi|_{M\times[\epsilon,U]}$ has only finitely many fixed point classes. We wish to describe $R(F)$.

Pick a basepoint $x_0 \in M$. Since $\Phi_0 = \text{id}$, we may assume that F induces the identity on G (see §2), so that the semiconjugacy classes of §2 are true conjugacy classes.

Let P be an isolated periodic orbit of period k and multiplicity m, where $\epsilon < k < U$. Let $x \in M$ be such that $(\Phi(x,t), k) \in P$ for some (hence all) t in $[0, k]$. The primitive loop is $\omega : [0, q] \to M$. Pick a path μ in M from x_0 to x. Let $C \subset G$ be the conjugacy class containing $g_C \equiv [\mu\omega^{-m}\mu^{-1}]$. By Proposition 2.1, $h \equiv [\mu\omega\mu^{-1}]$ determines an element of $H_1(Z(g_C))$ which we denote by $\{\omega\}$. We then associate with the orbit P the element $\iota(f^m, x)\{\omega\} \in H_1(Z(g_C)) \cong HH_1(\mathbb{Z}G)_C \subset HH_1(\mathbb{Z}G)$, where, as above, f is a first return map at $x \in P$.

Suppose the periodic set in $M \times (\epsilon, U)$ consists of finitely many periodic orbits P_i (of multiplicity m_i, period $m_i q_i$, and defining primitive loop $\omega_i : [0, q_i] \to M$). Let $C_i = \Psi(B(P_i))$ where $B(P_i)$ is the fixed point class of P_i (see §2 for the definition of Ψ). Then $\{\omega_i\} \in H_1(Z(g_{C_i}))$ and we may form what we call the *Nielsen–Fuller invariant:*

$$\Theta'(F) = \sum_i \iota(f_i^{m_i}, x_i)\{\omega_i\} \in \bigoplus_{C \in G_\phi} H_1(Z(g_C)) \cong HH_1(\mathbb{Z}G),$$

where f_i is a first return map at $x_i \in P_i$.

The similarity between this (non-transverse) invariant $\Theta'(F)$ and the transverse invariant $\Theta(F) = -R(F)$ (see 2.7) is intentional, for we have:

THEOREM 3.2. $\Theta'(F) = -R(F)$.

§4. The transverse Euler class

The classical Poincaré–Hopf index theorem asserts that if $\mathcal{X}$ is a smooth vector field on the compact oriented manifold M with only finitely many zeros then the global sum of the indices of $\mathcal{X}$ equals the Euler characteristic of M. The global sum of the indices of $\mathcal{X}$ can be viewed as a geometric definition of the Poincaré dual of the Euler class of the tangent bundle of M; on the other hand the Euler characteristic can be defined as an alternating sum of traces on homology (or cellular chains). We now proceed to formulate a "one–parameter" analogue of the Poincaré–Hopf index theorem.

Let $\mathcal{X}$ be a non-singular vector field on the closed connected oriented n–manifold M and let Φ be the flow on M determined $\mathcal{X}$. Let $\epsilon > 0$ be such that the restriction $F : M \times [-\epsilon, \epsilon] \to M$ of Φ contains no fixed points other than the points of $M \times \{0\}$. (Of course, all points $(x, 0)$ are fixed.)

The vector field defines an oriented trivial line subbundle, λ, of $\tau \equiv \tau(M)$, the tangent bundle of M. Let η be any $(n-1)$–dimensional subbundle of τ complementary to λ, e.g. if M is given a Riemannian metric, take η to be the normal bundle of the flow Φ. The Euler class of the vector bundle η, denoted by $e(\eta)$, lies in $H^{n-1}(M)$; see [**MS**].

The one–parameter Lefschetz class gives us a trace formula for $e(\eta)$:

THEOREM 4.1 (ONE–PARAMETER POINCARÉ–HOPF THEOREM). *$-L(F)$ is the Poincaré dual of $e(\eta)$.*

§5. Suspension flows

Consider the natural semiflow $\Phi\colon Y \times [0, \infty) \to Y$ on the mapping torus $Y = T(X, f)$ of a cellular map $f : X \to X$ which induces an isomorphism, θ, on fundamental groups. Changing notation, from now on let us write $H = \pi_1(X, v)$ and $G = \pi_1(Y, [v, 0])$. When we fix an integer ℓ and write $\Gamma = \Phi|_{Y\times[0,\ell+1]}$ we get an interesting result:

THEOREM 5.1. *$R(\Gamma) \in HH_1(\mathbb{Z}G)$ is represented by the Hochschild 1–cycle*

$$\sum_{n\geq 0}(-1)^{n+1}\,\mathrm{trace}\,\big((I - t[\tilde{f}_n]) \otimes \sum_{i=0}^{\ell-1}(t[\tilde{f}_n])^i\big).$$

If, in Theorem 5.1, we let $\ell \to \infty$, then the limit of the matrix $\sum_{i=0}^{\ell-1}(t[\tilde{f}_n])^i$ can be viewed as the inverse of $I - t[\tilde{f}_n]$ in the appropriate algebraic context, and then the whole formula becomes the Dennis trace of the element of K–theory represented by the finite alternating product of the invertible matrices $I - t[\tilde{f}_n]$, $n \geq 0$. This observation motivates much of what follows.

§6. K–theory and zeta functions

Consider a continuous self-map $f : X \to X$ of a finite complex. The "Lefschetz zeta function" of f is the formal power series $\zeta_f(t) = \exp\big(\sum_{n=1}^{\infty} \frac{1}{n} L(f^n)t^n\big) \in$

$\mathbb{Q}[[t]]$ where $L(f^n)$ is the Lefschetz number of the n-th iterate of f. Since we wish to take account of the full influence of the (typically non-commutative and infinite) fundamental group of X, we seek a "non-commutative" substitute for $\zeta_f(t)$. This requires some new algebra.

Let S be an associative ring with unit and $\theta : S \to S$ a ring homomorphism. We do not assume S is commutative; in typical applications S will be the integral group ring of a group H, and θ will be induced by an automorphism of H. Let (C, ∂) be a finitely generated chain complex of right S-modules such that each C_i is free with a given basis. Suppose $f_* : C \to C$ is a θ-*homomorphism*, i.e. a degree 0 homomorphism of the underlying graded abelian groups such that $f_i(mr) = f_i(m)\theta(r)$ for $m \in C_i$ and $r \in S$. The *Reidemeister trace* of f_*, which we denote by $R(f_*)$, is the element of $HH_0(S, S^\theta)$, the 0-th Hochschild homology of S with coefficients in the bimodule S^θ (see §1), represented by the alternating sum of traces: $\sum_j (-1)^j \operatorname{trace}[f_j]$ where $[f_j]$ is the matrix of f_j with respect to the given basis. For $n \geq 1$ the n-th iterate of f_*, denoted f_*^n, is a θ^n-homomorphism. Note that $R(f_*^n) \in HH_0(S, S^{\theta^n})$. Let $\bar{R}(f_*^n)$ denote the image of $R(f_*^n)$ in $HH_0(S, S^{\theta^n})_\theta \equiv \operatorname{coker}(\mathrm{id} - \theta_\#)$, the quotient group of co-invariants of the natural action induced by θ on $HH_0(S, S^{\theta^n})$. Geometric motivation for the passage from $R(f_*^n)$ to $\bar{R}(f_*^n)$ is given in §7. Also see [**J**, §**2**] for motivation.

DEFINITION. The *Lefschetz–Nielsen series* of a θ-homomorphism $f_* : C \to C$ is defined by $\mathrm{LN}(f_*) = \left(\bar{R}(f_*^n) \right)_{n=1}^\infty \in \prod_{n=1}^\infty HH_0(S, S^{\theta^n})_\theta$.

Define the θ-*twisted power series ring*, denoted by $S_\theta[[t]]$, as follows: elements of $S_\theta[[t]]$ are formal series $\sum_{i=0}^\infty u_i t^i$ where $u_i \in S$ and t is an indeterminate; and multiplication is defined by $(\sum_{i=0}^\infty u_i t^i)(\sum_{j=0}^\infty v_j t^j) = \sum_{k=0}^\infty (\sum_{i+j=k} u_i \theta^i(v_j)) t^k$.

Let I_{n_i} be an identity matrix of the same size as $[f_i]$. Since the matrix $I_{n_i} - t[f_i]$ is invertible over $S_\theta[[t]]$, we may regard $I_{n_i} - t[f_i]$ as an element of the infinite general linear group over $S_\theta[[t]]$ and so we can define $\Delta(f_*) \in K_1(S_\theta[[t]])$ as follows:

DEFINITION. $\Delta(f_*) \in K_1(S_\theta[[t]])$ is the element represented by the finite alternating product $\prod_{i \geq 0} (I_{n_i} - t[f_i])^{(-1)^{i+1}}$.

THEOREM 6.1 ("RATIONALITY" OF THE LEFSCHETZ–NIELSEN SERIES). *There is a natural homomorphism* $P_+ : HH_1(S_\theta[[t]]) \to \prod_{n=1}^\infty HH_0(S, S^{\theta^n})_\theta$ *such that* $\mathrm{LN}(f_*) = P_+ \,\mathrm{DT}(\Delta(f_*))$.

Thus $\Delta(f_*)$ (indeed its Dennis trace) determines the Lefschetz–Nielsen series of f_*. There is a "completed" version of the Hochschild homology of a twisted power series ring, denoted by $\widehat{HH}_*(S_\theta[[t]])$, and a natural homomorphism $HH_*(S_\theta[[t]]) \to \widehat{HH}_*(S_\theta[[t]])$. The homomorphism P_+ factors as $HH_1(S_\theta[[t]]) \to \widehat{HH}_1(S_\theta[[t]]) \xrightarrow{\hat{P}_+} \prod_{n=1}^\infty HH_0(S, S^{\theta^n})_\theta$. If the ring S is also a vector space over $\mathbb{Q}$ (so that division by a nonzero integer is possible) then we have the following sharpened form of the above theorem:

THEOREM 6.2 ("RATIONALITY", SECOND VERSION). *There is a natural right inverse,* Lg, *for* $\hat{P}_+$ *such that* $\mathrm{Lg}(\mathrm{LN}(f_*)) = \widehat{\mathrm{DT}}(\Delta(f_*))$ *where* $\widehat{\mathrm{DT}}$ *is the composite* $K_1(S_\theta[[t]]) \xrightarrow{\mathrm{DT}} HH_1(S_\theta[[t]]) \to \widehat{HH}_1(S_\theta[[t]])$.

The formula for Lg is reminiscent of a logarithm.

These theorems are called "rationality theorems" (akin to "rationality of the Lefschetz zeta function") as they show that the Lefschetz–Nielsen series can be computed from a finite alternating product of "characteristic polynomials" defining an element of $K_1(S_\theta[[t]])$. The Hochschild homology element $\mathrm{DT}(\Delta(f_*)) \in HH_1(S_\theta[[t]])$ should be thought of as a "non-commutative" substitute for the "Lefschetz zeta function" as we illustrate below.

Let R be a *commutative* ring and let $\rho : S \to M_m(R)$ be a ring homomorphism such that $\rho\theta = \rho$ where $M_m(R)$ is the R–algebra of $m \times m$ matrices over R. In typical applications S will be $\mathbb{Z}H$, θ will be induced by an automorphism of H and ρ will be induced by a representation $H \to GL_m(R)$. The ring homomorphism ρ induces a homomorphism $\rho_* : HH_*(S, S^{\theta^n}) \to HH_*(M_m(R)) \cong HH_*(R)$ where the Morita equivalence isomorphism $HH_*(M_m(R)) \cong HH_*(R)$ is given explicitly by the trace; furthermore, since $\rho\theta = \rho$, it follows that ρ_* factors through $HH_*(S, S^{\theta^n})_\theta$ yielding a homomorphism $HH_*(S, S^{\theta^n})_\theta \to HH_*(R)$ which will also be denoted by ρ_*.

Let $C' = C \otimes_\rho M_m(R)$. Then C' is a finitely generated complex of free right $M_m(R)$–modules; furthermore, the given basis for C_i determines a basis for C'_i. Since $f_* \otimes \mathrm{id} : C' \to C'$ is a homomorphism of graded right $M_m(R)$–modules, where $\mathrm{id} : M_m(R) \to M_m(R)$ is the identity, we may form its Reidemeister trace:

$$R(f_* \otimes \mathrm{id}) \in HH_0(M_m(R)) \cong HH_0(R) = R.$$

Explicitly, $R(f_* \otimes \mathrm{id}) = \sum_j (-1)^j \operatorname{trace} \rho([f_j]) \in R$ where $\rho([f_j])$ is viewed as an $mn_i \times mn_i$ matrix over R and n_i is the cardinality of the given basis of C_i. It is clear that $\rho_*(\bar{R}(f_*)) = R(f_* \otimes \mathrm{id})$. We write $L(f_*, \rho) \equiv R(f_* \otimes \mathrm{id})$. Following the terminology of [**F**], $L(f_*, \rho) \in R$ can be regarded as a "generalized Lefschetz number".

EXAMPLES 6.3. Let $f : X \to X$ be cellular inducing θ as before and let $S = \mathbb{Z}H$:

1. Let $C \equiv C_*(\tilde{X})$ be the cellular chain complex of $\tilde{X}$ (regarded as a right S–module complex) and let $\tilde{f}_* : C_*(\tilde{X}) \to C_*(\tilde{X})$ be the induced θ–homomorphism. Let $\rho : \mathbb{Z}H \to \mathbb{Z}$ be the augmentation homomorphism. Then $L(\tilde{f}_*, \rho) \in \mathbb{Z}$ is just the usual Lefschetz number, $L(f)$, of f (here $R = \mathbb{Z}$ and $m = 1$).
2. Let A be an abelian group and $\rho : H \to A$ a surjective homomorphism such that $\rho\theta = \rho$. Let $\rho : \mathbb{Z}H \to \mathbb{Z}A$ also denote the extension of $\rho : H \to A$ to a homomorphism of group rings. Suppose $Y \subset X$ is a subcomplex and that $f(Y) \subset Y$, i.e. f is a map of pairs. Let $\tilde{Y}$ be the inverse image of Y under the covering projection $\tilde{X} \to X$. Let $C \equiv C_*(\tilde{X}, \tilde{Y})$ be the relative cellular

chain complex of the pair $(\tilde{X}, \tilde{Y})$ (regarded as a right S–module complex) and let $\tilde{f}_*: C_*(\tilde{X}, \tilde{Y}) \to C_*(\tilde{X}, \tilde{Y})$ be the induced θ–homomorphism. Then $L(\tilde{f}_*, \rho) \in \mathbb{Z}A$ is the "generalized Lefschetz number" of [**F**,**§3**] (here $R = \mathbb{Z}A$, $m = 1$).

We wish to relate the invariant $\Delta(f_*)$ to the present discussion.

There is an extension of $\rho : S \to M_m(R)$ to a homomorphism $\bar{\rho} : S_\theta[[t]] \to M_m(R[[t]])$, where $R[[t]]$ is the commutative ring of formal power series over R, via the formula $\bar{\rho}(\sum_{j=0}^{\infty} u_j t^j) = \sum_{j=0}^{\infty} \rho(u_j) t^j$.

The matrices $I_{mn_i} - t\bar{\rho}([f_i])$ are invertible over $R[[t]]$ and so we can define $\Delta(f_*, \rho) \in K_1(R[[t]]))$ to be the element represented by the finite alternating product $\prod_{i \geq 0} (I_{mn_i} - t\bar{\rho}([f_i]))^{(-1)^{i+1}}$.

Let $\bar{\rho}_* : K_1(S_\theta[[t]]) \to K_1(R[[t]])$ be the composite

$$K_1(S_\theta[[t]]) \to K_1(M_m(R[[t]])) \xrightarrow{\cong} K_1(R[[t]])$$

where the first homomorphism is induced by $\bar{\rho}$: $S_\theta[[t]] \to M_m(R[[t]])$ and the second homomorphism is the Morita equivalence isomorphism. Clearly, $\bar{\rho}_*(\Delta(f_*)) = \Delta(f_*, \rho)$.

Recall that a *derivation* $D : S \to S$ of a ring S is a homomorphism of abelian groups such that $D(uv) = D(u)v + uD(v)$. For any commutative ring A there is a derivation D_t of $A[[t]]$ given by $D_t(\sum_{i=0}^{\infty} a_i t^i) = \sum_{i=1}^{\infty} i a_i t^{i-1})$. If A is a rational vector space then there is a formal integration homomorphism $I : A[[t]] \to A[[t]]$ defined by $I(\sum_{i=0}^{\infty} a_i t^i) = \sum_{i=0}^{\infty} \frac{a_i}{i+1} t^{i+1}$ which is a right inverse for D_t (i.e. $D_t I$ is the identity) and also $ID_t(u) = u - u_0$, $u = \sum_{i=0}^{\infty} u_i t^i$. Let $U_A \subset A[[t]]$ be the multiplicative subgroup of $A[[t]]$ consisting of power series of the form $1 + \sum_{i=1}^{\infty} u_i t^i$ and let $M_A \subset A[[t]]$ be the ideal of series of the form $\sum_{i=1}^{\infty} u_i t^i$. There is a formal logarithm which is a homomorphism $\log : U_A \to M_A$ defined by

$$\log(1 + \sum_{i=1}^{\infty} u_i t^i) = \sum_{n=1}^{\infty} \frac{(-1)^n}{n} (\sum_{i=1}^{\infty} u_i t^i)^n.$$

This logarithm has the property that $D_t \log(u) = D_t(u) u^{-1}$. The inverse to $\log : U_A \to M_A$ is the formal exponential $\exp : M_A \to U_A$ given by the formula:

$$\exp(\sum_{i=1}^{\infty} u_i t^i) = \sum_{n=0}^{\infty} \frac{1}{n!} (\sum_{i=1}^{\infty} u_i t^i)^n.$$

THEOREM 6.4. *Suppose that the commutative ring R is also a rational vector space. Then*

$$\det(\Delta(f_*, \rho)) = \prod_{i \geq 0} \det(I_{mn_i} - t\bar{\rho}([f_i]))^{(-1)^{i+1}} = \exp\left(\sum_{n=1}^{\infty} \frac{1}{n} L(f_*^n, \rho) t^n\right).$$

EXAMPLE. Let K be a field of characteristic zero, $S = R = K$, and suppose both ρ and θ are the identity $K \to K$. Then the conclusion of Theorem 6.4 is the familiar zeta function formula of [**M, §3**]:

$$\prod_{i \geq 0} \det(\mathrm{id} - tf_i)^{(-1)^{i+1}} = \exp\Big(\sum_{n=1}^{\infty} \tfrac{1}{n} L(f_*^n) t^n\Big).$$

EXAMPLE. Consider the situation of Example 6.3(2). Let $\rho' : \mathbb{Z}G \to \mathbb{Q}A$ be the composite of $\rho : \mathbb{Z}G \to \mathbb{Z}A$ and the inclusion $\mathbb{Z}A \hookrightarrow \mathbb{Q}A$. Then applying Theorem 6.4 with $S = \mathbb{Z}G$, $R = \mathbb{Q}A$, $C = C_*(\tilde{X}, \tilde{Y})$, $f_* = \tilde{g}_*$, and ρ' (in place of ρ) yields precisely the conclusion of [**F,Theorem 3**]; in particular, the series $\exp\big(\sum_{n=1}^{\infty} \frac{1}{n} L((\tilde{g}_*)^n, \rho')t^n\big)$ (denoted by $\tilde{\zeta}_{\rho'}(t)$ in [**F**]) is rational.

§7. The total Lefschetz–Nielsen invariant

In this and the next section, we apply the theory to suspension semiflows, deepening and enriching the preliminary results in §5. The special case of interest in dynamics is that of suspension *flows*, where X is an n–manifold, f is a diffeomorophism, and the mapping torus Y is an $(n+1)$–manifold supporting the suspension flow Φ. We note that in [$\mathbf{GN_2}$] we give the invariance theorems necessary to apply the "cellular" theory expounded here to that smooth case.

The *total Lefschetz–Nielsen invariant* of a continuous map $f\colon X \to X$, denoted by $\Delta(f)$, is defined to be $\Delta(\tilde{f}_*) \in K_1(\mathbb{Z}H_\theta[[t]])$. We saw in §6 that $\Delta(f)$ determines the classical Lefschetz zeta function of f and the Fried zeta functions associated with abelian covers of X; indeed Theorem 6.4 implies the rationality of these zeta functions. Moreover, this K_1–invariant, $\Delta(f)$, determines the Nielsen fixed point theory of all the iterates of f up to the ambiguity of passing to the co-invariants of the θ–action on Hochschild homology. Geometrically, this "ambiguity" is the difference between partitioning the fixed points of f into Nielsen classes and partitioning the corresponding orbits of the flow into classes in the sense of §2.

But there is another way of thinking about $\Delta(f)$, closely related to what we considered in §5. The group $G = \pi_1(Y)$ is isomorphic to the semidirect product $\langle H, t \mid t^{-1}ht = \theta(h), h \in H\rangle$. Consequently, we have ring "inclusions" $\mathbb{Z}H_\theta[[t]] \hookleftarrow \mathbb{Z}H_\theta[t] \hookrightarrow \mathbb{Z}G$. For each positive integer ℓ there is a sensible "truncation" homomorphism $\mu_\ell\colon HH_*(\mathbb{Z}H_\theta[[t]]) \to HH_*(\mathbb{Z}H_\theta[t])$ which ignores contributions from iterates of f beyond the ℓ–th. We have:

THEOREM 7.1. *$i_*\mu_\ell(\mathrm{DT}(\Delta(f))) = R(\Gamma) \in HH_1(\mathbb{Z}G)$ with Γ as in §5.*

One should interpret this as saying that $i_*\,\mathrm{DT}(\Delta(f))$ is a rigorous substitute for "$R(\Phi|_{Y\times[0,\infty)})$"; the latter is not rigorously defined because the one–parameter trace was defined only for homotopies parametrized by a compact interval.

§8. The Novikov ring and a "non-commutative" Alexander–Fried quotient

Just as with zeta functions in §7, our purpose here is to explain how our theory can elucidate the previously known "torsion" invariants of a suspension flow by placing them in their proper algebraic context.

We enlarge the twisted power series ring, $\mathbb{Z}H_\theta[[t]]$, to obtain the *Novikov ring*, $\widehat{\mathbb{Z}G}^+$, consisting of all twisted formal "meromorphic series", i.e. finitely many negative t-exponents permitted, with the same rule of twisting: $th = \theta(h)t$. It is the smallest ring which contains both $\mathbb{Z}H_\theta[[t]]$ and $\mathbb{Z}G$ as subrings.

PROPOSITION. *The finitely generated right $\widehat{\mathbb{Z}G}^+$-complex $C(\tilde{Y}) \otimes_{\mathbb{Z}G} \widehat{\mathbb{Z}G}^+$ is acyclic.*

Let $\tau(Y) \in K_1^{\pm G}(\widehat{\mathbb{Z}G}^+)$ denote the torsion of this acyclic chain complex, in the sense of Whitehead torsion [**C**]; i.e. the group of trivial units is $\pm G$. We have an "inclusion" $j\colon \mathbb{Z}H_\theta[[t]] \to \widehat{\mathbb{Z}G}^+$. The relationship between this torsion and the total Lefschetz–Nielsen invariant is:

THEOREM 8.1. $j_*(\Delta(f)) = -\tau(Y)$.

The torsion invariant $\tau(Y)$ can be thought of as a "non-commutative" generalization of the "Alexander invariant" of [**M**] or of the "Alexander quotient" of [**F**] as follows. The mapping torus structure on Y gives a natural homomorphism $\psi\colon G \to T$ where $T \equiv \langle t \rangle$ is the infinite cyclic group. Let P be a finitely generated abelian group and $\rho : G \to P$ a surjective homomorphism. Assume that ψ factors through ρ, i.e. there exists $\psi' : P \to T$ such that $\psi = \psi'\rho$. Let $\bar{Y}$ be the covering space of Y, corresponding to $\ker(\rho) \subset G$. Let $C_*(\bar{Y})$ be the cellular chain complex of $\bar{Y}$. We regard $C_*(\bar{Y})$ as a complex of right modules over the ring $\mathbb{Z}P$. Let $\mathbb{Z}P_N$ be the localization of $\mathbb{Z}P$ at its multiplicative group of non-zero divisors.

In [**F**], Fried shows that the complex $C' \equiv C_*(\bar{Y}) \otimes_{\mathbb{Z}P} \mathbb{Z}P_N$ is acyclic. He then defines the *Alexander quotient*, denoted by $\mathrm{ALEX}_P(Y)$, to be the element $-\tau(C') \in K_1^{\pm P}(\mathbb{Z}P_N)$ where $\tau(C')$ is the torsion of C'. Actually, Fried prefers to regard $\mathrm{ALEX}_P(Y)$ as an element of (units in $\mathbb{Z}P_N)/(\pm P)$ by applying the determinant, but the above formulation is better adapted to our purposes. The inclusion of rings $i : \mathbb{Z}P_N \hookrightarrow \widehat{\mathbb{Z}P}_N^+$ induces a homomorphism $i_* : K_1^{\pm P}(\mathbb{Z}P_N) \to K_1^{\pm P}(\widehat{\mathbb{Z}P}_N^+)$; ρ induces a homomorphism $\bar{\rho}\colon K_1^{\pm G}(S_\theta[[t]]) \to K_1^{\pm P}(\widehat{\mathbb{Z}P}_N^+)$.

THEOREM 8.2. $\bar{\rho}_*(\tau(Y)) = -i_*(\mathrm{ALEX}_P(Y))$.

Since i_* is injective, $\tau(Y)$ determines $\mathrm{ALEX}_P(Y)$ and so, by 8.1, $\Delta(f)$ determines $\mathrm{ALEX}_P(Y)$.

Our torsion invariant can be applied to obtain a new invariant, which we call the *non-commutative Alexander invariant*, for a fibered knot K in a homology 3–sphere. This invariant can be thought of as a "non-commutative" generalization of the (ideal in $\mathbb{Q}[t, t^{-1}]$ generated by the) Alexander polynomial of K.

In fact, the invariant $\tau(Y)$ is not only defined for mapping tori. All one needs is: a finite complex Y with fundamental group G, and a surjective homomorphism $\psi\colon G \to T$ (giving G a semidirect product structure) such that $C_*(\tilde{Y}) \otimes_{\mathbb{Z}G} \widehat{\mathbb{Z}G}^+$ is acyclic. In that case we denote the torsion invariant by $\tau(Y, \psi)$.

Even in this generality, if $h : Y \to Z$ is a homotopy equivalence, $\phi_Z\colon \pi_1(Z) \to T$ is a surjective homomorphism, and $\phi_Y = \phi_Z \circ h_*$ then there is a "difference formula" relating $\tau(Y, \psi_Y)$, $\tau(Z, \psi_Z)$ and the Whitehead torsion, $\tau(h)$, of h. When h is simple, we have $\tau(Z, \psi_Z) = h_*(\tau(Y, \psi_Y))$. Fried's Alexander quotient is also defined in this greater generality, provided the appropriate chain complex is acyclic; $\mathrm{ALEX}_P(Y)$ continues to be determined by $\tau(Y, \psi)$.

References

[C] M. M. Cohen, *A Course in Simple-Homotopy Theory*, Springer–Verlag, New York, 1973.

[Di_1] D. Dimovski, *One-parameter fixed point indices*, Pacific J. Math., to appear.

[Di_2] D. Dimovski, *One-parameter fixed point indices for periodic orbits*, these Proceedings.

[DG] D. Dimovski and R. Geoghegan, *One-parameter fixed point theory*, Forum Math. **2** (1990), 125–154.

[F] D. Fried, *Homological identities for closed orbits*, Invent. Math. **71** (1983), 419–442.

[Fu] F. B. Fuller, *An index of fixed point type for periodic orbits*, Amer. J. Math. **89** (1967), 133–148.

[GN_1] R. Geoghegan and A. Nicas, *Parametrized Lefschetz-Nielsen fixed point theory and Hochschild homology traces*, Amer. J. Math. (to appear).

[GN_2] R. Geoghegan and A. Nicas, *Trace and torsion in the theory of flows*, preprint.

[GN_3] R. Geoghegan and A. Nicas, *Higher Euler characteristics for groups*, in preparation.

[I] K. Igusa, *What happens to Hatcher and Wagoner's formula for $\pi_0 C(M)$ when the first Postnikov invariant is nontrivial?*, Algebraic K-theory, Number theory, Geometry and Analysis, Lecture notes in Math. vol. 1046, Springer–Verlag, New York, 1984, pp. 104–172.

[J] B. Jiang, *Nielsen theory for periodic orbits and applications to dynamical systems*, these Proceedings.

[M] J. Milnor, *Infinite cyclic coverings*, Conference on the Topology of Manifolds (Michigan State Univ., E. Lansing, Mich., 1967), Prindle, Weber & Schmidt, Boston, 1968, pp. 115–133.

[MS] J. Milnor and J. Stasheff, *Characteristic Classes*, Ann. of Math. Studies, No. 76, Princeton Univ. Press, Princeton, NJ, 1974.

Department of Mathematics, SUNY Binghamton, Binghamton, NY 13902–6000, USA

E-mail address: ross@math.binghamton.edu

Department of Mathematics, McMaster University, Hamilton, Ontario L8S 4K1, Canada

E-mail address: andy@icarus.math.mcmaster.ca

Contemporary Mathematics
Volume **152**, 1993

Recent Developments in Nielsen Theory and Discrete Groups

JANE GILMAN

ABSTRACT. We survey some recent developments in Riemann surfaces, Teichmüller theory, and discrete groups and their relation to Nielsen theory and dynamical systems. We discuss 1) the theory of abstract Fuchsian groups, convergence groups and Gabai's recent proof of the Nielsen realization conjecture, 2) discreteness criteria for subgroups of $PSL(2, C)$ and the iteration of commutator identities, and 3) open questions about pseudo-Anosov mapping-classes and product relations. While the paper is for the most part expository, some of the results on commutators and product relations are new.

Introduction

In 1976 Thurston classified surface diffeomorphisms. In 1978 I proved that Thurston's classification could be obtained using the Nielsen types of the lifts of the diffeomorphism to the unit disc. The ensuing renewed interest in Nielsen's three long papers on surface diffeomorphisms resulted in a number of papers making connections between Nielsen theory, dynamical systems, the complex analytic approach to Teichmüller theory and the topological approach to Teichmüller theory. This paper surveys some recent developments in Nielsen theory and discrete groups which have potential for additional interaction between a variety of fields.

1991 *Mathematics Subject Classification.* 20H10, 30F993, 34C35.

Key words and phrases. Mapping-class group, surface diffeomorphism, Nielsen realization, dynamical systems, pseudo-Anosov, discreteness, and commutator.

Research supported in part by NSF Grant No. DMS-9001881

The author thanks the Institute for Advanced Study for its hospitality while much of this work was being done.

This paper is in final form and no version of it will be submitted for publication elsewhere

The paper treats three main topics. The first topic is abstract Fuchsian groups, convergence groups and Gabai's proof of the Nielsen realization conjecture. Mathematicians working in Nielsen theory and dynamical systems should examine Gabai's techniques for possible applications. His techniques are very clever. He claims to give the proof that Nielsen meant to give and in this sense his techniques can be considered *new* classical techniques.

The second topic is discreteness criteria and the iteration of commutator inequalities. This is an area where dynamical systems experts might make a contribution to the theory of discrete groups. In recent years complex analysts have obtained some important results by iterating commutators. Although the results do make use of the language of dynamical systems, the dynamics of the situation is not completely understood. Dynamical systems might be able to furnish a fuller and more complete explanation of what is going on.

The last topic is some open problems in surface diffeomorphism. This is an area where complex analysts, topologists, dynamical systems experts or Nielsen theorists might make a contribution. Here *recent* means that the answers should by now be recent results or soon should be recent results. In other words these are questions which should be answerable.

Review of Thurston-Nielsen Theory

We begin with a review of Thurston's classification of surface diffeomorphisms and Nielsen theory and their connection. (See [**Th1**] and [**Gil1**] for complete details.)

Let S be a compact Riemann surface of genus $g \geq 2$ and t a diffeomorphism. A diffeomorphism is *reducible* if it fixes a set of simple closed curves on the surface. A diffeomorphism is *pseudo-Anosov* if it fixes a pair of transverse measured foliations. Thurston proved that a diffeomorphism of a surface which is not homotopic to a diffeomorphism of finite order is either homotopic to a reducible diffeomorphism or to a pseudo-Anosov diffeomorphism, but not both. For simplicity we call a diffeomorphism finite, reducible or pseudo-Anosov if its homotopy class contains such a map.

In order to make the connection with Nielsen theory more definitions are needed. Represent S as U/F where U is the unit disc and F is the group of covering transformations. We can look at $L(t) = \{\text{all lifts of } t \text{ to } U\}$ and if $T = \langle t \rangle$ is the group generated by t, then we can also look at $L(T) = \{\text{all lifts to } U \text{ of elements of } T\}$. While $L(t)$ is not a group, $L(T)$ is a group and F is a normal subgroup of $L(T)$ so that elements of $L(T)$ act on F by conjugation.

Every lift $h \in L(T)$ has a natural extension to the boundary of the unit disc. Nielsen studied this boundary action and he assigned to each lift a pair of integers, (V_h, U_h) which describes the boundary action. This pair of integers is called the *Nielsen type* of the lift h. Nielsen proved that a lift is determined uniquely by its boundary action. For simplicity we do not distinguish notationally between a lift and its extension to the boundary.

Since lifts act on F by conjugation, one can look at those elements of F fixed by a given lift. We let $N(h) = \{f \in F \mid hfh^{-1} = f\}$. $N(h)$ is called *the fixed element subgroup.* Nielsen proved that $N(h)$ is finitely generated. Let V_h be its rank. When a lift h is extended to the boundary of U, its isolated fixed points are alternately attracting and repelling (except in the case that $V_h = 1$). There may be an infinite number of boundary fixed points, but there are only a finite number of orbits under $N(h)$, the fixed element subgroup. U_h is defined to be the number of $N(h)$ orbits of isolated attracting boundary fixed points except when $V_h = 1$ and $N(h) = \langle f \rangle$ and h has precisely two fixed points on the boundary. In this latter case, case U_h is defined to be 0 and the type of h in this case is denoted either $(1,0)AR$ or $(1,0)N$ depending upon whether one of the fixed points is attracting and the other repelling or whether both are neutral.

As soon as we define V_h, it becomes clear that there is a connection between Nielsen theory and Thurston theory. Since the covering group F is isomorphic to the fundamental group of S, if $V_h \neq 0$ (so that h fixes an element f of F), then t must fix a curve on the surface (the curve that is the image of the fixed axis of f). Of course, the curve (the image of the axis) may not be a simple curve. It takes additional work to obtain a set of simple closed curves from a non-simple fixed curve, but this establishes the basic connection.

Roughly speaking, one can say that whether or not $V_h \neq 0$ for some lift determines whether t is reducible or not and whether or not $U_h \neq 0$ for some lift determines whether or not t is pseudo-Anosov or pseudo-Ansov-*like*, but additional definitions are needed to make this precise and accurate.

Teichmüller theorists often subdivide reducible mapping classes into two subcategories: A set of simple closed curves on a surface is a partition and a diffeomorphism that fixes a partition fixes a maximal partition. Removing the curves in a maximal partition divides the surface into a finite number of components which the diffeomorphism permutes. Each component is fixed by some smallest power of the diffeomorphism and the power of the map that fixes a component when restricted to that component is called a component map. A reducible diffeomorphism either has all component maps finite, in which case it is called *parabolic*, or it has at least one pseudo-Anosov component map, in which case it is called *pseudo-hyperbolic*. The names for these types of diffeomorphsims are chosen because there is an analogy between the way in which the mapping-class group acts on Teichmüller space and the way in which the group of Möbius transformations act on the unit disc [**B**]. It is of interest to note that Nielsen studied parabolic mapping classes. He called them mapping class of *algebraically finite type* [**N3**]. The accurate statement about U_h is that whether or not $U_h = 0$ governs whether or not t or one of its component maps is pseudo-Anosov.

Note that a hyperbolic fractional linear transformation is of type $(1,0)AR$ since hyperbolic fractional linear transformations fix two points on the boundary of the unit disc, one attracting and one repelling. Elliptic fractional linear transformations are of type $(0,0)$ with some power of type $(2g,0)$. To simplify the technicalities of this discussion we assume there are no parabolic fractional linear transformations.

The connection between Thurston theory and Nielsen theory can be formalized as follows:

THEOREM ([**Gil1**]). *Assume that S is compact.*
Let $L_0(T) = \{t \in L(T) \mid 1 - V_h - U_h < 0\}$.
Then t is

(i) *finite* $\Leftrightarrow$ *each $h \in L(T) - \{id\}$ is either of type $(1,0)AR$ or $(0,0)$.*
(ii) *pseudo-Anosov* $\Leftrightarrow$ $V_h = 0\ \forall\, h \in L_0(T)$ *and* $\exists\, h' \in L(T)$ *with* $U_{h'} \geq 2$.
(iii) *pseudo-hyperbolic (infinite reducible with at least one pseudo-Anosov component map)* $\Leftrightarrow$ $\exists\, h \in L_0(T)$ *with* $V_h \neq 0$ *and* $U_h \neq 0$.
(iv) *parabolic (infinite reducible with all component maps finite)* $\Leftrightarrow$ $U_h = 0\ \forall\, h \in L_0(T)$ *and* $\exists\, h \in L_0(T)$ *with* $V_h \neq 0$.

One thing to notice is that t is of finite order if and only if all lifts have the same type as fractional linear transformations (Möbius transformations). This motivates the following definitions.

DEFINITION. A lift h is called an *abstract hyperbolic* if h is of type $(1,0)AR$.

DEFINITION. A lift h is called an *abstract elliptic* if it is of type $(0,0)$ with some power of type $(2g,0)$. (Here g is the genus of S).

If G is a group of diffeomorphisms of S, one can consider $L(G)$, the group of all lifts to U of elements of G.

DEFINITION. If G is a group of homeomorphisms of S, $L(G)$ is called an *abstract Fuchsian group* if it contains only abstract hyperbolics and abstract elliptics.

One can prove

THEOREM ([**Gil4**]). *$L(G)$ is an abstract Fuchsian group if and only if G is a finite subgroup of the mapping-class group.*

This allows one to formulate the Nielsen realization problem as the problem of showing that an abstract Fuchsian group is quasiconformally conjugate to a Fuchsian group.

The Nielsen Realization Problem

The more standard formulation of the Nielsen Realization problem is to show that a diffeomorphism, t, of S that is finite up to homotopy can "be made" finite or to show that a group of homeomorphisms that are finite up to homotopy can "be made" finite. "Be made" finite means find a homotopic map or group that is finite. Equivalently the problem is to show that one can find a quasiconformal map φ such that φ conjugates $L(T)$ or

$$L(G) = \{\text{all lifts to } U \text{ of elements of } G\}$$

to a Fuchsian group. As mentioned before this is equivalent to showing that an abstract Fuchsian group is (quasiconformally conjugate to) a Fuchsian group.

Solutions to this problem have a long history. Nielsen showed in 1942 that cyclic groups are realizable. In 1948 Fenchel gave a proof for cyclic and solvable groups. His proof was different than Nielsen's. He applied the Smith fixed point theorem to the mapping-class group acting on Teichmüller space. In 1959 Kravetz proved that Teichmüller space is a straight space. The Nielsen realization for arbitrary groups was a corollary to this, but then Michelle Linch in the late sixties showed that Kravetz's proof that Teichmüller space was a straight space contained a gap. At this point a number of people working in Riemann surfaces and Teichmüller theory had published results that depended upon (Kravetz's proof of) the Nielsen realization. This naturally intensified interest in the validity of Nielsen realization. In 1971 Zieschang announced that Nielsen's methods extended to all groups, but he amended the announcement in 1974 by saying, that is, it extended to all groups for which Nielsen's original proof was correct. One needed to worry about the case when there was no simple axis for the group. In 1983 Kerchkoff gave a proof of the Nielsen realization for all groups using Thurston's theory of earthquakes. Most recently (1991), Gabai gave a proof for all groups which he claims is the proof that Nielsen meant to give. Cassen and Jungreis (1992) have also recently given a proof. The reason that topologists are interested in this is because of its connection with the Seifert Fibered Space Conjecture. (See [**CJ**], [**Ga1,2**], [**N2**], [**F**], [**Ker**],[**Kr**], [**Lin**], and [**Z1-3**].)

There are a number of equivalent characterizations of abstract Fuchsian groups. Gabai's proof of the Nielsen realization [**Ga2**] makes use of all of them. One important description is as a convergence group. The notion of a convergence group applies in a more general setting and Gabai's result is more general than the classical Nielsen realization theorem.

DEFINITION. Let X be a metric space and $G \subset \text{Homeo}(X)$. G is an X-*convergence group* if for all sequences of distinct elements there exists a subsequence g_i and points x and y in X such that g_i converges to x when restricted to $X - y$ and g_i^{-1} converges to y on $X - x$ and the convergence is uniform on compact subsets.

It is easy to establish that an S^1-convergence group is an abstract Fuchsian group. Here S^1 denotes the boundary of the unit disc. This convergence property was first studied by Nielsen [**N1**]. Convergence groups in higher dimensions have been studied by Tukia and Gehring and Martin among others ([**GeM0**], [**MT**],[**T1-3**]).

The Nielsen realization problem is, therefore, to prove that an S^1-convergence group is (quasiconformally conjugate to) a Fuchsian group.

A major difficulty in the proof of the realization lies with elliptics. Here is one way to view the problem with elliptics: One wants to associate to $L(G)$ a Fuchsian group, call it Möb(G). Since the elements of $L(G)$ are assumed to be abstract elliptics and abstract hyperbolics, the natural way to do this is to assign a hyperbolic to an abstract hyperbolic and an elliptic to an abstract elliptic. This is easy for abstract hyperbolics. If h is an abstract hyperbolic, associate to it the hyperbolic transformation with the same fixed points as h and with translation length $\frac{1}{n}$ times the translation length of f_h where n is the smallest integer such that h^n is in F and $f_h = h^n$. For an abstract elliptic element α of order n, one wants to associate an elliptic element e of order n. The question is, "Where should one put the fixed point of e?" An elliptic Möbius transformation always factors as the product of hyperbolics. Similarly, the abstract elliptic α can always be factored as the product of a hyperbolic, h_1, and an abstract hyperbolic, h_2. One can use the hyperbolic h_2' already associated to the abstract hyperbolic h_2, to give us $e = h_1 h_2'$. This then chooses a fixed point (possibly non-existant) for e. The problem with this method is that the choice of the fixed point or even the existence of a fixed point may depend upon the factorization of α. When there is a realization, all choices give the same group.

THEOREM ([**Gil4**]). *When F is a surface on which $L(G)$ is realizable, then Möb(G) is the correct group.*

An Outline of the Nielsen-Zieschang-Gabai Proof

Here is an outline of the Nielsen-Zieschang-Gabai proof of the Nielsen realization.

First find a simple axis.

DEFINITION. Let $f \in F$ be hyperbolic and let Ax_f be its axis, the non-euclidean geodesic that f fixes. Ax_f is simple with respect to $L(G)$ if $g(Ax_f) \cap Ax_f = \phi\ \forall g \in L(G)$ except when $g(Ax_f) = Ax_f$.

What Zieschang [**Z2**] proved was that Nielsen's proof was correct for any group that had a simple axis. In that case one could use the simple axis to cut up the unit disc into regions and straighten out the map or make it finite first on the boundary of the regions and then on the regions.

Zieschang observed that when there was no simple axis, the elliptic elements lay at the heart of the problem. In the absence of a simple axis the group $L(G)$ contains what Zieschang termed a semi-triangle group, a group that could be mapped onto a true triangle group with torsion free kernel.

Gabai's approach is to make a virtue out of elliptics. The idea is that an element in $L(G)$ is completely determined by its boundary behavior, so in some sense for any abstract elliptic one really does know where the fixed point interior to the unit disc is:

If there is no simple axis when one tries to cut up the interior of the disc into regions which one can handle, the images of an axes will cross each other and it is not clear how the crossing should go. For example, take six ordered points on the boundary of the disc in the clockwise order (a, b, c, a', b', c'). Connect a to a', b to b', and c to c' by simple curves. These three curves can intersect each other in three different ways and one needs to know which way to have them cross (Figure 1).

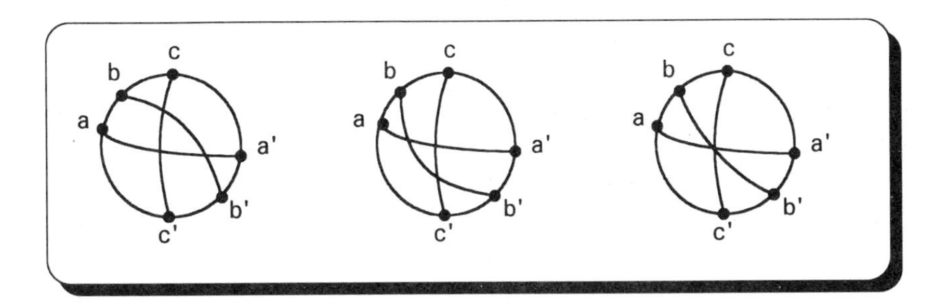

FIGURE 1. The three possible crossings

Gabai's idea is to use the elliptics as decision makers or markers. First, one can define the concept of an elliptic α lying to the *left*, *right*, or *on* a hyperbolic f. Namely, f has two boundary fixed points, an attracting fixed point P_f and a repelling one N_f. Write the order of α as $2n+1$ if it is odd and $2n$ if it is even. Then (the fixed point of) α is to the right of f if $\alpha^n(N_f) \in (N_f, P_f)$, and to the left if $\alpha^n(N_f) \in (P_f, N_f)$ and on f if $\alpha^n(N_f) = P_f$. Here the counterclockwise interval between the points X and Y on the boundary of U is written (X, Y).

Left, right and on are the first level of definitions. Gabai expands the definitions to include *close to* and *far* and builds extensive layers of similar definitions eventually proving that when there does not exist a simple axis, one can produce what he terms a *good elliptic-hyperbolic pair* which allows one to cut up the unit disc and straighten out the group action.

The technique is very clever and intricate, but conceptually straightforward. It should be examined for wider applications.

Discreteness Criteria

Let $X_1, \ldots, X_n$ be elements of $PSL(2, \mathbb{C})$ and let $G = \langle X_1, \ldots, X_n \rangle$ be the group they generate. The question that arises is "Can one find necessary and sufficient conditions for G to be discrete?"

The answer is that this is complicated. There is no simple sufficient condition. Even two generator subgroups of $PSL(2, \mathbb{R})$ require a messy algorithm. Purzitsky and Rosenberger developed the algebraic algorithm in a series of papers that appeared between 1972 and 1986. In 1991 Maskit and I gave a geometric algorithm, completing an incomplete treatment begun by Matelski. In addition we were able to explain the connection between the algebraic and the geometric algorithms. (See [[**GiMa**],[**Gil8**], [**Mat**], [**P1,2**],[**PR**], and [**R1-4**].)

There is an elegant necessary condition, which is Jørgensen's [**J**] inequality. Namely, if G is any discrete subgroup of $PSL(2,\mathbb{C})$, then any A or B in G satisfy $|\mathrm{tr}^2 A-4|+|\mathrm{tr}[A,B]-2| \geq 1$. Here tr denotes the trace and $[A,B] = ABA^{-1}B^{-1}$. Since we will need these terms later, note that we refer to $\mathrm{tr}^2 A - 4$ as the *trace parameter* and $\mathrm{tr}[A,B] - 2$ as the *commutator parameter.*

Since there is no good sufficient condition, sometimes it is replaced by iterating the process of taking commutators. There is a recurrent theme in recent literature: iterating commutators gives good results. But why it gives good results is only partially understood. So the question here is, "Does dynamical systems have a theory about when, why and what to iterate that could be brought to bear on this situation? Can dynamical systems explain in more detail than has yet been the case the possible outcomes under iteration?"

The process of iterating commutators means constructing sequences similar to the following one, which goes back to Jordan in 1878 [**Jord**]. Begin with X and Y. Let

$$\begin{aligned} Y_1 &= [X,Y] \\ Y_2 &= [X,[X,Y]] = [X,Y_1] \\ &\vdots \\ Y_{i+1} &= [X,Y_i] = [X,[X,...[X,Y]]...] \end{aligned}$$

Other variations are also used.

In 1957 Siegel obtained necessary conditions for discreteness using the trace parameter and the commutator parameter. However, he did not combine them into one inequality as Jørgensen did. Rather his results were of the form "if the trace parameter is greater than q, then the commutator parameter is at most...". In 1967 Leutbecher obtained an inequality for groups with a parabolic element and Shimuzu also obtained a similar result. In 1976 Jørgensen obtained his inequality. (See [**S**], [**Sh**], [**L**], and [**J**].)

In 1978 Brooks and Matelski iterated what they referred to as the Shimizu-Leutbecher ([**BrMat**]) sequence and applied Jørgensen's inequality to the iterates. They did not require the group to contain a parabolic. They connected this process with iterating a quadratic map. Dynamical systems experts may be familiar with their paper because it contains pictures of what has been referred to as the precursor of the Mandlebrot set. Brooks and Matelski do have a geometric explanation for what they are iterating. Thurston furnishes an algebraic explanation in [**Th2**] when he discusses groups with small elements and iterating the commutator.

Recently several different people have obtained variations on and improvements in Jørgensen's inequality including Gehring and Martin (1989) [**GeM1**], Gilman (1987) [**Gi6,7**] and Delan Tan (1989) [**Tan**]. Beginning in 1989 and still continuing, Gehring and Martin produced a series of important papers involving iterating of the commutator and trace parameters [**GeM1-7**]. While they do apply theory from dynamical systems to the process of iterating commutators, their emphasis is on obtaining new and sharp inequalities rather than on an exhaustive analysis of the dynamical system.

The algebraic explanation in Thurston is a standard argument. Namely, one computes that if A is close to the identity, that is, if $\|A - I\| \leq \varepsilon < 1$, then $[A, B]$ is within ε^2 of the identity (that is, $\|[A, B] - I\| \leq \varepsilon^2$). Here $\| \ \|$ denotes the matrix norm so that if $A = \begin{pmatrix} a & b \\ c & d \end{pmatrix}$, then $\|A\| = |a^2 + b^2 + c^2 + d^2|^{1/2}$.

This makes it clear that if one begins with an element of small trace, then iteration of commutators will produce a sequence of elements of decreasing trace. Since in a discrete group traces of element cannot become arbitrarily small, this gives information about discreteness.

However, the question remains in an arbitrary subgroup of $PSL(2, \mathbb{C})$ how do you find the initial element with small trace? Or what happens if you iterate the commutator process with an element of large trace? This is a dynamical system and one wonders whether there is a more detailed answer about what different things happen under iteration.

The partial geometric explanation is the following. For simplicity assume that A and B are loxodromic or elliptic so that they each fix non-euclidean straight lines, their axes in upper-half three space $\mathbb{H}^3$. Then $[A, B]$ gives a measurement of the hyperbolic distance between their axes [**BrMat**]. Iterating commutators iterates distances between axes and in a discrete group, the axes cannot accumulate.

Several sets of questions arise. One is roughly the same as with the algebraic case. How do we choose an initial set of axial distances to iterate? The second question follows from the observation that $[A, B]$ only gives an *indirect* measurement of the hyperbolic distance between their axes. So that iterating commutators only *indirectly* iterates distances between axes. The second question is, "What is the direct geometric meaning?"

To see that the commutator measures axial distance but only indirectly, let C be the transformation that moves the axis of A onto the axis of B. Then the axis of C is the common perpendicular to Ax_A and Ax_B and one can compute that

$$\operatorname{tr}^2(C^2) - 4 = \frac{\pm 16(\operatorname{tr}[A, B] - 2)}{(\operatorname{tr}^2 A - 4)(\operatorname{tr}^2 B - 4)}$$

([**Mey**]). This inequality is the explanation usually offered as the reason why the commutator parameter measures axial distance.

However, if we let CR be the cross ratio of the four fixed points of A and B (the ends of each of the axes on the sphere at infinity) one could also prove

PROPOSITION.
$$tr^2(C^2) - 4 = \frac{\pm 16CR}{(CR-1)^2}$$

PROOF. In [**Gil5**] it is shown that when A and B are hyperbolic elements of $PSL(2, \mathbb{R})$,
$$\mathrm{tr}[A, B] = 2 + \frac{f(CR)}{f(M_A)f(M_B)}$$
where $f(X) = \frac{X}{(X-1)^2}$ and M_X is the multiplier of the transformation with matrix X (so that $\mathrm{tr}\, X = \sqrt{M_X} + (\sqrt{M_X})^{-1}$). In [**GeM4**], it was shown that this formula actually holds for any loxodromic or elliptic elements of $PSL(2, \mathbb{C})$. Observe that $f(M_A) = \frac{1}{\mathrm{tr}^2 A - 4}$ to complete the proof.

This makes the commutator parameter appear superfluous. It is not superfluous and is the correct parameter to look at, but not because it directly measures axial distance.

Further evidence that the trace and the commutator parameters are the correct ones to look at comes from considering the case of arithmetic Kleinian groups. For such groups there is a standard construction of a quaternion algebra over the trace field (the trace field is the rationals with the traces of the elements of the group adjoined.) If 1, R, S, RS are the generators of the quaternion 100algebra, then from purely algebraic considerations $R^2 = \mathrm{tr}^2 A - 4$ and $S^2 = \mathrm{tr}[A, B] - 2$ (see [**Re**]). So the issue is not whether the commutator parameter is the correct one to consider, but merely to provide a clearer geometric understanding as to why it is the correct parameter.

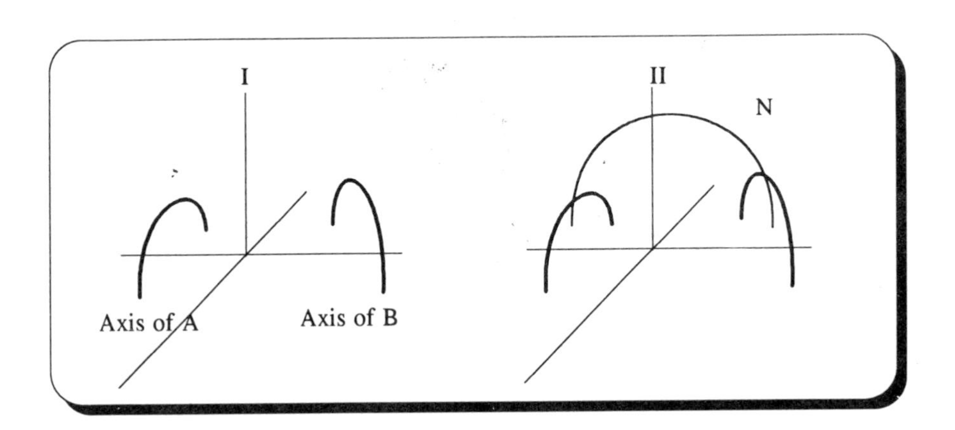

FIGURES 2(I) AND 2(II). Ax_A, Ax_B and N, their common perpendicular.

Here is a recipe for constructing the axis of $[A, B]$ when A and B are loxodromic with disjoint axes.

Let N be the common perpendicular to Ax_A and Ax_B (Figure 2.II). (N is the same as the axis of C.)

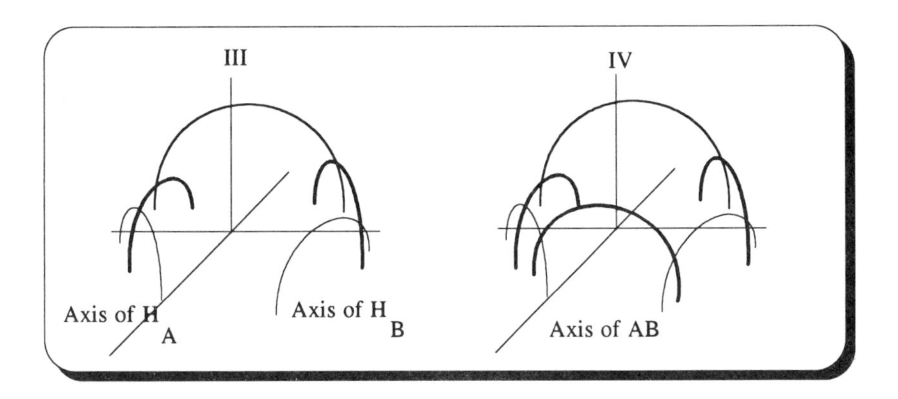

FIGURES 2(III) AND 2(IV). Add Ax_{H_A}, Ax_{H_B} and then add Ax_{AB}.

Let H_N be the half-turn about N. We can factor A has $H_N H_A$ and B as $H_B H_N$, where H_A and H_B are also half-turns. Let Ax_{H_A} and Ax_{H_B} be the axes of H_A and H_B, respectively (Figure 2.III). If Ax_{AB} is the axis of the product, AB, then one can show that Ax_{AB} is the common perpendicular of Ax_{H_A} and Ax_{H_B} (Figure 2.IV). (The lines Ax_A, Ax_B, N, Ax_{H_A}, Ax_{H_B} and Ax_{AB} bound a right-angled hexagon in hyperbolic 3-space.)

Let N_{AH_B} be the common perpendicular between the axes of A and H_B (Figure 2.V).

A can be factored as the product of half-turns about N_{AH_B} and another axis, call it N_A. The axis N_A is actually $A(H_A)$ (Figure 2.VI). $A = H_{N_{AH_B}} \cdot H_{N_A}$.

Also H_B can be factored as a product of half-turns about $H_{N_{AH_B}}$ and N_B where N_B is perpendicular to both N_{AH_B} and H_B and passes through their point of intersection (also Figure 2.VI).

The axis of the commutator of A and B is then the common normal to H_{N_A} and H_{N_B} (Figure 2.VII).

Although the construction does not actually locate the ends of the axis of the commutator, it suggests that the axis of the commutator is different than the axis of C.

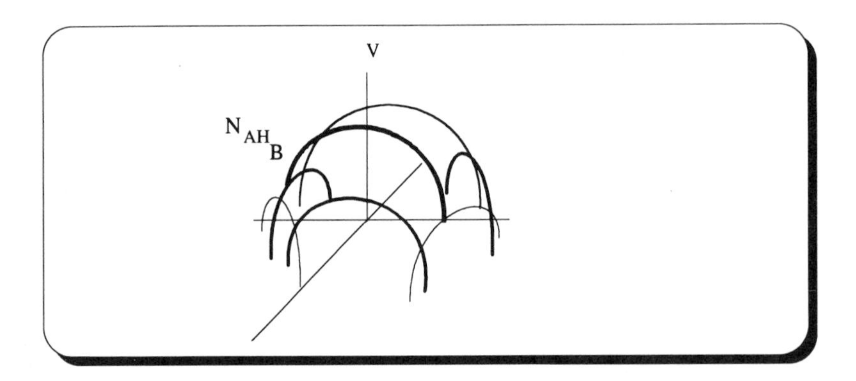

FIGURE 2(V). Add N_{AH_B} the common perpendicular to Ax_A and Ax_{H_B}.

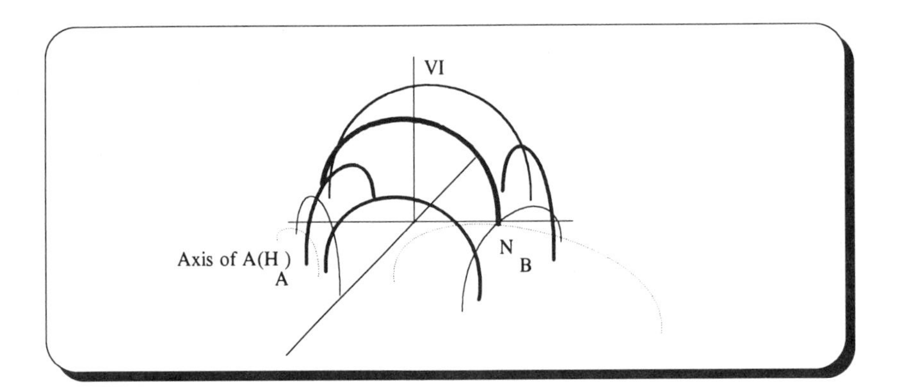

FIGURE 2(VI). Add $A(H_A)$ and N_B, the common perpendicular to N_{AH_B} and Ax_{H_B}.

This construction does not by itself help to clarify the situation. However, there is an alternate construction due to Fenchel, and it may be that combining or comparing the two constructions will shed some light on the situation.

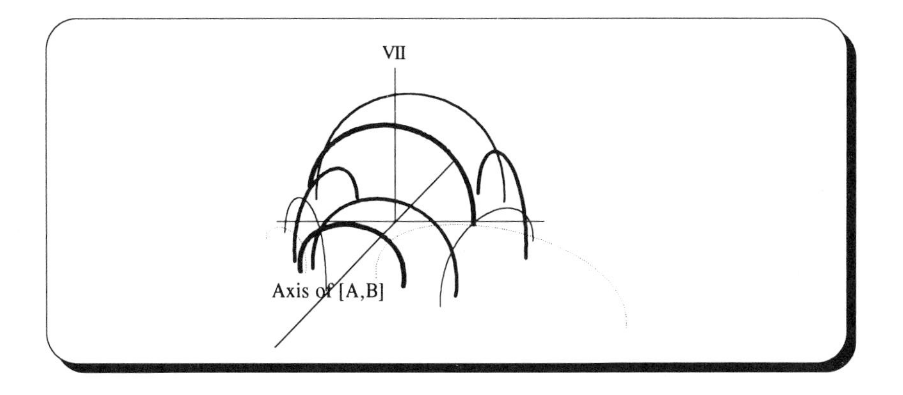

FIGURE 2(VII). Add $Ax_{[A,B]}$, the common perpendicular to $A(H_A)$ and N_B.

Questions About Surface Diffeomorphisms

In the course of his three papers on surface diffeomorphisms, Nielsen studied a number of examples of surface diffeomorphisms. One example that he returned to in a number of places is his example #13. Let $F = \langle a, b, c, d;\ [a, b][c, d] = 1\rangle$ so that if $S = U/F$ then S is a compact surface of genus 2. Let $t: S \to S$ be defined by requiring that some $h \in L(t)$ induce the automorphisms of F:

$$\begin{array}{ll} a \longmapsto c^{-1}a^{-1} & b \longmapsto b^{-1}a^{-1} \\ c \longmapsto b^{-1}a^{-1}d & d \longmapsto c^{-1}. \end{array}$$

Then it can be shown:

THEOREM (**[Gil2]**). *t is a pesudo-Anosov diffeomorphism with stretching factor λ of degree 4 satisfying*

$$x^4 + 2x^3 + x^2 + 2x + 1 = 0.$$

The fixed foliations each have two four-pronged singularities which t cyclically permutes.

This example is of interest for two reasons. First of all the fact that he kept returning to this example suggests that Nielsen understood something about the existence of pseudo-Anosov mapping classes and their importance. Secondly, the computations from Nielsen's papers are almost all of the computations that one needs to prove the theorem. The question arises as to whether Nielsen's computational methods can be codified into an algorithm for determining whether a diffeomorphism is reducible, pseudo-Anosov or finite beginning with the action of the diffeomorphism on the fundamental group and ending with the Nielsen types of the lifts. (The algorithm should also compute stretch factors, the number of components under a maximal partition, etc.)

The answer to the question should be yes. However, some of Nielsen's computations use ad hoc methods and questions about these methods need to be answered.

For example, Nielsen computes that both h^4 and $(ch)^4$ are of type $(0,4)$. He does this by computing the expansions of their boundary fixed points. Once he has found the boundary fixed points, it is evident from their expansions that h^4 and $(ch)^4$ are not conjugate. The formula that

$$\sum_{\substack{\text{equivalence classes of}\\ \text{lifts of } t}} 2(V_s - 1) + U_s \leq 4(g-1)$$

then implies that h^4 and $(ch)^4$ are representatives for all equivalence classes of lifts that contribute to the sum. (There are lifts of type $(0,2)$ but these do not contribute.) The question here is, once he knows that h^4 is of type $(0,4)$, how does he know that it will be productive to compute the type of $(ch)^4$? After all, $(ch)^4$ might have been equivalent to h^4, so that computing its type would have been unproductive. There are also questions about the method by which Nielsen computes the boundary fixed points of h^4. For example, $P_1 = \lim h^n(V_a) = \lim h^{-n}(V_b)$ where V_a denotes the attracting fixed point of a and V_b the attracting fixed point of b. Since $h^n(V_a)$ converges to P, from one side of the boundary of the disc and $h^{-n}(V_b)$ converges to P from the other side, the interval between V_a and V_b contains no other boundary fixed points of h^4. However, Nielsen does not compute $\lim h^n(V_c)$. Rather, he computes $\lim h^n(V_{d^{-1}a})$ and $\lim h^n(V_{c^{-1}b^{-1}a})$. Why does he choose the points $V_{d^{-1}a}$ and $V_{c^{-1}b^{-1}a}$ and not V_c?

Since Bestvinna and Handell [**BH**] have recently obtained an algorithm that determines the Thurston class of a diffeomorphism using train tracks, one might look at their method to see whether it can be translated into a method that uses the Nielsen types of the lifts.

An even more interesting question to answer is "What prompted Nielsen to write down this automorphism in the first place?" How is this automorphism related to known constructions of pseudo-Anosov diffeomorphisms, which are often as products of Dehn twists? What does it look like when factored as a product of Dehn twists?

While there are many results about the action on homology of non-pseudo-Ansosv diffeormorphisms, it has been suggested that with the exception of those pseudo-Anosov diffeomorphisms that induce the identity on homology, much can be gleaned from the homology action of even pseudo-Anosov diffeomorphisms.

Another question to ask is "What can one conclude about pseudo-Anosov diffeomorphisms from their action on homology".

Product Relations

The last question about surface diffeomorphisms is

Can one write out a complete multiplication table? That is, can one obtain necessary and sufficient conditions for each type of product relation to occur (e.g. for the product of two pseudo-Anosovs to be pseudo-Anosov, to be finite, etc)?

This problem has many pieces which vary in difficulty. Even finding examples of all possible product relations is complicated. The concept of reducible and irreducible is still valid for elements of finite order. It is shown in [**Gil3**] how to obtain examples of product relations for finite mapping classes in most cases.

For infinite product relations one can show

PROPOSITION. *There are examples of every possible product relation among elements of infinite order.*

PROOF:. Simply use a theorem of Kra's to construct examples. The theorem states:

THEOREM ([**Kr**]). *There is an isomorphism between $\pi_1(S_g, n)$, the fundamental group of a surface of genus g with n punctures, and subgroup of $M_{g,n+1}$, the mapping-class group of a surface of genus g with $n+1$ punctures. Under this isomorphism curves about punctures and simple closed curves are sent to parabolic mapping classes, non-simple, non-essential curves to pseudo-hyperbolic mapping-classes, and essential curves to pseudo-Anosov mapping-classes. (Note that an essential curve is one that intersects every non-trivial non-boundary curve.)*

For example, to obtain two pseudo-hyperbolics whose product is pseudo-Anosov, let S be a surface of genus 2 with fundamental group

$$\langle a_1, a_2, b_1, b_2;\ [a_1, b_1][a_2, b_2] = 1\rangle\,.$$

The curves $a_1a_2^{-1}$ and $b_1^{-1}b_2$ are non-simple and non-essential so they correspond to pseudo-hyperbolic mapping-classes on $S_{2,1}$. Their product, the curve $a_1a_2^{-1}b_1^{-1}b_2$, is essential and thus corresponds to a pseudo-Anosov mapping-class on $S_{2,1}$. (See Figures 3 and 4.)

For infinite by finite product relations some partial answers are known. A typical result is

THEOREM [**GG**]. *If S is compact and f and h are irreducible elements of the mapping-class group of finite but distinct orders, then $\langle f, h\rangle$ is of infinite order.*

This constructs two elements of finite order whose product is either infinite reducible or pseudo-Anosov. It would be nice to know which it is.

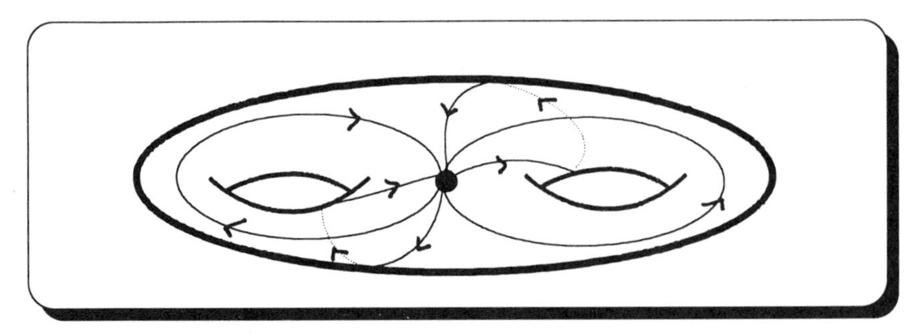

FIGURE 3. $a_1 a_2^{-1} b_1^{-1} b_2$ on S_2: pseudo-Anosov on $S_{2,1}$

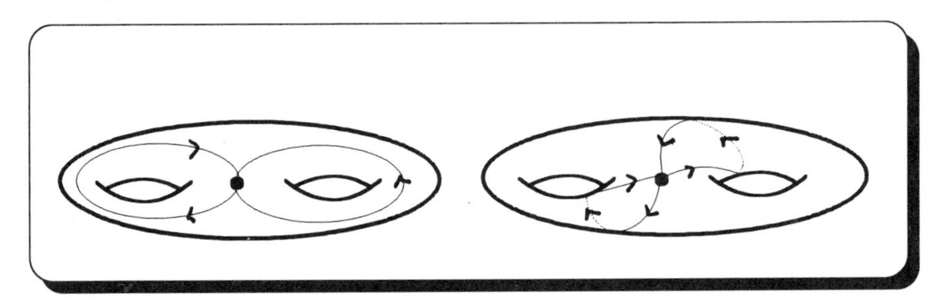

FIGURE 4. $a_1 a_2^{-1}$ and $b_1^{-1} b_2$ on S_2: both pseudo-hyperbolic on $S_{2,1}$

REFERENCES

[B] Bers, L., *An extremal problem for quasiconformal mapping and a theorem of Thurston*, Acta Math. **142** (1979), 223-274..

[BH] Bestvinna, M. and Handell, M., *Train tracks for surface diffeomorphisms*, preprint (1992).

[BrMat] Brooks, R. and Matelski, J. P., *The dynamics of 2-generator subgroups of $PSL(2,\mathbb{C})$*, Annals of Math. Studies **97** (1981), 65–72.

[F] Fenchel, W., *Estensioni gruppi descontinui e transformazioni periodiche delle surfice*, Rend. Acc. Lincei (Se. fis., mat. e nat.) **5** (1948), 326–329.

[Ga1] Gabai, David, *Convergence Groups are Fuchsian Groups*, Bull. AMS (1991).

[GA2] ———, *Convergence Groups are Fuchsian Groups*, Annals of Mathematics **136** (1992), 447-510.

[GeM0] Gehring, F. and Martin, G., *Discrete convergence groups*, Lecture Notes in Math. **1275** (1987), Springer.

[GeM1] ———, *Stability and extremality in Jørgensen's inequality*, Complex Variables **12** (1989), 277–282.

[GeM2] ———, *Iteration theory and inequalities for Kleinian groups*, Bull. AMS **21** (1989), 57–63.

[GeM3] ———, *The Matrix Chordal Norms of Möbius Transformations*, Complex Analysis (1988), 51–59.

[GeM4] ———, *Inequalities for Möbius Transformations and Discrete Groups*, Journal fur die reine und angewandte Mathematik **4l8** (1991), 31–7..

[GeM5] ———, *Commutators, Collars and the geometry of Möbius Groups* (to appear).

[GeM6] ———, *Axial distances in discrete Möbius groups* (to appear).

[GeM7] ———, *Discreteness in Kleinian groups and the iteration theory of quadratic mapping* (to appear).

[Gil1] Gilman, J., *On the Nielsen Type and the Classification of the Mapping-class Group*, Advances in Math. **40** (1981), 68–96.

[Gil2] ———, *Determining Thurston Classes Using Nielsen Types*, Trans. A.M.S. **272(2)** (1982), 669–675.

[Gil3] ———, *Structures of Elliptic Irreducible Subgroups of the Maping-class Group*, Proc. London Math. Soc. **47 (3)**, 27–42.

[Gil4] ———, *A Characterization of Finite Subgroups of the Mapping-class Group*, Proc. 1985 Alta Conference, Annals of Math. Studies, 1987, pp. 433–442.

[Gil5] ———, *Inequalities and discrete groups*, Canadian J. **XL(1)** (1988), 115–130.

[Gil6] ———, *A geometric approach to Jørgensen's inequality*, Advances in Math. **85 (2)** (1991), 193–197.

[Gil7] ———, *A geometric approach to the hyperbolic Jørgensen inequality*, Bull. A.M.S. **16(1)** (1987), 91–92.

[Gil8] ———, *Two-generator discrete subgroups of $PSL(2, R)$: the geometry of intersecting axes*, preprint.

[GG] Gilman, J and Gilman, R., *On the existence of cyclic surface kernels for pairs of Fuchsian groups*, J. London Math. Soc. **(2) 30** (1984), 451–464.

[GilMa] Gilman, J. and Maskit, B., *An algorithm for two-generator discrete groups*, Mich. Math. J. **38 (1)** (1991), 13–32..

[Jord] Jordan, C., *Mémoire sur les équations differentiells á integrale algébrique*, J. für Math. **84** (1878), 89–215.

[J] Jørgensen, T., *On discrete groups of Möbius transformations*, Amer. J. Math. **98** (1976), 739–749.

[Ker] Kerchkoff, S.P., *The Nielsen Realization Problem*, Annals of Math. **117** (1983), 235–265.

[Kr] Kra, Irwin, *On the Nielsen-Thurston-Bers type of some self-maps of Riemann surfaces*, Acta. Math. **146** (1981), 231–270.

[Kv] Kravetz, S., *On the geometry of teichmuller spaces and the structure of their modular group,*, Ann. Acad. Sci. Fenn. **Ser A VI 278** (1959), 1–35.

[L] Leutbecher, A., *Uber Spitzen diskontinuierlicher Gruppen von linear gebrochenen Transformationen*, Math. Z. **100** (1967), 183-200.

[Lin] Linch, Michele, *Thesis* (1972), Columbia University.

[MT] Martin, G. and Tukia, P., *Convergence and Möbius groups*, Holomorphic Functions and Moduli II, vol. 11, MSRI Publications, 1988, pp. 113–140.

[Mat] Matelski, Peter, *The classification of discrete 2-generator subgroups of $PSL(2, \mathbf{R})$*, Israeli J. of Math. **42** (1982), 309-317.

[Mey] Meyerhoff, R., *A lower bound for the volume of hyperbolic 3-rbifolds*, Duke Math. J. **57 (1)** (1988), 185–203.

[N1] Nielsen, J., *Untersuchungen zur Topologie der geschlossenen zweiseitigen Flachen*, Acta Math. **50** (1927), 189–358; Acta Math. **53**, 1–76; Acta Math **58**, 87–167.

[N2] ———, *Abbildungsklassen undlicher Ordnung*, Acta. Math. **75** (1942), 23–115.

[N3] ———, *Surface Transformation Classes of Algebraically Finite Type*, Math. Fys. Medd. Danske Vid. Selskd. **XXI 2** (1944), 1–89.

[P1] Purzitsky, N., *Two generator discrete free products*, Math. Z. **126** (1972), 209–223.

[P2] ———, *Real two-dimensional representations of two-generator free groups*, Math. Z. **127** (1972), 95–104.

[PR] Purzitsky, N. and Rosenberger, G., *Two generator Fuchsian roups of genus one*, Math. Z. **128** (1972), 245–251; *Correction*, Math. Z. **132** (1973), 261–262.

[Re] Reid, Alan, *A note on Trace Fields of Kleinian Groups*, Bulletin London Math Society **22** (1990), 349–352.

[R1] Rosenberger, G.,, *Fuchssche Gruppen, die freies Produkt zweier ztklischer Gruppen sind, un die Gleichung $x^2 + y^2 + z^2 = xyz$*, Math. Ann. **199** (1972), 213–228.

[R2] ———, *Von Untergruppen der Triangel gruppen*, Illinois J. Math. **22** (1978), 404–413.

[R3] ———, *Ein Bemerkung zu einer Arbeit von T. Jørgensen*, Math. Z. **165** (1979), 261–265.

[R4] ———, *All generating pairs of all two-generator Fuchsian groups*, Arch. Math. **46** (1986), 198–204.

[S] Siegel, C. L., *Uber enige Ungleichungen bei Bewgensgruppen in der nichteuklidishen Ebene*, Math. Ann. **133** (1957), 127-138.

[Sh] Shimuzu, H., *On groups operating on the product of the upper-half-planes.*, Ann. of Math. **(2) 79** (1963), 33–71.

[Tan] Tan, Delan, *On inequalities for discrete groups of Möbius transformations,*, Proc. AMS. (to appear).

[Th1] Thurston, W.P., *On the geometry and dynamics of surface diffeomorphisms.*, Bull. AMS **19 (2)** (1988), 417-431.

[Th2] ———, Lecture notes..

[T1] Tukia, P.,, *On two-dimensional groups,*, Ann. Acad. Sci. Fenn. **5** (1980), 73–78.

[T2] ———, *A quasiconformal group not isomorphic to a Möbius group,*, Ann. Acad. Sci. Fenn. **6** (1981), 149–160.

[T3] ———, *Homeomorphic conjugates of Fuchsian groups*, J. Reine. Agnew. Math. **391** (1988), 1–54.

[Z1] Zieschang, H.,, *On Extensions of fundamental groups of surfaces and related groups*, Bull. A.M.S. **77** (1971), 1116–1119.

[Z2] ———, *Addendum to On Extensions of fundamental groups of surfaces*, Bull. A.M.S. **80,** (1974), 366–367.

[Z3] ———, Springer LNM 875.

Mathematics Department, Rutgers Newark, NJ 07102

E-mail address: gilman@andromeda.rutgers.edu

Contemporary Mathematics
Volume **152**, 1993

Local Nielsen fixed point theory and the local generalized H-Lefschetz number

EVELYN L. HART

1. Introduction

Let X be a connected, finite dimensional, locally compact polyhedron. For example, X could be a manifold with or without boundary. Let U be an open, connected subset of X. For a map $f : U \to X$, we define the set of fixed points of f to be $\text{Fix}(f) = \{x \in U : f(x) = x\}$. Our goal is to estimate $\min(f) = \min\{|\text{Fix}(g)| : g \simeq f\}$. In other words, we would like to deform f to a new map having as few fixed points as possible. The goal of local Nielsen theory is to calculate a lower bound for $\min(f)$. This lower bound is the local H-Nielsen number of f, which is denoted by $N_H(f)$. Here H is a normal subgroup of $\pi_1(X)$. Different normal subgroups give different lower bounds, with the best approximation of $\min(f)$ occurring for the trivial subgroup, $H = 1$.

The classical Lefschetz number, which can sometimes be used to detect the existence of fixed points of a function f, is an alternating sum of traces of functions using information from the entire domain of f. As we will see, $N_H(f)$ is defined using neighborhoods of fixed points of f rather than using information involving all of U. The local generalized H-Lefschetz number $L_H(f; \widetilde{f}, \widetilde{\imath})$ is defined as an alternating sum of trace-like elements that are traces of functions defined on all of U. Also, $L_H(f; \widetilde{f}, \widetilde{\imath})$ provides information about $N_H(f)$. Thus $L_H(f; \widetilde{f}, \widetilde{\imath})$ provides a link between local Nielsen theory and Lefschetz trace theory.

We provide here, for readers unfamiliar with these topics, the background necessary for the rather technical results from [**4**] (joint work with J. Fares) and from [**7**] as well as a summary of those results. Local H-Nielsen fixed point theory is introduced in [**2**]. Note that if $U = X$ we have the classical Nielsen

---

1991 *Mathematics Subject Classification.* Primary 55M20.

This papers is in final form and no version of it will be submitted for publication elsewhere.

fixed point theory. Much of the work summarized here is a generalization of results for classical Nielsen theory by Fadell and Husseini in [**1**].

Local Nielsen theory is restricted to those maps f with $\mathrm{Fix}(f)$ compact in U. Similarly, in the definition of $\min(f)$, we require all homotopies $h : U \times I \to X$ to be admissible in the sense that they must have $\bigcup_{t \in I} \mathrm{Fix}(h_t)$ compact in U.

The calculation of $L_H(f; \widetilde{f}, \widetilde{\imath})$ is often difficult. An example for which $L_H(f; \widetilde{f}, \widetilde{\imath})$ can be computed appears in [**7**]. It involves maps $f : U \to L(p,q)$ defined on subspaces of lens spaces. In [**3**], Fadell and Husseini develop a process for calculating $L_H(f; \widetilde{f}, \widetilde{\imath})$ when X is a compact surface and $U = X$. The process works as long as certain orbits in a group action can be distinguished. train tracks. It is hoped that additional methods for calculating $L_H(f; \widetilde{f}, \widetilde{\imath})$, perhaps extensions of the method in [**3**] to other spaces, will be developed in the future. See [**8**], [**6**] and [**10**] for generalized Lefschetz numbers in other settings. For a study of the relationship between the local Nielsen number and the extension Nielsen number, see [**11**].

2. Background

Let X and U be as above with $f : U \to X$ and $\mathrm{Fix}(f)$ compact, and let H be any normal subgroup of the fundamental group $\pi_1(X)$. The set $\{f, U, X, H\}$ is the set of initial data we will study. In [**9**], only certain normal subgroups are allowed. Here, there is no restriction on the choice of normal subgroup.

The local H-Nielsen number of f can be defined algebraically, using covering transformations, or can be defined geometrically. For the geometric definition, we partition $\mathrm{Fix}(f)$ into equivalence classes. (See [**2**], [**8**] and [**9**].) Two fixed points x and y are equivalent if and only if there exists a path ω in U from x to y for which the loop $(f \circ \omega) * \omega^{-1}$ is in a loop class in H. The equivalence classes of fixed points are local H-Nielsen classes for f. Each class is assigned an integer called the local index of the class. (See [**5**].) A class of fixed points of f with non-zero index is called essential, because it cannot be removed by a deformation of f without introducing new fixed points. As in classical Nielsen theory, there is a one-to-one correspondence between the essential classes of f and the essential classes of g, whenever $g \simeq f$ via an admissible homotopy. The local H-Nielsen number of f, denoted by $N_H(f)$, is defined to be the number of essential local H-Nielsen classes. Thus $N_H(f)$ is invariant under admissible homotopy, and we have $N_H(f) \leq \min\{|\mathrm{Fix}(g)| : g \simeq f\}$.

The local index of a fixed point depends on the behavior of f on a neighborhood of the fixed point. We want to express $N_H(f)$ in terms of information about f on all of U. This motivates us to define the local generalized H-Lefschetz number in the next section.

If H_1 and H_2 are normal subgroups of $\pi_1(X)$ with $H_1 \subset H_2$, we have $N_{H_2}(f) \leq N_{H_1}(f)$. Thus our best lower bound for $\min(f)$ is $N_1(f)$, where 1 is the trivial subgroup. Unfortunately, it is often more difficult to calculate

$N_1(f)$ than $N_H(f)$ for some larger H.

The algebraic definition of $N_H(f)$ involves covering spaces. There exists a compact subset K of U large enough to contain all of the information from U that is required to determine local H-Nielsen classes. In other words, $\mathrm{Fix}(f) \subset \mathrm{int}(K)$, and two fixed points are equivalent in U if and only if they are also equivalent in K. We call such a K a stable subset of U. It is always possible to choose K with $K = \overline{\mathrm{int}(K)}$. Then whenever X is an n-manifold, K is an n-submanifold.

In order to study local H-Nielsen classes algebraically, we develop a connection between the classes of fixed points and the lifts of f to covering spaces of K and of X. We will see that the classes of fixed points can be described in terms of a group action called the Reidemeister action.

Let $\widetilde{X}$ be a regular covering space for X with $\widetilde{\pi}_X = \pi_1(X)/H$ the group of covering transformations. Let $\widetilde{K}$ be the universal covering space of K with $\widetilde{\pi}_K = \pi_1(K)$. (In [**4**], a more general setting is developed in which other covering spaces for K are permitted.)

There exist lifts of $f : K \to X$ and of the inclusion $i : K \hookrightarrow X$ making the following diagrams commute. Note that $\widetilde{\imath}$ is not an inclusion.

$$\begin{array}{ccc} \widetilde{K} & \xrightarrow{\tilde{f}} & \widetilde{X} \\ \downarrow \tilde{p}_K & & \downarrow \tilde{p}_X \\ K & \xrightarrow{f} & X \end{array} \qquad \begin{array}{ccc} \widetilde{K} & \xrightarrow{\tilde{\imath}} & \widetilde{X} \\ \downarrow \tilde{p}_K & & \downarrow \tilde{p}_X \\ K & \xrightarrow{i} & X \end{array}$$

Each lift of f to $\widetilde{K}$ may be written uniquely as $\alpha\widetilde{f}$ for some $\alpha \in \widetilde{\pi}_X$. An analogous statement is true for $\widetilde{\imath}$. The map $\widetilde{f}$ induces a homomorphism $\widetilde{\varphi} : \widetilde{\pi}_K \to \widetilde{\pi}_X$ given by $\widetilde{f}(\tau\widetilde{y}) = \widetilde{\varphi}(\tau)\widetilde{f}(\widetilde{y})$ for all $\tau \in \widetilde{\pi}_K$ and all $\widetilde{y} \in \widetilde{K}$. The lift $\widetilde{\imath}$ induces a homomorphism $\widetilde{\xi} : \widetilde{\pi}_K \to \widetilde{\pi}_X$ determined by the following formula. For all $\tau \in \widetilde{\pi}_K$ and $\widetilde{y} \in \widetilde{K}$, $\widetilde{\imath}(\tau\widetilde{y}) = \widetilde{\xi}(\tau)\widetilde{\imath}(\widetilde{y})$.

For any $\alpha \in \widetilde{\pi}_X$, points upstairs in $\widetilde{K}$ that are coincidence points for the lifts $\alpha\widetilde{f}$ and $\widetilde{\imath}$ are related to a class of fixed points of f downstairs. The set of coincidence points for $\alpha\widetilde{f}$ and $\widetilde{\imath}$ is $\mathrm{Coin}(\alpha\widetilde{f}, \widetilde{\imath}) = \left\{\widetilde{x} \in \widetilde{K} : \alpha\widetilde{f}(\widetilde{x}) = \widetilde{\imath}(\widetilde{x})\right\}$. We define N_H^α to be the image $\widetilde{p}_K\left(\mathrm{Coin}(\alpha\widetilde{f}, \widetilde{\imath})\right)$ in K. It can be shown that N_H^α equals either the empty set or a local H-Nielsen class of fixed points of f. (See [**2**] for a proof for the case $H = 1$.) The question remains: When do two of these images equal the same class of fixed points in K?

To answer that question, we consider a well-known group action. The Reidemeister action of $\widetilde{\pi}_K$ on $\widetilde{\pi}_X$ is given by the following equation. For any $\tau \in \widetilde{\pi}_K$ and any $\alpha \in \widetilde{\pi}_X$,

$$\tau \cdot \alpha = \widetilde{\xi}(\tau)\alpha\widetilde{\varphi}(\tau^{-1}).$$

Let $[\alpha]$ denote the orbit of α under the Reidemeister action, and let R denote the set of Reidemeister orbits. Let α and β be in $\widetilde{\pi}_X$. Whenever N_H^α and N_H^β are both non-empty we have $N_H^\alpha = N_H^\beta$ if and only if $[\alpha] = [\beta]$. Thus we can algebraically determine the number of classes of fixed points by determining the number of Reidemeister orbits that correspond to non-empty sets of coincidences in the covering space. Once again, the local H-Nielsen number $N_H(f)$ is the number of classes with non-zero index. Two difficulties remain. We do not always have a simple method for determining algebraically the index of a class of fixed points. Also, given two elements of $\widetilde{\pi}_X$, we do not always know whether they are in the same Reidemeister orbit.

The local generalized H-Lefschetz number provides information about the local Nielsen classes and their indices in a unified form.

3. The local generalized *H*-Lefschetz number

The following definition of $L_H(f;\widetilde{f},\widetilde{\imath})$, Theorem 3.1 and Corollary 3.1 appear in detail in **[4]** (joint work with J. Fares). Here we omit most of the technicalities that are needed to make the definition of $L_H(f;\widetilde{f},\widetilde{\imath})$ precise. There are triangulations and subdivisions involved which can obscure the basic ideas.

Let $C_q(\widetilde{K};\mathbb{Z})$ and $C_q(\widetilde{p}_X{}^{-1}(K);\mathbb{Z})$ be groups of oriented simplicial chains. Both of these chain groups are modules over the group ring $\mathbb{Z}[\widetilde{\pi}_X]$. There is a basis B_q of $C_q(\widetilde{K};\mathbb{Z})$ over $\mathbb{Z}[\widetilde{\pi}_X]$ for which $\widetilde{\imath}(B_q)$ is in one-to-one correspondence with B_q and for which $\widetilde{\imath}(B_q)$ is a basis for $C_q(\widetilde{p}_X{}^{-1}(K);\mathbb{Z})$ over $\mathbb{Z}[\widetilde{\pi}_X]$. The lift $\widetilde{f}$ induces a $\mathbb{Z}[\widetilde{\pi}_X]$-homomorphism

$$\Phi_q : C_q(\widetilde{K};\mathbb{Z}) \to C_q(\widetilde{p}_X{}^{-1}(K);\mathbb{Z}).$$

If we associate elements of B_q with the corresponding elements of $\widetilde{\imath}(B_q)$, we can define a trace-like element of $\mathbb{Z}[\widetilde{\pi}_X]$ for Φ_q. Let $\mathbb{Z}(R)$ be the free abelian group generated by the orbits of the Reidemeister action. Then we can define $T^R(\Phi_q) \subset \mathbb{Z}(R)$ to be the image of the trace of Φ_q when each element of $\widetilde{\pi}_X$ is sent to its orbit in R.

Finally, we define the local generalized H-Lefschetz number to be

$$L_H(f;\widetilde{f},\widetilde{\imath}) = \sum_q (-1)^q T^R(\Phi_q).$$

The connection between $L_H(f;\widetilde{f},\widetilde{\imath})$ and the local H-Nielsen number appears in Theorem 3.1.

To define $L_H(f;\widetilde{f},\widetilde{\imath})$ for a set of initial data, we have chosen a stable subset K of U, subdivisions of K and X and lifts of f and i. The information provided by $L_H(f;\widetilde{f},\widetilde{\imath})$ is essentially independent of these choices, even though the choices of $\widetilde{f}$ and $\widetilde{\imath}$ affect the Reidemeister action slightly. $L_H(f;\widetilde{f},\widetilde{\imath})$ is invariant under homotopy in the following sense. If $h : U \times I \to X$ is an admissible homotopy and if $\widetilde{h}_0$ and $\widetilde{h}_1$ are homotopic via a lift of h, then $L_H(h_0;\widetilde{h}_0,\widetilde{\imath}) = L_H(h_1;\widetilde{h}_1,\widetilde{\imath})$.

THEOREM 3.1. Given a set of initial data with X a connected, triangulable

n-manifold and $n \geq 3$, we have

$$L_H(f; \widetilde{f}, \widetilde{\imath}) = \sum_{\alpha \in W} i(N_H^\alpha)[\alpha] \in \mathbb{Z}(R),$$

where $W \subset \widetilde{\pi}_X$ contains exactly one element from each orbit of the Reidemeister action, and $i(N_H^\alpha)$ is the index of the local H-Nielsen class N_H^α. Thus, once $L_H(f; \widetilde{f}, \widetilde{\imath})$ is in this reduced form, $N_H(f)$ is the number of terms of $L_H(f; \widetilde{f}, \widetilde{\imath})$ with non-zero coefficient.

The sum of the coefficients in $L_H(f; \widetilde{f}, \widetilde{\imath})$ is equal to the classical local Lefschetz number as defined in [**5**]. In this theorem, K is automatically an n-submanifold of X, because K is chosen with $K = \overline{\text{int} K}$.

We would like to detect whether f is homotopic, via an admissible homotopy, to a map that is fixed point free. That is, we would like to know whether $\min(f) = 0$. For $H = 1$, a local obstruction is defined in [**2**] that detects whether f has this property. The preceding theorem combined with Theorem 5.12 of [**2**] provides the following generalization of the converse of the Lefschetz fixed point theorem.

COROLLARY 3.1. For a set of initial data with X a connected, triangulable n-manifold that is not simply connected, with $n \geq 3$ and $H = 1$, we have

$$L_1(f; \widetilde{f}, \widetilde{\imath}) = 0 \Leftrightarrow \min(f) = 0.$$

The following theorem appears in [**7**]. It is a generalization of work by Fadell and Husseini in [**1**], where the special case $U = X$ is considered. Note that we no longer require that X be a manifold. In [**7**], we define the lift index Λ that assigns an integer to each lift of f. In the special case $U = X$, the lift index is the same as the classical Lefschetz number of the lift.

THEOREM 3.2. Given a set of initial data for which each lift $\widetilde{f}$ is compact (that is, the closure of $\text{im}(\widetilde{f})$ is compact), let W be a subset of $\widetilde{\pi}_X$ containing exactly one element from each orbit of the Reidemeister action. If the isotropy subgroup $(\widetilde{\pi}_K)_\gamma$ for the Reidemeister action is finite for every $\gamma \in \widetilde{\pi}_X$, we have

$$L_H(f; \widetilde{f}, \widetilde{\imath}) = \sum_{\alpha \in W} \frac{\Lambda(\alpha \widetilde{f})}{|(\widetilde{\pi}_K)_\alpha|}[\alpha] = \sum_{\alpha \in W} i(N_H^\alpha)[\alpha] \in \mathbb{Z}(R).$$

Thus whenever $L_H(f; \widetilde{f}, \widetilde{\imath})$ is written in reduced form (with each Reidemeister orbit appearing at most once) the local H-Nielsen number $N_H(f)$ is the number of terms with coefficient different from zero.

Using this theorem, local Nielsen numbers are calculated in [**7**] for maps $f : K \to L(p, q)$, where K is a solid torus in a lens space.

REFERENCES

1. E. Fadell and S. Husseini, *Fixed point theory for non simply connected manifolds*, Topology **20** (1980), 53–92.

2. E. Fadell and S. Husseini, *Local fixed point index theory for non simply connected manifolds*, Illinois J. of Math. **25** (1981), 673–699.
3. E. Fadell and S. Husseini, *The Nielsen number on surfaces*, Contemp. Math. **21** (1983), 59–98.
4. J. Fares and E. Hart, *A generalized Lefschetz number for local Nielsen fixed point theory*, Top. & Appl. (to appear).
5. G. Fournier, *A simplicial approach to the fixed point index*, Fixed Point Theory, Proceedings, Lecture Notes in Math. vol. 886, Springer-Verlag, Berlin and New York, 1980) pp. 73–102.
6. R. Geoghegan and A. Nicas, *Parametrized Lefschetz-Nielsen fixed point theory and Hochschild homology traces*, Amer. J. Math. (to appear).
7. E. Hart, *Computation of the local generalized H-Lefschetz number*, preprint.
8. S. Husseini, *Generalized Lefschetz numbers*, Trans. Amer. Math. Soc. **272** (1982), 247–274.
9. D. McCord, *The converse of the Lefschetz fixed point theorem for surfaces and higher dimensional manifolds*, Ph.D. Thesis, University of Wisconsin-Madison, 1970.
10. B. Norton-Odenthal and P. Wong, *A relative generalized Lefschetz number*, Top. & Appl. (to appear).
11. P. Wong, *A note on the local and the extension Nielsen numbers*, Top. & Appl. **48** (1992), 207–213.

Department of Mathematics, Hope College, Holland, MI 49423
E-mail address: hart@hope.bitnet, hart@physics.hope.edu

Contemporary Mathematics
Volume **152**, 1993

Nielsen theory for periodic orbits and applications to dynamical systems

BOJU JIANG

ABSTRACT. A Nielsen theory for periodic orbits is presented that allows quantitative estimation of the number of periodic orbits for a self-map $f : X \to X$ of a compact polyhedron. Of special interest is the estimation of the asymptotic growth rate of the number of periodic orbits and the homotopy invariant lower estimation of the topological entropy, via a matrix representation of $\pi_1(T_f)$, where T_f is the mapping torus of f. As applications, some recent results in two-dimensional dynamical systems theory are improved.

Section 1. Introduction

Fixed point theory studies fixed points of a self-map f of a space X. Nielsen fixed point theory, in particular, is concerned with the properties of the fixed point set $\operatorname{Fix} f := \{x \in X \mid x = f(x)\}$ that are invariant under homotopies of the map f. When applied to the iterates of f, it provides information about the periodic points of f, e.g. lower bounds of the number of periodic points of various periods.

On the other hand, dynamical systems theory of maps studies the behavior of the orbits of a map f, i.e. sets of the form $O_f(x) := \{f^n(x) \mid 1 \le n < \infty\}$, among which the periodic ones play an important role. Clearly the topology of the space X and the homotopy properties of the map f would restrict the possible dynamical behavior of f. So it is no surprise that the two fields have a common interest and close interaction in the topological study of periodic points. In the early 1980's, long after the pioneering work [**Fu1**], [**Fu2**], researchers in both fields independently and almost simultaneously started to take up this study, see for example [**AF**], [**Fr1**], [**Fr2**], [**I**], [**J1**, §III.4], [**J2**], [**M1**], [**M2**].

1991 *Mathematics Subject Classification.* Primary 55M20; Secondary 57N05, 58C30, 58F20.

Partially supported by NSFC.
This paper is in final form and no version of it will be submitted for publication elsewhere.

An ideal testing ground for the homotopic theory of periodic points is the case of surface homeomorphisms. The Nielsen-Thurston classification theory has already shown that for generic surface homeomorphisms, the number of periodic points grows exponentially with the period. We would expect that a powerful periodic point theory can help to calculate or at least estimate the asymptotic growth rate. However, all the estimates obtained for the number of periodic points came short of confirming exponential growth. In other words, the topological periodic point theory was not able to match, let alone to improve, the relevant results of dynamical systems theory.

It is this gap that motivated the work [**J4**]. The present article is an exposition of the results in that paper and related work of the author. The aim is to introduce more algebraic tools into the theory so that new information on two-dimensional dynamical systems can be obtained. The exposition is divided into two sections. Section 2 sketches the theory and Section 3 describes applications. We shall emphasize the basic notions and illustrate the methods by concrete examples, without proofs.

In the theoretical section, we start with a review of Nielsen's notion of fixed point classes. Instead of counting periodic *points* of f (i.e. fixed points of f^n), we propose to count the periodic *orbits* of f and introduce the notion of periodic orbit classes. This slight shift of emphasis allows for a natural interpretation on the mapping torus T_f of f, and thus leads to familiar algebraic machinery.

Associated to matrix representations of the fundamental group $\pi_1(T_f)$, we introduce the notion of zeta functions of f. The zeta function is a formal power series that encodes periodic orbit information of all periods, hence also carries asymptotic information. On the other hand it is a rational function that is practically computable. So it is the key to the estimation of asymptotic invariants. Our zeta function differs from that of Fried [**Fr1**] in that matrix representations of $\pi_1(T_f)$ are used instead of f-invariant abelianizations of $\pi_1(X)$, so that non-abelian information can be better retained. This does make a difference in applications, as we will show in §3.4.

Homeomorphisms of compact surfaces and punctured surfaces are then discussed. The asymptotic Nielsen number is shown to be the largest stretching factor in the Thurston canonical form. Our method of estimation is of interest even for a pseudo-Anosov map, because it provides a practical way to estimate the stretching factor from the automorphism on the fundamental group.

In the section on applications, we pay more attention to 2-dimensional dynamical systems that are specified by certain braid data. This kind of systems was first studied by Matsuoka [**M1**] and has been much studied recently. The famous statement "period three implies chaos" for maps in one dimension is no longer true in two dimensions, but similar phenomena still persist in a more subtle form. Matsuoka observed that for planar homeomorphisms, a known periodic orbit can give rise to a braid and this data can tell much about the periodic orbits. He showed that 3-string braids almost always (with the only exception of certain simplest braids) guarantee the existence of infinitely many periodic orbits. We are able to strengthen his conclusion to positive topological entropy and exponential growth of periodic orbits that are typical of chaotic behavior, and even supply numerical lower estimates in various examples. This demonstrates the effectiveness of our theory.

Section 2. Nielsen theory for periodic orbits

2.1. Fixed point classes via the mapping torus approach.

The basis of Nielsen fixed point theory is the notion of a fixed point class.

Let X be a compact connected polyhedron, $f : X \to X$ be a map. The fixed point set $\operatorname{Fix} f := \{x \in X \mid x = f(x)\}$ splits into a disjoint union of *fixed point classes.* Two fixed points are in the same class if and only if they can be joined by a path which is homotopic (relative to end-points) to its own f-image. Each fixed point class $\mathbf{F}$ is an isolated subset of $\operatorname{Fix} f$, hence its *index* $\operatorname{ind}(\mathbf{F}, f) \in \mathbb{Z}$ is defined. A fixed point class is called *essential* if its index is non-zero. The number of essential fixed point classes is called the *Nielsen number* $N(f)$ of f. It is a homotopy invariant of f, so that every map homotopic to f must have at least $N(f)$ fixed points. (Cf. [**J1**, p.19].)

The notion of fixed point classes can be interpreted in terms of the mapping torus (see [**J2**]).

The *mapping torus* T_f of $f : X \to X$ is the space obtained from $X \times \mathbb{R}_+$ by identifying $(x, s+1)$ with $(f(x), s)$ for all $x \in X$, $s \in \mathbb{R}_+$, where $\mathbb{R}_+$ stands for the real interval $[0, \infty)$. On T_f there is a natural semi-flow ("sliding along the rays")

$$\varphi : T_f \times \mathbb{R}_+ \to T_f, \qquad \varphi_t(x, s) = (x, s+t) \text{ for all } t \geq 0.$$

A point $x \in X$ and a positive number $\tau > 0$ determine an orbit curve $\varphi_{(x,\tau)} := \{\varphi_t(x)\}_{0 \leq t \leq \tau}$ in T_f. We may identify X with the cross-section $X \times 0 \subset T_f$, then the map $f : X \to X$ is just the return map of the semi-flow φ.

Now, a point $x \in X$ is a fixed point of f if and only if the time-one orbit curve $\varphi_{(x,1)}$ is a closed curve. It turns out that $x, y \in \operatorname{Fix} f$ are in the same fixed point class if and only if the closed curves $\varphi_{(x,1)}$ and $\varphi_{(y,1)}$ are freely homotopic in T_f. (The term "free homotopy" means homotopy with no concern about base point.)

NOTATION. Let Γ be the fundamental group $\Gamma := \pi_1(T_f)$ and let Γ_c denote the set of conjugacy classes in Γ. Theoretically, it is better to regard Γ_c as the set of free homotopy classes of closed curves in T_f, so that it is independent of the base point of T_f. Let $\mathbb{Z}\Gamma$ be the integral group ring of Γ, and let $\mathbb{Z}\Gamma_c$ be the free abelian group with basis Γ_c. We use the bracket notation $\alpha \mapsto [\alpha]$ for both projections $\Gamma \to \Gamma_c$ and $\mathbb{Z}\Gamma \to \mathbb{Z}\Gamma_c$. In general, for any set S let $\mathbb{Z}S$ denote the free abelian group with the specified basis S. The *norm* in $\mathbb{Z}S$ is defined by $\|\sum_i k_i s_i\| := \sum_i |k_i| \in \mathbb{Z}$ when the s_i's in S are all different.

Let $\mathbf{F}$ be a fixed point class. Since for all $x \in \mathbf{F}$ the closed curves $\varphi_{(x,1)}$ are freely homotopic in T_f, they represent a well defined conjugacy class $[\varphi_{(x,1)}]$ in

Γ. This conjugacy class will be called the *coordinate* of $\mathbf{F}$ (in Γ), written

$$\mathrm{cd}_\Gamma(\mathbf{F}) = [\varphi_{(x,1)}] \in \Gamma_c.$$

We define the (generalized) *Lefschetz number*

$$L_\Gamma(f) := \sum_{\mathbf{F}} \mathrm{ind}(\mathbf{F}, f) \cdot \mathrm{cd}_\Gamma(\mathbf{F}) \quad \in \mathbb{Z}\Gamma_c, \tag{2.1}$$

the summation being over all fixed point classes $\mathbf{F}$ of f. Thus the Nielsen number $N(f)$ is the number of non-zero terms in $L_\Gamma(f)$, and the indices of the essential fixed point classes appear as the coefficients in $L_\Gamma(f)$, so that the norm $\|L_\Gamma(f)\|$ is the sum of absolute values of the indices of all the (essential) fixed point classes.

REMARK. $L_\Gamma(f)$ is, in spirit, the same as the classical invariant called the Reidemeister trace ([**R**], [**We**]) and later called the generalized Lefschetz number (e.g. [**FH**]), except that as coordinates we use ordinary conjugacy classes in $\pi_1(T_f)$ instead of skew conjugacy classes in $\pi_1(X)$, to make it algebraically more manageable.

2.2. Periodic orbit classes.

We intend to study the periodic points of f, i.e. the fixed points of the iterates of f.

We shall call $\mathrm{PP}f := \{(x,n) \in X \times \mathbb{N} \mid x = f^n(x)\}$ the *periodic point set* of f, where $\mathbb{N}$ denotes the set of natural numbers. A fixed point x of f^n is called an *n-point* of f, and its f-orbit $\{x, f(x), \ldots, f^{n-1}(x)\}$ an *n-orbit* of f. The latter is called a *primary* n-orbit if it consists of n distinct points, i.e. if n is the least period of the periodic point x.

A fixed point class $\mathbf{F}^n$ of f^n will be called an *n-point class* of f. One may interpret the n-point classes on the mapping torus T_{f^n}, with coordinates in the fundamental group $\Gamma_n := \pi_1(T_{f^n})$. But for various n these mapping tori are all different spaces and the Γ_n different groups. Rather, we propose to work on the same mapping torus T_f, and arrive at the notion of a periodic orbit class.

Observe that $(x,n) \in \mathrm{PP}f$, or equivalently $x \in \mathrm{Fix}\, f^n$, if and only if on the mapping torus T_f the time-n orbit curve $\varphi_{(x,n)}$ is a closed curve. We define $x, y \in \mathrm{Fix}\, f^n$ to be in the same *n-orbit class* if and only if $\varphi_{(x,n)}$ and $\varphi_{(y,n)}$ are in the same free homotopy class of closed curves in T_f. $\mathrm{Fix}\, f^n$ splits into a disjoint union of n-orbit classes.

Let $\mathbf{O}^n$ be an n-orbit class. If $x \in \mathbf{O}^n$, then its n-orbit $\{x, f(x), \cdots, f^{n-1}(x)\}$ is entirely in $\mathbf{O}^n$, because the closed orbit curves of these points are the same curve with different base points. Thus $\mathbf{O}^n$ is a union of n-orbits, hence the name "n-orbit class".

What is the relation between the notion of n-orbit classes and that of n-point classes? It turns out that [**J2**, §3] if $x \in \mathbf{O}^n$ and $\mathbf{F}^n$ is the n-point class containing x, then $\mathbf{O}^n = \mathbf{F}^n \cup f(\mathbf{F}^n) \cup \cdots \cup f^{n-1}(\mathbf{F}^n)$. So an n-orbit class is nothing but an f-orbit of n-point classes.

Thus the price we pay for dealing with a single mapping torus T_f is that we can no longer distinguish n-point classes individually, but only their f-orbits. This is quite agreeable for most purposes.

2.3. Coordinates and reducibility.

Since for all $x \in \mathbf{O}^n$ the closed curves $\varphi_{(x,n)}$ are freely homotopic in T_f, they represent a well defined conjugacy class $[\varphi_{(x,n)}]$ in Γ. This conjugacy class will be called the *coordinate* of $\mathbf{O}^n$ in Γ, written

$$\mathrm{cd}_\Gamma(\mathbf{O}^n) = [\varphi_{(x,n)}] \in \Gamma_c.$$

An important notion in the Nielsen theory for periodic orbits is that of reducibility. Suppose m is a factor of n and $m < n$. When the n-orbit class $\mathbf{O}^n$ contains an m-orbit class $\mathbf{O}^m$ then $\mathrm{cd}_\Gamma(\mathbf{O}^n)$ is the (n/m)-th power of $\mathrm{cd}_\Gamma(\mathbf{O}^m)$, because for $x \in \mathbf{O}^m$ the closed curve $\varphi_{(x,n)}$ is the closed curve $\varphi_{(x,m)}$ traced n/m times. This motivates the definition that the n-orbit class $\mathbf{O}^n$ is *reducible to period* m if $\mathrm{cd}_\Gamma(\mathbf{O}^n)$ has an (n/m)-th root, and that $\mathbf{O}^n$ is *irreducible* if $\mathrm{cd}_\Gamma(\mathbf{O}^n)$ is *primary* in the sense that it has no nontrivial root.

This notion of reducibility is consistent with that introduced in [**J1**]. An n-orbit class $\mathbf{O}^n$ is reducible to period m if and only if every n-point class $\mathbf{F}^n \subset \mathbf{O}^n$ is reducible to period m in the sense of [**J1**, Definition III.4.2].

2.4. Lefschetz numbers and n-orbit Nielsen numbers.

Every n-orbit class $\mathbf{O}^n$ is an isolated subset of $\mathrm{Fix}\, f^n$. Its *index* is $\mathrm{ind}(\mathbf{O}^n, f^n)$, the index of $\mathbf{O}^n$ with respect to f^n. An n-orbit class $\mathbf{O}^n$ is called *essential* if its index is non-zero.

For each natural number n, we define the (generalized) *Lefschetz number* (with respect to Γ)

$$L_\Gamma(f^n) := \sum_{\mathbf{O}^n} \mathrm{ind}(\mathbf{O}^n, f^n) \cdot \mathrm{cd}_\Gamma(\mathbf{O}^n) \quad \in \mathbb{Z}\Gamma_c, \tag{2.2}$$

the summation being over all n-orbit classes $\mathbf{O}^n$ of f.

The number of non-zero terms in $L_\Gamma(f^n)$ will be denoted $N_\Gamma(f^n)$, and called the *n-orbit Nielsen number* of f. It is the number of essential n-orbit classes, a lower bound for the number of n-orbits of f. The norm $\|L_\Gamma(f^n)\|$ is the sum of absolute values of the indices of all the (essential) n-orbit classes. It equals the sum of absolute values of the indices of all the (essential) n-point classes, because any two n-point classes contained in the same n-orbit class must have the same index. Hence $\|L_\Gamma(f^n)\| \geq N(f^n) \geq N_\Gamma(f^n)$.

Let $NI_\Gamma(f^n)$ be the number of non-zero primary terms in $L_\Gamma(f^n)$, called the *irreducible n-orbit Nielsen number* of f. It is the number of irreducible essential n-orbit classes, a lower bound for the number of primary n-orbits.

REMARK. For the readers who know the Nielsen type numbers defined in [**J1**, §III.4], we mention that $NI_\Gamma(f^n) = \frac{1}{n} NP_n(f)$.

2.5. Invariance properties.

The basic invariance properties, such as the homotopy invariance, the homotopy type invariance and the commutativity property, are similar (with similar proofs) to that for fixed points (cf. [**J1**, §§I.4–5]). For example, the

HOMOTOPY TYPE INVARIANCE. *Suppose $h : X \to X'$ is a homotopy equivalence. Suppose $f : X \to X$ and $f' : X' \to X'$ are maps such that the diagram*

$$\begin{array}{ccc} X & \xrightarrow{f} & X \\ {\scriptstyle h}\downarrow & & \downarrow{\scriptstyle h} \\ X' & \xrightarrow{f'} & X' \end{array}$$

commutes up to homotopy. Then $T_{f'}$ is homotopy equivalent to T_f, and when $\Gamma' = \pi_1(T_{f'})$ is suitably identified with $\Gamma = \pi_1(T_f)$, we have $L_\Gamma(f'^n) = L_\Gamma(f^n)$ for all n, hence also $N_\Gamma(f'^n) = N_\Gamma(f^n)$ and $NI_\Gamma(f'^n) = NI_\Gamma(f^n)$.

2.6. The trace formula for the Lefschetz numbers.

So far $L_\Gamma(f^n)$ is defined as a formal sum organizing the index and coordinate information of the periodic orbit classes. Its importance lies in its computability.

Pick a base point $v \in X$ and a path w from v to $f(v)$. Let $G := \pi_1(X, v)$ and let $f_G : G \to G$ be the composition

$$\pi_1(X,v) \xrightarrow{f_*} \pi_1\big(X, f(v)\big) \xrightarrow{w_*} \pi_1(X,v).$$

Let $p : \widetilde{X}, \tilde{v} \to X, v$ be the universal covering. The deck transformation group is identified with G. Let $\tilde{f} : \widetilde{X} \to \widetilde{X}$ be the lift of f such that the reference path w lifts to a path from $\tilde{v}$ to $\tilde{f}(\tilde{v})$. Then for every $g \in G$ we have $\tilde{f} \circ g = f_G(g) \circ \tilde{f}$ (cf. [**J1**, pp.24–25]).

Assume that X is a finite cell complex and $f : X \to X$ is a cellular map. Pick a cellular decomposition $\{e_j^d\}$ of X, the base point v being a 0-cell. It lifts to a G-invariant cellular structure on the universal covering $\widetilde{X}$. Choose an arbitrary lift $\tilde{e}_j^d$ for each e_j^d. These lifts constitute a free $\mathbb{Z}G$-basis for the cellular chain complex of $\widetilde{X}$. The lift $\tilde{f}$ of f is also a cellular map. In every dimension d, the cellular chain map $\tilde{f}$ gives rise to a $\mathbb{Z}G$-matrix $\widetilde{F}_d$ with respect to the above basis, i.e. $\widetilde{F}_d = (a_{ij})$ if $\tilde{f}(\tilde{e}_i^d) = \sum_i a_{ij}\tilde{e}_j^d$, $a_{ij} \in \mathbb{Z}G$.

For the mapping torus, take the base point v of X as the base point of T_f (recall that X is regarded as embedded in T_f). Let $\Gamma = \pi_1(T_f, v)$. By the van Kampen Theorem, Γ is obtained from G by adding a new generator z represented by the loop $\varphi_{(v,1)}w^{-1}$, and adding the relations $z^{-1}gz = f_G(g)$ for all $g \in G$:

$$\Gamma = \langle G, z \mid gz = zf_G(g) \text{ for all } g \in G \rangle. \tag{2.3}$$

Note that the homomorphism $G \to \Gamma$ induced by the inclusion $X \subset T_f$ is not necessarily injective.

In this notation, we can adapt the Reidemeister trace formula ([**R**], [**We**], see [**HJ**, §1] for an exposition) to our mapping torus setting and get a simple formula

$$L_\Gamma(f) = \sum_d (-1)^d \left[\operatorname{tr}(z\widetilde{F}_d)\right] \quad \in \mathbb{Z}\Gamma_c, \tag{2.4}$$

where $z\widetilde{F}_d$ is regarded as a matrix in $\mathbb{Z}\Gamma$. It generalizes to the

TRACE FORMULA FOR LEFSCHETZ NUMBERS. *For the Lefschetz numbers we have*

$$L_\Gamma(f^n) = \sum_d (-1)^d \left[\operatorname{tr}(z\widetilde{F}_d)^n\right] \quad \in \mathbb{Z}\Gamma_c. \tag{2.5}$$

EXAMPLE. (A bouquet of circles)

Let X be a bouquet of r circles with one 0-cell v and r 1-cells $a_1, \cdots, a_r$, and let $f : X \to X$ be a cellular map. The homomorphism $f_G : G \to G$ induced by f is determined by the images $a_i' := f_G(a_i)$, $i = 1, \cdots, r$. By (2.3), $\Gamma = \pi_1(T_f)$ has a presentation

$$\Gamma = \langle\, a_1, \cdots, a_r, z \mid a_i z = z a_i', \ i = 1, \cdots, r \,\rangle. \tag{2.6}$$

As pointed out in [**FH**], the matrices of the lifted chain map $\tilde{f}$ are

$$\tilde{F}_0 = (1),$$
$$\tilde{F}_1 = D := \left(\frac{\partial a_i'}{\partial a_j}\right),$$

where D is the Jacobian matrix in Fox calculus (see [**Bi**, §3.1] for an introduction). Then, by (2.4) and (2.5), in $\mathbb{Z}\Gamma_c$ we have

$$L_\Gamma(f) = [z] - \sum_{i=1}^{r} \left[z\frac{\partial a_i'}{\partial a_i}\right], \tag{2.7}$$

$$L_\Gamma(f^n) = [z^n] - [\operatorname{tr}(zD)^n]. \tag{2.8}$$

2.7. Twisted Lefschetz numbers and Lefschetz zeta function.

Suppose a group representation $\rho : \Gamma \to \mathrm{GL}_l(F)$ is given, where F is a field of characteristic 0. Then ρ extends to a ring representation $\rho : \mathbb{Z}\Gamma \to \mathcal{M}_{l\times l}(F)$, where $\mathcal{M}_{l\times l}(F)$ is the algebra of $l \times l$ matrices in F.

Define the *ρ-twisted Lefschetz number*

$$L_\rho(f^n) := \operatorname{tr}\left(L_\Gamma(f^n)\right)^\rho = \sum_{\mathbf{O}^n} \operatorname{ind}(\mathbf{O}^n, f^n) \cdot \operatorname{tr}\left(\operatorname{cd}_\Gamma(\mathbf{O}^n)\right)^\rho \quad \in F \tag{2.9}$$

for every $n \in \mathbb{N}$, the summation being over all n-orbit classes $\mathbf{O}^n$. It has the trace formula

$$L_\rho(f^n) = \sum_d (-1)^d \operatorname{tr}\left((z\widetilde{F}_d)^\rho\right)^n \quad \in F, \tag{2.10}$$

where for a $\mathbb{Z}\Gamma$-matrix A, its ρ-image A^ρ means the block matrix obtained from A by replacing each element a_{ij} with the $l \times l$ matrix a_{ij}^ρ.

We now define the (ρ-twisted) *Lefschetz zeta function* of f to be the formal power series

$$\zeta_\rho(f) := \exp \sum_n L_\rho(f^n) \frac{t^n}{n}. \tag{2.11}$$

It has constant term 1, so it is in the multiplicative subgroup $1 + tF[[t]]$ of the formal power series ring $F[[t]]$.

Clearly $\zeta_\rho(f)$ enjoys the same invariance properties as that of $L_\Gamma(f^n)$. As to its computation, we obtain from (2.10) the following

DETERMINANT FORMULA FOR THE LEFSCHETZ ZETA FUNCTION. *$\zeta_\rho(f)$ is a rational function in F.*

$$\zeta_\rho(f) = \prod_d \det\left(I - t(z\widetilde{F}_d)^\rho\right)^{(-1)^{d+1}} \quad \in F(t), \tag{2.12}$$

where I stands for suitable identity matrices.

By (2.9), (2.11) and the homotopy invariance, we have the

TWISTED VERSION OF THE LEFSCHETZ FIXED POINT THEOREM. *Let $f : X \to X$ be a map and $\rho : \pi_1(T_f) \to \mathrm{GL}_l(F)$ be a representation. If f is homotopic to a fixed point free map g, then $L_\rho(f) = 0$. If f is homotopic to a periodic point free map g, then $\zeta_\rho(f) = 1$.*

REMARK. When $F = \mathbb{Q}$ and $\rho : \Gamma \to \mathrm{GL}_1(\mathbb{Q}) = \mathbb{Q}$ is trivial (sending everything to 1), then $L_\rho(f) \in \mathbb{Z}$ is the ordinary Lefschetz number $L(f)$, and $\zeta_\rho(f)$ is the classical Lefschetz zeta function $\zeta(f) := \exp \sum_n L(f^n)t^n/n$ introduced by Weil (cf. [**Bt**]).

The following example provides a very useful recipe for computing $\zeta_\rho(f)$ on surfaces.

EXAMPLE. (Recipe for surfaces with boundary)

Let X be a surface with boundary, and $f : X \to X$ be a map. Suppose $\{a_1, \cdots, a_r\}$ is a free basis for $G = \pi_1(X)$, and suppose f induces the homomorphism $f_G : G \to G$ determined by the images $a_i' := f_G(a_i)$, $i = 1, \cdots, r$. Then X has the homotopy type of a bouquet X' of r circles corresponding to the basis elements, and f has the homotopy type of a map $f' : X' \to X'$ which induces the same homomorphism $G \to G$. By the homotopy type invariance of the invariants, we can replace f with f' in computations. Hence the formulas (2.6–8) are applicable.

When a representation $\rho : \Gamma \to \mathrm{GL}_l(F)$ is given, by (2.10) and (2.12) we have

$$L_\rho(f) = \mathrm{tr} z^\rho - \mathrm{tr}(zD)^\rho \quad \in F, \tag{2.13}$$

$$\zeta_\rho(f) = \frac{\det(I - t(zD)^\rho)}{\det(I - tz^\rho)} \quad \in F(t). \tag{2.14}$$

2.8. Asymptotic invariants.

The *growth rate* of a sequence $\{a_n\}$ of complex numbers is defined by

$$\operatorname*{Growth}_{n\to\infty} a_n := \max\left\{1, \limsup_{n\to\infty} |a_n|^{1/n}\right\} \tag{2.15}$$

which could be infinity. Note that $\operatorname{Growth} a_n = 1$ even if all $a_n = 0$. When $\operatorname{Growth} a_n > 1$, we say that the sequence *grows exponentially.*

We define the *asymptotic Nielsen number* of f to be the growth rate of the Nielsen numbers

$$N^\infty(f) := \operatorname*{Growth}_{n\to\infty} N(f^n) = \operatorname*{Growth}_{n\to\infty} N_\Gamma(f^n), \tag{2.16}$$

where the second equality is due to the obvious inequality $N_\Gamma(f^n) \leq N(f^n) \leq n \cdot N_\Gamma(f^n)$. And we define the *asymptotic irreducible Nielsen number* of f to be the growth rate of the irreducible Nielsen numbers

$$NI^\infty(f) := \operatorname*{Growth}_{n\to\infty} NI_\Gamma(f^n). \tag{2.17}$$

We also define the *asymptotic absolute Lefschetz number*

$$L^\infty(f) := \operatorname*{Growth}_{n\to\infty} \|L_\Gamma(f^n)\|. \tag{2.18}$$

Clearly $NI^\infty(f) \leq N^\infty(f) \leq L^\infty(f)$. All these asymptotic numbers are finite and share the invariance properties of $L_\Gamma(f^n)$.

It is interesting that in many important cases we actually have $NI^\infty(f) = N^\infty(f) = L^\infty(f)$. E.g. when X is a torus of any dimension, or when f is a homeomorphism of a surface X with $\chi(X) < 0$.

The asymptotic invariants measure the growth of the number of periodic orbits. In many applications, the estimation of these growth rates is more important than the estimation for a specific period n. The former is also often easier to do. Therefore in this paper we shall pay more attention to these asymptotic invariants.

A METHOD OF ESTIMATION. *Suppose $F = \mathbb{C}$ and $\rho : \Gamma \to \mathrm{U}(l)$ is a unitary representation. Let r be the minimum modulus of the zeros and poles of the rational function $\zeta_\rho(f)$. Then*

$$L^\infty(f) \geq \frac{1}{r}. \tag{2.19}$$

The proof is easy: We have $L_\rho(f^n) \in \mathbb{C}$ and $|L_\rho(f^n)| \leq l\|L_\Gamma(f^n)\|$, hence $L^\infty(f) \geq \operatorname{Growth} L_\rho(f^n)$. Then we know from complex analysis and (2.11) that $\operatorname{Growth} L_\rho(f^n)$ is the reciprocal of the radius of convergence of the function $\log \zeta_\rho(f)$, hence the formula.

The asymptotic invariants also provide lower bounds for the topological entropy, the most widely used measure of complexity for a map in the theory of dynamical systems (see **[Wa]** for an introduction).

ENTROPY THEOREM. ([**I**]) *Suppose X is a compact polyhedron, $f : X \to X$ is a map and $h(f)$ is its topological entropy. Then $h(f) \geq \log N^{\infty}(f)$.*

EXAMPLE. (Pseudo-Anosov maps, cf. [**FLP**, p.194])

Let X be a surface with $\chi(X) < 0$. Let f be a pseudo-Anosov homeomorphism with stretching factor $\lambda > 1$. Then

$$N^{\infty}(f) = \lambda, \qquad h(f) = \log \lambda. \tag{2.20}$$

EXAMPLE. (Linear maps of tori, cf. [**Wa**, p.203])

Let $T^k := \mathbb{R}^k/\mathbb{Z}^k$ be the k-dimensional torus. Let f be an endomorphism of T^k defined by an integer matrix A. Let $\lambda_1, \dots, \lambda_k$ be the eigenvalues of A. Then the ordinary Lefschetz number $L(f) = \det(I - A) = \prod_i (1 - \lambda_i)$ and

$$N^{\infty}(f) = \begin{cases} 1 & \text{if } L(f) = 0, \\ \prod_{|\lambda_i|>1} |\lambda_i| & \text{otherwise;} \end{cases} \qquad h(f) = \sum_{|\lambda_i|>1} \log |\lambda_i|. \tag{2.21}$$

2.9. Relative invariants mod a subpolyhedron.

Let X be a compact connected polyhedron as before, and A be a subpolyhedron. Let $f : X, A \to X, A$ be a self-map of the pair.

A fixed point x of f is *related to* A if there is a path c such that $c \simeq f \circ c : I, 0, 1 \to X, x, A$, where $\simeq$ means homotopic. A fixed point class $\mathbf{F}$ of f will be called a *fixed point class on* $X \setminus A$ if it is not related to A. The number of essential fixed point classes of f on $X \setminus A$ is called the *Nielsen number of the complement*, denoted $N(f; X \setminus A)$. It is a lower bound for the number of fixed points of f on $X \setminus A$, and it is invariant under homotopy of maps $X, A \to X, A$ ([**Z**], cf. [**S**, §2.3]). Obviously $N(f; X \setminus A) \leq N(f)$.

Under the mapping torus point of view, a fixed point x of f is related to A if and only if the corresponding closed orbit curve $\varphi_{(x,1)}$ in T_f is freely homotopic to a closed curve in $T_{f|A}$, the mapping torus of the restriction $f|A : A \to A$ naturally regarded as a subspace of T_f.

The Nielsen theory of periodic orbits for X developed above has a natural relative version for $X \setminus A$. A free homotopy class of closed curves in T_f (i.e. an element of Γ_c) will be called *related to* A if it contains a closed curve in $T_{f|A} \subset T_f$. An *n-orbit class of f on* $X \setminus A$ is defined to be an n-orbit class of f whose coordinate is not related to A. The *Nielsen number of the complement* $N_\Gamma(f^n; X \setminus A)$ is the number of essential n-orbit classes of f on $X \setminus A$. The *Lefschetz number of the complement* $L_\Gamma(f^n; X \setminus A) \in \mathbb{Z}\Gamma_c$ is obtained from $L_\Gamma(f^n)$ by deleting the terms related to A. Clearly $\|L_\Gamma(f^n; X \setminus A)\| \leq \|L_\Gamma(f^n)\|$.

Asymptotic relative invariants are defined in much the same way as in §2.8, such as

$$N^{\infty}(f; X \setminus A) := \operatorname*{Growth}_{n\to\infty} N(f^n; X \setminus A). \tag{2.22}$$

2.10. Surface homeomorphisms.

Let X be a compact surface with $\chi(X) < 0$, and let $f : X \to X$ be a homeomorphism. Based on Thurston's classification theory of surface homeomorphisms [**T**], a standard form for f is developed in [**JG**] with local models for its fixed point classes. Since all iterates of a standard map are still in standard form, we can obtain detailed knowledge of the periodic point classes. For example, we can see that "almost every" n-point class of f has index ± 1. A careful analysis leads to the following result.

ASYMPTOTIC INVARIANTS FOR SURFACE HOMEOMORPHISMS. *For a homeomorphism $f : X \to X$ of a surface X with $\chi(X) < 0$,*

$$NI^{\infty}(f) = N^{\infty}(f) = L^{\infty}(f) = \lambda, \tag{2.23}$$

where λ is the largest stretching factor of the pseudo-Anosov pieces in the Thurston canonical form of f ($\lambda := 1$ if there is no pseudo-Anosov piece).

Moreover, if A is a union of some components of the boundary ∂X of X and $f(A) = A$, then also

$$NI^{\infty}(f; X \setminus A) = N^{\infty}(f; X \setminus A) = L^{\infty}(f; X \setminus A) = \lambda. \tag{2.24}$$

2.11. Homeomorphisms of punctured surfaces.

Let X be a connected compact surface and let P be a nonempty finite set of points (punctures) in the interior of X. Let $f : X, P \to X, P$ be a homeomorphism. The (noncompact) surface $M := X \setminus P$ will be called a punctured surface, on which f restricts to a homeomorphism $f_M : M \to M$. In this section we assume that $|P| > \chi(X)$, where $|P|$ is the cardinality of P.

Let Y be the compactification of $M = X \setminus P$ obtained from the (smooth) surface X by blowing up each point of P into its circle of unit tangent vectors. Then Y is a compact surface with $\chi(Y) = \chi(X) - |P| < 0$, the added circles form a set $Q \subset \partial Y$, and $M = Y \setminus Q = X \setminus P$.

Generally speaking, the homeomorphism $f_M : Y \setminus Q \to Y \setminus Q$ is not extendable to a map of the pair Y, Q. So we use the following approximation procedure: We can always isotope f rel P to a homeomorphism $f' : X, P \to X, P$ that is differentiable on P. (There is a lot of freedom in this smoothing. When $p \in P$ is fixed under f, we can even make f' to be the identity on a neighborhood of p.) Let $g : Y, Q \to Y, Q$ be the blow-up of f', i.e. the homeomorphism extending $f'_M : Y \setminus Q \to Y \setminus Q$ to Q according to the differential of f' at P (cf. [**Bw**, §2]). Then the isotopy class of g is independent of the local smoothing f'. Since $G := \pi_1(Y) = \pi_1(M)$, we can identify the automorphism $g_G : \pi_1(Y) \to \pi_1(Y)$ with $f_G : \pi_1(M) \to \pi_1(M)$.

The Nielsen number of the complement $N(g; Y \setminus Q)$ is thus independent of the local smoothing f'. We define the *punctured Nielsen number* of f to be $N(f \setminus P) := N(g; Y \setminus Q)$. It is a lower bound for the number of fixed points of f on $M = X \setminus P$, and it is an isotopy invariant of f rel P.

REMARK. $N(f \setminus P)$ is not the same as $N(f; X \setminus P)$. The former is often a much better lower bound for the number of fixed points of f, as shown by the Example below. Note that the coordinates of the fixed point classes of f_M are in $\Gamma := \pi_1(T_g) = \pi_1(T_g \setminus T_{g|Q}) = \pi_1(T_f \setminus T_{f|P})$, which is not $\pi_1(T_f)$ but the fundamental group of the complement of the link $T_{f|P}$ in T_f.

Now generalize the above idea to the periodic orbits of f and define, for example,

$$N(f^n \setminus P) := N(g^n; Y \setminus Q), \qquad L_\Gamma(f^n \setminus P) := L_\Gamma(g^n; Y \setminus Q). \tag{2.25}$$

The first one is a lower bound for the number of n-points of f on $X \setminus P$.

The *asymptotic punctured Nielsen number* is defined as

$$N^\infty(f \setminus P) := \operatorname*{Growth}_{n\to\infty} N(f^n \setminus P). \tag{2.26}$$

ASYMPTOTIC INVARIANTS. *$N^\infty(f \setminus P)$ is the common growth rate of various punctured invariants:*

$$\begin{aligned} N^\infty(f \setminus P) &= \operatorname*{Growth}_{n\to\infty} N_\Gamma(f^n \setminus P) = \operatorname*{Growth}_{n\to\infty} NI_\Gamma(f^n \setminus P) \\ &= \operatorname*{Growth}_{n\to\infty} \|L_\Gamma(f^n \setminus P)\| = N^\infty(g). \end{aligned} \tag{2.27}$$

Consequently, as in §2.8, $N^\infty(f \setminus P)$ can also be estimated if we are given a unitary representation of $\Gamma = \pi_1(T_f \setminus T_{f|P})$.

ENTROPY THEOREM. *For any homeomorphism $f : X, P \to X, P$, we have*

$$h(f) \geq \log N^\infty(f \setminus P). \tag{2.28}$$

EXAMPLE. (The computational part of the Example in §3.3.)

Suppose X is a disk, P consists of three points in its interior. Then Y is a disk with three holes, and Q is the union of the boundary curves of the holes. We have $G = \pi_1(X \setminus P) = \pi_1(Y) = \langle a_1, a_2, a_3 \rangle$, the free group of rank 3.

Suppose $f : X, P \to X, P$ is a homeomorphism, and $g : Y, Q \to Y, Q$ is the blow-up of a local smoothing of f. Suppose f and g induce the automorphism $f_G : G \to G$ specified below by the images $a_i' := f_G(a_i)$. The Jacobian matrix D in Fox calculus is readily calculated.

$$\begin{cases} a_1' &= a_1 a_3 a_1^{-1}, \\ a_2' &= a_1, \\ a_3' &= a_3^{-1} a_2 a_3, \end{cases} \qquad D = \begin{pmatrix} 1 - a_1 a_3 a_1^{-1} & 0 & a_1 \\ 1 & 0 & 0 \\ 0 & a_3^{-1} & -a_3^{-1} + a_3^{-1} a_2 \end{pmatrix}. \tag{2.29}$$

According to the Example in §2.7, we have $\Gamma = \langle a_1, a_2, a_3, z \mid a_i z = z a_i',\ i = 1, 2, 3 \rangle$. An obvious way to get a representation of Γ is to abelianize and let $z \mapsto 1$.

We then get the projection $\Gamma \to H := \langle a \rangle$, all $a_i \mapsto a$. A U(1) representation ρ can be obtained by letting a to be a unimodular complex number. Thus

$$(zD)^\rho = \begin{pmatrix} 1-a & 0 & a \\ 1 & 0 & 0 \\ 0 & a^{-1} & 1-a^{-1} \end{pmatrix},$$

so that by (2.14),

$$\zeta_\rho(g) = \frac{\det(I - t(zD)^\rho)}{\det(I - tz^\rho)} = 1 - (1 - a - a^{-1})t + t^2. \tag{2.30}$$

Take $a = -1$, then we get the zeta function $\zeta_\rho(g) = 1 - 3t + t^2$ and its smallest root is $r = (3 - \sqrt{5})/2 < 2/5$. Hence, by (2.19), (2.23), (2.27) and (2.28), we get the estimates

$$NI^\infty(f \setminus P) = N^\infty(f \setminus P) = N^\infty(g) > \frac{5}{2}, \qquad h(f) > \log\frac{5}{2}. \tag{2.31}$$

Note that X is the disk, hence all $N(f^n) = 1$ and $N^\infty(f) = 1$. So the punctured Nielsen numbers do give better estimates.

Section 3. Applications to dynamical systems

We shall discuss some results in the recent literature of dynamical systems theory, and explain how they can be improved by use of the theory sketched in §2. For the sake of comparison, we shall use a box to mark the original part to be deleted, and use boldface for the added part.

All these applications are concerned with the growth rate of the number of periodic orbits for surface homeomorphisms.

Note that in dynamics, the "period" of a periodic point or periodic solution is always understood as its least period, i.e. we are only interested in primary periodic orbits. So $NI_\Gamma(f^n)$ is more important than $N_\Gamma(f^n)$.

These applications also serve to demonstrate certain features of our approach:

- smoothness not required;
- computational feasibility;
- better estimation is obtained by using $\|L_\Gamma(f^n)\|$ instead of $N(f^n)$;
- non-abelian information can be obtained via unitary representations.

3.1. Exponential growth without smoothness condition.

According to the Theorem in §2.10, if a surface homeomorphism has at least one pseudo-Anosov piece in its Thurston canonical form, then the number of its primary n-orbits grows exponentially. No smoothness assumption is needed.

For example, in the theorem of [**H**] below, the differentiability condition in the last sentence can be removed, and the conclusion can be strengthened to exponential growth.

THEOREM. *If $f : X \to X$ is an orientation reversing homeomorphism of a compact oriented surface of genus g, and if f has orbits with $g+2$ distinct odd periods, then $h(f) > 0$. Moreover,* if f is differentiable at the periodic points in question, then f has orbits with infinitely many distinct periods **the number of primary n-orbits of f grows exponentially in n.**

Thus, wherever the existence of pseudo-Anosov pieces is confirmed, the exponential growth of periodic orbits can be added to the conclusion of positive topological entropy. This remark applies e.g. to [**GST, Theorem**], [**LM**, Theorem 1] and [**P**, Theorem 2].

3.2. A general setting.

In many cases, exponential growth of the number of primary n-orbits can be shown without first proving the existence of pseudo-Anosov pieces, and even numerical estimates of the growth rate and the topological entropy can be obtained. We describe a general setting where this can be done.

Let X be a compact connected surface, and let $f : X \to X$ be a homeomorphism isotopic to the identity map of X. Suppose we are given a finite set $P \subset \text{int}X$ with $f(P) = P$, i.e. some periodic orbits of f are already known. We assume $r := |P| > \chi(X)$.

Suppose an isotopy $H = \{h_t\}_{t\in I} : id \simeq f : X \to X$ is given. The set $\mathcal{S} := \{(h_t(x), t) \in X \times I \mid x \in P\}$ is a geometric braid in $X \times I$ which represents a braid σ_P in the r-string braid group $\pi_1 B_{0,r}X$ of the surface X (cf. [**Bi**, p.6]).

Let M be the punctured surface $X \setminus P$ and let $f_M : M \to M$ be the restriction of f. Starting from the braid data σ_P, there is a classical procedure for computing the automorphism $f_G : \pi_1(M) \to \pi_1(M)$. (See [**Bi**, p.25] for the case $X = \mathbb{R}^2$ which is easily extended to any surface. The case of the torus T^2 is worked out in [**HJ**, §3].) Then representations of $\Gamma = \pi_1(T_f \setminus T_{f|P})$ can be found and Lefschetz zeta functions computed, so that the estimation method in §2.8 can be applied to obtain information on the dynamics of f.

The above setting is a generalization of the following dynamical problem first studied by Matsuoka [**M1**]:

Consider a differential equation on the plane

$$\frac{dx}{dt} = V(t, x), \qquad t \in \mathbb{R},\ x \in \mathbb{R}^2, \tag{3.1}$$

where $V_t := V(t, \cdot)$ are vector fields on $\mathbb{R}^2$, and V is C^1 on $\mathbb{R} \times \mathbb{R}^2$. We assume that:

(i) $V(t+1, x) = V(t, x)$ for all $t \in \mathbb{R}$, $x \in \mathbb{R}^2$. In other words, the vector field V_t is 1-periodic in the time variable t.

(ii) The general solution $x = \phi(t; x_0)$ to the equation (3.1) with initial condition $x|_{t=0} = x_0$ exists on $-\infty < t < \infty$ for any initial position $x_0 \in \mathbb{R}^2$. In other words, every solution extends forever in both directions of time.

(iii) $C = \{c_1(t), c_2(t), \cdots, c_r(t)\}$ is a given set of r periodic solutions of the equation, such that $\{c_1(0), c_2(0), \cdots, c_r(0)\} = \{c_1(1), c_2(1), \cdots, c_r(1)\}$.

This set C of known solutions gives rise to a geometric braid

$$\mathcal{S} = \{(c_i(t), t) \mid 0 \leq t \leq 1,\ 1 \leq i \leq r\} \tag{3.2}$$

in the space $\mathbb{R}^2 \times I$ which represents a braid σ_P in Artin's r-string braid group B_r. Matsuoka proposed to estimate the number of periodic solutions of (3.1) from the braid data σ_P, and showed that when $r = 3$, "genericly" it has an infinite number of periodic solutions (see §3.3 below). This problem arises naturally in the study of harmonic and subharmonic solutions of second order differential equations with periodic terms, as explained in [**M2**].

To fit this problem into the above general setting, we take X to be the sphere $S^2 = \mathbb{R}^2 \cup \{\infty\}$. Extend the general solution $x = \phi(t; x_0)$ to X by defining $\phi(t; \infty) = \infty$ for all t, and define the isotopy $H = \{h_t\} : id \simeq f : X \to X$ by $h_t(x) = \phi(t; x)$. Take $P = \{\infty, c_1(0), c_2(0), \cdots, c_r(0)\}$. Then periodic solutions of (3.1) correspond to periodic points of the map $f : X, P \to X, P$.

When the right hand side $V(t, x)$ of (3.1) is also periodic in one or both coordinates of $x \in \mathbb{R}^2$, the phase space can be reduced to the infinite cylinder or the torus respectively, cf. [**J3**] or [**HJ**].

3.3. The case of the thrice punctured plane.

Let us concentrate on the case of the three times punctured plane M, that is, the case where X is the sphere $\mathbb{R}^2 \cup \{\infty\}$, the isotopy $H = \{h_t\} : id \simeq f : X \to X$ keeps ∞ fixed, and P consists of ∞ and three points x_1, x_2, x_3 in the plane. The braid group involved is Artin's braid group B_3 with the standard presentation

$$B_3 = \langle \sigma_1, \sigma_2 \mid \sigma_1\sigma_2\sigma_1 = \sigma_2\sigma_1\sigma_2 \rangle, \tag{3.3}$$

the generators σ_1, σ_2 are depicted in Fig.1 and the relation is illustrated in the middle of Fig.2.

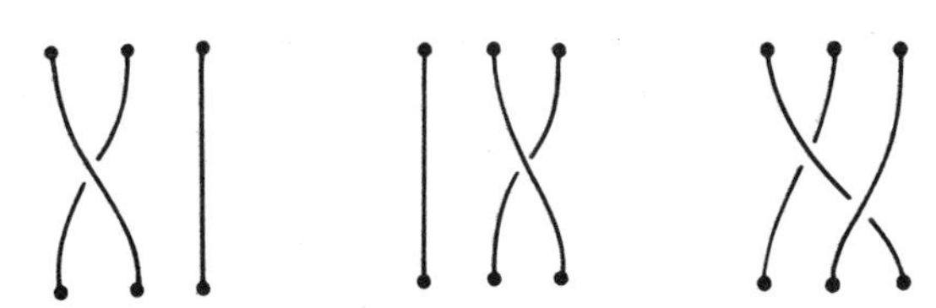

FIGURE 1. The braids σ_1, σ_2 and $\sigma_1\sigma_2^{-1}$ in B_3

The structure of the group B_3 is pretty well known. Its center is $Z = \langle (\sigma_1\sigma_2)^3 \rangle$, the infinite cyclic group generated by $(\sigma_1\sigma_2)^3 = (\sigma_1\sigma_2\sigma_1)^2$ (cf. [**Bi**, p.28]). This braid is a full twist corresponding to a 2π-rotation of the plane which has no effect on the map f. So we can work in the factor group B_3/Z.

The result obtained in [**M1**] (see also [**HJ**] and [**J3**]) can be formulated in the following way. Note that the only exceptional braids are the simplest ones that would be called twists rather than braids in non-mathematical language (cf. Fig.2).

THEOREM. *Suppose the braid* σ_P *is not conjugate to* σ_1^m, $(\sigma_1^2\sigma_2^2)^m$, $(\sigma_1\sigma_2)^m$ *or* $(\sigma_1\sigma_2\sigma_1)^m$ *in* B_3/Z, *for any integer* m. *Then* there are infinitely many periodic orbits. In addition, if all periodic orbits of f are hyperbolic, then *the number of* n*-orbits of* f *grows exponentially in* n.

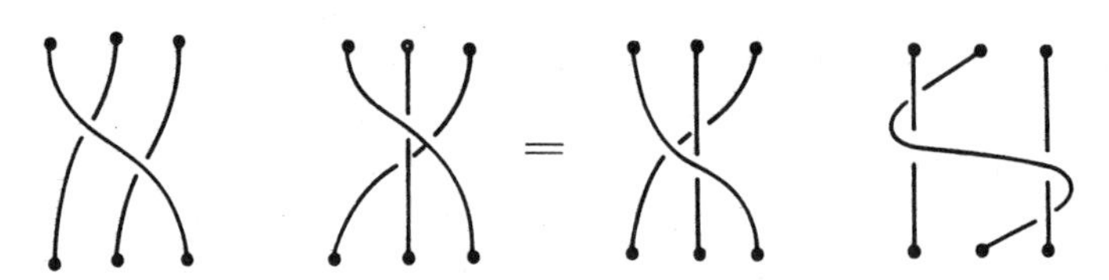

FIGURE 2. The braids $\sigma_1\sigma_2$, $\sigma_1\sigma_2\sigma_1 = \sigma_2\sigma_1\sigma_2$ and $\sigma_1^2\sigma_2^2$ in B_3

The argument in [**M1**] et al. was to count the number of terms in $L_G(f^n \setminus P)$, i.e. to estimate $N(f^n \setminus P)$ directly. The norm of $L_G(f^n \setminus P)$ was shown to grow exponentially. The purpose of the hyperbolicity condition was to link $N(f^n \setminus P)$ to $\|L_\Gamma(f^n \setminus P)\|$. Now that according to §2.10 we can estimate $N(f^n \setminus P)$ via $\|L_\Gamma(f^n \setminus P)\|$ anyway, that condition is no longer needed.

The algebraic technique used was the abelianization of Γ. The Fox calculus Jacobian matrix then turns out to be the classical Burau representation of the braid group (cf. [**Bi**, §3.2–3]), originally introduced for knot theory.

EXAMPLE.

A concrete example is the result of [**GST**] below, where we are able to supply a numerical estimate.

THEOREM. *Suppose* f *is an orientation preserving* differentiable *embedding of the disk, and* H *is an isotopy from the identity map to* f. *If* f *has a periodic orbit of period 3 whose image under* H *represents the braid* $\sigma_P = \sigma_1\sigma_2^{-1} \in B_3$, *then* f *has an periodic orbit of period* n *for every integer* n. **Moreover, the growth rate of the number of such periodic orbits is greater than** $\frac{5}{2}$, **and the topological entropy** $h(f) > \log\frac{5}{2}$.

Although this theorem refers to embeddings of the disk rather than homeomorphisms of the plane, the isotopy H of the embedding f of the disk can be extended to an isotopy of a homeomorphism of the whole plane. We can even make the extended H be the identity outside of some large circle. All the additional fixed points and periodic points created by this extension are related to the point at infinity, hence are not counted in our approach (§2.11). So we are still within the current context.

The automorphism $f_G : G \to G$ can be read off from the picture of the braid σ_P by pulling the loop down the braid. (The free generators a_1, a_2, a_3 of

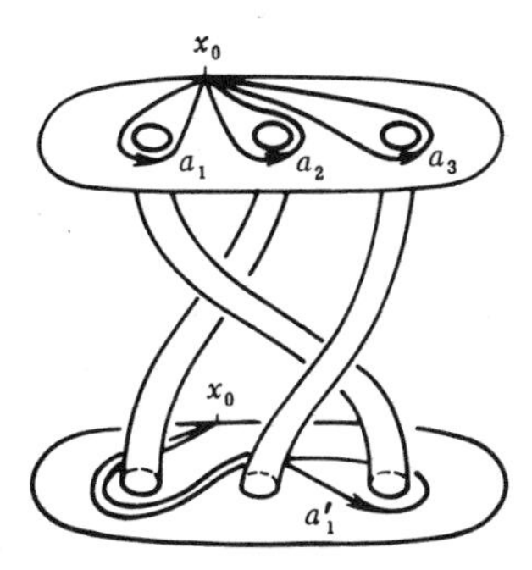

FIGURE 3. The elements a_1, a_2, a_3 and a'_1 in G

$G = \pi_1(M)$ are shown on the top on Fig.3, from left to right, and the bottom shows $a'_1 = f_G(a_1)$ which is a_1 pulled down. Note that the base point on the outer boundary is pulled straight downward.) The result is

$$\begin{cases} f_G(a_1) &= a_1 a_3 a_1^{-1}, \\ f_G(a_2) &= a_1, \\ f_G(a_3) &= a_3^{-1} a_2 a_3, \end{cases}$$

the same as the first part of (2.29). Hence the conclusion in boldface follows from the Example in §2.11.

As another concrete example, we mention that for the braid $\sigma_P = \sigma_2\sigma_1^2\sigma_2^2$ discussed in [**J3**], the same method leads to $\zeta_\rho = 1 - 4t + t^2$ and the estimate $N^\infty(f \setminus P) \geq 2 + \sqrt{3} > \frac{7}{2}$.

3.4. An example on the torus.

Homeomorphisms of the torus T^2 isotopic to the identity are studied in [**LM**], where the authors asked (p. 118) whether one can find a positive lower bound for the topological entropy of such a homeomorphism as a function of the rotation types for its periodic orbits. We choose their Example 2 (for which their theorem shows $h(f) > 0$) to test the potential of our method. It is easily seen that their original construction of the map fits into the following description via braids.

Think of $X = T^2$ as $\mathbb{R}^2/\mathbb{Z}^2$ and let the three points $x_1 = (0,0)$, $x_2 = (\frac{1}{3}, \frac{1}{3})$ and $x_3 = (\frac{2}{3}, \frac{2}{3})$ constitute the set P. The map $f : X, P \to X, P$ is isotopic to the identity via an isotopy $H = \{h_t\}_{t\in I} : id \simeq f : X \to X$ whose restriction on P is given by

$$\begin{cases} h_t(x_1) = x_1, \quad h_t(x_2) = x_2 + (2t, 0), \quad h_t(x_3) = x_3, & \text{if } 0 \leq t \leq \dfrac{1}{2}, \\ h_t(x_1) = x_1, \quad h_t(x_2) = x_2, \quad h_t(x_3) = x_3 + (0, 2t), & \text{if } \dfrac{1}{2} \leq t \leq 1. \end{cases}$$

Let $M = X \setminus P$ and $f_M : M \to M$ be the restriction of f. Then M is the thrice punctured torus. Its fundamental group $G = \langle a_1, a_2, b_1, b_2 \rangle$ is a free group of rank 4, with generators shown on the left of Fig.4. The braid σ_P determined by the isotopy H is $\sigma_P = \lambda_2\mu_3$, where λ_2 is the braid keeping x_1, x_3 fixed while moving x_2 once along a longitude curve, and μ_3 is the one keeping x_1, x_2 fixed while moving x_3 once along a meridian curve (see the right of Fig.4).

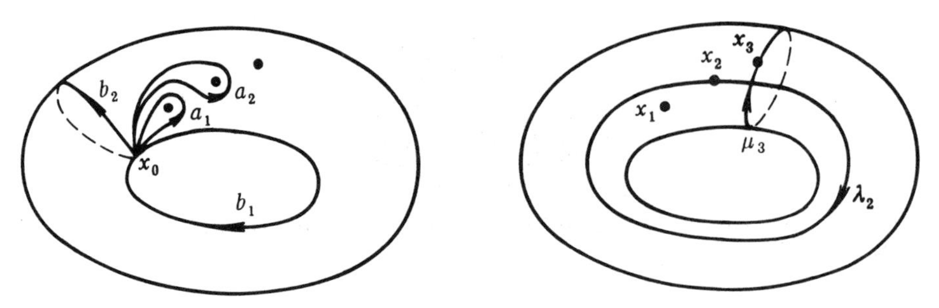

FIGURE 4. Basis $\{a_1, a_2, b_1, b_2\}$ and braids λ_2, μ_3 on T^2

By means of the formulas in [**HJ**, p.118], we calculate that the automorphism $f_G : G \to G$ is

$$(3.4) \qquad f_G : \begin{cases} a_1 \mapsto a_1' = a_1, \\ a_2 \mapsto a_2' = (a_2b_2b_1b_2^{-1})a_2(a_2b_2b_1b_2^{-1})^{-1}, \\ b_1 \mapsto b_1' = a_1a_2b_2b_1b_2^{-1}, \\ b_2 \mapsto b_2' = a_2b_2. \end{cases}$$

It is routine to compute the Fox calculus Jacobian matrix D.

Now $\Gamma = \langle a_i, b_i, z \mid a_iz = za_i', b_iz = zb_i' \text{ for } i = 1, 2\rangle$. The abelianization that worked so well for the plane does not work here, because when Γ is abelianized, the nature of (3.4) forces both a_1, a_2 to become 1, so that any U(1) representation can only give the trivial estimate $N^\infty(f \setminus P) \geq 1$. Thus we have to look for nonabelian representations of Γ. Fortunately a simple representation of Γ into the multiplicative group of unimodular quaternions works.

$$(3.5) \qquad \rho : \Gamma \to \mathrm{Sp}(1); \quad z \mapsto i, \quad a_1 \mapsto -1, \quad a_2 \mapsto -1, \quad b_1 \mapsto j, \quad b_2 \mapsto k.$$

But quaternions form a skew-field, not a field. We have to identify Sp(1) with SU(2) via the correspondence

$$(3.6) \quad 1 \mapsto \begin{pmatrix} 1 & 0 \\ 0 & 1 \end{pmatrix}, \quad i \mapsto \begin{pmatrix} 0 & -1 \\ 1 & 0 \end{pmatrix}, \quad j \mapsto \begin{pmatrix} i & 0 \\ 0 & -i \end{pmatrix}, \quad k \mapsto \begin{pmatrix} 0 & i \\ i & 0 \end{pmatrix}.$$

Then we get $\zeta_\rho = (1+t)^2[(1-t+t^2)^2 + 3t^2]$, from which follow the estimates $N^\infty(f \setminus P) > 2.3$ and $h(f) > \log 2.3$.

This example shows that on the torus, such estimations are quite accessible for concrete homeomorphisms. But to find a useful representation of Γ could be a challenge when applied to a family of homeomorphisms specified by qualitative conditions such as the rotation set.

REFERENCES

[AF] Asimov, D., Franks, J., *Unremovable closed orbits*, Geometric Dynamics (J. Palis Jr., ed.), Lecture Notes in Math. vol. 1007, Springer-Verlag, Berlin, Heidelberg, New York, 1983, pp. 22–29.

[Bi] Birman, J.S., *Braids, Links, and Mapping Class Groups*, Ann. Math. Studies vol. 82, Princeton Univ. Press, Princeton, 1974.

[Bt] Bott, R., *On the shape of a curve*, Advances in Math. **16** (1975), 23–38.

[Bw] Bowen, R., *Entropy and the fundamental group*, The Structure of Attractors in Dynamical Systems (N.G. Markley et al., eds.), Lecture Notes in Math. vol. 668, Springer-Verlag, Berlin, Heidelberg, New York, 1978, pp. 21–29.

[FH] Fadell, E., Husseini, S., *The Nielsen number on surfaces*, Topological Methods in Nonlinear Functional Analysis (S.P. Singh et al., eds.), Contemp. Math. vol. 21, Amer. Math. Soc., Providence, 1983, pp. 59–98.

[FLP] Fathi, A., Laudenbach, F., Poénaru, V., *Travaux de Thurston sur les surfaces, Séminaire Orsay*, Astérisque vol.66–67, Soc. Math. France, Paris, 1979.

[Fr1] Fried, D., *Periodic points and twisted coefficients*, Geometric Dynamics (J. Palis Jr., ed.), Lecture Notes in Math. vol. 1007, Springer-Verlag, Berlin, Heidelberg, New York, 1983, pp. 261–293.

[Fr2] Fried, D., *Homological identities for closed orbits*, Invent. Math. **71** (1983), 419–442.

[Fu1] Fuller, F.B., *The treatment of periodic orbits by the methods of fixed point theory*, Bull. AMS **72** (1966), 838–840.

[Fu2] Fuller, F.B., *An index of fixed point type for periodic orbits*, Amer. J. Math. **89** (1967), 133–148.

[GST] Gambaudo, J.-M., van Strien, S., Tresser, C., *Vers un ordre de Sarkovskii pour les plongements du disque préservant l'orientation*, C. R. Acad. Sci. Paris Série I **310** (1990), 291–294.

[H] Handel, M., *The entropy of orientation-reversing homeomorphisms of surfaces*, Topology **21** (1982), 291–296.

[HJ] Huang, H.-H., Jiang, B.-J., *Braids and periodic solutions*, Topological Fixed Point Theory and Applications, Proceedings, Tianjin 1988 (B. Jiang, ed.), Lecture Notes in Math. vol. 1411, Springer-Verlag, Berlin, Heidelberg, New York, 1989, pp. 107–123.

[I] Ivanov, N.V., *Entropy and the Nielsen numbers*, Soviet Math. Dokl. **26** (1982), 63–66.

[J1] Jiang, B., *Lectures on Nielsen Fixed Point Theory*, Contemp. Math. vol. 14, Amer. Math. Soc., Providence, 1983.

[J2] Jiang, B., *A characterization of fixed point classes*, Fixed Point Theory and its Applications (R.F. Brown, ed.), Contemp. Math. vol. 72, Amer. Math. Soc., Providence, 1988, pp. 157–160.

[J3] Jiang, B., *Periodic orbits on surfaces via Nielsen fixed point theory*, Topology — Hawaii (K.H. Dovermann, ed.), World Scientific Publishing Co., Singapore, 1991.

[J4] Jiang, B., *Estimation of the number of periodic orbits*, preprint.

[JG] Jiang, B., Guo, J., *Fixed points of surface diffeomorphisms*, Pacific J. Math. (to appear).

[LM] Llibre, J., MacKay, R.S., *Rotation vectors and entropy for homeomorphisms of the torus isotopic to the identity*, Ergod. Th. Dyn. Sys. **11** (1991), 115–128.

[M1] Matsuoka, T., *The number and linking of periodic solutions of periodic systems*, Invent. Math. **70** (1983), 319–340.

[M2] Matsuoka, T., *Waveform in dynamical systems of ordinary differential equations*, Japan. J. Appl. Math. **1** (1984), 417–434.

[P] Pollicott, M., *Rotation sets for homeomorphisms and homology*, Tran. AMS **331** (1992), 881–894.

[R] Reidemeister, K., *Automorphismen von Homotopiekettenringen*, Math. Ann. **112** (1936), 586–593.

[S] Schirmer, H., *A survey of relative Nielsen fixed point theory*, these Proceedings.

[T] Thurston, W.P., *On the geometry and dynamics of diffeomorphisms of surfaces*, Bull. AMS **19** (1988), 417–431.

[Wa] Walters, P., *An Introduction to Ergodic Theory*, Springer, New York, 1982.

[We] Wecken, F., *Fixpunktklassen, II*, Math. Ann. **118** (1942), 216–234.
[Z] Zhao X.-Z., *A relative Nielsen number for the complement*, Topological Fixed Point Theory and Applications, Proceedings, Tianjin 1988 (B. Jiang, ed.), Lecture Notes in Math. vol. 1411, Springer-Verlag, Berlin, Heidelberg, New York, 1989, pp. 189–199.

DEPARTMENT OF MATHEMATICS, PEKING UNIVERSITY, BEIJING 100871, CHINA

Contemporary Mathematics
Volume **152**, 1993

Genericity of Homeomorphisms with Connected Stable and Unstable Sets

JORGE LEWOWICZ AND JUAN TOLOSA

1. Introduction

Let $f: M \to M$ be a homeomorphism of a smooth compact connected riemannian manifold M. For $\varepsilon > 0$ and $x \in M$, call

$$S_\varepsilon(x) = \{y \in M: \operatorname{dist}(f^n(x), f^n(y)) \leqslant \varepsilon,\ n \geqslant 0\}$$

and

$$U_\varepsilon(x) = \{y \in M: \operatorname{dist}(f^n(x), f^n(y)) \leqslant \varepsilon,\ n \leqslant 0\}$$

the ε-stable and the ε-unstable sets of x.

Stable and unstable sets are basic elements of the dynamical structure of f and a fundamental tool to face problems of classification of dynamical systems under conjugacy. See, for instance, [**F**], for the case of Anosov diffeomorphisms, and [**H**], [**L**] in connection with the topological equivalence of expansive homeomorphisms of surfaces. In case f is expansive, for any $x \in M$ these stable and unstable sets contain non-trivial (infinite) connected pieces.

In this paper we obtain some general results on the existence of such connected pieces at each $x \in M$ (Proposition 1.1). When x is a periodic point that is not a repeller (attractor) it is easy to show that $S_\varepsilon(x)$ ($U_\varepsilon(x)$) contains such a piece, for any $\varepsilon > 0$. However, points x in a minimal set may have trivial $S_\varepsilon(x)$ and $U_\varepsilon(x)$ for small $\varepsilon > 0$. Consider the Denjoy map of S^1, i.e., take a rotation of S^1 by an angle $2\pi\alpha$, where α is irrational, and replace the points of a dense orbit $\{x_n, n \in \mathbb{Z}\}$ by intervals of size decreasing with $|n|$, in order to get a new space also homeomorphic to S^1. The Denjoy transformation may be defined by assigning to each point that was not on the added intervals, the former image under the rotation, and mapping linearly the interval we put instead of x_n into the one replacing x_{n+1}, $n \in \mathbb{Z}$. Any two points that do not lie in the same

1991 *Mathematics Subject Classification.* Primary 54H20, 58F12; Secondary 58F13, 58F15.
This paper is in final form and no version of it will be submitted for publication elsewhere

added interval will be, under some positive and some negative iteration, at a distance larger than the length of the interval replacing x_0. In fact, between these two points we find, in the original rotation, positive and negative iterates of x_0. Thus, for any x in the Denjoy minimal set which is not an endpoint of some added interval, $S_\varepsilon(x) = U_\varepsilon(x) = \{x\}$ if $\varepsilon > 0$ is small enough.

For each endpoint of the interval replacing x_0, which also belongs to the Denjoy minimal set, there is a non-trivial connected set that is at the same time the ε-stable and the ε-unstable set (ε small) of it. Moreover, this connected set has the property that its diameter decreases under positive and negative iteration. Proposition 1.1 shows that if for arbitrarily small ε no limit point of f has an ε-stable (ε-unstable) set with this property, then for each $x \in M$ which is not a periodic repeller (attractor) there is a non-trivial connected set included in $S_\varepsilon(x)$ (respectively $U_\varepsilon(x)$). Theorem 2.2 shows that this property on the limit set of f is C^0-generic; thus, for f in a a C^0-residual subset of Homeo(M) (the space of homeomorphisms of M with the usual topology) and each $x \in M$, $S_\varepsilon(x)$ and $U_\varepsilon(x)$ contain non-trivial connected pieces.

Let now dim $M = 2$; the description of local stable and unstable sets and the classification results of [**H**] and [**L**] are based chiefly on the existence of those connected pieces, and on the fact that two such pieces meet at most at one point. As a matter of fact, the same description of local stable and unstable sets may be obtained, even for non-expansive f, at points where the above mentioned properties of these stable and unstable pieces hold for them and for neighbouring points. This is the case, for instance, if we take the homeomorphipsm of S^2 defined, after identifying x with $-x$ on T^2, applying the usual linear Anosov map $\begin{pmatrix} 2 & 1 \\ 1 & 1 \end{pmatrix}$ of T^2. We get a non-expansive homeomorphism of S^2 for which, except at the image of branch points under the canonical projection, the stable and unstable sets are topological manifolds.

We show that for f in a C^0-residual set of Homeo(M), if f has a compact connected non-trivial attractor $C \subset \Omega(f)$, these connected pieces $C_\varepsilon(x) \subset S_\varepsilon(x)$ and $D_\varepsilon(x) \subset U_\varepsilon(x)$ meet only at x, provided $x \in C$. Then the arguments in [**L**] permit us to show that $D_\varepsilon(x)$ is an arc.

Thus, for almost all attractors, the connected unstable sets of each of its points are arcs.

Finally, we want to thank the participants of the Seminar of Dynamical Systems of the IMERL[1] for useful conversations on these topics.

1. General results

Let M be a compact connected smooth riemannian manifold and f a homeomorphism of M onto M. Let $\omega(f)$ ($\alpha(f)$) denote the set of all ω-limit points

[1] Instituto de Matemática y Estadística Rafael Laguardia, Montevideo, Uruguay.

(α-limit points) of f, i.e.,

$$\omega(f) = \bigcup_{x \in M} \omega(x), \qquad \alpha(f) = \bigcup_{x \in M} \alpha(x),$$

where $\omega(x)$ ($\alpha(x)$) is the set of ω-limit points (respectively α-limit points) of x. Obviously, $\omega(f) \bigcup \alpha(f) \subset \Omega(f)$; as usual, $\Omega(f)$ stands for the set of non-wandering points of f.

PROPOSITION 1.1. *Assume there is a sequence of positive numbers ρ_m, $\rho_m \to 0$, such that for any $\rho = \rho_m$ there is no point $x \in \omega(f)$ and no connected set C containing x such that dist$(f^n(x), f^n(y)) \leqslant \rho$ for every $y \in C$ and all $n \in \mathbb{Z}$ and that dist$(x, z) = \rho$ for some $z \in C$. Then, for any $\varepsilon > 0$ and $x \in M$, there is a compact connected set $C_\varepsilon(x)$, $x \in C_\varepsilon(x) \neq \{x\}$, such that dist$(f^n(x), f^n(y)) \leqslant \varepsilon$ for every $y \in C_\varepsilon(x)$ and all $n \geqslant 0$, unless x is a periodic repeller.*

PROOF. Let ε be a positive number and $x \in M$; we shall first assume that for some m with $\rho_m = \rho < \varepsilon$, f has the following property: for each $k = 1, 2, \ldots$ there is an $n_k > 0$ such that, for some ν, $0 \leqslant \nu \leqslant n_k$, $f^{-\nu}(B_{1/k}(f^{n_k}(x)))$ is not included in $B_\rho(f^{n_k - \nu}(x))$. (For $\sigma > 0$ and $x \in M$, $B_\sigma(x)$ denotes the ball $\{y \in M \colon \mathrm{dist}(x, y) \leqslant \sigma\}$.)

Let $y \in B_{1/k}(f^{n_k}(x))$ be such that $f^{-\nu}(y) \notin B_\rho(f^{n_k - \nu}(x))$. Take an arc joining $f^{n_k}(x)$ to y within $B_{1/k}(f^{n_k}(x))$, say, $a_k(t)$, $0 \leqslant t \leqslant 1$, $a_k(0) = f^{n_k}(x)$, $a_k(1) = y$, and let t_k^* be the supremum of those $t \in [0, 1]$ such that $f^{-\nu}(a_k([0, t]))$ is contained in the interior of $B_\rho(f^{n_k - \nu}(x))$ for every ν, $0 \leqslant \nu \leqslant n_k$. Thus, for these ν,

$$f^{-\nu}(a_k([0, t_k^*])) \subset B_\rho(f^{n_k - \nu}(x)),$$

and for some ν_k, $0 \leqslant \nu_k \leqslant n_k$,

$$f^{-\nu_k}(a_k([0, t_k^*])) \bigcap \partial B_\rho(f^{n_k - \nu_k}(x)) \neq \varnothing.$$

For each k, we choose then such a ν_k and show that $\lim_{k \to \infty} \nu_k = +\infty$. In fact, if some sub-sequence of ν_k were bounded, we would have that for infinitely many k, $\nu_k = N$ for some fixed N, and therefore f^N would map sets of diameter at least ρ onto sets of arbitrarily small diameter, which is absurd.

On the other hand, if $n_k - \nu_k$ were unbounded, we would be able to find $y \in \omega_f(x)$ and a compact connected set C, $y \in C$, $\mathrm{dist}(y, z) = \rho$ for some $z \in C$, such that $\mathrm{dist}(f^n(y), f^n(u)) \leqslant \rho$ for every $u \in C$ and all $n \in \mathbb{Z}$, in contradiction with the hypothesis of the proposition. Indeed, such a connected set C may be obtained as follows: assume that $f^{n_k - \nu_k}(x)$ converges to, say, y (the construction is the same in case we have to replace $n_k - \nu_k$ by a convergent sub-sequence) and take

$$C = \bigcap_{k=1}^{\infty} \mathrm{clos}\left(\bigcup_{j \geqslant k} f^{-\nu_j}(a_j([0, t_j^*])) \right).$$

Thus, $n_k - \nu_k$ is bounded, and therefore the arcs $f^{-n_k}(a_k([0, t_k^*]))$ have diameters bounded away from zero. The set

$$C_\varepsilon(x) = \bigcap_{k=1}^{\infty} \operatorname{clos}\left(\bigcup_{j=k}^{\infty} f^{-n_j}(a_j([0, t_j^*])) \right)$$

clearly satisfies the requirements of the thesis of the proposition.

Let us suppose now that the assumption made in the first paragraph of this proof does not hold. Then, for every m with $\rho_m = \rho < \varepsilon$, there is a $k = k(m) > 0$ such that for every $n \geqslant 0$

$$f^{-\nu}\left(B_{1/k}(f^n(x))\right) \subset B_\rho(f^{n-\nu}(x))$$

if $0 \leqslant \nu \leqslant n$. Consequently, for any $y = \omega_f(x)$, we have that

$$f^n(B_{1/k}(y)) \subset B_\rho(f^n(y)), \quad n \leqslant 0.$$

As $\rho_m \to 0$, we get that $\omega_f(x)$ is uniformly Lyapunov stable in the past. Take $y \in \omega_f(x)$; since $\operatorname{dist}(x, \operatorname{clos}(\{f^n(y) : n \in \mathbb{Z}\})) > 0$ contradicts the stability in the past of $\omega_f(x)$, it follows that $x \in \omega_f(x)$ and that $\omega_f(x)$ is a minimal set which, because of its uniform stability properties, consists of almost-periodic motions ([**NS**], p. 390).

If x is not a periodic point, we choose $\rho_m < \varepsilon$, the corresponding $k_m = k$, and a point $y \in B_{1/k}(x)$, $y \neq x$, $y \in \omega_f(x)$. Join x to y through an arc contained in $B_{1/k}(x)$, and let $n_i \to -\infty$ be a sequence of negative integers such that $f^{n_i}(x) \to x$. Since, on account of the uniform stability of $\omega_f(x)$ in both senses, the diameters of the n_i-iterates of this arc are bounded away from zero, we take $C_\varepsilon(x)$ as the usual intersection, for $i \geqslant 0$, of the closures of the unions of the n_j-iterates of the arc, $j \geqslant i$. Clearly, $C_\varepsilon(x)$ satisfies the required properties.

If x is periodic and the diameters of $f^n(B_{1/k}(x))$, $n \leqslant 0$, do not tend to zero, the previous arguments apply and allow us to construct a set $C_\varepsilon(x)$ as required. This completes the proof of the proposition. □

REMARK. The same arguments prove that, unless x is a periodic attractor, there exists a compact connected set $D_\varepsilon(x)$, $x \in D_\varepsilon(x) \neq \{x\}$, such that $\operatorname{dist}(f^n(x), f^n(y)) \leqslant \varepsilon$ for every $y \in D_\varepsilon(x)$ and $n \leqslant 0$.

2. Generic properties

Let M and f be as before. Call

$$S_\varepsilon^f(x) = \{y \in M : \operatorname{dist}(f^n(x), f^n(y)) \leqslant \varepsilon, \ n \geqslant 0\}$$

and

$$U_\varepsilon^f(x) = \{y \in M : \operatorname{dist}(f^n(x), f^n(y)) \leqslant \varepsilon, \ n \leqslant 0\}.$$

Let f satisfy Axiom A[2]. Since every basic set of f is isolated and $f|_{\Omega(f)}$ is expansive, we may choose $\varepsilon = \varepsilon_f > 0$ such that $S_\varepsilon(x) \bigcap U_\varepsilon(x) = \{x\}$ for every $x \in \Omega(f)$. For every f satisfying Axiom A we choose once and for all such an ε_f.

LEMMA 2.1. *Let f satisfy Axiom A and the strong transversality condition. Then, for every $m = 1, 2, \dots$, there is a C^0-neighbourhood $\mathcal{U}(f, m)$ of f such that if $g \in \mathcal{U}(f, m)$, for any $x \in \Omega(f)$ we have that*

$$\left(S^g_\varepsilon(x) - S^g_{\varepsilon/m}(x)\right) \bigcap U^g_\varepsilon(x) = \varnothing,$$

where $\varepsilon = \varepsilon_f$.

PROOF. Arguing by contradiction, let us assume that for some $m > 0$ there is a sequence $\{g_\nu\}$ of homeomorphisms of M that converges to f in the C^0-topology and such that for each $\nu = 1, 2, \dots$ there exists $x_\nu \in \Omega(g_\nu)$, and

$$y_\nu \in \left(S^{g_\nu}_\varepsilon(x_\nu) - S^{g_\nu}_{\varepsilon/m}(x_\nu)\right) \bigcap U^{g_\nu}_\varepsilon(x_\nu).$$

For these ν, let h_ν denote a semi-conjugacy between f and g_ν, i.e., a continuous map of M onto M, such that $f \circ h_\nu = h_\nu \circ g_\nu$. Furthermore, let the h_ν converge in the C^0-topology to the identity map of M [**Hu**].

We have that $\operatorname{dist}(g^n_\nu(x_\nu), g^n_\nu(y_\nu)) \leqslant \varepsilon$ for every $n \in \mathbb{Z}$ and that

$$\operatorname{dist}(g^{n_\nu}_\nu(x_\nu), g^{n_\nu}_\nu(y_\nu)) > \varepsilon/m$$

for some $n_\nu > 0$. Let $z_\nu = g^{n_\nu}_\nu(x_\nu)$, $u_\nu = g^{n_\nu}_\nu(y_\nu)$, and call (z_∞, u_∞) a limit pair of (z_ν, u_ν). Clearly, $z_\infty \neq u_\infty$ and $\operatorname{dist}(f^n(z_\infty), f^n(u_\infty) \leqslant \varepsilon$, $n \in \mathbb{Z}$. But since $z_\nu \in \Omega(g_\nu)$, $h_\nu(\Omega(g_\nu)) \subset \Omega(f)$, and $\operatorname{dist}(z_\nu, h_\nu(z_\nu)) \to 0$, we get that $h_\nu(z_\nu) \to z_\infty$, $z_\infty \in \Omega(f)$, and $u_\infty \in S^f_\varepsilon(z) \bigcap U^f_\varepsilon(z)$; a contradiction. □

THEOREM 2.2. *There is a C^0-residual set Σ such that if $g \in \Sigma$, $\varepsilon > 0$, and $x \in M$, then $S^g_\varepsilon(x)$ $(U^g_\varepsilon(x))$ contains a compact connected set $C_\varepsilon(x)$ $(D_\varepsilon(x))$, $x \in C_\varepsilon(x) \neq \{x\}$ $(x \in D_\varepsilon(x) \neq \{x\})$.*

PROOF. For f satisfying Axiom A and the strong transversality condition, take the chosen $\varepsilon_f > 0$ such that for $x \in \Omega(f)$, $S^f_\varepsilon(x) \bigcap U^f_\varepsilon(x) = \{x\}$, where, as before, $\varepsilon = \varepsilon_f > 0$. Let m be a positive integer and let $\mathcal{U}(f, m)$ be the C^0-neighbourhood of f given by Lemma 2.1. Call $\mathcal{N}_m$ the union of the $\mathcal{U}(f, m)$ for all f satisfying the above mentioned conditions; then $\Sigma = \bigcap_{m=1}^\infty \mathcal{N}_m$ is a C^0-residual set [**S**].

If $g \in \Sigma$, g belongs to some $\mathcal{U}(f, m)$ for each m. Choose $\rho_m > 0$, $\varepsilon/m < \rho_m < \varepsilon$ $(\varepsilon = \varepsilon_f)$, in such a way that $\lim \rho_m \to 0$ when $m \to \infty$.

Then, if $x \in \Omega(g)$ and C is a compact connected set containing x such that $\operatorname{dist}(g^n(x), g^n(y)) \leqslant \rho_m$ for every $y \in C$ and all $n \in \mathbb{Z}$, we have that

$$\operatorname{dist}(x, z) < \rho_m$$

for any $z \in C$, for otherwise $z \in (S^g_\varepsilon(x) - S^g_{\varepsilon/m}(x)) \bigcap U^g_\varepsilon(x)$, in contradiction

[2] Recall that this means that $\Omega(f)$ is hyperbolic and the set of periodic points of f is dense in $\Omega(f)$, see [**Sm**].

with Lemma 2.1. Therefore the thesis of the theorem follows from Proposition 1.1 and the fact that there is a C^0-residual subset of Homeo(M) such that each homeomorphism in this set has no periodic attractors or repellors [**PPSS**]. □

3. The size of stable and unstable sets

In [L1], Section 1, it is shown that if f is an Anosov diffeomorphism of a compact connected riemannian manifold, there exists a positive integer m such that either $\|(f^m)'_x u\| \geqslant 2\|u\|$ or $\|(f^{-m})'_x u\| \geqslant 2\|u\|$, for each $u \in T_xM$ and every $x \in M$. But the arguments there also show the existence of such an m, with the same property, on the restriction of a diffeomorphism f of M to a compact f-invariant hyperbolic subset C of M. Let $A: \bigcup_{x\in C} T_xM \to \mathbb{R}$ be the positive quadratic form defined, for $u \in T_xM$, $x \in C$, by $A(u) = \sum_{i=0}^{m-1}\sum_{j=0}^{m-1} \|(f^{-i+j})'_x u\|^2$. Then it is easy to check that

$$A(f'_x u) - 2A(u) + A((f^{-1})'_x u) = \|(f^m)'_x u\|^2 - 2\|u\|^2 + \|(f^{-m})'_x u\|^2,$$

which is positive for every $u \in T_xM$, $\|u\| \neq 0$, $x \in C$. Because of the continuity of f' and $(f^{-1})'$, this quadratic form satisfies the same properties on a neighbourhood of C, and moreover, we can define for x, y in some neighbourhood of C a function $V(x,y) = A(u)$, where $\exp_x u = y$, provided y is close enough to x. For some $\alpha > 0$ we will have, again on account of the continuity of f' and $(f^{-1})'$, that if x, y belong to this neighbourhood and $0 < \operatorname{dist}(x,y) < \alpha$,

$$V(f(x), f(y)) - 2V(x,y) + V(f^{-1}(x), f^{-1}(y)) > 0.$$

Now let f satisfy Axiom A, B_i, $i = 1, \dots, r$, being its basic sets, $\Omega(f) = \bigcup_{i=1}^{r} B_i$. Let $\rho > 0$, $\rho < \min_{i\neq j}(\min_{x_i\in B_i, x_j\in B_j} \operatorname{dist}(x_i, x_j))$, and $\alpha > 0$ be so that on $\{x \in M: \operatorname{dist}(x, \Omega(f)) < \rho\}$ we may define a quadratic function V with the above mentioned property for $0 < \operatorname{dist}(x,y) < \alpha$. Let $\varepsilon > 0$, and let $k > 0$ and $\delta > 0$ be chosen so that $V(x,y) \leqslant k$ implies $\operatorname{dist}(x,y) \leqslant \varepsilon$; and $\operatorname{dist}(x,y) > \delta$ if $V(x,y) \geqslant k$. Let $\mathcal{U} = \mathcal{U}(f,\rho,k)$ be a C^0-neighbourhood of f such that for $g \in \mathcal{U}$

$$\left\{x \in M: \operatorname{dist}(x, \Omega(g)) \leqslant \frac{\rho}{2}\right\} \subset \{x \in M: \operatorname{dist}(x, \Omega(f)) < \rho\}$$

and that if $\operatorname{dist}(x, B) < \rho$, $\operatorname{dist}(y, B) < \rho$, for B a basic set of f that is not a periodic repeller, and if $0 < \operatorname{dist}(x,y) < \alpha$, we have

i) $V(g^{-1}(x), g^{-1}(y)) > k$ for some y with $V(x,y) = k$.

ii) $V(g(x), g(y)) - 2V(x,y) + V(g^{-1}(x), g^{-1}(y)) > 0$ for every y with $V(x,y) \geqslant k$.

Let g be a homeomorphism of M and let $C_\varepsilon(x)$ and $D_\varepsilon(x)$ denote the connected components containing x of the g-stable set $S_\varepsilon(x)$ and the g-unstable set $U_\varepsilon(x)$ of x respectively.

LEMMA 3.1. *Let $g \in \mathcal{U}(f,\rho,k)$. Assume that for some $x \in M$ and some f-basic set B that is not a periodic repeller (attractor) we have that $\operatorname{dist}(g^n(x), B) \leqslant$*

$\rho/2$ for $n \geqslant 0$ ($n \leqslant 0$). Then $C_\varepsilon(x)$ (respectively $D_\varepsilon(x)$) contains a point y such that $dist(x,y) = \delta$.

PROOF. We prove that $C_\varepsilon(x)$ contains such a y, arguing by contradiction. Assume then that each connected set joining x to $\partial V_k(x)$, where

$$V_k(x) = \{y \in M : V(x,y) \leqslant k\},$$

contains a point y such that $g^n(y) \notin V_k(g^n(x))$ for some $n > 0$. Because of the compactness of $V_k(x)$, we may assume that all those n are less than some $N > 0$. Choose $\nu > N$; for some $z \in \partial V_k(g^\nu(x))$ we have $g^{-1}(z) \notin V_k(g^{\nu-1}(x))$, because of i). Join $g^\nu(x)$ to z through an arc $a: [0,1] \to V_k(g^\nu(x))$. Let t^* be the supremum of those t for which

$$g^{-n}(a[0,t]) \subset V_k(g^{\nu-n}(x)),$$

$0 \leqslant n \leqslant \nu$. Then, because of the contradiction assumption, for some p, $0 < p < \nu$, $g^{-p}(a(t^*)) \in \partial V_k(g^{\nu-p}(x))$, and, at the same time, $g^{-n}(a(t^*)) \in V_k(g^{\nu-n}(x))$ for $0 \leqslant n \leqslant \nu$, which is absurd on account of ii). The second part of the result is proved similarly. □

For f satisfying Axiom A and the strong transversality property, and for $n > 0$, choose ρ_n, $0 < \rho_n < 1/n$, and $k_n > 0$, $\text{dist}(x,y) < 1/n$ if $V(x,y) \leqslant k_n$, in order that each $g \in \mathcal{U}(f, \rho_n, k_n)$ fulfills conditions i) and ii).

Denote $\mathcal{V}(f,n)$ a C^0-neighborhood of f included in $\mathcal{U}(f,\rho_n,k_n)$ such that for $g \in \mathcal{V}(f,n)$ there is a semi-conjugacy h, $f \circ h = h \circ g$, such that $\text{dist}(x, h(x)) < 1/n$ for $x \in M$. Let $\mathcal{N}_n$ be the union, for f satisfying Axiom A and the strong transversality condition, of the $\mathcal{V}(f,n)$. Then $\Sigma = \bigcap_{n>0} \mathcal{N}_n$ is C^0-residual.

PROPOSITION 3.2. *Let $g \in \Sigma$ and let C be a nontrivial compact connected g-invariant subset of M, $C \subset \Omega(g)$. Assume, moreover, that C is an attractor, i.e., there exists a neighbourhood U of C, $g(clos\, U) \subset int\, U$, such that $\bigcap_{n\geqslant 0} g^n(U) = C$. Then for each $\varepsilon > 0$ the diameters of the $C_\varepsilon(x)$ are bounded away from zero on U. The diameters of the $D_\varepsilon(x)$ are bounded away from zero on C.*

PROOF. Let $g \in \Sigma$. Given $\varepsilon > 0$, choose n such that $1/n < \varepsilon$, $2/n < \text{dist}(C, M-U)$, and $2/n < \text{diam}\, C$. Then $g \in \mathcal{V}(f,n)$ for some f with the Axiom A and strong transversality properties. Since $C \subset \Omega(g)$ we have $h(C) \subset \Omega(f)$, and as $h(C)$ is connected, it is included in some basic set B of f. Since $2/n < \text{dist}(C, M-U)$, there is a neighbourhood W of $h(C)$ such that $\text{clos}(h^{-1}(W)) \subset U$. ¿From this remark it follows that $B = h(C)$ is an attractor for f. Then Lemma 3.1 applies and allows us to obtain easily the thesis of the proposition since B, being connected and infinite ($2/n < \text{diam}\, C$ and $\text{dist}(x, h(x)) < 1/n$ imply $\text{diam}\, B > 0$), cannot be the orbit of a periodic point. □

4. Dim $M = 2$

Let $g: M \to M$, $g \in \Sigma$, have a compact connected attractor $C \subset \Omega(g)$, $\text{diam}\, C = d > 0$, and let U be an open neighbourhood of C such that $\bigcap_{n\geqslant 0} g^n(U)$

$= C$, and ρ a positive number so that the ball of radius 10ρ centered at each $x \in M$ is homeomorphic to a disk in $\mathbb{R}^2$ and that $\{x \in M\colon \mathrm{dist}(x, C) \leqslant 10\rho\} \subset U$.

Fix ε, $0 < \varepsilon < \rho/10$, and let $\sigma > 0$, $\sigma < \varepsilon$, be such that if $\mathrm{dist}(x, C) < \rho/10$, $C_\varepsilon(x)$ contains a point y_0 with $\mathrm{dist}(x, y_0) = \sigma$, and that for $x \in C$, $D_\varepsilon(x)$ contains a z_0 satisfying $\mathrm{dist}(x, z_0) = \sigma$.

Since C is an attractor, $D_\varepsilon(x) \subset C$ for each $x \in C$; indeed, $\mathrm{dist}(g^{-n}(x), g^{-n}(z)) \leqslant \varepsilon$, $n \geqslant 0$, for $z \in D_\varepsilon(x)$, implies $z \in \bigcap_{n\geqslant 0} g^n(U) = C$.

Let $f\colon M \to M$ satisfying Axiom A and strong transversality be such that g is semi-conjugate to f via a continuous onto map h, $f \circ h = h \circ g$, such that $\mathrm{dist}(x, h(x)) < \sigma/2$ for every $x \in M$. Now, $h(C) = B$ is an attracting basic set of f as we have shown before; moreover, $h^{-1}(h(C)) = C$, for otherwise there would exist $y \notin C$ such that $h(g^{-n}(y)) = f^{-n}(h(y)) \in B$ for $n \geqslant 0$; thus, $g^{-n}(y) \in U$ for $n \geqslant 0$, which is absurd.

For f satisfying Axiom A and $x \in \Omega(f)$ we will denote, as usual,

$$W^s(x) = \{y \in M\colon \mathrm{dist}(f^n(x), f^n(y)) \to 0 \text{ as } n \to \infty\}$$

the stable manifold of f at x. The unstable manifold $W^u(x)$ is defined similarly, replacing ∞ by $-\infty$. For $\varepsilon > 0$,

$$W^s_\varepsilon(x) = \{y \in W^s(x)\colon \mathrm{dist}(f^n(x), f^n(y)) \leqslant \varepsilon,\ n \geqslant 0\};$$

$W^u_\varepsilon(x)$ is defined analogously.

LEMMA 4.1. *For every $x \in C$ we have $h(D_\varepsilon(x)) \subset W^u(h(x))$.*

PROOF. For $x \in C$, consider $h(D_\varepsilon(x)) \subset B$, and let β, γ be the endpoints of the maximum arc containing $h(x) = \xi$ and included in $h(D_\varepsilon(x)) \bigcap W^u(\xi)$. Construct a neighbourhood of this maximum arc by taking a very close but strictly larger arc of $W^u(\xi)$ with endpoints $\beta' \notin h(D_\varepsilon(x))$ to the left of β and $\gamma' \notin h(D_\varepsilon(x))$ to the right of γ and by tracing through each point η of this new arc the local stable manifolds $W^s_{\theta'}(\eta)$, where θ' is chosen so small that the neighbourhood r constructed in this way is homeomorphic to a rectangle $b \times c$, where $b, c \subset \mathbb{R}$ are intervals, b homeomorphic to the arc with endpoints β', γ' and c homeomorphic to $W^s_{\theta'}(\xi)$. We may assume that through each point ζ included in B and in this rectangle we may trace $W^u_\theta(\zeta)$ for some θ, $0 < \theta < \varepsilon$, and that this arc meets $W^s_\theta(\xi)$: if this were not the case, we might take negative iterates of f in order to get that the maximum arc with endpoints β, γ becomes small enough to apply the local product structure on B.

If $h(D_\varepsilon(x))$ does not coincide with this maximum arc, we may assume that the rectangle is so small that $h(D_\varepsilon(x))$ contains some points in the exterior of the rectangle. Thus, the connected component of $h(D_\varepsilon(x)) \bigcap r$ containing ξ must reach the boundary of the rectangle. Through each ζ that belongs to this connected component we trace $W^u_\theta(\zeta)$ and find the intersection $W^s_\theta(\xi) \bigcap W^u_\theta(\zeta)$.

We claim that the range of the mapping $\zeta \to W^s_\theta(\xi) \bigcap W^u_\theta(\zeta)$ is $\{\xi\}$, which is absurd.

If not, we would get a non-trivial subarc δ of $W^s_\theta(\xi)$ contained in B. Since the unstable arc through ξ is also included in B, we obtain that B contains open sets. Since stable manifolds of points in int $B = B^0$ are also included in B^0 as it is easy to show inasmuch as this happens for stable manifolds of the interior periodic points, we get, on account of the local product structure on B, that $\partial B^0 = \varnothing$, i.e., $B = M$. But this implies that f is Anosov; on the other hand, this arc δ on $W^s_\theta(\xi)$ has the property that, for $n \geqslant 0$, $f^{-n}(\delta)$ is contained in a disk of radius ρ, which is impossible.

Thus, $h(D_\varepsilon(x))$ coincides with an arc of $W^u(h(x))$. □

LEMMA 4.2. *If $x \in C$ then $h(C_\varepsilon(x)) \subset W^s(h(x))$ and $h(D_\varepsilon(x)) \bigcap h(C_\varepsilon(x)) = \{h(x)\}$.*

PROOF. Let now β, γ denote the endpoints of the maximum arc $\beta\gamma$ containing $h(x) = \xi$ of $h(C_\varepsilon(x)) \bigcap W^s(\xi)$. We iterate f forwards in order to get $f^n(\beta\gamma) \subset W^s_\theta(f^n(\xi))$, for some $n > 0$, where θ, $0 < \theta < \varepsilon$, is such that $W^s_\theta(\zeta) \bigcap W^u_\theta(f^n(\xi)) = \{\zeta\}$ for $\zeta \in W^u_\theta(f^n(\xi))$. If $h(C_\varepsilon(x))$ had other points than those of $\beta\gamma$, then, as in the previous lemma, we would get, projecting through $W^s_\theta(\zeta)$ into $W^u_\theta(f^n(\xi))$, a non-trivial arc on $W^u_\theta(f^n(\xi))$ whose forward iterates have diameter less than $4\varepsilon < \rho$. Let η be an f-periodic point in B so close to $f^n(\xi)$ that by projection through $W^s_\theta(\zeta)$, for ζ in that arc, we get another non-trivial arc δ, $\delta \subset W^u(\eta)$; the diameter of $f^n(\delta)$, $n \geqslant 0$, is this time less than $6\varepsilon < \rho$. The unstable manifold through η can be obtained as $\bigcup_{k\geqslant 0} f^{k\mu}(\delta)$, μ being the period of η. Let τ be an accumulation point of $W^u(\eta)$. This implies that $W^s_\theta(\tau)$ meets twice $W^u(\eta)$, and we get, therefore, a disk of radius 2ρ centered at τ containing another disk D bordered by an unstable arc and a stable one. Now we finish the proof of both assertions of the lemma by showing that this is impossible. Since at the border of D the diameter is less than 4ρ, for some $n > 0$, $f^n(D)$ is so close to B that we may define on $f^n(D)$, and consequently on D, a stable vector field that never vanishes. Take half stable manifolds entering D and starting on the unstable border of D. Since no half stable manifold can neither stay in the interior of D nor meet the stable border of D, we get that the continuous map that sends a point on the unstable border of D to the first point where the half stable manifold through it meets again this unstable border, has a fixed point, which is absurd. □

LEMMA 4.3. *$C_\varepsilon(x) \bigcap D_\varepsilon(x) = \{x\}$ for $x \in C$.*

PROOF. Let $y \in C_\varepsilon(x) \bigcap D_\varepsilon(x)$ and suppose that $\operatorname{dist}(x, y) > 0$. Choose f satisfying Axiom A and strong transversality, such that $f \circ h = h \circ g$ and $2 \operatorname{dist}(x, h(x)) < \operatorname{dist}(x, y)$. Then $h(x) \neq h(y) \in h(C_\varepsilon(x)) \bigcap h(D_\varepsilon(x))$. □

COROLLARY 4.4. *Let $x \in C$ and $y \in C_\varepsilon(x)$ $(D_\varepsilon(x))$. Then $diam(g^n(C_\varepsilon(x))) \to 0$ $(diam(g^n(D_\varepsilon(x))) \to 0)$ when $n \to +\infty$ (respectively $n \to -\infty$).*

PROOF. Otherwise we would get a point $z \in \omega(x)$ and a non-trivial connected set containing z and included in $C_\varepsilon(z) \bigcap D_\varepsilon(z)$. □

PROPOSITION 4.5. *For $x \in C$, $D_\varepsilon(x)$ is compact, connected and locally connected.*

PROOF. It follows from the previous corollary that given ε', $0 < \varepsilon' < \varepsilon$, there exists $\delta > 0$ such that, if $y \in C_\varepsilon(x)$ ($D_\varepsilon(x)$) and $\text{dist}(x, y) < \delta$, then $y \in C_{\varepsilon'}(x)$ (respectively $y \in D_{\varepsilon'}(x)$). The proof of the proposition is now the same as that of Corollary 2.4 (p. 121) of [**L**]. □

THEOREM 4.6. *If* $\dim M = 2$ *then there is a C^0-residual set Σ of Homeo(M) such that if $g \in \Sigma$ has a compact connected attractor $C \subset \Omega(g)$, diam$C > 0$, then there exists $\varepsilon > 0$ such that if $\varepsilon < \varepsilon_0$, $x \in C$, the connected component $D_\varepsilon(x)$ containing x of $U_\varepsilon(x)$ is a homeomorphic image of an interval. Furthermore,* $\lim_{n \to -\infty} dist(g^n x, g^n y) = 0$ *for $y \in D_\varepsilon(x)$.*

PROOF. Choose ε as in the second paragraph of this section, and let $\varepsilon < \varepsilon$. Since, by the previous proposition, $D_\varepsilon(x)$ is locally connected, any two points may be joined by an arc within $D_\varepsilon(x)$. Assume that for some $\sigma > 0$, $\sigma < \varepsilon$, there are three arcs a, b, c in $D_\varepsilon(x)$ with origin x, joining x to $\partial B_\sigma(x)$ and such that $a \bigcap b = b \bigcap c = a \bigcap c = \{x\}$.

Take f satisfying Axiom A and strong transversality, and semi-conjugate to g via h, where $\text{dist}(x, h(x)) < \delta$ for $x \in M$; here $\delta > 0$ is chosen so small that the endpoints of each of these arcs are at a distance no less than 10δ to the other two arcs. Then we cannot have that the h-image of an endpoint of some arc lies on the h- image of the other arcs; but this is impossible.

This argument also proves that an interior point of an arc like, say, a, cannot be joined to another point of $D_\varepsilon(x)$ through an arc within $D_\varepsilon(x)$ that meets a only at that point. This proves the first assertion; on account of Corollary 4.4, this completes the proof. □

REFERENCES

[F] J. Franks, *Anosov Diffeomorphisms*, Proceedings of the Symposium in Pure Mathematics, vol. 14, 1970, pp. 61–94.

[H] K. Hiraide, *Expansive homeomorphisms of compact surfaces are Pseudo-Anosov*, Osaka J. Math. **27** (1990), 117–162.

[Hu] M. Hurley, *Combined structural and topological stability are equivalent to Axiom A and the strong transversality condition*, Ergod. Th. and Dynam. Sys. **4** (1984), 81–88.

[L] J. Lewowicz, *Expansive homeomorphisms of surfaces*, Bol. Soc. Bras. Mat. **20** (1989), 113–133.

[L1] J. Lewowicz, *Lyapunov functions and topological stability*, J. of Differential Equations **38** (1980), 192–209.

[NS] V. V. Nemytskii and V. V. Stepanov, *Qualitative theory of differential equations*, Princeton University Press, Princeton, New Jersey, 1960.

[S] M. Shub, *Structurally stable diffeomorphisms are dense*, Bull. Amer. Math. Soc. **78** (1972), 817–819.

[Sm] S. Smale, *Differentiable Dynamical Systems*, Bull. Amer. Math. Soc. **73** (1967), 747–817.

[PPSS] J. Palis, C. Pugh, M. Shub, and D. Sullivan, *Genericity theorems in topological dynamics*, Dynamical Systems, Warwick (1974), Lecture Notes in Math., vol. 468, Springer-Verlag, Berlin and New York, 1975, pp. 241–250.

IMERL, Facultad de Ingeniería, Universidad de la República, Av. Julio Herrera y Reissig 565, Montevideo, Uruguay

NAMS, Stockton State College, Pomona, NJ 08240

E-mail address: jtolosa@vax002.stockton.edu

Contemporary Mathematics
Volume **152**, 1993

Lefschetz numbers for periodic points

JAUME LLIBRE

ABSTRACT. We develop a modified Lefschetz number for analysing if a given period belongs to the set of periods of a self-map. We apply it to study of the periodic points of some classes of maps: transversal maps on compact manifolds, Lie group endomorphisms, torus maps and transversal graph maps.

1. Introduction and statement of the main results

We develop in this paper a modified Lefschetz number for analysing if a given period belongs to the set of periods of a self-map. Essentially we work with the Lefschetz numbers for periodic points instead of the usual Lefschetz number for fixed points. As these techniques are homological they are computable and they apply equally to all maps in an appropriate homotopy class. The modified Lefschetz numbers are applied to study the set of periods of several classes of self-maps.

For simplicity of exposition all spaces considered here will be compact manifolds or finite graphs, and thus they admit an index theory (see, for instance [**6**]).

Let $f : X \to X$ be a continuous map. A *fixed point* of f is a point x of X such that $f(x) = x$. Denote the totality of fixed points by $\text{Fix}(f)$. The point $x \in X$ is *periodic with period* m if $x \in \text{Fix}(f^m)$ but $x \notin \text{Fix}(f^k)$ for all $k = 1, \dots, m-1$. Let $\text{Per}(f)$ denote the set of all periods of periodic points of f.

Let M be a compact manifold of dimension n. A continuous map $f : M \to M$ induces endomorphisms $f_{*k} : H_k(M; \mathbb{Q}) \to H_k(M; \mathbb{Q})$ (for $k = 0, 1, \dots, n$) on

1991 *Mathematics Subject Classification.* Primary 58F20.
The author was supported in part by DGICYT Grant PB90-0695.
This paper is in final form and no version of it will be submitted for publication elsewhere.

the rational homology groups of M. The *Lefschetz number* of f is defined by

$$L(f) = \sum_{k=0}^{n} (-1)^k \mathrm{trace}(f_{*k}).$$

By the renowned Lefschetz fixed point theorem: if $L(f) \neq 0$ then f has fixed points (see, for instance [6]). Of course, we can consider the Lefschetz number of f^m but (in general) it is not true that if $L(f^m) \neq 0$ then f has periodic points of period m. As it is well-known a fixed point of f^m need not have period m, so it will be useful to have a method for detecting the difference between "real" and "false" periodic points of period m (i.e., points having period some proper divisor of m). Thus, for instance, the map $f(\theta) = -\theta$ on $\mathbb{S}^1$ satisfies that $\mathrm{Per}(f) = \{1, 2\}$ and $L(f^m)$ is equal to 2 if m is odd, and 0 otherwise.

Let $f : X \to X$ be a map and suppose that the sets $\mathrm{Fix}(f^m)$ are finite for all $m \in \mathbb{N}$ (as usual $\mathbb{N}$ denotes the set of all natural numbers, here 0 is *not* a natural number). If the cardinality of a set A is denoted by $\#A$, we define

$$F_m = \#\mathrm{Fix}(f^m)$$

and

$$P_m = \#\{x \in \mathrm{Fix}(f^m) : x \notin \mathrm{Fix}(f^k) \text{ for } k = 1, \ldots, m-1\}.$$

Then

$$F_m = \sum_{d|m} P_d, \tag{1.1}$$

where $\sum_{d|m}$ denotes the sum over all positive divisors d of m. If we consider the *Moebius function* $\mu(m)$ defined by

$$\mu(m) = \begin{cases} 1 & \text{if } m = 1, \\ 0 & \text{if } k^2 | m \text{ for some } k \in \mathbb{N}, \\ (-1)^r & \text{if } m = p_1 \cdot \ldots \cdot p_r \text{ distinct prime factors,} \end{cases}$$

then

$$P_m = \sum_{d|m} \mu(d) F_{\frac{m}{d}} \tag{1.2}$$

for every $m \in \mathbb{N}$ (see, for instance [27]). Thus, the sequences of integers $\{F_m\}$ and $\{P_m\}$ satisfy the formulas (1.1) and (1.2).

Let $f : M \to M$ be a continuous map on the compact manifold M. For every $m \in \mathbb{N}$ we define the *Lefschetz number of period* m, $l(f^m)$, as follows

$$l(f^m) = \sum_{d|m} \mu(d) L(f^{\frac{m}{d}}). \tag{1.3}$$

Therefore

$$L(f^m) = \sum_{d|m} l(f^d).$$

The Lefschetz number of period m will become interesting after showing that for many classes of maps we have: if $l(f^m) \neq 0$ then $m \in \text{Per}(f)$. Dold [**9**] showed that for any $m \in \mathbb{N}$ if $\text{Fix}(f^m)$ is compact then m divides $l(f^m)$. Other authors like Halpern [**14**] or Heath, Piccinini and You [**15**] have introduced a similar definition for Nielsen numbers instead of Lefschetz numbers.

A C^1 map $f : M \to M$ defined on a compact C^1 differentiable manifold is called *transversal* if $f(M) \subset \text{Int}(M)$ and if for all $m \in \mathbb{N}$ at each point $x \in \text{Fix}(f^m)$ we have $\det(I - df^m(x)) \neq 0$, i.e. 1 is not an eigenvalue of $df^m(x)$. Notice that if f is transversal then for all $m \in \mathbb{N}$ the graph of f^m intersets transversally the diagonal $\{(y, y) : y \in M\}$ at each point (x, x) such that $x \in \text{Fix}(f^m)$.

Let $f : M \to M$ be a transversal map. Then the fixed points of f^m are isolated. Since M is compact, $\#\text{Fix}(f^m)$ is finite for every $m \in \mathbb{N}$.

Since f is transversal it is well-known that if x is a fixed point of f^m then the index of f^m at x, $\text{index}(f^m, x)$, is $(-1)^{u_+(x)}$ where $u_+(x)$ (respectively $u_-(x)$) denotes the number of real eigenvalues of $df^m(x)$ which are strictly greater than 1 (respectively less than -1). Then we have

$$L(f^m) = \sum_{x \in \text{Fix}(f^m)} \text{index}(f^m, x) = \sum_{x \in \text{Fix}(f^m)} (-1)^{u_+(x)}, \tag{1.4}$$

for more details see again [**6**] and [**18**].

We classify the points of f of period m as follows

$$\begin{aligned}
\widetilde{EE}_m &= \{x \text{ point of period } m : u_+(x) \text{ and } u_-(x) \text{ are even}\},\\
\widetilde{EO}_m &= \{x \text{ point of period } m : u_+(x) \text{ is even and } u_-(x) \text{ is odd}\},\\
\widetilde{OE}_m &= \{x \text{ point of period } m : u_+(x) \text{ is odd and } u_-(x) \text{ is even}\},\\
\widetilde{OO}_m &= \{x \text{ point of period } m : u_+(x) \text{ and } u_-(x) \text{ are odd}\};
\end{aligned}$$

and we denote by $\widetilde{ee}_m, \widetilde{eo}_m, \widetilde{oe}_m$ and $\widetilde{oo}_m$ the cardinals of the sets $\widetilde{EE}_m$, $\widetilde{EO}_m$, $\widetilde{OE}_m$ and $\widetilde{OO}_m$, respectively.

Periodic points of transversal maps have been studied by several authors: Franks [**10, 12**], Matsuoka [**23**], Matsuoka and Shiraki [**24**], Casasayas, Llibre and Nunes [**7,8**], Llibre and Swanson [**22**], etc. We present the following result on the periodic points of transversal maps.

THEOREM A. *Let f be a transversal map.*

(a) *If m is odd then*

$$\widetilde{ee}_m + \widetilde{eo}_m = \widetilde{oe}_m + \widetilde{oo}_m + l(f^m).$$

(b) *If m is even then*

$$\widetilde{ee}_m + \widetilde{eo}_m + 2\widetilde{oo}_{\frac{m}{2}} = \widetilde{oe}_m + \widetilde{oo}_m + 2\widetilde{eo}_{\frac{m}{2}} + l(f^m).$$

Theorem A will be proved in Section 2. From Theorem A if follows immediately:

COROLLARY B. *Let f be a transversal map. Suppose that $l(f^m) \neq 0$ for some $m \in \mathbb{N}$.*

(a) *If m is odd then $m \in Per(f)$.*
(b) *If m is even then $\{m/2, m\} \cap Per(f) \neq \emptyset$.*

As an example of application of Corollary B consider a transversal map f on the n-sphere of degree $D \notin \{-1, 0, 1\}$. Since

$$l(f^m) = \sum_{d|m} \mu(d)[1 + (-1)^n D^{\frac{m}{d}}] \neq 0,$$

from Corollary B it follows easily that $\{n \in \mathbb{N} : n \text{ is odd}\} \subset \text{Per}(f)$. For more results on transversal maps on spheres see [**7**].

Now the Lefschetz numbers for periodic points are applied to study the set of periods of Lie group endomorphisms and of continuous self-maps on tori. Several authors have studied these two topics, see for instance Halpern [**13**], Nakaoka [**26**], Alsedà, Baldwin, Llibre, Swanson and Szlenk [**1**] and Keppelmann [**19**].

THEOREM C. *Let G be a Lie group and Γ a discrete normal subgroup such that G/Γ is compact. Let $F : G \to G$ be a Lie group endomorphism such that $F(\Gamma) \subset \Gamma$ and suppose that the eigenvalues of $dF(e) : T_e(G) \to T_e(G)$ contains no roots of unity (where e is the identity element of G). Let $\widetilde{F} : G/\Gamma \to G/\Gamma$ be the Lie group endomorphism induced by F. If $l(\widetilde{F}^{p^k}) \neq 0$ for every prime number p and for every $k \in \mathbb{N}$, then $Per(\widetilde{F}) = \mathbb{N}$.*

THEOREM D. *Let $f : \mathbb{T}^n \to \mathbb{T}^n$ be a continuous map on the n-torus, and suppose that the endomorphism f_{*1} (induced by f on the first rational homological group of $\mathbb{T}^n$) has no roots of unity as eigenvalues. If $l(f^{p^k}) \neq 0$ for every prime number p and for every $k \in \mathbb{N}$, then $Per(f) = \mathbb{N}$.*

From Theorem D it follows easily (see for more details Section 4):

COROLLARY E. *Let $f : \mathbb{T}^n \to \mathbb{T}^n$ be a continuous map. If the sequence $|L(f^m)|$ is strictly increasing, then $Per(f) = \mathbb{N}$.*

This corollary was proved in [**1**].

In Section 3 we study the set of periods of the Lie group endomorphisms and prove Theorem C. These results are used in Section 4 together with the Nielsen fixed point theory to prove Theorem D.

Finally, to show that these kind of results can be extended to spaces different from manifolds we consider the following class of self-maps on graphs.

We say that a finite graph G is *proper* if the valence of all its vertices is different from two (see Section 5 for precise definitions). Roughly speaking all the vertices of a proper graph are endpoints, branching points or isolated vertices.

Let G be a proper graph and let V be the set of its vertices. Then the map $f : G \to G$ is a *transversal graph map* if the following conditions hold: (1) f

is continuous, (2) f is C^1 in $G \setminus V$, (3) f has no periodic points in V, (4) if $f^n(x) = x$ then $df^n(x) \neq 1$.

Periodic points for continuous self-maps on finite graphs have been studied by several authors, see for instance Imrich [**16**], Imrich and Kalinowski [**17**], Alsedà, Llibre and Misiurewicz [**2**], Baldwin [**4**], Llibre [**20**], Messano [**25**], Blokh [**5**] and Llibre and Misiurewicz [**21**].

For finite graphs we can define the Lefschetz numbers of period m as in (1.3).

THEOREM F. *Let f be a transversal graph map and suppose that $l(f^m) \neq 0$ for some $m \in \mathbb{N}$.*

(a) *If m is odd then $m \in Per(f)$.*
(b) *If m is even then $\{m/2, m\} \cap Per(f) \neq \emptyset$.*
(c) *If f satisfies that $df^k(x) > -1$ for every $k \in \mathbb{N}$ and every $x \in Fix(f^k)$, then $m \in Per(f)$.*

Theorem F will be proved in Section 5.

2. Transversal maps

The main goal of this section is to prove Theorem A.

Taking advantage of the notation introduced in [**23**] and in [**22**], for a fixed $m \in \mathbb{N}$ we classify the fixed points of f^m as follows

$$\begin{aligned} EE_m &= \{x \in \mathrm{Fix}(f^m) : u_+(x) \text{ and } u_-(x) \text{ are even}\}, \\ EO_m &= \{x \in \mathrm{Fix}(f^m) : u_+(x) \text{ is even and } u_-(x) \text{ is odd}\}, \\ OE_m &= \{x \in \mathrm{Fix}(f^m) : u_+(x) \text{ is odd and } u_-(x) \text{ is even}\}, \\ OO_m &= \{x \in \mathrm{Fix}(f^m) : u_+(x) \text{ and } u_-(x) \text{ are odd}\}. \end{aligned}$$

We denote by ee_m, eo_m, oe_m and oo_m the cardinals of the sets EE_m, EO_m, OE_m and OO_m, respectively.

LEMMA 2.1. *Let f be transversal map. Then for every $m \in \mathbb{N}$ we have*

$$ee_m + eo_m = oe_m + oo_m + L(f^m).$$

PROOF. From (1.4) we have

$$\begin{aligned} L(f^m) &= \sum_{x \in \mathrm{Fix}(f^m)} (-1)^{u_+(x)} \\ &= ee_m + eo_m - oe_m - oo_m. \end{aligned}$$ □

The next lemma was proved in [**22**] for transversal maps on surfaces, but its proof extends to transversal maps on manifolds.

LEMMA 2.2. *Let f be a transversal map. Then the following equalities hold:*

$$ee_m = \widetilde{ee}_m + oo_{\frac{m}{2}} - \sum_{\substack{d|m \\ d>1}} \mu(d) ee_{\frac{m}{d}},$$

$$eo_m = \widetilde{eo}_m - \sum_{\substack{d|m \\ d>1 \\ d \text{ odd}}} \mu(d) eo_{\frac{m}{d}},$$

$$oe_m = \widetilde{oe}_m + eo_{\frac{m}{2}} - \sum_{\substack{d|m \\ d>1}} \mu(d) oe_{\frac{m}{d}},$$

$$oo_m = \widetilde{oo}_m - \sum_{\substack{d|m \\ d>1 \\ d \text{ odd}}} \mu(d) oo_{\frac{m}{d}}.$$

PROOF. Note that if $x \in EE_m$ (respectively OE_m), then $x \in EE_{km}$ (respectively OE_{km}) for every $k \in \mathbb{N}$. But if $x \in EO_m$ (respectively OO_m), then $x \in EO_{km}$ (respectively OO_{km}) if k is odd, and $x \in OE_{km}$ (respectively EE_{km}) if k is even.

Then the fixed points of EE_m can be either periodic points of period m of $\widetilde{EE}_m$; or periodic points of $\widetilde{EE}_r$ of period r a proper divisor of m; or, if m is even, periodic points of $\widetilde{OO}_s$ of period s a divisor of m such that m/s is even. In short, we have

$$ee_m = \widetilde{ee}_m + oo_{\frac{m}{2}} - \sum_{\substack{d|m \\ d>1}} \mu(d) ee_{\frac{m}{d}},$$

because

$$\sum_{1 \le i \le k} ee_{\frac{m}{p_i}} - \sum_{1 \le i < j \le k} ee_{\frac{m}{p_i p_j}} + \cdots + (-1)^{k+1} ee_{\frac{m}{p_1 \cdot \cdots \cdot p_k}} = - \sum_{\substack{d|m \\ d>1}} \mu(d) ee_{\frac{m}{d}}$$

is the number of periodic points $x \in EE_m$ whose period is some proper divisor r of m and $x \in \widetilde{EE}_r$, being $m = p_1^{\alpha_1} \cdot \ldots \cdot p_k^{\alpha_k}$ with $p_1, \ldots, p_k$ distinct primes; and if m is even $oo_{m/2}$ is the number of periodic points $x \in EE_m$ whose period is a proper divisor s of m such that m/s is even and $x \in \widetilde{OO}_s$.

The other three equalities of the lemma follow in a similar fashion. $\square$

We are interested in studying the set of periods of f. To this purpose it is useful to have information on the whole sequence $\{L(f^m)\}_{m \in \mathbb{N}}$ of the Lefschetz numbers of all the iterates of f. The *Lefschetz zeta function of f* defined as

$$Z_f(t) = \exp \left(\sum_{m=1}^{+\infty} \frac{L(f^m)}{m} t^m \right)$$

is a generating function for that sequence, and it may be computed independently through

$$Z_f(t) = \prod_{k=0}^{\dim M} \det(I_{n_k} - tf_{*k})^{(-1)^{k+1}},$$

where $n_k = \dim H_k(M;\mathbb{Q})$, I_{n_k} is the $n_k \times n_k$ identity matrix and we take $\det(I_{n_k} - tf_{*k}) = 1$ if $n_k = 0$, see for more details [**11**].

For transversal maps the Lefschetz zeta function may be related in a simple way with its set of periodic orbits. Given γ a periodic orbit of f of period $p(\gamma)$ and $x \in \gamma$, we define $u_+(\gamma) = u_+(x)$ and $u_-(\gamma) = u_-(x)$ (it is easy to check that they are well-defined). With this notation, we have the following proposition due to Franks [**10**], which is one of the basic results that we will need.

PROPOSITION 2.3. *Let $f : M \to M$ be a transversal map. Then*

$$Z_f(t) = \prod_{\gamma} \left(1 - (-1)^{u_-(\gamma)} t^{p(\gamma)}\right)^{(-1)^{u_+(\gamma)+u_-(\gamma)+1}},$$

where γ goes over all the periodic orbits of f.

From the definitions of $l(f^m)$ and $Z_f(t)$ we get the following well-known formal relation

$$Z_f(t) = \prod_{m=1}^{+\infty} (1 - t^m)^{-\frac{l(f^m)}{m}}, \tag{2.1}$$

for more details see, for instance [**3**].

PROOF OF THEOREM A. Lef f be a transversal map. A periodic orbit of f and type $\widetilde{EE}_m$, $\widetilde{EO}_m$, $\widetilde{OE}_m$ or $\widetilde{OO}_m$ contributes to the product of Proposition 2.3 in the factor $(1 - t^m)^{-1}$, $(1 + t^m)$, $(1 - t^m)$ or $(1 + t^m)^{-1}$ respectively. Therefore if we write the product of Proposition 2.3 by substituting $(1 + t^m)$ by $(1 - t^{2m})/(1 - t^m)$, we obtain

$$Z_f(t) = \prod_{m=1}^{+\infty} (1 - t^m)^{\frac{\widetilde{oe}_m + \widetilde{oo}_m - \widetilde{ee}_m - \widetilde{eo}_m + 2(\widetilde{eo}_{m/2} - \widetilde{oo}_{m/2})}{m}}.$$

Of course, in the above formula $\widetilde{eo}_{m/2}$ and $\widetilde{oo}_{m/2}$ are zero if m is odd. Now from (2.1) it follows the theorem. □

3. Lie group endomorphisms

A *Lie group* G is a smooth differentiable manifold which is also endowed with a group structure such that the map $G \times G \to G$ defined by $(a, b) \longmapsto a \circ b^{-1}$ is smooth. Here we denote by $\circ$ the group operation on G. A map $F : G \to G$ is a *Lie group endomorphism* if F is both smooth and a group endomorphism of the abstract group. We denote by e the identity element of G.

The following result was proved in [**1**] for a Lie torus endomorphism, but its proof extends to any Lie group.

LEMMA 3.1. *Suppose $F : G \to G$ is a Lie group endomorphism. If F has periodic points of relatively prime periods k and j, then F has a periodic point of period kj.*

PROOF. Let x and y be periodic points of period k and j, respectively. Set $z = x \circ y$. Clearly, $F^{kj}(z) = z$. Therefore, z is a periodic point of F of period a divisor r of kj. Thus, either $r = kj$ and we are done, or there is a prime p such that p divides kj and r divides kj/p. Since k and j are relatively prime, p divides only one of them. So, either k or j divides kj/p but not both. Suppose j divides kj/p. Then

$$x \circ y = z = F^{\frac{kj}{p}}(z) = F^{\frac{kj}{p}}(x) \circ F^{\frac{kj}{p}}(y) = F^{\frac{kj}{p} \pmod k}(x) \circ y.$$

Hence $F^{\frac{kj}{p} \pmod k}(x) = x$, which is a contradiction because k does not divide kj/p. □

The following result was proved by Halpern [**13**] for a Lie torus endomorphism and generalized by Nakaoka [**26**] to a Lie group endomorphism.

PROPOSITION 3.2. *Let G be a Lie group and Γ a discrete subgroup such that G/Γ is compact. Let $F : G \to G$ be a Lie group endomorphism such that $F(\Gamma) \subset \Gamma$ and 1 is not an eigenvalue of $dF(e) : T_e(G) \to T_e(G)$. Then, for the map $\widetilde{F} : G/\Gamma \to G/\Gamma$ induced by F, we have*

$$\#Fix(\widetilde{F}) = |L(\widetilde{F})|.$$

Proposition 3.2 follows from Proposition 1 of [**26**]. Notice that we do not need in the hypotheses of Proposition 3.2 that $\text{Fix}(f) = \{e\}$ as in Proposition 1 of [**26**]. This is due to the fact that our proposition does not state that the Nielsen number of $\widetilde{F}$ is equal to $\#\text{Fix}(\widetilde{F})$ as it is stated in Proposition 1 of [**26**].

REMARK 3.3. *The key point in the proof of Proposition 3.2 is to show that each fixed point x of $\widetilde{F}$ is isolated, and that its index is constant equal to the sign of $det(I - d\widetilde{F}(x)) = det(I - dF(e))$.*

Remark 3.3 will also play a key role in the proof of Theorem C.

PROOF OF THEOREM C. Suppose that the topological dimension of G is n, and let $\lambda_1, \dots, \lambda_n$ be all the eigenvalues of $dF(e)$ listed with their multipicities. Since $\det(I - dF^m(e)) = \prod_{i=1}^n (1 - \lambda_i^m)$ for every $m \in \mathbb{N}$, and $\lambda_1, \dots, \lambda_n$ contains no roots of unity, it follows that $\det(I - dF^m(e)) \neq 0$ for all $m \in \mathbb{N}$. Therefore, by Remark 3.3, the map $\widetilde{F}$ is transversal.

Since $\widetilde{F}$ is transversal, if $m \in \mathbb{N}$ is odd and $l(\widetilde{F}^m) \neq 0$ then by Theorem A(a) we get statement (a). In particular, if $p > 2$ is prime and $l(\widetilde{F}^{p^k}) \neq 0$ for all $k \in \mathbb{N}$, we have that $\{p^k : k \in \mathbb{N}\} \subset \text{Per}(\widetilde{F})$.

Let $m \in \mathbb{N}$. Since $\det(I - dF^m(e)) = \prod_{i=1}^n (1 - \lambda_i^m)$, by Remark 3.3 it follows that if a fixed point of $\widetilde{F}^m$ belongs to one of the sets EE_m, EO_m, OE_m, OO_m, then all the fixed points of $\widetilde{F}^m$ belongs to the same set.

Let $p = 2$. From Theorem A it follows that

$$\widetilde{ee}_{2^k} + \widetilde{eo}_{2^k} + 2\widetilde{oo}_{2^{k-1}} = \widetilde{oe}_{2^k} + \widetilde{oo}_{2^k} + 2\widetilde{eo}_{2^{k-1}} + l(\widetilde{F}^{2^k}) \tag{3.1}$$

for every $k \in \mathbb{N}$. Since all the fixed points of $\widetilde{F}^{2^k}$ belong to a unique set $EE_{2^k}, EO_{2^k}, OE_{2^k}, OO_{2^k}$, from Lemma 2.2, equality (3.1) and Lemma 2.1, for every $k \in \mathbb{N}$ we obtain: if $EO_{2^k} \cup OE_{2^k} \cup OO_{2^k} = \emptyset$ then

$$\begin{aligned} \widetilde{ee}_{2^k} &= ee_{2^k} - ee_{2^{k-1}} - oo_{2^{k-1}}, \\ \widetilde{ee}_{2^k} &= -2\widetilde{oo}_{2^{k-1}} + l(\widetilde{F}^{2^k}), \\ ee_{2^k} &= L(\widetilde{F}^{2^k}); \end{aligned} \tag{3.2}$$

if $EE_{2^k} \cup OE_{2^k} \cup OO_{2^k} = \emptyset$ then

$$\begin{aligned} \widetilde{eo}_{2^k} &= eo_{2^k}, \\ \widetilde{eo}_{2^k} &= l(\widetilde{F}^{2^k}), \\ eo_{2^k} &= L(\widetilde{F}^{2^k}); \end{aligned} \tag{3.3}$$

if $EE_{2^k} \cup EO_{2^k} \cup OO_{2^k} = \emptyset$ then

$$\begin{aligned} \widetilde{oe}_{2^k} &= oe_{2^k} - oe_{2^{k-1}} - eo_{2^{k-1}}, \\ \widetilde{oe}_{2^k} &= -2\widetilde{eo}_{2^{k-1}} - l(\widetilde{F}^{2^k}), \\ oe_{2^k} &= -L(\widetilde{F}^{2^k}); \end{aligned} \tag{3.4}$$

if $EE_{2^k} \cup EO_{2^k} \cup OE_{2^k} = \emptyset$ then

$$\begin{aligned} \widetilde{oo}_{2^k} &= oo_{2^k}, \\ \widetilde{oo}_{2^k} &= -l(\widetilde{F}^{2^k}), \\ oo_{2^k} &= -L(\widetilde{F}^{2^k}). \end{aligned} \tag{3.5}$$

Since $F(e) = e$, we have that $\widetilde{F}$ has fixed points. Since there are no roots of unity between the eigenvalues of $dF(e)$, -1 is not an eigenvalue of $dF^m(e)$ for all $m \in \mathbb{N}$. Therefore, since $\widetilde{F}^m$ has fixed points, by applying Proposition 3.2 to F^m, it follows that $|L(\widetilde{F}^m)| = \#\text{Fix}(\widetilde{F}^m) \neq 0$ for all $m \in \mathbb{N}$.

If $EE_{2^k} \cup OE_{2^k} \cup OO_{2^k} = \emptyset$ or $EE_{2^k} \cup EO_{2^k} \cup OE_{2^k} = \emptyset$, since $l(\widetilde{F}^{2^k}) = L(\widetilde{F}^{2^k}) - L(\widetilde{F}^{2^{k-1}})$, from (3.3) or (3.5) it follows that $L(\widetilde{F}^{2^{k-1}}) = 0$, in contradiction with the fact that $L(\widetilde{F}^m) \neq 0$ for all $m \in \mathbb{N}$. Hence, since for every $k \in \mathbb{N}$ there is at most a unique set $EE_{2^k}, EO_{2^k}, OE_{2^k}, OO_{2^k}$ non-empty, we have $EO_{2^k} = OO_{2^k} = \emptyset$ for every $k \in \mathbb{N}$. Then equalities (3.2) and (3.4) become: if $EO_{2^k} \cup OE_{2^k} \cup OO_{2^k} = \emptyset$ then

$$\begin{aligned} \widetilde{ee}_{2^k} &= ee_{2^k} - ee_{2^{k-1}}, \\ \widetilde{ee}_{2^k} &= l(\widetilde{F}^{2^k}), \\ ee_{2^k} &= L(\widetilde{F}^{2^k}); \end{aligned}$$

if $EE_{2^k} \cup EO_{2^k} \cup OO_{2^k} = \emptyset$ then

$$\begin{aligned} \widetilde{oe}_{2^k} &= oe_{2^k} - oe_{2^{k-1}}, \\ \widetilde{oe}_{2^k} &= -l(\widetilde{F}^{2^k}), \\ oe_{2^k} &= -L(\widetilde{F}^{2^k}). \end{aligned}$$

So, if $l(\widetilde{F}^{2^k}) \neq 0$, either $\widetilde{ee}_{2^k} \neq 0$ or $\widetilde{oe}_{2^k} \neq 0$. Hence $\{2^k : k \in \mathbb{N}\} \subset \mathrm{Per}(\widetilde{F})$. In short, from Theorem A(a) we have that $\{p^k : k \in \mathbb{N}\} \subset \mathrm{Per}(\widetilde{F})$ for every prime natural number p. Let $m \in \mathbb{N}$. If $m = 1$ we are down (because $F(e) = e$ and consequently $1 \in \mathrm{Per}(\widetilde{F})$). Suppose $m > 1$. We write $m = p_1^{\alpha_1} \cdot \ldots \cdot p_r^{\alpha_r}$ where $p_1, \ldots, p_r$ are distinct primes and $\alpha_i \geq 1$ for $i = 1, \ldots, r$. By the inductive use of Lemma 3.1 applied to $\widetilde{F}$, we get $m \in \mathrm{Per}(\widetilde{F})$. □

4. Continuous maps on tori

Let $f : \mathbb{T}^n \to \mathbb{T}^n$ be a continuous map. The *Nielsen number* $N(f)$ of f is defined as follows. First an equivalence relation $\sim$ is defined on the set $\mathrm{Fix}(f)$. Two fixed points x, y are equivalent, $x \sim y$, provided there is a path γ in $\mathbb{T}^n$ from x to y such that $f \circ \gamma$ and γ are homotopic and the homotopy fixes endpoints. The set of equivalence classes $\mathrm{Fix}(f)/\sim$ is known to be finite and each equivalence class is compact. Each of these equivalence classes will be called a *fixed point class*. Using a fixed point index, we may assign an index $i_f(C)$ to each fixed point class C. A *essential class* is a fixed point class C such that $i_f(C) \neq 0$. Then the Nielsen number is the number of essential classes. For more details on the Nielsen number see [**18**].

Notice that from the definition of Nielsen number, f has at least $N(f)$ fixed points. The main difference between the Lefschetz and the Nielsen numbers is that, in general, the Lefschetz number only gives existence of fixed points, while the Nielsen number provides a lower bound for the number of fixed points.

The following result applied to our maps explains the importance of the Nielsen number in fixed point theory. *If $f : \mathbb{T}^n \to \mathbb{T}^n$ is a continuous map, then each map $g : \mathbb{T}^n \to \mathbb{T}^n$ homotopic to f has at least $N(f)$ fixed points. Furthermore, $N(f) = N(g)$.*

Let $f : \mathbb{T}^n \to \mathbb{T}^n$ be a continuous map and let A be an $n \times n$ matrix representative of the induced homology endomorphism $f_{*1} : H_1(T^n; \mathbb{Q}) \to H_1(T^n; \mathbb{Q})$. In what follows A will be called a *matrix associated* to f. Then f is homotopic to the map $f_A : \mathbb{T}^n \to \mathbb{T}^n$ covered by the linear map $A : \mathbb{R}^n \to \mathbb{R}^n$ (for more details see [**13**]). Then, by using Nielsen numbers in [**1**] (see Corollary 3.5) or in [**15**] (see Proposition 2.2 and Theorem 3.7) the following result is proved.

PROPOSITION 4.1. *Let $f : \mathbb{T}^n \to \mathbb{T}^n$ be a continuous map and let A be a matrix associated to f. If no eigenvalue of A is a root of unity, then Per(f_A) $\subset$ Per(f)*

PROOF OF THEOREM D. Let $f : \mathbb{T}^n \to \mathbb{T}^n$ be a continuous map and let A be a matrix associated to f. Since the space $\mathbb{R}^n$ with the euclidean addition group structure is an abelian Lie group, $\mathbb{Z}^n$ is clearly a discrete subgroup of $\mathbb{R}^n$ and A has no roots of unity as eigenvalues, we can apply Theorem C to the continuous map $f_A : \mathbb{R}^n/\mathbb{Z}^n \to \mathbb{R}^n/\mathbb{Z}^n$ induced by the Lie group endomorphism $A : \mathbb{R}^n \to \mathbb{R}^n$.

Now suppose that $l(f^{p^k}) \neq 0$ for every prime number p and for every $k \in \mathbb{N}$. Then $l(f_A^{p^k}) \neq 0$ for every prime number p and for every $k \in \mathbb{N}$. By Theorem C, $\mathrm{Per}(f_A) = \mathbb{N}$. Again from Proposition 4.1 it follows that $\mathrm{Per}(f_A) = \mathbb{N} \subset \mathrm{Per}(f)$. $\square$

PROOF OF COROLLARY E. Let $f : \mathbb{T}^n \to \mathbb{T}^n$ be a continuous map and let A be a matrix associated to f. Since $L(f^m) = \det(I - A^m)$ (see for instance [**13**]) and $|L(f^m)|$ is strictly increasing, it follows that no eigenvalue of A is a root of unity. Furthermore, since $l(f^{p^k}) = L(f^{p^k}) - L(f^{p^{k-1}})$ if p is prime and $k \in \mathbb{N}$, and $|L(f^{p^k})| > |L(f^{p^{k-1}})|$, we have that $l(f^{p^k}) \neq 0$. Hence, we are in the hypotheses of Theorem D, consequently $\mathrm{Per}(f) = \mathbb{N}$. $\square$

5. Transversal graph maps

A *finite graph* will be a finite CW-complex of dimension one. That is, a finite graph G is a Hausdorff space which has a finite subspace V (points of V are called *vertices*) such that $G \setminus V$ is the disjoint union of a finite number of open subsets $e_1, \ldots, e_k$ called *edges*, and each e_i is homeomorphic to an open interval of the real line.

Note that a finite graph is compact, since it is the union of a finite number of compact subsets (the closed edges $\overline{e}_i$ and the vertices). It may be either connected or disconnected, and it may have isolated vertices.

The *valence* of a vertex is the number of edges with the vertex as an endpoint. We say that a finite graph is *proper* if the valence of all its vertices is distinct from two.

A graph map $f : G \to G$ induces endomorphisms $f_{*k} : H_k(G;\mathbb{Q}) \to H_k(G;\mathbb{Q})$ (for $k = 0, 1$) on the rational homology groups of G, where $H_0(G;\mathbb{Q}) \approx \mathbb{Q}^c$, $H_1(G;\mathbb{Q}) \approx \mathbb{Q}^d$, and c and d are the number of connected components of G and the number of independent loops of G respectively. A *loop* is a subset of G homeomorphic to a circle. The endomorphisms f_{*0} and f_{*1} are represented by integer matrices. Then the Lefschetz number of f is

$$L(f) = \mathrm{trace}(f_{*0}) - \mathrm{trace}(f_{*1}).$$

Let $f : G \to G$ be a transversal graph map (see the definition in the introduction). Then condition (4) implies that if $f^m(x) = x$ then $df^m(x) \neq \pm 1$, because if $df^m(x) = -1$ then $f^{2m}(x) = x$ and $df^{2m}(x) = 1$, in contradiction with (4).

So, for every $m \in \mathbb{N}$ we have

$$\begin{aligned}
OE_m &= \{x \in \mathrm{Fix}(f^m) : df^m(x) > 1\},\\
EE_m &= \{x \in \mathrm{Fix}(f^m) : -1 < df^m(x) < 1\},\\
EO_m &= \{x \in \mathrm{Fix}(f^m) : df^m(x) < -1\},\\
OO_m &= \emptyset.
\end{aligned}$$

Now we can extend the lemmas and theorems of Section 2 to any transversal graph map by using exactly the same arguments. In particular, statements (a) and (b) of Theorem A give the corresponding ones of Theorem F. Furthermore, in the hypotheses of statement (c) of Theorem F we have $EO_m = \emptyset$. Therefore, since always $OO_m = \emptyset$, from Theorem A we have

$$\widetilde{ee}_m = \widetilde{oe}_m + l(f^m).$$

Then, if $l(f^m) \neq 0$ it follows that at least one of the numbers $\widetilde{ee}_m$ or $\widetilde{oe}_m$ is non-zero, consequently $m \in \mathrm{Per}(f)$. Hence, Theorem F is proved. □

REFERENCES

1. L. Alsedà, S. Baldwin, J. Llibre, R. Swanson and W. Szlenk, *Miminal sets of periods for torus maps via Nielsen numbers*, Pacific Journal of Math. (to appear).
2. L. Alsedà, J. Llibre and M. Misiurewicz, *Periodic orbits of maps of Y*, Trans. Amer. Math. Soc. **313** (1989), 475–538.
3. I.K. Babenko and S.A. Bogatyi, *The behavior of the index of periodic points under iterations of a mapping*, Math. USSR Izvestiya **38** (1992), 1–26.
4. S. Baldwin, *An extension of Šarkovskiĭ's Theorem to the n-od*, Ergod. Th. & Dynam. Sys. **11** (1991), 249–271.
5. A. M. Blokh, *The spectral decomposition, periods of cycles and Misiurewicz conjecture for graph maps*, preprint, 1990.
6. R.F. Brown, *The Lefschetz fixed point theorem*, Scott, Foresman and Company, Glenview, IL, 1971.
7. J. Casasayas, J. Llibre and A. Nunes, *Periodic orbits of transversal maps*, preprint, Univ. Autònoma de Barcelona, 1991.
8. J. Casasayas, J. Llibre and A. Nunes, *Periods and Lefschetz zeta functions*, Pacific Journal of Math. (to appear).
9. A. Dold, *Fixed point indices of iterated maps*, Invent. math. **74** (1983), 419–435.
10. J. Franks, *Some smooth maps with infinitely many hyperbolic periodic points*, Trans. Amer. Math. Soc. **226** (1977), 175–179.
11. J. Franks, *Homology and dynamical systems*, CBMS Regional Conf. Ser. in Math., no. 49, Amer. Math. Soc., Providence, R.I., 1982.
12. J. Franks, *Period doubling and the Lefschetz formula*, Trans. Amer. Math. Soc. **287** (1985), 275–283.
13. B. Halpern, *Periodic points on tori*, Pacific Journal of Math. **83** (1979), 117–133.
14. B. Halpern, *Nielsen type numbers for periodic points*, (unpublished).
15. P. Heath, R. Piccinini and C.Y. You, *Nielsen type numbers for periodic points I*, Springer Lecture Notes in Math. **1411** (1989), 88–106.
16. W. Imrich, *Periodic points of small periods of continuous mappings of trees*, Ann. Discrete Math. **27** (1985), 443–446.
17. W. Imrich and R. Kalinowski, *Periodic points of continuous mappings of trees*, Ann. Discrete Math. **27** (1985), 447–460.
18. B. Jiang, *Lectures on Nielsen fixed point theory*, Contemporary Mathematics **14**, Amer. Math. Soc., 1983.

19. E.C. Keppelmann, *Periodic points on nilmanifolds and solvmanifolds*, Ph. D., University of Wisconsin - Madison, 1991.
20. J. Llibre, *Periodic points of one dimensional maps*, European Conference on Iteration Theory, ECIT 89, World Scientific, Singapore (1991), 194–198.
21. J. Llibre and M. Misiurewicz, *Horseshoes, entropy and periods for graph maps*, to appear in Topology.
22. J. Llibre and R. Swanson, *Periodic points for transversal maps on surfaces*, preprint, Univ. Autònoma de Barcelona, 1991.
23. T. Matsuoka, *The number of periodic points of smooth maps*, Ergod. Th. & Dynam. Sys. **9** (1989), 153–163.
24. T. Matsuoka and H. Shiraki, *Smooth maps with finitely many periodic points*, Mem. Fac. Sci. Kochi. Univ. (Math.) **11** (1990), 1-6.
25. B. Messano, *Continuous functions from an arcwise connected tree into itself: periodic points, global convergence, plus-global convergence*, Ricerche di Matem. **38** (1989), 199–205.
26. M. Nakaoka, *Periodic points on nilmanifolds*, in Manifolds & Lie Groups (Notre Dame, 1980), Prog. Math. Series, vol. **14**, Birkhauser (1981), 315-324.
27. I. Niven and H.S. Zuckerman, *An introduction to the theory of numbers*, fourth edition, John Wiley & Sons, New York, 1980.

DEPARTAMENT DE MATEMÀTIQUES, UNIVERSITAT AUTÒNOMA DE BARCELONA, BELLATERRA, BARCELONA 08193, SPAIN

E-mail address: imat0@cc.uab.es

Contemporary Mathematics
Volume **152**, 1993

The Burau representation of the braid group and the Nielsen-Thurston classification

TAKASHI MATSUOKA

Abstract. We prove a formula which concerns the computation of the Burau representation of the braid group. We apply this formula to the problem of determining whether a given isotopy class of homeomorphisms on a punctured disk contains a pseudo-Anosov component.

1. Introduction

The Burau representation is a matrix representation of the braid group introduced by W. Burau in 1936. This representation is one of the fundamental tools in low dimensional topology, and it has also proved to be important in fixed point theory and the theory of dynamical systems. Indeed, it is known that if f is an orientation-preserving homeomorphism on a punctured disk, then the fixed point indices of some homological fixed point classes of f coincide with the coefficients of the trace of the reduced Burau matrix of the braid b corresponding to the map f up to sign. (See Theorem A in the next section.) Thus by computing the Burau matrices for powers of the braid b, we can obtain an information concerning the existence of fixed points and periodic points of f together with their homological properties.

In this paper, we prove a formula concerning the Burau representation, which reduces the computation of the Burau matrices and their traces to the summation of some polynomials and hence make the computation simpler. In the case of 3-braids, this formula has been already obtained by the author in [20, p.429] , and has been used to get some results on the number of periodic solutions for 2-dimensional, time-periodic systems of ordinary differential equations in [20], [21].

In Section 4, our formula will be applied to the classification problem of homeomorphisms on punctured disks. According to the Nielsen-Thurston classification theory [9],[13],[25], any isotopy class of homeomorphisms on a compact surface can be built up by gluing together a finite number of building blocks, each of which is periodic or pseudo-Anosov. We consider the problem of deciding whether a given isotopy class has a pseudo-Anosov component or not. This problem is

1991 Mathematics Subject Classification 55M20, 57M99, 58F99. This paper is in final form and no version of it will be submitted for publication elsewhere .

important in the theory of dynamical systems, since if an isotopy class has a pseudo-Anosov component, then every homeomorphism in this class must have dynamical complexity, e.g., it has positive topological entropy and an infinite number of periodic points. When the surface is a punctured disk, several authors have obtained sufficient conditions under which an isotopy class has a pseudo-Anosov component(for instance, Boyland[4], Kobayashi[18]). Also, algorithms have been found which determine all of the components, and in particular, solve this problem by Benardete, Gutierrez, and Nitecki[1], Franks and Misiurewicz[11].

Here, we restrict ourselves to some family of isotopy classes, and exhibit explicitely all the isotopy classes within this family having no pseudo-Anosov components. To get this result, we do not use the algorithms mentioned above. Instead we compute the Burau matrices by using our formula, and use it to analyze the structure of the set of periodic points. The Burau representation is known to be unfaithful in general[22], nevertheless our method is applicable within this family.

2. The Burau representation and fixed points

Let B_n denote the braid group on n strings. (For the definition of the braid group, see Birman[2], Moran [23].) For $i = 1, \ldots, n-1$, let σ_i be the n-braid represented by n strings shown in Fig. 1. Then the braid group B_n has a presentation with generators $\sigma_1, \ldots, \sigma_{n-1}$ and defining relations:[2, 1.8]

$$\sigma_i\sigma_j = \sigma_j\sigma_i \qquad \text{if } \mid i-j \mid > 1,$$
$$\sigma_i\sigma_{i+1}\sigma_i = \sigma_{i+1}\sigma_i\sigma_{i+1} \qquad \text{for } i = 1, \ldots, n-2.$$

If $n \geq 3$, the center $Z(B_n)$ of the group B_n is the infinite cyclic subgroup genereated by the full twist braid $\theta_n = (\sigma_1\sigma_2\cdots\sigma_{n-1})^n$ (see [2 ,Theorem 1.84]).

We define the *reduced Burau representation*

$$R : B_n \to GL_{n-1}(\mathbf{Z}[t, t^{-1}]),$$

where $\mathbf{Z}[t, t^{-1}]$ is the ring of integer polynomials in the variable t and its inverse, by the following rules:

$$R(bb') = R(b)R(b'). \tag{1}$$

$$R(\sigma_1) = \left(\begin{array}{cc|c} -t & 1 & 0 \\ 0 & 1 & \\ \hline 0 & & I_{n-3} \end{array}\right), \quad R(\sigma_{n-1}) = \left(\begin{array}{c|cc} I_{n-3} & & 0 \\ \hline & 1 & 0 \\ 0 & t & -t \end{array}\right), \tag{2}$$

$$R(\sigma_i) = \left(\begin{array}{c|ccc|c} I_{i-2} & & 0 & & 0 \\ \hline & 1 & 0 & 0 & \\ 0 & t & -t & 1 & 0 \\ & 0 & 0 & 1 & \\ \hline 0 & & 0 & & I_{n-i-2} \end{array}\right) \qquad \text{for } i = 2, \ldots, n-2,$$

where I_k denotes the identity matrix of size k. Notice that $R(\theta_n) = t^n I_{n-1}$.

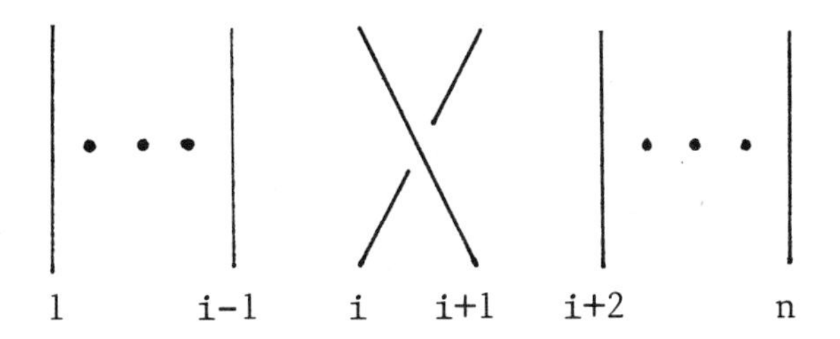

Figure 1. The braid σ_i

In the rest of this section, we shall explain the known relationship between the Burau representation and fixed points of homeomorphisms on punctured disks. First recall that the braid group is closely related to the group of isotopy classes of self-homeomorphoisms on punctured disks as follows: Let D_n denote an n-punctured disk, i.e., D_n is the 2-dimensional disk D with n disjoint open disks removed. Let $\mathrm{Homeo}^+(D_n, \partial D)$ be the group of orientation- preserving self-homeomorphims $f : (D_n, \partial D) \to (D_n, \partial D)$, where ∂D is the outer boundary of D_n, and let $\pi_0(\mathrm{Homeo}^+(D_n, \partial D))$ be the group of isotopy classes in $\mathrm{Homeo}^+(D_n, \partial D)$. Then it is known that the group $\pi_0(\mathrm{Homeo}^+(D_n, \partial D))$ is isomorphic to the quotient group $B_n/Z(B_n)$ via the isomorphism defined as follows(cf. [14]). Suppose an element f of $\mathrm{Homeo}^+(D_n, \partial D)$ is given. Then there exists an isotopy $F = \{f_t\} : D \to D$ such that f_0 is the identity and $f_1 |_{D_n} = f$. Let $D(1), \ldots, D(n)$ be the components of $D - D_n$, and let A_i be the image of $D(i) \times [0,1]$ under the isotopy $F : D \times [0,1] \to D \times [0,1]$. Shrinking each of these solid tubes $A_1, \ldots, A_n$ to a string, we obtain an n-braid which is denoted by $b(f)$. See Fig. 2. Since this braid $b(f)$ is well defined by the isotopy class $[f]$ up to composition with full twists, the correspondence which sends $[f]$ to the coset $[b(f)] \in B_n/Z(B_n)$ is an isomorphism.

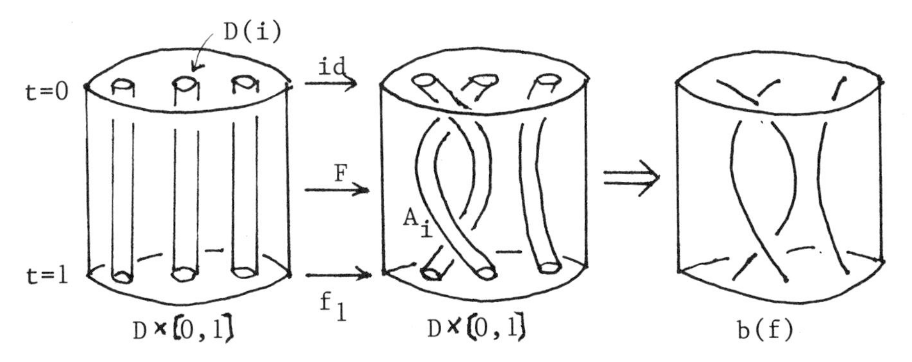

Figure 2.

Now we fix an isotopy $\{f_t\} : D \to D$. For a fixed point x of f, we define the linking number $\mathrm{lk}(x)$ of x with the braid $b(f)$ as follows. If we identify $(x, 0)$ with $(x, 1)$ for all $x \in D$ in the space $D \times [0,1]$, we obtain a solid torus $D \times S^1$, and under this identification the braid $b(f) \subset D \times [0,1]$ becomes a link denoted by $\hat{b}(f)$. Then define $\mathrm{lk}(x)$ as the linking number of the closed curve $\{(f_t(x), t) \mid 0 \leq t \leq 1\} \subset D \times S^1$ with the link $\hat{b}(f)$. For example, if the curve $f_t(x)$ go through the space $D \times [0,1]$ as in Fig. 3, then $\mathrm{lk}(x) = 2$.

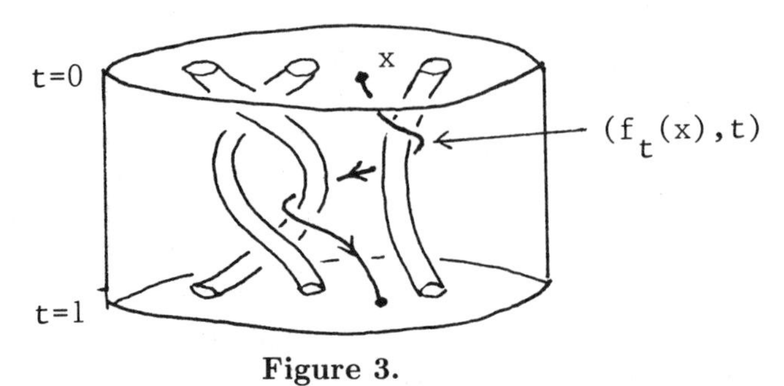

Figure 3.

Using the notion of linking number, we can divide the fixed point set $\mathrm{Fix} f$ as $\mathrm{Fix} f = \bigcup_{i \in \mathbf{Z}} \mathrm{Fix}_i f$, where

$$\mathrm{Fix}_i f = \{x \in \mathrm{Fix} f \mid \mathrm{lk}(x) = i\}.$$

The relationship between the Burau representation and the fixed point indices of f is stated in the following theorem, which shows that the fixed point indices $\mathrm{ind}(f, \mathrm{Fix}_i f)$ are determined by computing the reduced Burau matrix of the braid $b(f)$ (see [5],[17] for the definition of fixed point index).

Theorem A. *Let $f \in \mathrm{Homeo}^+(D_n, \partial D)$. Then*

$$\sum_{i \in \mathbf{Z}} \mathrm{ind}(f, \mathrm{Fix}_i f) t^i = -\mathrm{tr} R(b(f)).$$

This theorem is contained implicitely in [19 ,Prop.2], and has been fully described in Huang and Jiang [15, Sect.2 (D)]. They obtained Theorem A as a direct consequence of the trace formula for the generalized Lefschetz numbers proved by Fadell and Husseini[7], [16], Fried[12, Theorem 1] and of the computation of the generalized Lefschetz number for surface maps carried out by Fadell and Husseini[8]. In the case where f has a 0-dimensional , hyperbolic chain recurrent set, this theorem can be derived easily from Franks' theorem[10].

Remark. In [20], [21], the author adopted an alternative definition of the reduced Burau representation, which is secified by (1) and by sending each generator σ_i to the transpose $R(\sigma_i)^T$ of the matrix $R(\sigma_i)$, instead of $R(\sigma_i)$ itself. See [6, pp. 159-160] for the proof that this representation actually becomes a reduction of the Burau representation.

3. Computation of the Burau matrices

In this section, we prove a formula concerning the reduced Burau matrices and their traces. To state the result, we need a special expression for braids. Let $\rho_n = \sigma_1\sigma_2 \cdots \sigma_{n-1}$. Note that $\theta_n = (\rho_n)^n$. We shall fix an integer $n \geq 3$(in the case of $n = 2$, it is easy to compute the Burau matrices) and write ρ_n and θ_n simply as ρ and θ respectively. For a positive integer i, let $\rho(i) = \rho\sigma_1^i$. For a finite sequence $I = (i_1, \ldots, i_d)$ of positive integers, let $\rho(I) = \rho(i_1) \cdots \rho(i_d)$.

Proposition 1. *For any $b \in B_n$, there exist an integer m and a finite sequence $I = (i_1, \dots, i_d)$ of positive integers such that b is conjugate to the braid $\theta^m \rho(I)$.*

Proof. For $b, b' \in B_n$, we write $b \sim b'$ if $b' = \theta^m b$ for some $m \in \mathbf{Z}$. Since $\rho\sigma_i = \sigma_{i+1}\rho$, we have

$$\sigma_i = \rho^{i-1}\sigma_1\rho^{1-i} \sim \rho^{i-1}\sigma_1\rho^{n+1-i}, \tag{3}$$

$$\rho^{-1} \sim \rho^{-1}\rho^n = \rho^{-1}(\rho\sigma_1)^{n-1} = \sigma_1\rho(1)^{n-2}. \tag{4}$$

Hence for any $j \in \mathbf{N}$ less than n,

$$\rho^j \sim (\rho^{-1})^{n-j} = (\sigma_1\rho(1)^{n-2})^{n-j}. \tag{5}$$

Let $b \in B_n$. Then $b \sim \sigma_{j_1} \cdots \sigma_{j_r}$ for some $1 \le j_1, \dots, j_r \le n-1$, because

$$\sigma_i^{-1} \sim \rho^{i-1}\sigma_1^{-1}\rho^{n-i+1} = \rho^{i-1}(\sigma_2 \cdots \sigma_{n-1})\rho^{n-i},$$

which is a positive word. Then by (3), $b \sim b'$ for some b' which is a product of σ_1 and $\rho^j (1 \le j \le n-1)$. Therefore, by (5), b' is conjugate to a product of $\rho(i), i \in \mathbf{N}$, and the result follows. *q.e.d.*

Remark. The expression given in Proposition 1 is not unique. For instance,we have $\rho(i, \underbrace{1, \dots, 1}_{\text{n-2}}, j) = \theta\rho(i+j-1)$.

For $n, d \in \mathbf{N}$, denote by $\mathcal{A}_{n,d}$ the set of sequences $A = (a_1, \dots, a_d)$ of length d which satisfy the following conditions:

(i) $a_i = 0, 1, 2$ for each $i = 1, \dots, d$,
(ii) $A \neq 0$ i.e., $a_i \neq 0$ for some i,
(iii) if $a_i = 1$, then $a_{i+k} \neq 0$ for some k with $1 \le k \le n-2$,
(iv) if $a_i = 2$, then $d \ge n-1, a_{i+1} = \cdots = a_{i+n-2} = 0$, and $a_{i+n-1} \neq 0$.

In (iii),(iv), the subindices of a's are taken modulo d , i.e., $a_j (j > d)$ means $a_{j'}$, where $1 \le j' \le d, j \equiv j' \bmod d$. For example, $(2, 0, 0)$ satisfies the condition (iv) for $n = 4$, because $a_2 = a_3 = 0, a_4 (= a_1) = 2 \neq 0$. Condition (iii) (resp. (iv)) is equivalent to say that at most $(n-3)$ (resp. exactly $(n-2)$) 0's follow after 1 (resp. 2). Clearly $\mathcal{A}_{n,d} \subset \{0, 1, 2\}^d$.

Example.

$$\mathcal{A}_{4,3} = \{(1,1,1), (1,1,0), (1,0,1), (0,1,1), (2,0,0), (0,2,0), (0,0,2)\},$$

$$\begin{aligned}\mathcal{A}_{4,4} = \{&(1,1,1,1), (1,1,1,0), (1,1,0,1), (1,0,1,1), (0,1,1,1), (1,0,1,0),\\ &(0,1,0,1), (2,0,0,1), (0,0,1,2), (0,1,2,0), (1,2,0,0)\}.\end{aligned}$$

For $i \in \mathbf{N}$, define polynomials $P_l(i), l = 0, 1, 2$ by

$$P_0(i) = t, \qquad P_2(i) = (-t)^{i+1}, \tag{6}$$

$$P_1(i) = \begin{cases} -\sum_{k=2}^{i}(-t)^k & \text{if } i \ge 2, \\ 0 & \text{if } i = 1, \end{cases}$$

For $A \in \{0, 1, 2\}^d$ and $I = (i_1, \dots, i_d) \in \mathbf{N}^d$, let $P_A(I) = P_{a_1}(i_1) \cdots P_{a_d}(i_d)$.

Theorem 1. *If m is an integer and $I = (i_1, \dots, i_d)$ is a finite sequence of positive integers, then*

$$\mathrm{tr}R(\theta^m \rho(I)) = t^{mn} \sum_{A \in \mathcal{A}_{n,d}} P_A(I).$$

Proof. For $s = 1, \dots, d$, let $\mathcal{A}'_s$ be the set of $A = (a_1, \dots, a_s) \in \{0,1,2\}^s$ satisfying the following two conditions:

(i) if $a_i = 1$, then there exists an l with $1 \leq l \leq n-2$ such that $a_{i+l} \neq 0$ or $i + l > s$,

(ii) if $a_i = 2$, then $i + n - 2 \leq s, a_{i+1} = \cdots = a_{i+n-2} = 0$, and $a_{i+n-1} \neq 0$ or $i + n - 1 = s + 1$.

Let

$$\mathcal{B}_{0,s} = \{(a_1, \dots, a_s) \in \mathcal{A}'_s \mid a_1 \neq 0\},$$

$$\mathcal{B}_{k,s} = \{(a_1, \dots, a_s) \in \mathcal{A}'_s \mid a_1 = 0, \max\{l \mid a_1 = \ldots = a_l = 0\} \leq k\}, \quad k \geq 1,$$

$$\mathcal{C}^l_j = \{(l, \underbrace{0, \dots, 0}_{\text{j-1}})\}, \quad j \geq 1, l = 1, 2.$$

Notice that

$$\mathcal{B}_{k,s} - \mathcal{B}_{k-1,s} = \begin{cases} \underbrace{\{0, \dots, 0\}}_{k} \times \mathcal{B}_{0,s-k} & \text{if } k \leq s, \\ \emptyset & \text{if } k > s. \end{cases}$$

For a subset $\mathcal{D}$ of $\{0,1,2\}^d$ define

$$S(\mathcal{D}; I) = \sum_{A \in \mathcal{D}} P_A(I). \tag{7}$$

For $0 \leq k \leq n-2, 0 \leq j \leq d, l = 1, 2$, define polynomials $\phi_l(k, j; I) \in \mathbf{Z}[t, t^{-1}]$ by

$$\phi_l(k, j; I) = \begin{cases} S(\mathcal{B}_{k,d-j} \times \mathcal{C}^l_j; I) & \text{if } 1 \leq j \leq d-1, \\ S(\mathcal{C}^l_d; I) & \text{if } j = d \text{ and } k = 0, \\ 0 & \text{if } j = d \text{ and } k \geq 1, \text{ or if } j = 0. \end{cases}$$

Also, let

$$\phi_1(k, -1; I) = 0, \quad \phi_2(k, -1; I) = S(\mathcal{B}_{k,d}), \quad \phi_2(0, d+1; I) = 0. \tag{8}$$

Define

$$\psi(k, j; I) = \phi_1(k, j-2; I) - \phi_2(k, j-2; I) + \phi_2(k, j-1; I). \tag{9}$$

Lemma. *Let $r_{i,j}(I)$ be the $i-j$ entry of the matrix $R(\rho(I))$. Then we have:*

$$r_{i,j}(I) = \begin{cases} -t^{1-j}\psi(0,j;I) & \text{if } i=1, j \le d+2, \\ t^{i-j}\psi(n-i,j;I) & \text{if } i \ge 2, j \le d+1, \\ 1 & \text{if } j \ge d+2, i = j-d, \\ 0 & \text{otherwise.} \end{cases}$$

Proof. We prove this lemma by induction on d. Since

$$R(\rho(i)) = \begin{pmatrix} P_1(i) & (-t)^i & U(i) & 0 & \cdots & 0 \\ -t^2 & 0 & 1 & & & \vdots \\ -t^3 & & 0 & 1 & & \vdots \\ \vdots & \vdots & & \ddots & \ddots & 0 \\ \vdots & & & & \ddots & 1 \\ -t^{n-1} & 0 & & \cdots & & 0 \end{pmatrix}, \tag{10}$$

where $U(i) = (-t)^{i-1} - t^{-2}P_1(i)$, this lemma holds for $d=1$. Assume the result holds for some $d \ge 1$. Suppose $I = (i_0, \dots, i_d)$ is a sequence of positive integers of length $d+1$. We set $I' = (i_1, \dots, i_d)$, and

$$r_{i,j} = r_{i,j}(I), \quad r'_{i,j} = r_{i,j}(I').$$

Let

$$\phi_l(k,j) = \phi_l(k,j;I), \quad \phi'_l(k,j) = \phi_l(k,j;I'),$$
$$\psi_l(k,j) = \psi_l(k,j;I), \quad \psi'_l(k,j) = \psi_l(k,j;I'), \quad S_l = P_l(i_0).$$

For $j < d$, it is easy to verify the following equalities:

$$\phi_l(k,j) = \begin{cases} S_1(\phi'_l(0,j) + \phi'_l(n-3,j)) \\ \qquad + S_2(\phi'_l(n-2,j) - \phi'_l(n-3,j)) & \text{if } k=0, \\ S_0\phi'_l(0,j) & \text{if } k=1, \\ S_0(\phi'_l(0,j) + \phi'_l(k-1,j)) & \text{if } k \ge 2. \end{cases} \tag{11}$$

Also, we have

$$\phi_l(k,d) = \begin{cases} S_1\phi'_l(0,d) & \text{if } k=0, \\ S_0\phi'_l(0,d) & \text{if } k \ge 1, \end{cases} \tag{12}$$

$$\phi_l(0,d+1) = t^d S_l. \tag{13}$$

(11),(12) imply that if $j \le d+1$, then we have:

$$\psi(k,j) = \begin{cases} S_1(\psi'(0,j) + \psi'(n-3,j)) \\ \qquad + S_2(\psi'(n-2,j) - \psi'(n-3,j)) & \text{if } k=0, \\ S_0\psi'(0,j) & \text{if } k=1, \\ S_0(\psi'(0,j) + \psi'(k-1,j)) & \text{if } k \ge 2, \end{cases} \tag{14}$$

Also we have

$$\psi(k, d+2) = S_0\psi'(0, d+2) \quad \text{for } k \geq 1. \tag{15}$$

First consider $r_{1,j}$. Notice that by (10),

$$r_{1,j} = S_1 r'_{1,j} - t^{-1}S_2 r'_{2,j} - t^{-2}(S_1 - S_2)r'_{3,j}. \tag{16}$$

Therefore for $j \leq d+1$, we have by (14) and the induction hypothesis,

$$\begin{aligned} r_{1,j} &= -t^{1-j}(S_1\psi'(0,j) + S_2\psi'(n-2,j) + (S_1 - S_2)\psi'(n-3,j)) \\ &= -t^{1-j}\psi(0,j). \end{aligned}$$

Also,by (8),(12),(13),(16), we have

$$r_{1,d+2} = -t^{-d-1}\psi(0, d+2), \quad r_{1,d+3} = -t^{-d-2}\psi(0, d+3).$$

Therefore the lemma holds for I in case of $i = 1$. In the case of $i \geq 2$, since by (10), $r_{i,j} = -t^i r'_{1,j} + r'_{i+1,j}$ for $i \leq n-2$ and $r_{n-1,j} = -t^{n-1}r'_{1,j}$, we can verify the lemma for I and $j \leq d+2$ by (14),(15). It is clear that $r_{j-d-1,j} = 1$ for $j \geq d+3$, and that all the $r_{i,j}$' s which have not been computed are zero. Thus we have proved the lemma for $d+1$. *q.e.d.*

Now we compute the trace of $R(\rho(I))$. Let $\xi = \min\{d+1, n-1\}$, and let

$$\delta = \begin{cases} S(\bigcup_{i=3}^{n}(\underbrace{\{0, \ldots, 0\}}_{n-i+1} \times \mathcal{B}_{0,d-n+1} \times \mathcal{C}_{i-2}^{2}) & \text{if } d+1 \geq n, \\ 0 & \text{if } d+1 \leq n-1. \end{cases}$$

Then by Lemma,

$$\begin{aligned} \mathrm{tr}R(\rho(I)) &= -\psi(0,1;I) + \sum_{i=2}^{\xi}\psi(n-i,i;I) \\ &= S(\mathcal{B}_{0,d};I) + \phi_2(n-\xi, \xi-1;I) + \sum_{i=3}^{\xi}\phi_1(n-i,i-2;I) \\ &\quad + \sum_{i=3}^{\xi}\{\phi_2(n-i+1,i-2;I) - \phi_2(n-i,i-2;I)\} \\ &= S(\mathcal{B}_{0,d};I) + S(\bigcup_{i=3}^{\xi}\mathcal{B}_{n-i,d-i+2} \times \mathcal{C}_{i-2}^{1};I) + \delta \\ &= S(\mathcal{A}_{n,d};I) = \sum_{A \in \mathcal{A}_{n,d}} P_A(I). \end{aligned}$$

Also, $R(\theta^m) = t^{mn}I_{n-1}$. Thus the proof is completed.

4. The Nielsen-Thurston classification

In this section, we apply the formula on the Burau representation obtained in Sect. 3 to the classification problem of homeomorphisms on punctured disks.

We first recall the Nielsen-Thurston classification theory for homeomorphisms on compact surfaces briefly. Let M be a compact surface. A homeomorphism $f : M \to M$ is called *pseudo-Anosov* if there are a pair of transverse measured foliations $\mathcal{F}^s$ and $\mathcal{F}^u$ and a real number $\lambda > 1$ such that $f(\mathcal{F}^s) = \lambda^{-1}\mathcal{F}^s, f(\mathcal{F}^u) = \lambda\mathcal{F}^u$. The map f is *reducible* by a union $C = \bigcup_{i=1}^m C_i$ of disjoint simple closed curves, if each C_i is not homotopic to either a point or a component of ∂M, C_i and C_j are not homotopic for $i \neq j$, and if $f(C) = C$. f is *irreducible*, if f is not reducible.

Theorem B. [9],[13],[25] *If ϕ is an isotopy class of a homeomorphism on a compact surface M, then either*

(i) *ϕ contains a periodic homeomorphism,*
(ii) *ϕ contains a pseudo-Anosov homeomorphism, or*
(iii) *ϕ contains a homeomorphism f_0 which is reducible by a union C, and moreover if we denote the closures of the components of $M - C$ by $M_1, \dots, M_k$, then for each i, the isotopy class of $f_0^{\mu_i}$ restricted to M_i satisfies (i)or (ii),where μ_i is the least positive integer such that $f_0^{\mu_i}(M_i) = M_i$.*

We call an isotopy class $[f_0^{\mu_i} \mid M_i]$ in (iii) a *component* of the class ϕ. Theorem B means that any isotopy class on a compact surface can be built up by gluing together periodic and pseudo-Anosov components.

We consider here the problem of determining whether a given isotopy class contains a pseudo-Anosov component, in the case where M is an n- punctured disk and f preserves orientation. We shall assume $n \geq 3$.

As we saw in Sect. 3, any braid b is expressed as $b \equiv \rho(I)$ modulo cojugacy and the center by some sequence I of positive integers. We shall restrict ourselves to a subclass of the braids which are expressed by I contained in the set $\mathcal{I}_n$ defined as follows. If $J = (j_1, \dots, j_s)$ and $K = (k_1, \dots, k_t)$ are sequences of integers, let JK denote the sequence $(j_1, \dots, j_s, k_1, \dots, k_t)$. Also, write J^l for $JJ\cdots J$(l times), where $l \in \mathbf{N}$. We denote by $\sim$ the equivalence relation in the set of finite sequences of positive integers given by the relation

$$(i_1, i_2, \dots, i_s) \sim (i_t, i_{t+1}, \dots, i_s, i_1, i_2, \dots, i_{t-1}),$$

where $2 \leq t \leq s$. We denote by $\mathcal{I}_n$ the set of the equivalence classes $[I]$ of finite sequences $I = (i_1, \dots, i_d)$ of positive integers satisfying that

(i) I is non-repetitive,i.e., there is no sequence J with $I = J^l$ for some $l \geq 2$,
(ii) The sum of any consecutive (n-2) elements of I is greater or equal to n, i.e., for each $k = 1, \dots, d, \sum_{j=k}^{k+n-3} i_j \geq n$, where the index j is taken modulo d,i.e., if $j > d$ then i_j means $i_{j'}$,where $1 \leq j' \leq d, j \equiv j' \mod d$.

For example, $[(1,3,1,4)] \in \mathcal{I}_4$, but $[1,3,1,2]$ and $[(1,3,1,3)]$ are not contained in $\mathcal{I}_4$, because $[1,3,1,2]$ has two consecutive elements 1 and 2 whose sum is less than 4, and $[(1,3,1,3)]$ is repetitive.

For a sequence I, let $\phi_I \in \pi_0(\mathrm{Homeo}^+(D_n, \partial D))$ denote the isotopy class which corresponds to the braid $\rho(I)$. In the rest of this paper, we shall write the equivalence class $[I]$ simply as I, if there is no danger of ambiguity. Let

$$\mathcal{I}_n^* = \{I \in \mathcal{I}_n \mid \phi_I \text{ has no pseudo-Anosov components}\}.$$

This definition makes sense, because if two sequences I and I' are equivalent, then ϕ_I has no pseudo-Anosov component if and only if so does $\phi_{I'}$. We shall find explicitely all of the members of $\mathcal{I}_n^*$ in Theorem 2 below. For $n \in \mathbf{N}$, define

$$\Gamma_n = \{(j_1, \dots, j_m) \mid 2 \leq m \leq n-1, 1 \leq j_1, \dots, j_m \leq n-2, \\ \sum_{i=1}^{m} j_i = (m-1)(n-1)\}.$$

We shall identify two elements

$$(j_1, \dots, j_m) \text{ and } (j_t, \dots, j_m, j_1, \dots, j_{t-1})$$

of Γ_n for $2 \leq t \leq m$. For $j \in \mathbf{N}$, define two sequences α_j, β_j by

$$\alpha_j = (\underbrace{1, \dots, 1}_{j-1}, 3) \qquad \beta_j = (\underbrace{1, \dots, 1}_{j-1}, 2).$$

For $J = (j_1, \dots, j_m) \in \Gamma_n$ and ν with $0 \leq \nu \leq \frac{m-1}{2}$, we define an element $I_\nu(J)$ of $\mathcal{I}_n$ as follows. If $\nu \geq 1$, we define a sequence $(l_1, l_2, \dots, l_{2m-1}, l_{2m})$ of positive integers of length $2m$ by

$$l_{2i} = \nu(n-1) - \sum_{a=1}^{\nu} j_{i+a}, \quad l_{2i-1} = j_i - l_{2i} \quad \text{for } 1 \leq i \leq m. \tag{17}$$

Define a sequence $I'_\nu(J) = \omega_1 \cdots \omega_m$ by

$$\omega_i = \begin{cases} \alpha_{j_i} & \text{if } \nu = 0, \\ \beta_{l_{2i-1}} \beta_{l_{2i}} & \text{if } \nu \geq 1. \end{cases} \tag{18}$$

If $I'_\nu(J)$ has no repetition, then $I'_\nu(J) \in \mathcal{I}_n$, and we put $I_\nu(J) = I'_\nu(J)$. If $I'_\nu(J)$ is repetitive, let $I_\nu(J)$ be the non-repetitive sequence K with $I'_\nu(J) = K^s$ for some $s \geq 2$. Recall that $I_\nu(J)$ may be identified with its equivalence class.

Example. Let $n = 5$. Then $\Gamma_5 = \{(1,3), (2,2), (2,3,3), (3,3,3,3)\}$. Consider the case of $J = (2,3,3), \nu = 1$. Then $(l_1, \dots, l_6) = (1,1,2,1,1,2)$ and by (18), $I'_1(2,3,3) = (\beta_1^2\beta_2)^2$. Hence $I_1(2,3,3) = \beta_1^2\beta_2 = (2,2,1,2) = (1,2,2,2)$. Also, in the similar manner, it is shown that

$$I_0(1,3) = (1,1,3,3), \quad I_0(2,2) = (1,3), \quad I_0(2,3,3) = (1,3,1,1,3,1,1,3),$$

$$I_0(3,3,3,3) = (1,1,3), \quad I_1(3,3,3,3) = (1,2,2).$$

The following theorem exhibits the elements of $\mathcal{I}_n^*$.

Theorem 2. $\mathcal{I}_n^* = \{I_\nu(J) \mid J \in \Gamma_n, 0 \le \nu \le \frac{m-1}{2}\}$.

Example. If $n = 5$, by Theorem 2, $\mathcal{I}_5^*$ consists of the following 6 elements:

$$(1,3), (1,1,3), (1,1,3,3), (1,3,1,1,3,1,1,3), (1,2,2), (1,2,2,2).$$

Since each sequence $I_\nu(J)$ consists only of 1's and 2's, or only of 1's and 3's, we have

Corollary. *Let $I = (i_1, \dots, i_d) \in \mathcal{I}_n$. If there is a k with $i_k \ge 4$, or if there are k and l with $i_k = 2, i_l = 3$, then ϕ_I has a pseudo-Anosov component.*

We now apply Theorem 2 to some isotopy classes derived from the Smale's Horse-shoe diffeomorphism [24]. Let $H : D \to D$ be the Smale's horse-shoe diffeomorphism. H maps a rectangle R in D to a horseshoe-shaped region $H(R)$ as depicted in Fig. 4. If x is a periodic point of H of least period n, then we obtain an n-punctured disk D_x by adding a circle to each puncture of $D - \gamma$, where γ is the orbit containing x, and an orientation-preserving homeomorphism $H_x : D_x \to D_x$ in the standard way (cf. [3]). It is well known that the set Per$H$ of the periodic points of H is contained in $A \cup B$, where A, B are rectangles shown in Fig. 4, and that there is a one-to-one correspondence ω between Per$H$ and the set W of non-repetitive words in the letters $\{a, b\}$ of finite length. ω is defined by sending a periodic point x of length n to $\alpha_0 \cdots \alpha_{n-1}$, where

$$\alpha_i = \begin{cases} a & \text{if } H^i(x) \in A, \\ b & \text{if } H^i(x) \in B. \end{cases}$$

For $\alpha \in W$, let $H_\alpha = H_{\omega^{-1}(\alpha)}$. Then it can be shown that $b(H_{a^k b^3}) = \rho(2) \in B_{k+3}$ with respect to some isotopy. Since $(2) \in \mathcal{I}_{k+3} - \mathcal{I}_{k+3}^*$, by Theorem 2, the isotopy class of $H_{a^k b^3}$ has a pseudo-Anosov component. Also, we have $b(H_{ab^{2k+1}}) = \rho(\underbrace{1, \dots, 1}_{k-1}, 2) \in B_{2k+2}$. Since

$$(\underbrace{1, \dots, 1}_{k-1}, 2) = I_k(\underbrace{2k, \dots, 2k}_{2k+1}) \in \mathcal{I}_{2k+2}^*,$$

the isotopy class of $H_{ab^{2k+1}}$ has no pseudo-Anosov components.

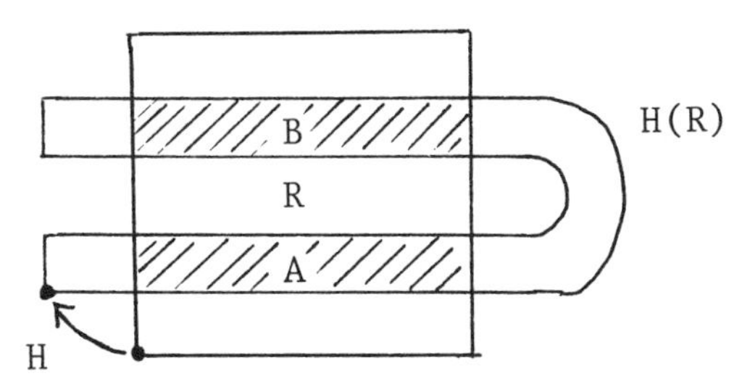

Figure 4. The horse-shoe diffeomorphism H

5. Proof of Theorem 2

The proof is divided into two parts.

(I) First we show that $\mathcal{I}_n^* \subset \{I_\nu(J) \mid \nu, J\}$. Suppose $A = (a_1, \dots, a_d) \in \mathcal{A}_{n,d}$, and $I = (i_1, \dots, i_d) \in \mathcal{I}_n$. We define a partition of $\{1, \dots, d\}$ into subsets $A(0), A(1), A(2)$ by

$$A(l) = \{k = 1, \dots, d \mid a_k = l\} \qquad \text{for } l = 0, 1, 2.$$

Let $\bar{A}(2) = \bigcup_{k \in A(2)} \{k, k+1, \dots, k+n-2\}$, and $A'(0) = A(0) - \bar{A}(2)$. Notice that $A(0) = A'(0) \cup (\bar{A}(2) - A(2))$. Define

$$\mid I \mid = i_1 + \cdots + i_d, \quad \mid I \mid_k = i_{k+1} + \cdots + i_{k+n-2},$$

$$\gamma_l = \sum_{k \in A(l)} (i_k + 1), \qquad \gamma = \gamma_1 + \gamma_2.$$

Define polynomials $\Phi_l (l = 1, 2, 3)$ by

$$\Phi_1 = \prod_{\substack{k \in A(1) \\ i_k \geq 3}} \sum_{j=0}^{i_k - 2} (-t)^{-j}, \qquad \Phi_2 = \prod_{k \in A(2)} t^{n-1-|I|_k}, \tag{19}$$

$$\Phi_3 = \prod_{k \in A'(0)} t^{1-i_k}.$$

Then we have :

Lemma 1. *If $A \in \mathcal{A}_{n,d}$, then*

(i) *$P_A(I) = (-1)^\gamma t^{|I|} \Phi_1 \Phi_2 \Phi_3$, and*

(ii) *if $P_A(I) \neq 0$, then*

$$\deg P_A(I) = \mid I \mid - \sum_{k \in A(2)} (\mid I \mid_k - n + 1) - \sum_{k \in A'(0)} (i_k - 1).$$

In particular, $\deg P_A(I) \leq \mid I \mid$.

Proof. (i) By (6), we have $P_A(I) = \Pi_1 \Pi_2 \Pi_3$, where

$$\Pi_1 = \prod_{k \in A(1)} (-\sum_{j=2}^{i_k} (-t)^j), \qquad \Pi_2 = \prod_{k \in A(2)} (-t)^{i_k+1} t^{n-2},$$

$$\Pi_3 = t^{\sharp A'(0)} \qquad (\sharp \text{ denotes the cardinality}).$$

Since

$$\Pi_1 = (-1)^{\gamma_1} \prod_{k \in A(1)} t^{i_k} \Phi_1,$$

$$\Pi_2 = (-1)^{\gamma_2} \prod_{k \in \bar{A}(2)} t^{i_k}\Phi_2, \qquad \Pi_3 = (\prod_{k\in A'(0)} t^{i_k})\Phi_3,$$

the equality holds.

(ii)If $P_A(I) \neq 0$, then $\deg P_A(I) = \mid I \mid - \sum_{l=1}^{3} \deg \Phi_l$. Hence, the conclusion follows from the followings:

$$\deg \Phi_1 = 0, \quad \deg \Phi_2 = \sum_{k\in A(2)} (n-1- \mid I \mid_k), \quad \deg \Phi_3 = \sum_{k \in A'(0)} (1 - i_k).$$

q.e.d.

Lemma 1 (ii) implies that $\deg \mathrm{tr} R(\rho(I))) = \deg S(\mathcal{A}_{n,d}; I) = \mid I \mid$. For a subset $\mathcal{D}$ of $\mathcal{A}_{n,d}$, define integers $r_c(\mathcal{D}; I)$ by

$$S(\mathcal{D}; I) = \sum_{c=0}^{\infty} r_c(\mathcal{D}; I) t^{|I|-c}.$$

We write $r_c(\mathcal{D})$ for $r_c(\mathcal{D}; I)$, and $r_c(A)$ for $r_c(\{A\})$, and let $r_c(I) = r_c(\mathcal{A}_{n,d}; I)$. Then $r_1(I)$ is computed as follows:

Lemma 2.

$$r_1(I) = (-1)^{\gamma+1} \sharp \{k \mid i_k \geq 2, \mid I \mid_{k-n+1} > n\}.$$

Proof. Let $A \in \mathcal{A}_{n,d}$. Define $A"(0) = A'(0) \cap \{k \mid i_k \geq 2\}$. If $\sharp A(2) \geq 2$, then $\deg P_A(I) \leq \mid I \mid -2$, so $r_1(A) = 0$. Suppose $A(2) = \emptyset$. Then $\Phi_2 = 1$, and Φ_3 is a monomial, so

$$r_1(A) = \begin{cases} -\sharp\{k \mid i_k \geq 3\}(-1)^{\gamma} & \text{if } A"(0) = \emptyset, \\ (-1)^{\gamma} & \text{if } A"(0) = \{k_0\}, i_{k_0} = 2 \text{ for some } k_0, \\ 0 & \text{otherwise.} \end{cases}$$

Theorefore

$$r_1(\{A \mid A(2) = \emptyset\}) = -\sharp\{k \mid i_k \geq 2\}(-1)^{\gamma}. \tag{20}$$

Suppose $A(2) = \{k_0\}$ for some k_0. Then

$$r_1(A) = \begin{cases} (-1)^{\gamma} & \text{if } \mid I \mid_{k_0} = n, A"(0) = \emptyset, i_{k_0+n-1} \geq 2, \\ 0 & \text{otherwise.} \end{cases}$$

Theorefore $r_1(\{A \mid \sharp A(2) = 1\}) = (-1)^{\gamma} \sharp \mathcal{H}$, where $\mathcal{H}$ is the set of $A \in \mathcal{A}_{n,d}$ such that $A(2) = \{k_0\}$ for some k_0 with $|I|_{k_0} = n, i_{k_0} + n - 1 \geq 2$ and that $A"(0) = \emptyset$. Let $H = \{k \mid i_k \geq 2, \mid I \mid_{k-n+1} = n\}$. Then a map $\tau : \mathcal{H} \to H$ is defined by $\tau(A) = k_0 + n - 1$, where $\{k_0\} = A(2)$. Since this map τ is a bijection, $\sharp\mathcal{H} = \sharp H$. Hence

$$r_1(\{A \mid \sharp A(2) = 1\}) = (-1)^{\gamma} \sharp H. \tag{21}$$

By (20),(21), we complete the proof. *q.e.d.*

Define three sets $\Delta_n^+, \Delta_n^0, \Delta_n$ of sequences of even length as follows:

$$\begin{aligned}
\Delta_n^+ &= \{(l_1,\dots,l_{2m}) \mid m \in \mathbf{N}, l_k > 0, l_k + l_{k+1} \le n-2, \\
&\qquad \text{and } l_k + l_{k+1} + l_{k+2} \ge n-1 \text{ for any } k\}, \\
\Delta_n^0 &= \{(l_1,\dots,l_{2m}) \mid m \in \mathbf{N}, l_k = 0 \text{ for even } k, \\
&\qquad 0 < l_k \le n-2, l_k + l_{k+2} \ge n-1 \text{ for odd } k\}, \\
\Delta_n &= \Delta_n^+ \cup \Delta_n^0.
\end{aligned}$$

For integers $l > 0, l' \ge 0$, let

$$\omega_{l,l'} = \begin{cases} \alpha_l & \text{if } l' = 0, \\ \beta_l \beta_{l'} & \text{if } l' > 0. \end{cases} \tag{22}$$

For $L = (l_1,\dots,l_{2m}) \in \Delta_n$, let

$$I(L) = \omega_{l_1 l_2} \cdots \omega_{l_{2m-1} l_{2m}}. \tag{23}$$

Lemma 3. *Let $I \in \mathcal{I}_n$. Then $r_1(I) = 0$, if and only if there exists a sequence $L \in \Delta_n$ such that $I = I(L)$.*

Proof. Suppose $r_1(I) = 0$. Suppose $1 \le k \le d$ and $i_k \ge 2$. Then by Lemma 2, $\mid I \mid_{k-n+1} = n$, and hence

$$\begin{aligned}
(i_{k-n+1},\dots,i_{k-1}) = (1,\dots,1,3,1,\dots,1) \qquad &\text{or} \\
(1,\dots,1,2,1,\dots,1,2,1,\dots,1)&.
\end{aligned}$$

This implies that $I = I(L)$ for some $L \in \Delta_n$. Conversely, suppose $L \in \Delta_n$. Then for any k with $i_k \ge 2$, we have $\mid I(L) \mid_{k-n+1} = n$ and by Lemma 2, $r_1(I(L)) = 0$. *q.e.d.*

Given an $L = (l_1,\dots,l_{2m}) \in \Delta_n$, let $\mid L \mid = l_1 + \cdots + l_{2m}$, and for $c \in \mathbf{N}$, define subsets $N_c(L), N_c^+(L)$ of $\{1,\dots,2m\}$ by

$$N_c(L) = \{s \in \{1,\dots,2m\} \mid \sum_{t=0}^{2c} l_{s-t} \ge c(n-1)\},$$

$$N_c^+(L) = \{s \in \{1,\dots,2m\} \mid \sum_{t=0}^{2c} l_{s-t} > c(n-1)\}.$$

Notice that $N_c(L) \subset N_{c-1}^+(L)$ for any $c \in \mathbf{N}$.

Lemma 4. *Let $c \in \mathbf{N}$, and $L \in \Delta_n$. Assume $|L| > c(n-1)$, and $r_{c'}(I(L)) = 0$ for any $1 \leq c' < c$. Then $r_c(I(L)) = \sharp N_c(L) - \sharp N^+_{c-1}(L)$.*

Proof. Since $N_1(L) = N^+_0(L)$ for any $L \in \Delta_n$ and $r_1(I(L)) = 0$ by Lemma 3, the conclusion holds for $c = 1$. Assume $c \geq 2$ and this lemma holds for $1, \ldots, c-1$. For $c' \leq c$, let $N_{c'} = N_{c'}(L), N^+_{c'} = N^+_{c'}(L), r_{c'} = r_{c'}(I(L))$. Let $I = I(L)$. Suppose $r_1 = \cdots = r_{c-1} = 0$. Then

$$N_{c'} = N^+_{c'-1} \text{ for any } 0 \leq c' \leq c-1. \tag{24}$$

Suppose $I(L) = (i_1, \ldots, i_d)$, and let $k(s) = l_1 + \cdots + l_s$. Then $i_{k(s)} \geq 2$ if $l_s > 0$. Suppose A is an element of $\mathcal{A}_{n,d}$ with $S(A) \neq 0$. For $k \in A(2)$, let

$$Q(A,k) = \max\{s \mid s = 1, \ldots, 2m, l_s > 0, k(s) \in \{k+1, \ldots, k+n-2\}\},$$

$$Q_A = \{Q(A,k) \mid k \in A(2)\},$$

$$\mathcal{Q} = \{Q_A \mid A \in \mathcal{A}_{n,d}, S(A) \neq 0\}.$$

Define

$$\mathcal{G} = \{A \in \mathcal{A}_{n,d} \mid S(A) \neq 0, A"(0) = \emptyset\}.$$

Then the correspondence $A \mapsto Q_A$ gives a bijection from $\mathcal{G}$ to $\mathcal{Q}$. For $Q \in \mathcal{Q}$, we denote by A_Q the unique element of $\mathcal{G}$ which corresponds to Q. For each $Q \in \mathcal{Q}$, define a subset $\mathcal{P}_Q$ of the power set Power$\{1, \ldots 2m\}$, by

$$\mathcal{P}_Q = \{P \subset \{1, \ldots, 2m\} - \bar{A}_Q(2) \mid l_s > 0, \{s-1, s+1\} \cap P = \emptyset \text{ for any } s \in P\}.$$

Note that if $L \in \Delta^0_n$, then

$$\mathcal{P}_Q = \text{Power}(\{1, 3, \ldots, 2m-1\} - \bar{A}_Q(2)).$$

If $L \in \Delta^+_n$, then for any $Q \in \mathcal{Q}$ and any $P \in \mathcal{P}_Q$, there is a unique A such that $Q_A = Q, \{s \mid i_{k(s)} = 2, k(s) \in A(0)\} = P$. We denote this A by$A(P,Q)$.Clearly

$$S(A(P,Q)) = (-1)^{\sharp P} t^{|I| - \sharp(P \cup Q)}. \tag{25}$$

If $L \in \Delta^0_n$, then for any A with $S(A) \neq 0$, by (19), we have

$$S(A) = t^{|I|} \prod_{P \in \mathcal{P}_{Q_A}} (1 - t^{-1})^{\sharp P} t^{-\sharp Q_A}. \tag{26}$$

Therefore by (25), (26), for any $L \in \Delta_n$,

$$\begin{aligned} \text{tr} R(\rho(I)) &= \sum_{Q \in \mathcal{Q}} \sum_{A : Q_A = Q} S(A) \\ &= \sum_{Q \in \mathcal{Q}} \sum_{P \in \mathcal{P}_Q} (-1)^{\sharp P} t^{|I| - \sharp P \cup Q}. \end{aligned}$$

and hence,

$$r_c = \sum_{Q \in \mathcal{Q}} \sum_{\substack{P \in \mathcal{P}_Q \\ \sharp P = c - \sharp Q}} (-1)^{\sharp P}. \tag{27}$$

Let $E_{s,c} = \{s, s+2, \ldots, s+2(c-1)\}, E'_{s,c} = E_{s,c} - \{s\}$, and let Λ_0 (resp. Λ_1)) denotes the set of pairs (Q, P) which satisfy that

$$Q \in \mathcal{Q}, P \in \mathcal{P}_Q, \sharp P + \sharp Q = c, \sharp P \text{ is even (resp. odd)},$$
$$(Q, P) \neq (E_{s,c}, \emptyset), (E'_{s,c}, \{s\}) \text{ for any } s.$$

Notice that

$$E_{s,c} \in \mathcal{Q} \text{ and } \mathcal{P}_{E_{s,c}} \ni \emptyset \text{ if and only if } s + 2c - 1 \in N_c, \tag{28}$$

$$E'_{s,c} \in \mathcal{Q} \text{ and } \mathcal{P}_{E'_{s,c}} \ni \{s\} \text{ if and only if } s + 2c - 1 \in N^+_{c-1}.$$

We claim that $\sharp\Lambda_0 = \sharp\Lambda_1$. To show this, we define a map $\Theta : \Lambda_0 \to \Lambda_1$ as follows: Let $k_0 = \min\{k(s) \mid s \in P \cup Q, s-2 \text{ is not in } P \cup Q\}$. Then for some $p \geq 0$,$\{k_0 + 2, \ldots, k_0 + 2p\} \subset Q$, and $k_0 + 2p + 2$ is not contained in Q. In the case of $k_0 \in Q$, clearly $(Q - \{k_0\}, P \cup \{k_0\}) \in \Lambda_1$. We let $\Theta(P, Q)$ be this element. If $k_0 \in P$, then $k_0 + 2p + 1 \in N^+_p$, because if $k_0 + 2p + 1$ is not contained in N^+_p, then k_0 must be in $\bar{A}(2)$, which is a contradiction. Since $p \leq c - 1$, by (24) $k_0 + 2p + 1 \in N_{p+1}$. Therefore $(Q \cup \{k_0\}, P - \{k_0\}) \in \Lambda_1$. We set $\Theta(P, Q) = (Q \cup \{k_0\}, P - \{k_0\})$ in this case. Clearly Θ is a bijection , and so, we have $\sharp\Lambda_0 = \sharp\Lambda_1$ Therefore by (26), (27),

$$\begin{aligned} r_c = & \sum_{(Q,P)=(E_{s,c},\emptyset)} (-1)^{\sharp P} + \sum_{(Q,P)=(E'_{s,c},\{s\})} (-1)^{\sharp P} \\ & + \sum_{(Q,P)\in\Lambda_0} (-1)^{\sharp P} + \sum_{(Q,P)\in\Lambda_1} (-1)^{\sharp P} \\ = & \sharp N_c - \sharp N^+_{c-1} + \sharp\Lambda_0 - \sharp\Lambda_1 \\ = & \sharp N_c - \sharp N^+_{c-1}. \end{aligned}$$

q.e.d.

Suppose $I \in \mathcal{I}_n$ and ϕ_I has only periodic components. Let f_0, μ_i be as defined in Theorem B. Let μ be the least common multiple of $\mu_1, \ldots, \mu_k$, and let $g = f_0^\mu$. Since D_n is compact, we have a finite partition:

$$\text{Fix} g = \text{Fix}_{i_1} g \cup \cdots \cup \text{Fix}_{i_s} g$$

for some $i_1 < \cdots < i_s$. We can assume $\text{Fix} g^n = \text{Fix} g$. Indeed, if we choose the Thurston's canonocal form for the map f_0 in Theorem B (iii), then this property holds. Therefore since $\text{lk}(x, g^n) = n\text{lk}(x, g)$, we have $\text{Fix}_{i_j} g = \text{Fix}_{ni_j} g^n$. Therefore by Theorem A,

$$\text{tr}R(\rho(I^{\mu n})) = -\sum_{j=1}^{s} \text{ind}(\text{Fix}_{i_j} g) t^{ni_j},$$

and in particular, $r_c(I^{\mu n}) = 0$ for $1 \leq c \leq n-1$. By Lemma 3, $I^{\mu n} = I(L)$ for some $L = (l_1, \ldots, l_{2q}) \in \Delta_n$. Let $N_c^0 = N_c - N_c^+$. By Lemma 4, $N_c = N_{c-1}^+$ for $1 \leq c \leq n-1$. Also,$N_{n-1} = \emptyset$, because for any k, $\sum_{s=0}^{2(n-1)} l_{k-s} \leq n(n-2) < (n-1)^2$. Therefore, for any $0 \leq c \leq n-2$,

$$\begin{aligned} N_c &= N_c^0 \cup N_c^+ = \cdots = N_c^0 \cup N_{c+1}^0 \cup \cdots \cup N_{n-2}^0 \cup N_{n-1} \\ &= N_c^0 \cup \cdots \cup N_{n-2}^0. \end{aligned}$$

Since if $k \in N_c^0$ then $k - 2 \in N_{c-1}^+ = N_c = N_c^0 \cup \cdots \cup N_{n-2}^0$, we have

$$\{2, 4, \ldots, 2q\} = N_\nu^0, \{1, 3, \ldots, 2q-1\} = N_{\nu'}^0$$

for some $0 \leq \nu, \nu' \leq n-2$. We can assume $\nu \leq \nu'$ by letting $L = (l_2, \ldots, l_{2m}, l_1)$ if necessary. Let $m = \nu + \nu' + 1$, $L' = (l_1, \ldots, l_{2m})$. Then for any k, $\sum_{j=k}^{k+2m-1} l_j = \sum_{j=k}^{k+2\nu} l_j + \sum_{j=k+2\nu+1}^{k+2m-1} l_j = \nu(n-1) + \nu'(n-1) = (m-1)(n-1)$, so we have $l_k = l_{k+2m}$. Therefore m devides q, and $L = (L')^{\frac{q}{m}}$. Define $J = (j_1, \ldots, j_m)$ by $j_k = l_{2k-1} + l_{2k}$. Then $J \in \Gamma_n$, and $I_\nu{}'(J)$ coincides with $I(L')$, and therefore $I^{\mu n} = I(L) = (I_\nu'(J))^{\frac{q}{m}}$. Hence $I^{\mu n}$ is a power of $I_\nu(J)$. Since I is non-repetitive, we have $I = I_\nu(J)$.

(II) We now show that every $I_\nu(J)$ is an element of $\mathcal{I}_n^*$.

Lemma 5. *Let J, J' be elements of Γ_n having the same length m. Then $\rho(I_\nu(J))$ and $\rho(I_\nu(J'))$ are conjugate for any $0 \leq \nu \leq \frac{m-1}{2}$.*

Proof. For $1 \leq k \leq n$, let $B_n^{(k)}$ denote the subgroup of B_n generated by $\sigma_1, \ldots, \sigma_{k-1}$. For $0 \leq E \leq n-k$, define a homomorphism $\phi_E : B_n^{(k)} \to B_n$ by $\phi_E(\sigma_i) = \sigma_{i+E}$. Since $\rho\sigma_i = \sigma_{i+1}\rho$, we have $\phi_E(b) = \rho^E b \rho^{-E}$ for any $b \in B_n^{(k)}$.

For $1 \leq i, j \leq n-1$, let $\sigma(i,j) = \sigma_i\sigma_{i+1}\cdots\sigma_j$ if $i \leq j$, $\sigma(i,j) = e$ if $i > j$.Let

$$e(l, l') = n - 1 - l - l',$$

$$\xi(l, l') = \sigma(n - l - l', n - l')(= \phi_{e(l,l')}\sigma(1, l+1)),$$

$$\eta(l) = \sigma(2, n-1-l)^{-1}.$$

Let $\rho(\beta_0)$ mean σ_1. Then $\rho(\alpha_l) = \rho(\beta_l)\rho(\beta_0)$. Since $\theta^{-1} = (\sigma_1\rho)^{1-n}, (\sigma_1\rho)^k\sigma_1 = \begin{cases} \sigma(1, k+1)\rho^k & \text{if } k \geq 0, \\ \rho^k\sigma(2, -k)^{-1} & \text{if } k < 0 \end{cases}$, we have for any $0 \leq l \leq n-2$,

$$\sigma_1\rho(\beta_l)\sigma_1^{-1} = (\sigma_1\rho)^l\sigma_1 = \sigma(1, l+1)\rho^l,$$

$$\theta^{-1}\sigma_1\rho(\beta_l)\sigma_1^{-1} = (\sigma_1\rho)^{l+1-n}\sigma_1 = \rho^{l+1-n}\eta(l).$$

Hence we have for $l \geq 0, l' > 1$, with $l + l' \leq n-2$,

$$\theta^{-1}\sigma_1\rho(\beta_l)\rho(\beta_{l'})\sigma_1^{-1} = \sigma(1, l+1)\rho^{-e(l,l')}\eta(l') = \rho^{-e(l,l')}\xi(l, l')\eta(l'). \tag{29}$$

Let $J \in \Gamma_n, 0 \leq \nu \leq \frac{m-1}{2}$. Let $L_\nu(J) = (l_1, \ldots, l_{2m})$ be the sequence defined by (17) if $\nu \geq 1$, and be $(j_1, 0, j_3, 0, \ldots, j_m, 0)$ if $\nu = 0$. Then

$$\rho(I'_\nu(J)) = \rho(\beta_{l_1}) \cdots \rho(\beta_{l_{2m}}). \tag{30}$$

Define $\nu_1 = m - \nu - 1, \nu_2 = \nu$, and $l_i^1 = l_i$ for $1 \leq i \leq 2\nu_1 + 1$, $l_i^2 = l_{i+2\nu+1}$ for $1 \leq i \leq 2\nu_2 + 1$. Let $a = 1$ or 2. For $i = 1, \ldots, \nu_a$, let

$$e_i^a = e(l_{2i}^a, l_{2i+1}^a), \quad E_i^a = e_{i+1}^a + \cdots + e_{\nu_a}^a,$$

$$\xi_i^a = \xi(l_{2i}^a, l_{2i+1}^a), \quad \eta_i^a = \eta(l_{2i+1}^a)$$

Note that $E_0^a = l_1^a$. Let

$$A_i^a = \begin{cases} \phi_{E_i^a}(\xi_i^a \eta_i^a) & \text{if } 1 \leq i \leq \nu_a, \\ \sigma(1, l_1^{3-a} + 1) & \text{if } i = \nu_a + 1, \end{cases}$$

$$A^a = \begin{cases} A_2^a \cdots A_{\nu_a+1}^a & \text{if } \nu_a \geq 1, \\ e & \text{if } \nu_a = 0. \end{cases} \tag{31}$$

If we set $\mu = l_1 + l_2 + 1$, then

$$A_1^a \in B_n^{(\mu)} - B_n^{(\mu-1)}, A^a \in B_n^{(\mu-1)}.$$

We have by (29),(30),(31)

$$\begin{aligned} \rho(I'_\nu(J)) &= \prod_{a=1}^{2} \{\rho(\beta_{l_1^a}) \prod_{i=1}^{\nu_a} \rho(\beta_{l_{2i}^a}) \rho(\beta_{l_{2i+1}^a})\} \\ &= \prod_{a=1}^{2} \{\sigma(1, l_1^a + 1) \rho^{l_1^a} \prod_{i=1}^{\nu_a} \rho^{-e_i^a} \xi_i^a \eta_i^a\} \\ &= \prod_{a=1}^{2} \{\sigma(1, l_1^a + 1) \prod_{i=1}^{\nu_a} \rho^{E_i^a} \xi_i^a \eta_i^a \rho^{-E_i^a})\} \\ &= \prod_{a=1}^{2} \{\sigma(1, l_1^a + 1) \prod_{i=1}^{\nu_a} A_i^1\} \sim A_1^1 A_1 A_2^1 A_2, \end{aligned}$$

where $\sim$ means conjugacy.

For $J = (j_1, \ldots, j_m), J' = (j'_1, \ldots, j'_m) \in \Gamma_n$, we write $J \sim J'$ if for some $1 \leq k_0 \leq m$, $j'_{k_0} = j_{k_0} + \epsilon, j'_{k_0+1} = j_{k_0+1} - \epsilon, j'_k = j_k$ for any $k \neq k_0, k_0 + 1$, where $\epsilon = 1, -1$. We shall prove that

$$\text{if } J \sim J', \text{ then } \rho(I_\nu(J)) \sim \rho(I_\nu(J')). \tag{32}$$

We can assume that $j_2 \geq 1$, and $J' = (j_1 + 1, j_2 - 1, j_3, \ldots, j_m)$. Then $L_\nu(J')$ coincides with $(l'_1, \ldots, l'_{2m})$ defined by $l'_i = l_i + \delta_i - \delta'_i$, where $\delta_2 = \delta_{2\nu_1+3} = 1, \delta_i =$

0 for $i \neq 2, 2\nu_1+3$, $\delta'_3 = \delta'_{2\nu_1+4} = 1, \delta'_i = 0$ for $i \neq 3, 2\nu_1+4$. This implies that, if we let

$$A'^a_1 = \sigma(l^a_1+1, \mu+1)\sigma(\mu_a, \mu)^{-1},$$

where

$$\mu_a = \begin{cases} \mu & \text{if } \nu = 0 \text{ and } a = 1, \\ \mu + l^a_3 - n + 2 & \text{otherwise,} \end{cases}$$

then we have $\rho(I'_\nu(J')) \sim A'^1_1 A_1 A'^2_1 A_2$. Since

$$A'^a_1 = \sigma(l^a_1+1, \mu+1)\sigma_\mu^{-1}\sigma(\mu_a, \mu-1)^{-1} = \sigma_{\mu+1}^{-1}\sigma(l^a_1+1, \mu)\sigma(\mu_a, \mu-1)^{-1}\sigma_{\mu+1}$$
$$= \sigma_{\mu+1}^{-1} A^a_1 \sigma_{\mu+1},$$

and $A_a \in B_n^{(\mu-1)}$, we have $\rho(I'_\nu(J')) \sim \prod_{a=1}^2 \sigma_{\mu+1}^{-1} A^a_1 \sigma_{\mu+1} A_a = \rho(I'_\nu(J))$, and hence (32) has been proved. Now let $J, J' \in \Gamma_n$ have the same length m. Then there are some sequences $J(1), \ldots, J(s)$ such that $J \sim J(1) \sim \cdots \sim J(s) \sim J'$. Hence by (32), we have $\rho(I_\nu(J)) \sim \rho(I_\nu(J'))$. *q.e.d.*

Let $J_m = (m-1, \underbrace{n-2, \ldots, n-2}_{m-1}) \in \Gamma_n$. Notice that $L_\nu(J_m) = (l_1, \ldots, l_{2m})$, where

$$l_k = \begin{cases} \nu, & \text{if } k \text{ is even and } 2 \leq k \leq 2\nu'+2, \\ n-2-\nu, & \text{if } k \text{ is odd and } 3 \leq k \leq 2\nu'+1, \\ \nu', & \text{if } k = 1, \text{or if } k \text{ is odd and } k \geq 2\nu'+3, \\ n-2-\nu', & \text{if } k \text{ is even and } k \geq 2\nu'+4. \end{cases}$$

Therefore, if we let $\tau(k,l) = \sigma(k, k+l)\sigma(k, k+l-1)^{-1}$ for $k \geq 1, l \geq 0, k+l \leq m$, then $\rho(I'_\nu(J_m)) = \rho(I(L_\nu(J_m)))$ is conjugate to the n-btraid

$$b(\nu, m) = \tau(\nu'+1, \nu) \cdots \tau(3, \nu)\tau(2, \nu)\sigma(1, \nu+1)$$
$$\tau(\nu+1, \nu') \cdots \tau(3, \nu')\tau(2, \nu')\sigma(1, \nu'+1).$$

(For example, $b(1,4)$ is depicted in Fig. 5.) It is easy to see that $b(\nu, m)$ coincides with the n-braid

$$\theta_{m+1}\{\sigma(1,\nu)^{-1}\sigma(\nu+2, m)^{-1}\}^m,$$

which clearly corresponds to an isotopy class having only periodic components. This shows that $I_\nu(J_m)$ is in $\mathcal{I}^*_n$, and hence by Lemma 5, every $I_\nu(J)$ is in $\mathcal{I}^*_n$. Thus we have completed the proof of Theorem 2.

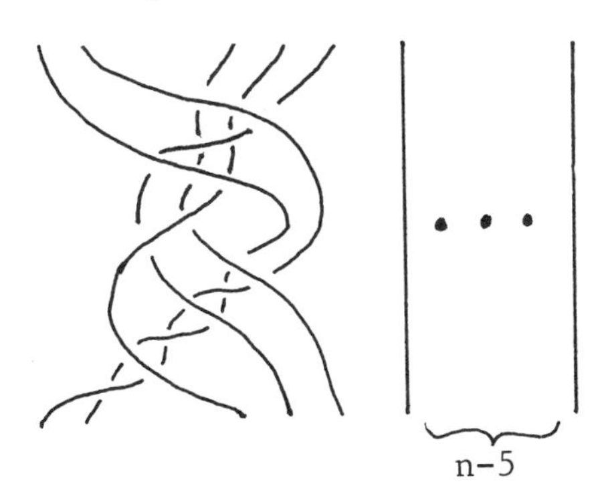

Figure 5. The braid $b(1,4) \in B_n$

References

1. D. Benardete, M. Gutierrez, Z. Nitecki, *Braids and the Nielsen-Thurston classification*, preprint.
2. J. S. Birman, *Braids,links,and mapping class groups*, Ann. of Math. Studies Vol. 82, Princeton Univ. Press, Princeton, 1974.
3. R. Bowen, *Entropy and the fundamental group*, The structure of attractors in dynamical systems, Lecture Notes in Math. Vol. 668, Springer-Verlag, Berlin, New York, 1978, pp. 21–29.
4. P. Boyland, *Braid types and a topological method of proving positive entropy*, preprint.
5. R. Brown, *The Lefschetz fixed point theorem*, Scott-Foresman, Chicago, 1971.
6. G. Burde and H. Zieschang, *Knots*, De Gruyter Studies in Math. Vol. 5, de Gruyter, Berlin, New York, 1985.
7. E. Fadell and S. Husseini, *Fixed point theory for non simply connected manifolds*, Topology **20** (1981), 53–92.
8. E. Fadell S. Husseini, *The Nielsen number on surfaces*, Topological Methods in Nonlinear Functional Analysis (S.P.Singh et al., eds.), Contemp. Math. Vol 21, Amer. Math. Soc, Providence, 1983, pp. 59-98.
9. A. Fathi, F. Laudenbach and V. Poénaru, *Travaux de Thurston sur les surfaces*, Astérisque **66-67** (1979).
10. J. Franks, *Knots,links,and symbolic dynamics*, Ann. of Math. **113** (1981), 529-552.
11. J. Franks , M. Misiurewicz, *Cycles for disk homeomorphisms and thick trees*, preprint.
12. D. Fried, *Periodic points and twisted coefficients*, Geometric Dynamics (J. Palis Jr., ed.), Lecture Notes in Math. Vol. 1007, Springer-Verlag, Berlin, Heidelberg, New York, 1983, pp. 261- 293.
13. M. Handel and W. P. Thurston, *New proofs of some results of Nielsen*, Adv. Math. **56** (1985), 173-191.
14. V. Hansen, *Braids and coverings: selected topics*, London Math. Soc. Stud. Texts Vol. 18, Cambridge Univ. Press, 1989.
15. H.-H. Huang B.-J. Jiang, *Braids and periodic solutions*, Topological Fixed Point Theory and Applications (B. Jiang, ed.), Lecture Notes in Math. Vol. 1411, Springer-Verlag, Berlin, Heidelberg, New York, 1989, pp. 107-123.
16. S. Husseini, *Generalized Lefschetz numbers*, Trans. Amer. Math. Soc. **272** (1982), 247-274.
17. B. Jiang, *Lectures on Nielsen fixed point theory*, Contemp. Math. Vol. 14, Amer. Math. Soc., Providence, 1983.
18. T. Kobayashi, *Links of homeomorphisms of a disk and topological entropy*, Invent. Math. **80** (1985), 153-159.
19. T. Matsuoka, *The number and linking of periodic solutions of periodic systems*, Invent. Math. **70** (1983), 319-340.
20. T. Matsuoka, *Waveform in the dynamical study of ordinary differential equations*, Japan J. Appl. Math. **1** (1984), 417-434.
21. T. Matsuoka, *The number and linking of periodic solutions of non-dissipative systems*, J. Differential Equations **76** (1988), 190–201.
22. J. A. Moody, *The Burau representation of the braid group B_n is unfaithful for large n*, Bull. Amer. Math. Soc. **25** (1991), 379-384.
23. S. Moran, *The mathematical theory of knots and braids:an introduction*, North-Holland Math. Studies Vol. 82, North-Holland, Amsterdam, 1983.
24. S. Smale, *Differentiable dynamical systems*, Bull. Amer. Math. Soc. **73** (1967), 747–817.
25. W. P. Thurston, *On the geometry and dynamics of diffeomorphisms of surfaces*, Bull. Amer. Math. Soc. **19** (1988), 417-431.

Department of Mathematics , Naruto University of Education , Naruto, 772 Japan

Contemporary Mathematics
Volume **152**, 1993

Computing Nielsen numbers

CHRISTOPHER K. MCCORD

ABSTRACT. For Nielsen numbers to be truly useful in applied problems, they should be computable. Since it is generally impossible to compute Nielsen numbers directly from the definition, other methods of computation must be developed. In this paper, some of the existing computational methods are surveyed. All of these methods rely on a combination of homological and fundamental group information for their application.

1. The Problem

Consider a discrete dynamical system given by iteration of a function $f : X \to X$. From one point of view, analysing this system means identifying its essential dynamical features (fixed points, periodic points, recurrent sets, etc.), the way in which those features interact (say through intersections of stable and unstable sets) *and* the persistance of those features under perturbation of the system. Typically, one wishes to identify the "minimal" structure that persists under all perturbations. Certainly one of the basic ingredients in such a description is a count of the number of fixed points of the system, and the persistence of that count under perturbation. That is, the minimum number of fixed points in the homotopy class of a map

$$MF(f) = \min\{\#Fix(g)\,|\,g \simeq f\},$$

is a number of interest. And it is the fact that the Nielsen number $N(f)$ is an (almost always sharp) lower bound on $MF(f)$ that makes computation of the Nielsen number a matter of interest.

Unfortunately, Nielsen numbers cannot in general be computed from the definition. The definition requires that the fixed point set $Fix(f)$ be partitioned into fixed point classes, and that the fixed point index of each class be computed. Then $N(f)$ is the number of classes with nonzero index. Except in the simplest

1991 *Mathematics Subject Classification.* Primary 55M20.

This paper is in final form and no version of it will be submitted for publication elsewhere.

cases (some of which will be described below), this cannot be done. Because it is both desirable and difficult to compute Nielsen numbers, a great deal of work in Nielsen theory has been devoted to the question of computation. The results of this work can be broadly divided into two categories: theorems which relate Nielsen numbers to other (hopefully, more computable) invariants; and theorems which relate Nielsen numbers of different maps to one another. Examples of the former are the Jiang conditions and the theorems which relate Nielsen numbers and Lefschetz numbers of maps; an example of the latter is the Nielsen number product formula for fibrations.

In this paper, I will briefly survey some of these efforts. This survey is not meant to be encyclopedic, but rather has some definite limits and a definite point of view. I am interested in identifying conditions on the space X which guarantee that, for all self maps $f : X \to X$, there is an algorithm which computes $N(f)$. Moreover, this algorithm should only require information about the space and the map that can actually be obtained in practice. For instance, an algorithm that requires knowledge of $\pi_n(X)$ and $f_{n\#} : \pi_n(X) \to \pi_n(X)$ for all n is asking for more information than can usually be produced. On the other hand, if an algorithm requires only knowledge of $\pi_1(X)$, $H_*(X;\mathbb{Q})$, and the corresponding homomorphisms $f_{1\#}$ and f_*, then there is a realistic chance of obtaining that information and applying the algorithm.

Even if an algorithm can be found for a class of spaces, this does not mean that Nielsen numbers on those spaces have been made truly computable. There remains the question of implementing the algorithm. While there are some very real questions here that must ultimately be dealt with, I will not address them in this paper. For the purposes of this paper, a class of spaces can be thought of as "Nielsen-computable" if there is an algorithm for computing Nielsen numbers which requires as data only homological and fundamental group information, and which works for all self maps. While Lefschetz numbers need not be the only homological information used, in practice, it is Lefschetz numbers that arise the most naturally. In most of the situations considered here, the calculations will reduce to a calculation of Lefschetz numbers, though computing $N(f)$ may often involve computing Lefschetz numbers of maps other than f.

Of course, Nielsen numbers are not computed for their own sake. They are computed because $MF(f) \geq N(f)$ for all self maps on compact polyhedra. Further, if X is polyhedra without local separating points, and is not a surface with negative Euler characteristic, then $MF(f) = N(f)$. So for such spaces, computing $N(f)$ exactly computes $MF(f)$. On the other hand, if no algorithm for computing $N(f)$ exists, lower bounds for $N(f)$ (which are in turn lower bounds for $MF(f)$, and so for $\#Fix(f)$) may still be of interest.

We begin with consideration of two simple classes of spaces: spaces with finite fundamental groups, and tori. This discussion will also serve to fix the notation used throughout the paper. In §3, the Jiang condition and Jiang spaces are examined. The next class of spaces, infrasolvmanifolds, will require more development. The topology of infrasolvmanifolds is described in §4, and the Nielsen theory in §5. The last section suggests some directions for future investigations.

2. Some Simple Examples

If X is a compact connected polyhedra, and $f : X \to X$ is a self map, then Nielsen theory begins by partitioning $Fix(f)$ into fixed point classes by the equivalence relation $x \sim y$ if there exists a path ω from x to y with $f\omega \simeq \omega$. A fixed point class S is *essential* if its fixed point index $Ind(f, S) \neq 0$. The set of essential fixed point classes will be denoted $\mathcal{E}(f)$, and its cardinality is of course the Nielsen number $N(f)$.

Suppose X is simply connected. Then for any two points $x, y \in Fix(f)$ and any path ω connecting them, $f\omega \simeq \omega$. Thus all of $Fix(f)$ forms a single fixed point class, whose index is $L(f)$. That is, if X is simply connected, then $N(f) = 1$ if $L(f) \neq 0$, and $N(f) = 0$ if $L(f) = 0$.

The next step up from this trivial case is the case when $\pi_1(X)$ is finite. In this case, the universal cover $\tilde{X}$ is also compact, and every $f : X \to X$ lifts to some $\tilde{f} : \tilde{X} \to \tilde{X}$. If $\mathcal{D} \cong \pi_1(X)$ is the group of deck transformations, then every $\beta\tilde{f}$, $\beta \in \mathcal{D}$ is also a lift of f. Since the Nielsen numbers of the maps $\beta\tilde{f}$ are all trivial to compute, $N(f)$ can be computed if the fixed point classes of f can be related to the fixed point classes of the lifts. But this relation is well-known [**11**]:

$$Fix f = \bigcup_{\beta \in \mathcal{D}} p(Fix(\beta\tilde{f}))$$

with each $p(Fix(\beta\tilde{f}))$ a (possibly empty) fixed point class of f, and with $p(Fix(\beta_1\tilde{f})) = p(Fix(\beta_2\tilde{f}))$ if and only if $\beta_2\tilde{f} = \gamma\beta_1\tilde{f}\gamma^{-1}$ for some $\gamma \in \mathcal{D}$. It will be convenient to use the isomorphism between $\mathcal{D}$ and $\pi_1(X)$ to reformulate this in terms of $\pi_1(X)$. We therefore introduce the Reidemeister classes $\mathcal{R}(f) = \pi_1(X)/\sim$, where $\omega_1 \sim \omega_2$ if and only if $\omega_2 = \gamma\omega_1 f_\#(\gamma^{-1})$ for some $\gamma \in \pi_1(X)$. The Reidemeister number $R(f)$ is the number of Reidemeister classes. If $\mathcal{T} \subset \pi_1(X)$ is a set of representatives of $\mathcal{R}(f)$ in $\pi_1(X)$, then

$$Fix(f) = \bigsqcup_{\beta \in \mathcal{T}} p(Fix(\beta\tilde{f}).$$

Each $Fix(\beta\tilde{f})$ is a covering space of the fixed point class $S = p(Fix(\beta\tilde{f})$. To describe this covering, we introduce the fix subgroup

$$F_\#(f, S) = \{\omega \in \pi_1(X, s) | f_\#(\omega) = \omega\},$$

where $s \in S$. This subgroup depends on the fixed point class S, but not on the point s chosen. This subgroup appears repeatedly, and plays a crucial role in computational issues. In this instance, $F_\#(f, S)$ is the covering group of $Fix(\beta\tilde{f})$ over S. From this, the fixed point index of S can be computed:

$$L(\beta\tilde{f}) = Ind(\beta\tilde{f}, Fix(\beta\tilde{f})) = |F_\#(f, S)| Ind(f, S),$$

so

$$Ind(f,S) \neq 0 \Leftrightarrow L(\beta\tilde{f}) \neq 0 \Leftrightarrow N(\beta\tilde{f}) \neq 0.$$

Thus $N(f) = \#\{[\beta] \in \mathcal{R}(f) | L(\beta\tilde{f}) \neq 0\}$, and we have

ALGORITHM 2.1 ($N(f)$ FOR SPACES WITH FINITE FUNDAMENTAL GROUP). *If X is a compact polyhedron with finite fundamental group, then computation of the Nielsen number $N(f)$ for a self map $f : X \to X$ requires knowledge of:*

- *the fundamental group $\pi_1(X)$ and its action on $H_*(\tilde{X};\mathbb{Q})$;*
- *the fundamental group homomorphism $f_\# : \pi_1(X) \to \pi_1(X)$;*
- *the homology homomorphism $\tilde{f} : H_*(\tilde{X};\mathbb{Q}) \to H_*(\tilde{X};\mathbb{Q})$ for some lift $\tilde{f} : \tilde{X} \to \tilde{X}$ of f.*

Then $N(f)$ is computed by the following procedure:

(i) *From $\pi_1(X)$ and $f_\#$, find the Reidemeister classes and select a transveral $\mathcal{T}$ of $\mathcal{R}(f)$ in $\pi_1(X)$.*

(ii) *For each $\beta \in \mathcal{T}$, use $\beta_* : H_*(\tilde{X};\mathbb{Q}) \to H_*(\tilde{X};\mathbb{Q})$ and $\tilde{f}_*$ to compute $L(\beta\tilde{f}) = \sum_n (-1)^n tr(\beta_{n*}\tilde{f}_{n*})$.*

(iii) *$N(f)$ is the number of nonzero Lefschetz numbers so found.*

Note that if $\pi_1(X)$ acts trivially on $H_*(\tilde{X};\mathbb{Q})$, then $N(f) = R(f)$ if $L(\tilde{f}) \neq 0$, and $N(f) = 0$ if $L(\tilde{f}) = 0$. We will see a similar result again when we consider Jiang spaces in the next section.

Spaces with finite fundamental group are perhaps the one class of spaces for which the Nielsen number can be computed more or less directly from the definition. We turn now to another class of spaces, tori, where the calculation does not follow immediately from the defintion, but is nonetheless particularly simple. It will also serve as an important model for more complicated calculations later.

There is a one-to-one correspondence between homotopy classes of maps $[T^n, T^n]$ and fundamental group homomorphisms $Hom(\mathbb{Z}^n, \mathbb{Z}^n) \cong gl_n(\mathbb{Z})$. That is, every homotopy class is represented by an $n \times n$ integer matrix. We will assume throughout this section that f is a homomorphism, which lifts to integer matrix $F : \mathbb{R}^n \to \mathbb{R}^n$. The relation between Nielsen numbers and Lefschetz numbers on tori is

THEOREM 2.1 ([6]). *Every fixed point class has the same index, which is either $1, 0$, or -1.*

It follows immediately from this that $N(f) = |L(f)|$ for all $f : T^n \to T^n$. Moreover, it is well-known that for a torus map $L(f) = \det(I - f_{1*})$, where f_{1*} is the induced map on $H_1(T^n;\mathbb{Q})$. That is, $N(f) = \det(I - f_{1*})$, which is as computable a formula as one could ask for.

ALGORITHM 2.2 ($N(f)$ FOR TORI). *If T^n is a torus, then computation of the Nielsen number $N(f)$ for a self map $f : T^n \to T^n$ requires knowledge of a* *matrix representative F of the fundamental group homomorphism $f_\# : \pi_1(T^n) \to \pi_1(T^n)$. Then $N(f)$ is computed by*

$$N(f) = |L(f)| = |\det(I - F)|.$$

EXAMPLE 2.1. Suppose $f : T^2 \to T^2$ is the torus homomorphisms represented by the map

$$F = \begin{bmatrix} 2 & -2 \\ -2 & 0 \end{bmatrix}.$$

Then $\det(I - F) = -5$, so there are five coincidence classes, each with index -1, and $N(f) = |L(f)| = 5$. There are five fixed points, each of which forms a distinct fixed point class. For each class, the fundamental group homomorphism $f_\#$ is also represented by F, so $F_\#(f, S) = Fix(F) = \ker(I - F) = 0$.

On the other hand, for the torus homomorphism $g : T^2 \to T^2$ represented by the matrix

$$G = \begin{bmatrix} 3 & 2 \\ 2 & 3 \end{bmatrix},$$

then $N(g) = L(g) = \det(I - G) = 0$, so all coincidence classes have index 0, and g is homotopic to a map which is fixed point free. But the map g is not fixed point free: $Fix(g) = \{\theta_1 + \theta_2 = 1\} \cup \{\theta_1 + \theta_2 = 1/2\}$. The two circles are each a fixed point class with index 0. The fixed points can be removed by adding a small irrational translation of the form $(\theta_0, -\theta_0)$ to g. For each class, $F_\#(f, S) = Fix(F) = \ker(I - G) = \langle e_1 - e_2 \rangle \cong \mathbb{Z}$, where e_1, e_2 are the standard generators of the fundamental group.

These two examples are typical. If $f : T^n \to T^n$ is the unique torus homomorphism which represents its homotopy class, then either $Fix(f)$ consists of $N(f)$ points, each forming a distinct class and each having index ± 1 and trivial fix subgroup; or $Fix(f)$ consists of a disjoint union of sub-tori, with each sub-torus a fixed point class of index 0 and fix subgroup equal to the kernel of $I - F$. In the latter case, perturbing the map by an irrational rotation along the sub-torus will remove all fixed points.

3. Jiang Spaces

One of the earliest steps towards computability in Nielsen theory was the development by Jiang **[10]** in the 1960's of the condition now known as the *Jiang condition* – a sufficient condition for computing the Nielsen number of a map in terms of the Lefschetz and Reidemeister numbers of the map. A space with the property that the Jiang condition is satisfied for all self maps is referred to as a *Jiang space*.

Given a map $f : X \to X$ and $x \in \mathrm{Fix}(f)$, define

$$J(f) = \{\omega \in \pi = \pi_1(X, x) \mid \omega = [H(x, -)] \text{ for some } H : f \simeq f\}.$$

Some of the important properties of $J(f)$ are:

(i) $J(f)$ is independent of the base point chosen (up to conjugation).

(ii) $J(f) \subseteq \mathbb{C}_\pi(f_\#(\pi))$, the centralizer of $f_\#(\pi)$ in π, with equality if X is aspherical.
(iii) $J(X) = J(id) \subseteq J(f)$ for all f.
(iv) If $f_\#(\pi) \subseteq J(f)$, then all fixed point classes have the same index.

It is this last condition, $f_\#(\pi) \subseteq J(f)$, that is known as the Jiang condition. In order for X to be a Jiang space, that is, for the Jiang condition to be satisfied for all maps, it is necessary and sufficient that the condition be satified for the identity map. If $J(X) = \pi$, then for every f we have $f_\#(\pi) \subseteq \pi = J(X) \subseteq J(f)$. Note that $J(X) \subseteq \mathbb{Z}(\pi)$, the center of π, so X can only be a Jiang space if π is abelian. Thus many spaces of interest (e.g. surfaces with negative Euler characteristic) are not Jiang spaces. However, several significant classes of spaces are known to be Jiang spaces: H-spaces, including compact topological groups; homogenous spaces G/G_0, where G is a compact topological group and G_0 is a compact connected subgroup; principal G-bundles over a simply-connected base[**23**]; and generalized lens spaces. The question of determining necessary and sufficient conditions for a space to be a Jiang space continues to receive considerable attention: the reader is referred to [**21, 22, 23, 26**].

The significance of the Jiang condition is that, since all fixed point classes have the same index, they are either all essential or all inessential. Further, the sum of the fixed point indices (i.e. the Lefschetz number) is zero if and only if all fixed point classes are inessential. Thus the Nielsen number can be computed as follows:

ALGORITHM 3.1 ($N(f)$ FOR JIANG SPACES). *Suppose X is a compact polyhedron with $J(X) = \pi_1(X)$. Then computation of the Nielsen number $N(f)$ for a self map $f : X \to X$ requires knowledge of the fundamental group endomorphism $f_\# : \pi_1(X) \to \pi_1(X)$ and the homology endomorphism $f_* : H_*(X;\mathbb{Q}) \to H_*(X;\mathbb{Q})$. Then $N(f)$ is computed as follows:*
(i) *From $f_\#$, compute $coker(1 - f_\#)$.*
(ii) *From f_*, compute $L(f) = \sum(-1)^n tr(f_{n*})$.*
(iii) *Then*

$$N(f) = \begin{cases} |coker(1 - f_\#)| & L(f) \neq 0 \\ 0 & L(f) = 0 \end{cases}$$

That is, $N(f) = R(f)$ if all fixed point classes have the same nonzero index. But if $J(f) = \pi_1(X)$ (and, in particular, π_1 is abelian), it follows that $R(f) = |coker(1 - f_\#)|$. Of course, since π_1 is abelian, the Hurewicz homomorphism is an isomorphism, and $f_\#$ can be equally well thought of as the endomorphism f_{1*} on $H_1(X;\mathbb{Z})$.

4. Nilmanifolds, Solvmanifolds & Infrasolvmanifolds

It would not be too gross a distortion of history to say that the above results represent the state of the art, circa 1980. Shortly thereafter, two papers appeared which significantly advanced the ability to find algorithms for Nielsen numbers: Anosov's theorem that $N(f) = |L(f)|$ for self maps of nilmanifolds [**5**]; and You's determination of necessary and sufficient conditions for the Nielsen product formula for fibrations [**28**]. Anosov's result was reproved using the product formula by Fadell and Husseini [**8**]. This reformulation pointed the way to a variety of results [**12, 13, 14, 16, 18, 19**] that have produced algorithms for a broad class of spaces known as infrasolvmanifolds. In this section, we will consider nilmanifolds and solvmanifolds, infranilmanifolds and infrasolvmanifolds. We will develop briefly the structures of these spaces. There is a considerable literature on the topology of these manifolds. Some of the most important references are: for nilmanifolds, Mal'cev's elegant and exhaustive treatment [**15**]; for solvmanifolds, Mostow [**20**], Wang [**27**], and a series of studies by Auslander, the most relevant of which are [**2, 3**] ([**3**] contains an extensive bibliography). Auslander should also be mentioned for infranilmanifolds [**1**]. Infrasolvmanifolds appear most often in the literature in connection with flat riemannian manifolds; we mention in particular [**4, 7, 9**].

There two descriptions of infrasolvmanifolds: one constructive, and one algebraic. The constructive approach is modeled on the construction of the torus T^n as $\mathbb{R}^n/\mathbb{Z}^n$. To generalize this construction, think of $\mathbb{R}^n$ as a simply connected abelian Lie group, and $\mathbb{Z}^n$ as a discrete subgroup of it. To define infrasolvmanifolds constructively, take a solvable connected simply connected Lie group S. Consider the Lie group $G = S \rtimes K$, where K is a compact subgroup of $Aut(S)$. G acts on S by $(s, \alpha) \cdot s' = s\alpha(s')$. If $\pi \subset G$ is a torsion-free subgroup with finite projection Φ onto K, then $M = \pi \setminus S$ is an infrasolvmanifold. M is connected, and is compact if and only if π is uniform in $S \rtimes K$, or equivalently if and only if Γ is uniform in S.

If Φ is solvable, then M is a homogeneous space of a solvable simply connected (not necessarily connected) Lie group, and is called a *solvmanifold.* In this case M can also be constructed as a homogeneous space of a solvable connected (not necessarily simply connected) Lie group, or as a homogenous space of a solvable connected simply connected Lie group with π not necessarily discrete. It is not in general possible to construct a solvmanifold as a quotient of a solvable connected simply connected Lie group by a discrete subgroup. The Klein bottle, for example, is a solvmanifold which admits no such construction. Solvmanifolds that can be so formed (i.e. solvmanifolds with $\Phi = 0$) are called *special solvmanifolds.* If S is a nilpotent Lie group, M is called an *infranilmanifold*; if we further require $\Phi = 0$ (i.e. $M = S/\pi$, with S a connected simply connected nilpotent Lie group and π a discrete subgroup), then M is a *nilmanifold.* Similarly, if S is abelian, M is an *infratorus* or *infrabelian manifold.* And of course, if S is abelian and $\pi \subset S$, we recover the familiar construction of the torus.

One obvious consequence of this construction is that every infrasolvmanifold has a finite regular cover by a solvmanifold (indeed, by a special solvmanifold), and similarly, every infranilmanifold (resp. infrabelian manifold) has a finite regular cover by a nilmanifold (resp. torus). The other topological construction of central importance is the *Mostow fibration* [**20**]: every solvmanifold M has a (nonorientable) fibration

$$N \to M \to T$$

with N a nilmanifold and T a torus.

EXAMPLE 4.1. Let U_n be the matrix group of $n \times n$ upper triangular unipotent matrices, and let $U_n(\mathbb{Z})$ be the subgroup of integer matrices. U_n is nilpotent and simply connected (in fact, homeomorphic to Euclidean space of dimension $\frac{n(n-1)}{2}$). Then $U_n/U_n(\mathbb{Z})$ is a nilmanifold. This in fact the typical example. A consequence of Mal'cev's analysis of nilmanifolds is that every nilmanifold has the form S/π, where S embeds as a simply connected subgroup of some U_n. As an example of an infranilmanifold, form the automorphism $\alpha : U_n \to U_n$ by conjugation in $GL_n(\mathbb{R})$ by $diag(-1, 1, \ldots, 1)$. Form $U_n \rtimes \mathbb{Z}_2$ in the obvious way. Let $\pi \subset U_n(\mathbb{Z}) \rtimes \mathbb{Z}_2$ be the subgroup generated by all $E_{i,j}$ with $1 \leq i < j \leq n$ except for $E_{n-1,n}$, and by $(E_{n-1,n}, \alpha)$ (where $E_{i,j} \in U_n$ is the upper triangular unipotent matrix with i, j entry equal to 1 and all other upper triangular entries 0). Then U_n/π is an infranilmanifold. In a similar way, the Klein bottle is constructed by letting $diag(-1, 1)$ act on $\mathbb{R}^2$. That is the Klein bottle can be viewed as an infranilmanifold (in fact, an infratorus) as well as a solvmanifold. When viewed as a solvmanifold, its Mostow fibration is the familiar fibration $S^1 \to K \to S^1$.

The basis of the algebraic approach is that infrasolvmanifolds are compact $K(\pi, 1)$'s, so $\pi = \pi_1(M)$ is finitely generated and torsion free, and M is classified up to homeomorphism by π. That is, two infrasolvmanifolds are homeomorphic if and only if their fundamental groups are isomorphic. For a group π to admit a compact $K(\pi, 1)$, π must be torsion-free finitely generated. To be the fundamental group of a nilmanifold, π must also be nilpotent. This exactly characterizes nilmanifolds: M is a compact nilmanifold if and only if M is an aspherical manifold with nilpotent fundamental group. Alternatively, a group π is the fundamental group of a nilmanifold if and only if π is finitely generated torsion-free nilpotent. The corresponding statement for solvmanifolds, that a group is the fundamental group of a solvmanifold if and only if it is finitely generated torsion-free solvable, is false. For a solvmanifold must have a Mostow fibration, which implies that the fundamental group must admit a factorization

$$1 \to \Gamma \to \pi \to \Lambda \to 0$$

with Γ torsion-free nilpotent and Λ torsion-free abelian. A group that admits such a factorization is referred to as a *strongly torsion-free $\mathcal{S}$-group*. (It follows immediately that such a group is solvable). This condition characterizes solvmanifolds: π is the fundamental group of a solvmanifold if and only if π is a strongly torsion-free $\mathcal{S}$-group.

To characterize fundamental groups of infrasolvmanifolds, recall that for any property P of groups, a group G is *virtually* P if there is a normal subgroup of finite index H which has property P. Similarly, a group is *poly-* P if there is a normal series $\{G_i\}$ for G such that each subquotient G_i/G_{i+1} has property P. In particular, G is polycyclic if there is a normal series with $G_i/G_{i+1} \cong \mathbb{Z}$. A group is the fundamental group of an infrasolvmanifold if and only if it is a virtually polycyclic group, or equivalently, a poly-{cyclic or finite} group [**4**]. Of course, π is the fundamental group of an infrasolvmanifold (resp. infranilmanifold, infratorus) if and only if π has a normal subgroup Γ of finite index which is a strongly torsion-free $\mathcal{S}$-group (resp. torsion-free nilpotent group, torsion-free abelian group).

Of course, torsion-free abelian, nilpotent and solvable groups have a great deal of structure, which we must exploit to produce the Nielsen-Lefschetz formulas. As the essense of our Nielsen formulas will be reduction to the torus case, we are interested in "finding" tori related to these manifolds – that is, finding torsion-free abelian groups related to the fundamental group. To begin with, suppose π is torsion-free nilpotent, and consider the commutator series $\pi^0 = \pi, \pi^j = [\pi^{j-1}, \pi]$. Clearly, $\pi^{j+1} \subset \pi^j$ and $\pi^j \lhd \pi$. As π is nilpotent, this series descends to the identity: $\pi^{N+1} = [\pi, \pi^N] = 1$ for some N, so π^N is central. The successive quotients π^j/π^{j-1} are abelian, but not necessarily torsion-free. We therefore introduce the idea of the *completion* of a subgroup. If $H \lhd G$, then the completion of H in G, denoted $C(H)$, is the minimal normal subgroup of G that contains H and has $G/C(H)$ torsion-free. In general, H might have infinite index in $C(H)$, but for the groups we will be concerned with, the index $[C(H) : H]$ will be finite. Now, let Γ_j be defined by $\Gamma_1 = \pi$ and $\Gamma_j = C([\pi, \Gamma_{j-1}])$. Because π embeds as a uniform subgroup of a nilpotent Lie group G, we can show that $\Gamma_j = C(\pi^{j+1}) = G^{j+1} \cap \pi$. In particular, the successive quotients $\Lambda_j = \Gamma_j/\Gamma_{j+1}$ are torsion-free abelian and Γ_N is central.

The structure we require for strongly torsion-free $\mathcal{S}$-groups is an extension of this. If π is a strongly torsion-free $\mathcal{S}$-group, then the factorization

$$1 \to \Gamma \to \pi \to \Lambda \to 0$$

with Γ nilpotent and Λ torsion-free abelian means that $[\pi, \pi]$ is nilpotent, as is $\Gamma_1 = C([\pi, \pi])$. Let $\Lambda_0 = \pi/\Gamma_1$, and let Γ_j, Λ_j be derived as indicated above from the commutator series of Γ_1. (Note that this gives a polycyclic decomposition of π.) Each Γ_j is normal in π, so the conjugation action of π on itself defines an action on each $\Lambda_j = \Gamma_j/\Gamma_{j+1}$. Moreover, as Γ_{j+1} contains $[\Gamma_1, \Gamma_j]$, Γ_1 acts trivially on Λ_j. Thus there is a well-defined action of Λ_0 on each Λ_j. These actions will all be trivial if and only if π is nilpotent – that is, if and only if $\Lambda_0 = 0$. We will refer to the groups Λ_j and the action of Λ_0 on the Λ_j as the *linearization of* π.

Certain classes of solvmanifolds can be usefully distinguished by the linearizations of their fundamental groups – in particular, by the eigenvalues of the action of Λ_0 on the Λ_j. A solvmanifold S is said to be *rotational* if all eigenvalues of these actions lie on the unit circle (and so are all roots of unity). In contrast, S is *exponential* if the only action eigenvalue on the unit circle is 1, and is a $\mathcal{NR}$-solvmanifold ($\mathcal{NR}$ for "non-rotational") if no root of unity other than 1 appears as an eigenvalue. A solvmanifold is rotational if and only if it is an infranilmanifold; it is both rotational and exponential (i.e. has only 1 as an eigenvalue) if and only if it is a nilmanifold.

If π is virtually polycyclic, then the solvradical π_S of π, the maximal normal solvable subgroup of π, is a subgroup of finite index. Since the solvradical $\pi_S = \Gamma_0$, its commutator subgroup Γ_1 and the commutator series Γ_j of Γ_1 are all characteristic subgroups of π, the conjugation action of π on Γ_0 induces actions on the subgroups Γ_j, and so on the subquotients Λ_j. Γ_1 acts trivially on $\Lambda_j, j \geq 1$, and Γ_0 acts trivially on Λ_0, so there are actions of π/Γ_0 on Λ_0 and of π/Γ_1 on each $\Lambda_j, j \geq 1$. We refer to these actions as the linearization of π. Note that the restriction of the action of π/Γ_1 to $\Lambda_0 = \Gamma_0/\Gamma_1$ gives the linearization of Γ_0 described above. If π is infranilpotent, we make the same construction, only using the nilradical π_N (the maximal normal nilpotent subgroup) instead of the solvradical. In this case, $\Lambda_0 = 0$ and the only actions present are the actions of π/π_N on each $\Gamma_j, j \geq 1$.

One of the reasons for working with the linearizations just outlined is that they are particularly well-behaved with respect to maps. If π_1, π_2 are strongly torsion-free $\mathcal{S}$-groups, and $\phi : \pi_1 \to \pi_2$ is a homomorphism, then clearly $\phi([\pi_1, \pi_1]) \subset [\pi_2, \pi_2]$. More importantly, this containment is preserved by completions: $\phi(\Gamma_{1,1}) \subset \Gamma_{2,1}$. Similarly, ϕ maps the commutator series for $\Gamma_{1,1}$ into the commutator series for $\Gamma_{2,1}$, and this containment is preserved by completions: $\phi(\pi_1^j) \subset \pi_2^j$, and $\phi(\Gamma_{1,j}) \subset \Gamma_{2,j}$. Thus there are in turn well-defined maps $\bar{\phi}_j : \Lambda_{1j} \to \Lambda_{2j}$. Moreover, these maps respect the actions of Λ_0 on the Λ_j. If $A_j : \Lambda_{10} \to Aut(\Lambda_{1j})$ and $B_j : \Lambda_{20} \to Aut(\Lambda_{2j})$ are the homomorphisms expressing the actions, then $\bar{\phi}_j A_j(\lambda) = B_j(\bar{\phi}_0 \lambda)\bar{\phi}_j$ for all $\lambda \in \Lambda_1$. We refer to the maps $\bar{\phi}_j$ as the linearization of ϕ.

If π is virtually polycyclic, the situation is not quite so nice. The solvradical π_S and nilradical π_N are characteristic, but are not necessarily preserved by all endomorphisms $\phi : \pi \to \pi$. And if π_S is not preserved by ϕ, there can be no guarantee that the linearization of π_S is preserved either. We must therefore take a slightly different approach to decomposing ϕ. Since $\phi(\pi_S)$ and $\phi(\Gamma_j)$ may not be contained in π_S and Γ_j, we choose a decomposition of π so that the stages of this new decomposition map under ϕ into the corresponding stages of the decomposition generated by π_S and its commutator series. Namely, let $\Gamma_j' = \Gamma_j \bigcap \phi^{-1}(\Gamma_j)$. Then $\Lambda_j' = \Gamma_j/\Gamma_{j+1}$ is torsion-free abelian, and the restriction of ϕ to Γ_j' induces a well-defined map $\bar{\phi}_j : \Lambda_j' \to \Lambda_j$. Similarly, the inclusion homomorphism $\Gamma_j' \to \Gamma_j$ induces an injection $\bar{\iota}_j : \Lambda_j' \to \Lambda_j$.

So far, these have all been purely algebraic constructions on the fundamental group level. But, as infrasolvmanifolds are aspherical, all fundamental group constructions can be realized topologically.

LEMMA 4.1. *Suppose M is an infrasolvmanifold, with $\pi = \pi_1(M)$. Then*

(i) *If $1 \to \Gamma \to \pi \to \Lambda \to 1$ is a factorization of π with Λ torsion-free, then there exists a fibration $F \to M \to B$ with $\pi_1(F) = \Gamma$ and $\pi_1(B) = \Lambda$.*

(ii) *If Γ is abelian, then the actions of Λ on Γ induced by conjugation and induced by the fibration are equal. In particular, the fibration is orientable if and only if Γ is central.*

(iii) *If $\phi : \pi \to \pi$ is a homomorphism, then there is a map $f : M \to M$ with $f_\# = \phi$.*

(iv) *If $\Gamma_i \to \pi_i \to \Lambda_i$, $i = 1, 2$, are two factorizations as in (i) and $\phi : \pi_1 \to \pi_2$ has $\phi(\Gamma_1) \subseteq \Gamma_2$, then there is a commutative diagram*

$$\begin{array}{ccccc} F_1 & \to & M_1 & \to & B_1 \\ \downarrow & & \downarrow f & & \downarrow \bar{f} \\ F_2 & \to & M_2 & \to & B_2 \end{array}$$

with $f_\# = \phi$ and $\bar{f}_\# = \bar{\phi}$.

(v) *If $\Gamma \lhd \pi$ has finite index, then there exists a regular finite covering $\tilde{M} \xrightarrow{p} M$ with $im(p_\#) = \Gamma$.*

(vi) *If $\Gamma_i \lhd \pi_i$ has finite index and $\phi : \pi_1 \to \pi_2$ has $\phi(\Gamma_1) \subseteq \Gamma_2$, then there exists a commutative covering space diagram*

$$\begin{array}{ccc} \tilde{M}_1 & \xrightarrow{\tilde{f}} & \tilde{M}_2 \\ \downarrow p_1 & & \downarrow p_2 \\ M & \xrightarrow{f} & M \end{array}$$

with $f_\# = \phi$ and $\tilde{f}_\# = \phi\,|_{\Gamma_1}$. If $\phi|_{\Gamma_1}$ is injective, then $\tilde{f}$ is a covering projection.

That is, every self map of a nilmanifold can (up to homotopy) be factored through a sequence of orientable fibrations

$$\begin{array}{ccccc} T_j & \to & N_j & \to & N_{j-1} \\ \downarrow \bar{f}_j & & \downarrow f_j & & \downarrow f_{j-1} \\ T_j & \to & N_j & \to & N_j' \end{array}$$

where T_j is a torus with fundamental group Λ_j and N_j is a nilmanifold with fundamental group π/Γ_j. Similarly, every self map of a solvmanifold can be factored through a (nonorientable) fibration

$$\begin{array}{ccccc} N & \to & S & \to & T_0 \\ \downarrow & & \downarrow f & & \downarrow \bar{f}_0 \\ N & \to & S & \to & T_0 \end{array}$$

where N is a nilmanifold with fundamental group Γ_1 and T_0 is a torus with fundamental group Λ_0. The map can be further factored into $\bar{f}_j : T_j \to T_j$ by factoring $f|_N$.

Finally, every self map $f : M \to M$ of an infrasolvmanifold can be lifted to a map $\tilde{f} : S_1 \to S_2$ of solvmanifolds, where $\pi_1(S_2) = \Gamma_0$, the solvradical of $\pi_1(M)$, and $\pi_1(S_1) = \Gamma_0 \bigcap f_{\#}^{-1}(\Gamma_0)$. The solvmanifolds S_1 and S_2 are finite regular covers of M, and the identity map $id : M \to M$ lifts to a covering projection $p : S_1 \to S_2$. That is, there is a commutative diagram

$$\begin{array}{ccc} S_1 & \xrightarrow{\tilde{f},p} & S_2 \\ \downarrow p_1 & & \downarrow p_2 \\ M & \xrightarrow{f,id} & M \end{array}$$

known as the *lifting diagram* of f.

5. Nielsen Numbers on Infrasolvmanifolds

The work of the previous section shows that nilmanifolds and solvmanifolds can be decomposed into tori through a sequence of fibrations. The crucial distinction between nilmanifolds and solvmanifolds is that these fibrations can all be chosen to be orientable if the manifold is a nilmanifold; but cannot all be orientable if it is a solvmanifold. Since the computation of Nielsen numbers on tori is so straightforward, all that is required to compute Nielsen numbers on nilmanifolds and solvmanifolds is knowledge of how Nielsen numbers behave with respect to fibrations. This too is well-known [**28**], and turns out to be particularly simple in our setting.

If $F \to E \xrightarrow{p} B$ is a fibration, a map $f : E \to E$ is a *fiber-preserving map* if there exists a map $\bar{f} : B \to B$ such that $\bar{f} \circ p = p \circ f$. Then for every $b \in B$ we can define $F_b = p^{-1}(b)$ and $f_b = f\mid_{F_b} : F_b \to F_{\bar{f}(b)}$. The general problem in Nielsen theory for fiber-preserving maps is to determine the relationship between $N(f)$, $N(\bar{f})$ and the various $N(f_b)$ as b ranges over $Fix(\bar{f})$. For convenience, assume that all of the spaces F, E and B are Wecken spaces (i.e. spaces on which $N(f) = MF(f)$ for all self maps).

To begin with, we can assume that all non-essential fixed point classes of $\bar{f}$ have been removed, and that all essential classes consist of a single point. (If not, homotope $\bar{f}$ to a map $\bar{f}'$which satisfies these conditions. Then use the homotopy lifting property of the fibration to find a map f' homotopic to f which covers $\bar{f}'$.) Then every class in $\mathcal{R}(f)$ is contained in a single fiber F_b over some $b \in Fix(\bar{f})$. That is, $Fix(f)$ projects to $Fix(\bar{f})$, and $p|_{Fix(f)}$ maps fixed point classes to fixed point classes. Now, we can modify each f_b (as b ranges over $Fix(\bar{f})$) so that all non-essential fixed point classes of f_b have been removed, and that all essential classes consist of a single point.

At this point, we have maps which minimize the number of fixed points *in the fiber-preserving homotopy class.* But if we consider maps homotopic to f which are not fiber-preserving (or which are homotopic to f through non-fiber-preserving maps), we may be able to reduce the number of fixed points. That is, $N(f) \leq \sum_{\mathcal{E}(\bar{f})} N(f_b)$, but equality does not necessarily hold. Moreover, the right hand side is expressed as a sum, rather than as a the product $N(\bar{f})N(f_b)$, because the fibration is not assumed to be orientable, and so the maps f_b on different fibers may not be homotopic.

Two things can go wrong to prevent the equality $N(f) = \sum_{\mathcal{E}(\bar{f})} N(f_b)$ from holding: points in $Fix(f_b)$ which are in distinct fixed point classes for the maps restricted to the fiber may be in the same fixed point class of f; and fixed point classes which are essential in the fiber may not be essential in the total space. The first of these is controlled entirely by the fundamental group.

Suppose $x, y \in Fix(f)$ are in the same fixed point class of f. Then there exists a path ω in E with $f\omega \simeq \omega$ in E. But then $p(x), p(y)$ are in $Fix(\bar{f})$, and $p\omega$ is a path from $p(x)$ to $p(y)$ with $\bar{f}(p\omega) \simeq p\omega$. That is, $p(x)$ and $p(y)$ are in the same fixed point class of $\bar{f}$. Since we assumed that fixed point classes of $\bar{f}$ are singletons, $p(x) = p(y)$, and we conclude that each fixed point class of f must lie in a single fiber. That is, if $S_x = S_x(f) \in \mathcal{R}(f)$ denotes the fixed point class (in E) of x, then S_x is a union of f_b-fixed point classes, where $b = p(x)$. In that case, $p\omega$ is then a loop at b with $\bar{f}_{\#}[p\omega] = [p\omega]$. This suggests that we may be able to compute the cardinality of S_x via the fundamental group. This is made precise by the following

LEMMA 5.1. [**28**, Corollary 5.4] *If $x \in Fix(f), b = p(x)$ and $\pi_1(F_{\bar{f}(b)}, f(x)) \to \pi_1(E, f(x))$ is injective, then the number of f_b-fixed point classes in S_x is*

$$[F_{\#}(\bar{f}, b) : p_{\#}F_{\#}(f, x)].$$

The details of the proof are rather technical, but the intuitive idea is clear: as indicated above, f_b-fixed point classes in S_x generate elements of $F_{\#}(\bar{f}, p(x))$. In fact, every element of $F_{\#}(\bar{f}, p(x))$ can be so realized. But some of the elements generated may have been generated trivially, that is, by projecting (via $p_{\#}$), elements of $F_{\#}(f, x)$. Can any class in S_x generate an element of $p_{\#}F_{\#}(f, x)$, other than x itself? Such a class y exists only if there exists a path c in F_b from x to y with $fc \simeq c$ in E. This is equivalent to requiring that the loop $c^{-1} * fc$ which lies in $F_{\bar{f}(b)}$, be contractible in E but not in $F_{\bar{f}(b)}$. The injectivity condition precludes this, so the quotient $F_{\#}(\bar{f}, b)/p_{\#}F_{\#}(f, x)$ exactly counts the number of classes in S_x

A sufficient condition then, for each f-fixed point class to consist of a single f_b-fixed point class, is that $\pi_1(F) \to \pi_1(E)$ is injective, and that $F_{\#}(\bar{f}, b) = 0$ for all $b \in Fix(\bar{f})$. In this case, the second question – whether essential f_b-classes must be essential (f, g)-classes – is easily answered. If each f-fixed point class consists of a single f_b-fixed point class (in our case, a single point x), then

$$Ind(f; x) = Ind(\bar{f}; p(x))Ind(f_b; x),$$

and x is essential as an f-fixed point class if and only if it is essential as an f_b-class and $p(x)$ is an essential $\bar{f}$-class.

Thus we have, as a special case of You's product formula:

THEOREM 5.1. *Suppose* $F \to E \xrightarrow{p} B$ *is a fibration such that* $\pi_1(F) \to \pi_1(E)$ *is injective. If* $f : E \to E$ *is a fiber-preserving map such that* $F_\#(\bar{f}, b) = 0$ *for all* b *in essential* $(\bar{f})$*-fixed point classes, then*

(i) *If* $N(\bar{f}) = 0$, *then* $N(f) = 0$.

(ii) *If* $N(\bar{f}) \neq 0$, *then*

$$N(f) = \textstyle\sum_{\mathcal{E}(\bar{f})} N(f_b) \qquad L(f) = \textstyle\sum_{\mathcal{E}(\bar{f})} Ind(\bar{f}; b) L(f_b).$$

(iii) *If the fibration is orientable, then all restrictions* f_b *of* f *to fibers are homotopic. Then* $N(f) = N(\bar{f})N(f_b)$ *and* $L(f,g) = L(\bar{f})L(f_b)$.

Note that orientability of the fibrations is, in some sense, not very important. Orientability of $p : E \to B$ guarantees that all restrictions of the maps to fibers have the same homotopy type, so that $N(f_b)$ and $L(f_b)$ are independent of b. We will see in our applications of the product formula below that this independence can sometimes be achieved without assuming orientability.

A crucial step in doing so is to determine how the maps on the fibers over the various $\bar{f}$-fixed point classes are related to one another. To do so, fix a "reference class" b_0 in $Fix(\bar{f})$, and a path c_b from b_0 to each $b \in Fix(\bar{f})$. Let τ_c be the fiber-translation map $\tau_c : F_0 \to F_b$. Then the path $\bar{f}c$ induces a fiber-translation map $\tau_{\bar{f}c}$ from F_0 to F_b. τ_c and $\tau_{\bar{f}c}$ are homotopy equivalences, which we may treat without loss as homeomorphisms. More importantly, they form a homotopy-commutative diagram

$$\begin{array}{ccc} F_0 & \xrightarrow{f_0} & F_0 \\ \downarrow \tau_c & & \downarrow \tau_{\bar{f}c} \\ F_b & \xrightarrow{f_b} & F_b \end{array}$$

In particular, if ω is a loop at b_0, then $f_0 \circ \tau_\omega \simeq \tau_{\bar{f}\omega} \circ f_0$.

We can further rewrite the expression for f_b as

$$\begin{aligned} f_b &\simeq \tau_{\bar{f}c} \circ f_0 \circ \tau_c^{-1} \\ &= \tau_c \circ \tau_c^{-1} \circ \tau_{\bar{f}c} \circ f_0 \circ \tau_c^{-1} \\ &= \tau_c \circ \tau_{\bar{f}c * c^{-1}} \circ f_0 \circ \tau_c^{-1} \end{aligned}$$

Then the homotopy-type invariance of Nielsen and Lefschetz numbers implies that $N(f_b) = N(\tau_{\bar{f}c*c^{-1}} f_0)$ and $L(f_b) = L(\tau_{\bar{f}c*c^{-1}} f_0)$. The $\bar{f}$-commutativity of f_0 indicated above imply that these quantities are independent of the path c chosen. Thus the sums in 5.1 can be rewritten as $N(f) = \sum N(\tau_{\bar{f}c*c^{-1}} f_0)$ and $L(f) = \sum L(\tau_{\bar{f}c*c^{-1}} f_0)$. If $\pi_1(B)$ is abelian, these sums range over a transversal in $\pi_1(B)$ of $\operatorname{coker}(id - f_\#)$.

To apply theorem 5.1 to the nilmanifold and solvmanifold fibrations, we need to know that, in all of the fibrations, $\pi_1(F) \to \pi_1(E)$ is injective and $F_\#(f, S) = 0$ for all essential fixed point classes of all maps. The first condition is trivial, as all of the spaces are aspherical. The second was shown in §2 to be true for all maps on tori. A simple diagram chase then gives

PROPOSITION 5.1 ([**8, 16**]). *If S is an essential fixed point class of a self map f on a solvmanifold, then $F_{\#}(f, S) = 0$.*

With this, the product formula can be applied to the sequence of orientable fibrations of a nilmanifold generated by the commutator series of its fundamental group. The result is:

ALGORITHM 5.1 ($N(f)$ FOR NILMANIFOLDS). *Suppose N is a nilmanifold. Then computation of the Nielsen number $N(f)$ for a self map $f : N \to N$ requires knowledge of the linearization of f; that is, the matrices $F_i : \Lambda_i \to \Lambda_i$ derived from the commutator series of $\pi_1(N)$. Then $N(f)$ is computed by*

$$N(f) = \prod_{i=0}^{N} |\det(F_i - I)| \,.$$

Alternatively, if the homology endomorphism $f_* : H_*(N; \mathbb{Q}) \to H_*(N; \mathbb{Q})$ is known, then the Nielsen number can be computed by

$$N(f) = |L(f)| = \left| \sum_n (-1)^n tr(f_{n*}) \right|$$

To extend this result to solvmanifolds, we use the Mostow fibration to factor the solvmanifold into a nilmanifold and a torus, then apply the previous result to the nilmanifold. Since the Mostow fibration is a non-orientable fibration, we must expect a summation, instead of a product formula.

ALGORITHM 5.2 ($N(f)$ FOR SOLVMANIFOLDS). *Suppose S is a solvmanifold. Then computation of the Nielsen number $N(f)$ for a self map $f : S \to S$ requires knowledge of the linearizations of $\pi_1(S)$ and $f_{\#}$:*

- *The polycyclic decomposition $\{\Lambda_i\}_{i=0}^{N}$ of $\pi_1(S)$ derived from the commutator series*
- *The matrices A_{ij} that express the action of each element e_j of some basis of Λ_0 on each Λ_i*
- *The matrix representatives $F_i : \Lambda_i \to \Lambda_i$.*

Then $N(f)$ is computed as follows:

(i) *Compute $coker(F_0 - I)$ and choose a set of representatives $\mathcal{T} \subset \Lambda_0$ of $coker(F_0 - I)$. Write each $v \in \mathcal{T}$ as $v = \sum n_j(v) e_j$*

(ii) *For each $i = 1, \ldots, N$ and each $v \in \mathcal{T}$, compute $A_i(v) = \prod_j A_{ij}^{n_j(v)}$*

(iii) *Then*

$$N(f) = \sum_{v \in \mathcal{T}} \prod_{i=1}^{N} |\det(A_i(v) F_i - I)| \,.$$

Note that, if no A_{ij} has a root of unity as an eigenvalue (i.e. S is an $\mathcal{NR}$-solvmanifold), then the action matrices may be omitted from the formula [**12**], which becomes

$$N(f) = |L(f)| = \prod_{i=0}^{N} |\det(F_i - I)|\,.$$

Of course, in this case, as in the nilmanifold case, $N(f)$ may then also be computed from the homology endomorphism f_*.

EXAMPLE 5.1. Consider the Klein bottle K, with fundamental group $\mathbb{Z} \rtimes \mathbb{Z}$. Clearly, $\Lambda_0 = \Lambda_1 = \mathbb{Z}$, with Λ_0 acting on Λ_1 by $A(n) = (-1)^n$. Any $f : K \to K$ has linearization $F_0 n = an, F_1 m = bm$. The commutativity condition $F_1 A(n) = A(F_0 n) F_1$ simply becomes $-b = (-1)^a b$, so we must either have $b = 0$ or a odd. If $b = 0$, then $\det(A(n)F_1 - I) = -1$ for all n and $N(f) = |\det(F_0 - I)| = |a-1|$. If $b \neq 0$, then a is odd and $\det(F_0 - I) = a - 1$ is even. Thus any transversal of $coker(F_0 - I)$ will consist of an even number of integers, half of which will be even, the other half odd. The even integers will each contribute $|b-1|$ to the sum; the odd integers will contribute $|-b-1|$ to the sum. Thus, the sum which computes $N(f)$ is $\frac{1}{2}|a-1|\,(|b-1| + |b+1|)$. Thus, for the Klein bottle at least, the algorithm produces a fairly simple formula for the Nielsen number:

$$N(f) = \frac{|a-1|}{2}\,(|b-1| + |b+1|)\,.$$

The next step, producing an algorithm for infrasolvmanifolds, is more complicated. The reason for this is that, in order to reduce infrasolvmanifold calculations to the solvmanifold calculations just described, we must employ liftings. But in so doing, the lifting condition may force us to lift $f : M \to M$ to $\tilde{f} : \tilde{M}_1 \to \tilde{M}_2$ with $\tilde{M}_1 \neq \tilde{M}_2$ and lift the identity map on M to a covering projection $p : \tilde{M}_1 \to \tilde{M}_2$. Then fixed points of f are covered not by fixed points, but by coincidences of $\tilde{f}$ and p. This requires us to move from Nielsen fixed point theory to Nielsen coincidence theory. While coincidence theory proceeds (for the most part) by generalizing the fixed point theory "in the obvious way," the ultimate results we require are more involved and less elegant than the corresponding fixed point theory results.

Of course, we don't need the full strength of coincidence theory here. We are in a rather special case in which the spaces involved, M_1 and M_2, are compact solvmanifolds of the same dimension, with M_1 a finite cover of M_2, and one of the two maps involved is a covering projection. This case has been studied in [**17, 18, 25**], and we will content ourselves here with a summary of those results and the algorithm they imply.

Suppose M is a compact infrasolvmanifold and $f : M \to M$ is a self map. If $\pi = \pi_1(M)$ and $\Gamma_2 \lhd \pi$ is a solvable normal subgroup of finite index in π, then we define $\Gamma_1 = \Gamma_2 \bigcap f_{\#}^{-1}(\Gamma_2)$, and construct manifolds $\tilde{M}_1, \tilde{M}_2$ such that $\Gamma_i = \pi_1(\tilde{M}_i)$. Let $p : \tilde{M}_1 \to \tilde{M}_2$ denote the covering projection, and let $\Phi_i = \pi/\Gamma_i$. Of course, Φ_2 acts on $\tilde{M}_2$ via covering transformations, so if $\tilde{f} : \tilde{M}_1 \to \tilde{M}_2$ is a lift of f, then so is $\beta\tilde{f}$ for all $\beta \in \Phi_2$. The essential result of

[18] is that

$$N(f) = \frac{1}{|\Phi_1|} \sum_{\beta \in \Phi_2} N(\beta \tilde{f}, p)$$

where $N(\beta \tilde{f}, p)$ is the Nielsen coincidence number of $\beta \tilde{f}$ and p. It remains to show how $N(\beta \tilde{f}, p)$ is computed. This is described in detail in [**17, 25**]. In broad outline, it proceeds in the same manner as the computation of Nielsen fixed point numbers on solvmanifolds: find fibrations of $\tilde{M}_1$ and $\tilde{M}_2$ so that f and p are (up to homotopy) fiber-preserving. A succesion of such fibrations are constructed, with the base of each a torus. The induced maps on the tori may be taken to be homomorphisms, whose Nielsen coincidence number is computed from the determinant of their difference. All of this combines to produce the following

ALGORITHM 5.3 ($N(f)$ FOR INFRASOLVMANIFOLDS). *Suppose M is an infrasolvmanifold. Then computation of the Nielsen number $N(f)$ for a self map $f : M \to M$ requires knowledge of $\pi = \pi_1(M)$ and $f_\# : \pi \to \pi$. Then $N(f)$ is computed as follows*

(i) *Find the solvradical π_S of π. Let $\Gamma_2 = \pi_S, \Gamma_1 = \Gamma_2 \bigcap f_\#^{-1}(\Gamma_2)$ and $\Phi_i = \pi/\Gamma_i$.*

(ii) *Find $\Gamma_{21} = C([\Gamma_2, \Gamma_2]), \Gamma_{2j} = C([\Gamma_{2j-1}, \Gamma_{21}])$.*

(iii) *Find $\Gamma_{1j} = \Gamma_{2j} \bigcap f_\#^{-1}(\Gamma_{2j})$.*

(iv) *If any Γ_{1j} does not have finite index in Γ_{2j}, the Nielsen number of f is then 0.*

(v) *If every Γ_{1j} has finite index in Γ_{2j}, proceed with the computation of the linearizations. Compute $\Lambda_{ij} = \Gamma_{ij}/\Gamma_{ij+1}$. Choose bases for each Λ_{2j} and find matrix representatives $\{B_j(\beta)\}$ for the action of $\beta \in \Phi_2$ on each Λ_{2j}. Find matrix representatives $\{B_{jk}\}$ for the action of the basis elements e_k of Λ_{20} on each Λ_{2j}.*

(vi) *Find the linearizations of $f_\# : \Gamma_1 \to \Gamma_2$ and $p_\# : \Gamma_1 \to \Gamma_2$. That is, choose a basis for each Λ_{1j} and find matrix representatives $F_j, P_j : \Lambda_{1j} \to \Lambda_{2j}$.*

(vii) *For each $\beta \in \Phi_2$, compute $coker(B_0(\beta)F_0 - P_0)$ and choose a set of representatives $\mathcal{T}_\beta \subset \Lambda_{20}$ of $coker(B_0(\beta)F_0 - P_0)$. Write each $v \in \mathcal{T}_\beta$ as $v = \sum n_k(v) e_k$*

(viii) *For each $j = 1, \ldots, N$ and each $v \in \mathcal{T}_\beta$, compute $B_j(\beta, v) = B_j(\beta) \circ \left(\prod_k B_{jk}^{n_k(v)}\right)$*

(ix) *Then*

$$N(f) = \frac{1}{|\Phi_1|} \sum_{\beta \in \Phi_2} \sum_{v \in \mathcal{T}_\beta} \prod_{j=1}^{N} |\det(B_j(\beta, v)F_j - P_j)|.$$

If M is an infranilmanifold and the nilradical is used in place of the solvradical, then the matrices B_{jk} are all trivial and $B_j(\beta, v) = B_j(\beta)$. The sum over $\mathcal{T}_\beta$

then drops out, and

$$N(f) = \frac{1}{|\Phi_1|} \sum_{\beta \in \Phi_2}' \prod_{j=0}^{N} |\det(B_j(\beta)F_j - P_j)|.$$

6. Conclusion

So where do we go next? What spaces are likely candidates for new algorithms for Nielsen numbers? We can draw some hints for new directions by looking for common features in the spaces just described. The spaces considered in this survey were:

- Spaces with finite fundamental group.
- Jiang spaces, which necessarily have abelian fundamental group.
- Infrasolvmanifolds, which have poly{cyclic or finite} fundamental group.

All of these have well-behaved fundamental groups, and for all of them, it is the structure of the fundamental group that allows the analysis to proceed. At the opposite extreme, surfaces with negative Euler characteristic, which are among the most difficult spaces to compute Nielsen numbers on, have fundamental groups which contain free groups. This suggests that the most promising spaces to consider are those with amenable fundamental groups [**24**]. It is "almost true" that aspherical manifolds with amenable fundamental group are infrasolvmanifolds. A promising direction then would be to try to extend the results on infrasolvmanifolds to aspherical manifolds with amenable fundamental group. If this is to be done in the same "bootstrap" manner that has carried us from tori to infrasolvmanifolds, then it is likely that more Nielsen-theoretic tools (along the lines of the Nielsen theory for fibrations and finite covers) will have to be developed.

Another promising direction is to consider how Nielsen theory behaves with respect to other constructions: joins, wedge products, mapping cones and cylinders, etc. If the Nielsen theory of such constructions is tractable, then a space so constructed from Nielsen-computable spaces should itself be Nielsen-computable. For example, there is some evidence that there is an algorithm for Nielsen numbers on wedges of tori and spheres. Admittedly, such wedge spaces would not represent a very significant addition to the catalogue of computable spaces, but it does suggest a direction of growth for Nielsen theory.

References

1. L. Auslander, *Bieberbach's theorems on space groups and discrete uniform subgroups of Lie groups*, Ann. of Math. **71** (1960), 579–590.
2. ———, *Discrete uniform subgroups of solvable Lie groups*, Trans. Amer. Math. Soc. **99** (1961), 398–402.
3. ———, *An exposition of the structure of solvmanifolds. I: Algebraic theory*, Bull. Amer. Math. Soc. **79** (1973), 227–261.
4. L. Auslander, F. E. A. Johnson, *On a conjecture of C. T. C. Wall*, J. London Math. Soc. (2) **14** (1976), 331–332.

5. D. V. Anosov, *The Nielsen number of maps of nil-manifolds*, Russian Math. Surveys **40** (1985), 149–150.
6. R. Brooks, R. Brown, J. Pak, and D. Taylor, *Nielsen numbers of maps of tori*, Proc. Amer. Math Soc. **52** (1975), 398–400.
7. L. Charlap, *Bieberbach Groups and Flat Manifolds*, Springer-Verlag, New York and Berlin, 1986.
8. E. Fadell and S. Husseini, *On a theorem of Anosov on Nielsen numbers for nilmanifolds.* Nonlinear Functional Analysis and its Applications (Maratea,1985), NATO Adv. Sci. Inst. Ser. C: Math. Phys. Sci., **173** , Reidel Dordrecht, Boston, Mass., 1986, 47–53.
9. F. T. Farrell, L. E. Jones, *Classical Aspherical Manifolds* CBMS Lecture Notes, # 75, Amer. Math. Soc., Providence, R.I. 1990
10. B. Jiang, *Estimation of Nielsen numbers*, Acta Math. Sinica **14** (1964), 304–312.
11. ———, *Lectures on Nielsen fixed point theory*, Contemp. Math. **14** Amer. Math. Soc., Providence, RI 1983.
12. E. Keppelmann and C. McCord, *The Anosov theorem for exponential solvmanifolds*, Pac. J. Math. (to appear).
13. S. Kwasik and K. B. Lee, *The Nielsen numbers of homotopically periodic maps of infranil-manifolds*, J. London Math. Soc.(2) **38** (1988), 544–554.
14. K. B. Lee, *Nielsen numbers of periodic maps on solvmanifolds*, Proc. Amer. Math. Soc. **116** (1992), 575–579.
15. A. Mal'cev, *On a class of homogeneous spaces*, Amer. Math. Soc. Transl. (2) **39** (1951), 276–307.
16. C. McCord, *Nielsen numbers and Lefschetz numbers on solvmanifolds*, Pac. J. Math. **147** (1991), 153–164.
17. ———, *Lefschetz and Nielsen coincidence numbers on nilmanifolds and solvmanifolds*, Topology & Appl. **43** (1992), 249–261.
18. ———, *Estimating Nielsen numbers on infrasolvmanifolds*, Pac. J. Math. **154** (1992), 345–368.
19. ———, *Nielsen numbers for homotopically periodic maps on infrasolvmanifolds*, Proc. Amer. Math. Soc. (to appear).
20. G.D. Mostow, *Factor spaces for solvable groups*, Ann. of Math. **60** (1954), 1–27.
21. J. Oprea, *A homotopical Conner-Raymond theorem and a question of Gottlieb*, Can. Math. Bull. **33** (1990), 219–229.
22. J. Oprea, J. Pak, *Principal bundles over tori and maps which induce the identity on homotopy*, Top. & Appl. (to appear).
23. J. Pak, *On the fibered Jiang spaces*, Contemp. Math. **72** (1988), 179–181.
24. A. Paterson, *Amenability* Mathematical Surveys and Monographs **29** American Mathematical Society: Providence, R.I. 1988.
25. A. Rarivoson, *Calculation of the Nielsen coincidence number on solvmanifolds* Ph.D. Thesis, University of Cincinnati, 1993.
26. J. Siegel, *G-spaces, H-spaces and W-spaces*, Pac. J. Math. **31** (1969), 209–214.
27. H.-C. Wang, *Discrete subgroups of solvable Lie groups* I, Ann. of Math. **64** (1956), 1–19.
28. C.-Y. You, *Fixed point classes of a fiber map*, Pac. J. Math. **100** (1982), 217–241.

INSTITUTE FOR DYNAMICS, DEPARTMENT OF MATHEMATICS, UNIVERSITY OF CINCINNATI, CINCINNATI, OHIO 45221-0025

E-mail address: mccord@cmccord.csm.uc.edu, chris.mccord@uc.edu

Contemporary Mathematics
Volume **152**, 1993

The Structure of Isolated Invariant Sets and the Conley Index

KONSTANTIN MISCHAIKOW

1. Introduction

> " ... many significant properites of the flow are reflected in the existence of isolating neighborhoods, or perhaps more accurately, in the companion isolated invariant set... This is true in some generality of those properties which are stable under perturbation" *C. Conley* **[2]**

Coarsely put, *if it is observable, then it can be isolated.* If one accepts this premise, then it is natural to try to study dynamical systems in terms of those structures and properties which can be detected through isolating neighborhoods. To a large extent this is the basis for Conley's theory of dynamical systems.

Recall that given a flow $\varphi : R \times X \to X$ on a locally compact space X an *isolating neighborhood* is a compact set N such that its maximal invariant set

$$\mathrm{Inv}(N, \varphi) := \{x \in N \mid \varphi(R, x) \subset N\} \subset int(N)$$

and S is an *isolated invariant set* if $S = Inv(N)$ for some isolating neighborhood N. From the definition it is easy to see that isolating neighborhoods are robust objects, i.e. under any sufficiently small continuous perturbation of the flow an isolating neighborhood remains an isolating neighborhood. After a little reflection, one can convince oneself that given a "typical" flow, the typical compact set is in fact an isolating neighborhood. Thus isolating neighborhoods are nice objects; one can hope to find them and once found they have a tendency to stay put. Unfortunately, as is remarked above, the object of interest is really the companion isolated invariant set. Thus, the fundamental problem with Conley's approach is:

1991 *Mathematics Subject Classification.* 34C35,54H20.
Research was supported in part by NSF Grant DMS-9101412.
This paper is in final form and no version of it will be submitted for publication elsewhere.

N is what you have, but Inv(N) is what you want.

The purpose of this paper is to present results, most of them recent, which indicate that the Conley index of $\text{Inv}(N)$ can be used to obtain information concerning the dynamics on $\text{Inv}(N)$. In particular, there are two types of results which shall be presented:

(i) basic existence results, i.e. there exist fixed points or periodic orbits in $\text{Inv}(N)$,

(ii) lower bounds on the complexity and structure of the dynamics over all of $\text{Inv}(N)$.

Obviously, isolated invariant sets can be extremely complicated objects. Thus it is reasonable to ask whether there is a natural means of decomposing them, while preserving the perturbation properities of the corresponding isolating neighborhoods. This leads to Conley's idea of a *Morse decomposition* of an invariant set S, which is denoted by

$$\mathcal{M}(S) = \{M(p) \mid p \in \mathcal{P}\}.$$

A Morse decompositon consists of at most a finite number of disjoint compact invariant subsets of S which contain the chain recurrent set of S. These individual isolated invariant subsets $M(p)$ are called *Morse sets*, and the remaining portion $S \setminus \bigcup M(p)$ is referred to as the set of connecting orbits.

As a decomposition $\mathcal{M}(S)$ is very robust with regards to perturbations, though this is not meant to imply that as sets the Morse sets remain unchanged. However, given this decomposition one is naturally led to the following goals.

Q1 *Describe the structure of the Morse sets.*

Q2 *Describe the structure of the set of connecting orbits.*

In this setting it is natural to view **Q1** as a "local" question and **Q2** as a "global" question.

Before entering into any details, a brief discussion of what is meant by "describe" is necessary. Non-linear systems can exhibit extremely complicated dynamics, and the structure of the dynamics can change dramatically as a function of natural parameters. A generally accepted means of describing these structures is to prove the existence of a conjugacy with respect to a well understood or more easily computable system, e.g., the invariant set of the Smale horseshoe and the shift dynamics on bi-infinite sequences of two symbols. While from a qualitative and theoretical point of view such a description is quite desirable, in practice there are at least two serious drawbacks to this approach. First, starting with a specific evolution equation obtaining a rigorous proof of the desired conjugacy can be extremely difficult. Second, even if a conjugacy for a fixed parameter value is given, determining the range for which the conjugacy is valid or how the conjugacy changes as a function of the parameter is a daunting task. These drawbacks can be a serious handicap. For many models arising from the sciences and engineering, not only are the parameters not known to great precision, but often the class of potential non-linearities is large. Therefore one is faced with the task of providing a rigorous description of the dynamics for classes of equations for which one expects (or for which the numerics indicates) a wide range of behaviors.

Given the problem of trying to describe a potentially diverse set of dynamical structures simultaneously, it seems reasonable to forsake the idea of establishing a conjugacy with a simpler system. Instead, the more modest goal will be to obtain a semi-conjugacy; in other words to establish a minimal dynamic structure which is present for all dynamics over a specified range of parameter values. It is hoped that the advantages and disadvantages of this approach will be made clearer in later sections of this paper.

Section 2 provides a very brief description of the Conley index, primarily as a means of introducing the notation which will be used throughout this paper. Section 3 describes existence results for fixed points and periodic results. Section 4 shows that the existence of chaotic dynamics can be obtained via index information. And finally, in Section 5 we turn to questions about global dynamics.

A word of warning, this is not intended to be a comprehensive survey article, in part because there are too many basic results which have not yet been proven. Furthermore, in many cases the the most general result is not stated, and hence, readers who need more details are strongly recommended to examine the original sources for the results.

Acknowledgements: Much of what is described in this paper arises from joint work with C. McCord, M. Mrozek, and T. Gedeon. Numerous discussions with them helped form the basis of what is presented here, namely the author's interpretation of these results and the author's belief of what constitutes interesting future questions.

2. The Conley Index

A quick review of the Conley index is presented in this section. The goal is to introduce the notation which will be used in the later sections, rather than to explain the ideas. Basic references for the index theory are [**1, 7, 8, 15, 17, 18, 19, 20, 22**]. The Conley index is a topological invariant of isolated invariant sets. Originally defined for flows it has been extended to case of maps. We shall make use of the index in both settings.

Remark: For simplicity of presentation, throughout this paper is will be assumed that the space X on which the flow or map is defined is a local compact, metric ANR.

Given a flow $\varphi : R \times X \to X$, an *index pair* for the isolated invariant set S is a compact pair (N, L) with $L \subset N$ such that:

(i) $\overline{N \setminus L}$ is an isolating neighborhood for S and $S \subset int(N \setminus L)$.

(ii) L is *positively invariant* in N, i.e. if $x \in L$ and $\varphi([0,T],x) \subset N$, then $\varphi([0,T],x) \subset L$.

(iii) L is an *exit set* for N, i.e. if $x \in N$ and $\varphi(T,x) \notin N$, then $\varphi(t,x) \in L$ for some $0 < t < T$.

An index pair is *regular* if, in addition, $\tau : N \to [0,\infty)$ defined by

$$\tau(x) = \begin{cases} \sup\{t > 0 | x \cdot [0,t] \subset N \setminus L\} & \text{if } x \in N \setminus L \\ 0 & \text{if } x \in L \end{cases}$$

is a continuous function. Thus, for a regular index pair, L is a neighborhood deformation retract (along flow lines) in N. Regular index pairs always exist and, furthermore, the homotopy type of the quotient space N/L is independent of the index pair chosen. Therefore, the homotopy type of N/L is an invariant of S; it is called the Conley index of S and is denoted by $h(S)$.

Ordering regular index pairs by inclusion results in an inverse system of index pairs $\{H^*(N_\alpha, L_\alpha)\}$, with the inclusion induced cohomology map $H^*(N_\alpha, L_\alpha) \to H^*(N_\beta, L_\beta)$ an isomorphism for every $\beta < \alpha$. The inverse limit of this system, denoted $CH^*(S)$, is the cohomology Conley index of S. Since each bonding map in the system is an isomorphism, $CH^*(S) \cong H^*(N_\alpha, L_\alpha)$ for every α. That is, the cohomology of any index pair represents the cohomology Conley index.

In the case of normally hyperbolic invariant sets the Conley index is easy to compute.

PROPOSITION 2.1. *If S is a normally hyperbolic invariant set for a flow on a manifold with orientable unstable manifold of (normal) dimension u, then $CH^{q+u}(S) \cong H^q(S)$.*

Turning now to maps, let $f : X \to X$ be a continuous function. The definition of an index pair, while similar in spirit, is slightly more technical. Let $\tilde{N}$ be an isolating neighborhood and let $S = \text{Inv}(\tilde{N})$. (N, L) is a compact pair with $L \subset N$ is an *index pair* for S if:

(i) $\text{Inv}(\tilde{N}) \subset int(N \setminus L)$;

(ii) given $x \in N$ or $x \in L$ and $f(x) \in \tilde{N}$, then $f(x) \in N$ or $f(x) \in L$, respectively;

(iii) given $x \in N$ and $f(x) \notin \tilde{N}$, then $x \in L$.

Unlike the case for flows the homology of the pair (N, L) is *not*, in general, an invariant of S. To obtain the index, we begin with the observation that by excision, the inclusion map $i : (N, L) \to (N \cup f(L), L \cup f(L))$ induces an isomorphism on the Alexander–Spanier cohomology of the pairs. The composition

$$H^*(N, L) \xrightarrow{(i^*)^{-1}} H^*(N \cup f(L), L \cup f(L)) \xrightarrow{f} H^*(N, L)$$

is called the *index map* for f and denoted by F. The generalized kernel of F is defined by

$$GK(F) := \{v \in H^*(N, L) \mid F^n(v) = 0 \text{ for some } n > 0\}.$$

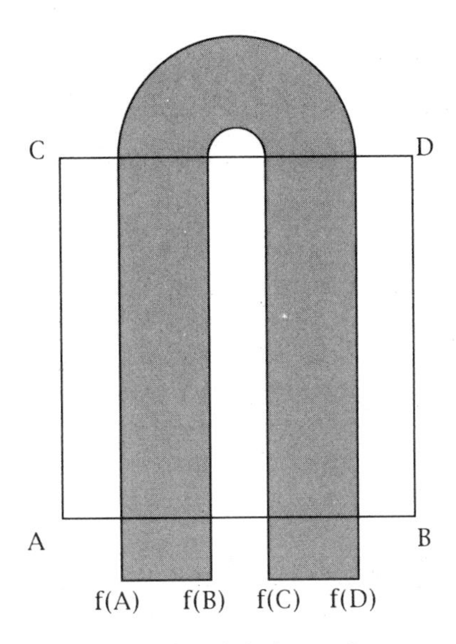

FIGURE 1. Smale's horseshoe map.

Let

$$CH^*(S) := \frac{H^*(N,L)}{GK(F)},$$

and observe that F induces an isomorphism (in our setting $H^*(N,L)$ is finite dimensional) on $CH^*(S)$ which is denoted by $\chi^*(S)$. The Conley index of S is this pair, i.e.

$$\mathrm{Con}(S) := (CH^*(S), \chi^*(S)).$$

This is an invariant of S **[15]**.

The following result follows directly from the definition.

PROPOSITION 2.2. *Let S be a hyperbolic fixed point with unstable manifold of dimension n, then*

$$Con^k(S; Z_2) = \begin{cases} (Z_2, Id) & \text{if } k = n \\ 0 & \text{otherwise} \end{cases}$$

A more interesting calculation is the following, where the Conley index of the Smale horseshoe is computed. The horseshoe map is shown in Figure 2. An index pair (N, L) for $\mathrm{Inv}(N)$ is shown in Figure 2, along with chains α and β which generate $H_*(N, L; Z_2)$. Observe that

$$f_* = \begin{bmatrix} 1 & 1 \\ 1 & 1 \end{bmatrix}$$

and hence, $f_*^2 = 0$, i.e.

$$\mathrm{Con}^*(\mathrm{Inv}(N); Z_2) \approx 0.$$

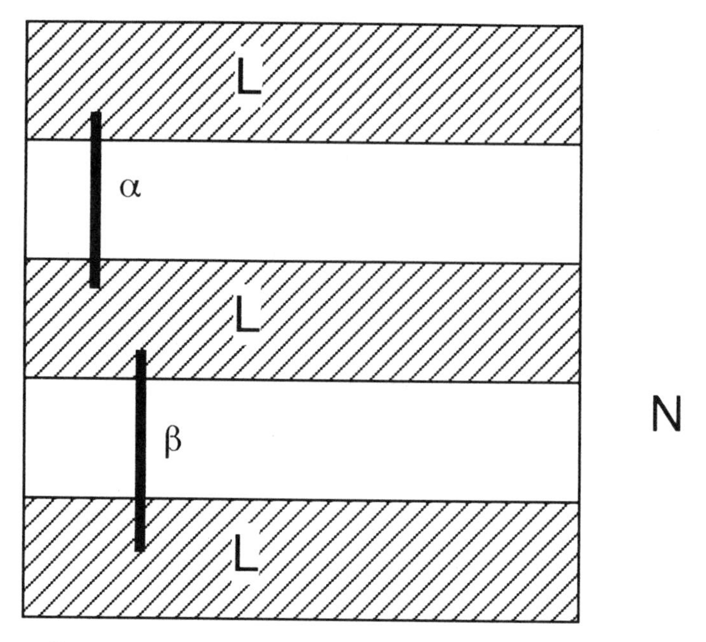

FIGURE 2. An index pair for the horseshoe.

3. Simple Dynamics

Fixed point theory is a well developed subject. Because of its importance both very general and problem-specific techniques exist. Thus, one should not expect too much from a fixed point theorem based only on the Conley index; it is easy to imagine that any result which can be proven with such a theorem can also be obtained via a more standard approach. However, we begin with a fixed point result because it is an easily understood example which demonstrates both the potential and the limitations of the Conley index theory.

Let us begin by considering the ideal result. Let N be an isolating neighborhood for a flow φ and assume that

$$CH^k(\mathrm{Inv}(N)) \approx \begin{cases} Z & \text{if } k = n \\ 0 & \text{otherwise,} \end{cases}$$

i.e. $\mathrm{Inv}(N)$ has the index of a hyperbolic fixed point with unstable manifold of dimension n. Then, $\mathrm{Inv}(N)$ contains a fixed point with an n-dimensional unstable manifold. Observe that we are asking for *both* an existence result and a "stability" result. As is demonstrated in Figure 3, such a theorem is impossible. A mere existence result does, however, exist.

THEOREM 3.1. [McCord [**7**]] *If*

$$CH^*(Inv(N)) \approx Z$$

then $Inv(N)$ contains a fixed point.

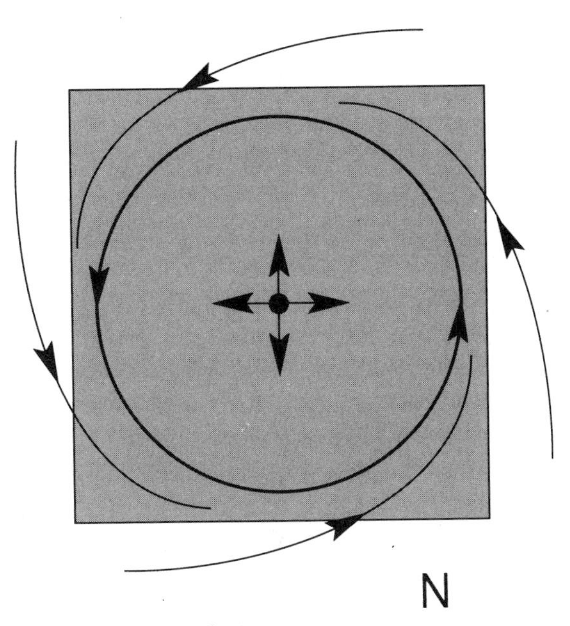

FIGURE 3. $h(\text{Inv}(N)) \sim \Sigma^0$ but the only fixed point has index Σ^2.

The proof of this theorem is a straightforward application of the Lefschetz fixed point theorem once one shows, as was done in [**7**], that CH is a homology theory. While, as was suggested above, this theorem may be of limited interest, the implications of the proof are important. McCord showed that CH can be viewed as a functor from the category of isolated invariant sets to the category of graded abelian groups. This theorem shows that some of the dynamic properties of the invariant set are preserved by this functor, i.e. the algebra indicates the existence of fixed points. The functorial view of the index has been most fully described in [**17**], and much of what follows in this and other the sections can be viewed as an attempt to understand what dynamic properties are preserved by these functors and how to recover them.

The subtlety of this problem can already be seen in the existence question for periodic orbits. A naive approach would begin with the following conjecture. Given an isolating neighborhood N with the index of a hyperbolic periodic orbit, then there exists a periodic orbit in $\text{Inv}(N)$. However, as the example shown in Figure 3 indicates, this obviously false. The work of Fuller [**3**] and Schweitzer [**23**] shows that at a minimum one needs to assume the existence of a return map on the isolating neighborhood. To make these comments precise, let Ξ be a local section of the flow φ. Ξ is a *Poincare section* for N if $\Xi \cap N$ is closed and for any $x \in N$, there exist times $t_x^- < 0 < t_x^+$ such that

$$\varphi(t_x^\pm, x) \in \Xi.$$

This leads to the following result.

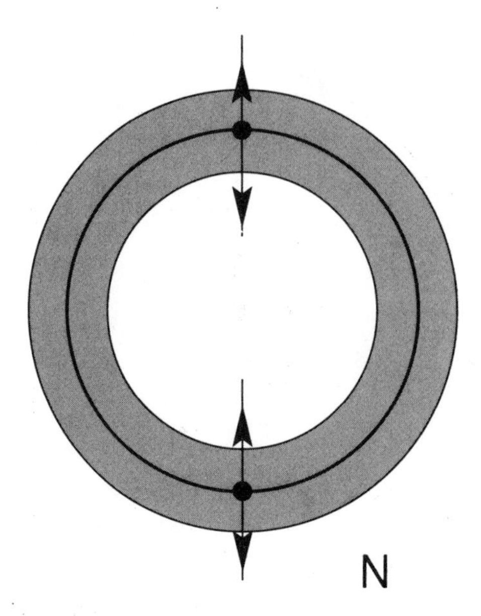

FIGURE 4. $h(\text{Inv}(N)) \sim \Sigma^1 \vee \Sigma^2$ but there is no periodic orbit contained in $\text{Inv}(N)$.

THEOREM 3.2. [C. McCord, K. Mischaikow, and M. Mrozek [**10**]] *Assume N has a Poincare section and*

$$CH^k(Inv(N), F) \approx \begin{cases} F & \textit{if } k = n, n+1 \\ 0 & \textit{otherwise} \end{cases}$$

Then, $Inv(N)$ contains a periodic orbit.

sketch of proof: Let $S = \text{Inv}(N)$. The proof is done in five steps.
(Step 1) The existence of a Poincaré section Ξ allows one to define a continuous first return map which is roughly of the form

$$f : \Xi \to \Xi.$$

Observe that one can view $\Xi \cap N$ as an isolating neighborhood for the discrete dynamical system generated by $f : \Xi \to \Xi$. Let $S_\Xi = \text{Inv}(N \cap \Xi, f)$. The first step is to relate the index of the map $CH^*(\Xi \cap N, f)$ with the index of the flow $CH^*(N, \varphi)$. Using the forward images of the section one can construct a Meyer-Vietoris decomposition of N and the corresponding long exact sequence. Written in terms of the Conley indexes one obtains the following long exact sequence.

$$\to CH^k(S, \varphi) \to CH^k(S_\Xi, f) \stackrel{id - \chi^k}{\to} CH^k(S_\Xi, f) \to CH^k(S, \varphi) \to .$$

(Step 2) Let $s_k = \text{dim} CH^k(S)$, $d_k = dimker(id - \chi^k)$ and observe that $s_k =$

$d_k + d_{k-1}$. By hypothesis

$$s_k = \begin{cases} 1 & \text{if } k = n, n+1 \\ 0 & \text{otherwise} \end{cases}$$

This implies that

$$d_k = \begin{cases} 1 & \text{if } k = n \\ 0 & \text{otherwise} \end{cases}$$

(Step 3) Recall that the *zeta function* for $f : \Xi \to \Xi$ is given by

$$\zeta_{(\Xi,f)}(t) := \exp\left(\sum_{m=0}^{\infty} \frac{i(\Xi, f^m)}{m} t^m\right)$$

where $i(\Xi, f^m)$ is the fixed point index of f^m on Ξ. The important observation is that $\zeta_{(\Xi,f)}(t) \neq 1$ implies that f has a periodic point in Ξ.

(Step 4) While the definition is step 3 makes is clear how the zeta function is related to the existence of periodic orbits, it is not of much use with regards to computation. However, M. Mrozek **[16]** has shown that the zeta function is rational, in particular

$$\zeta_{(\Xi,f)}(t) = \prod_{n=0}^{\infty} [\det(\mathrm{Id} - t\chi^n)]^{(-1)^{n+1}} \tag{1}$$

$$= \frac{\det(\mathrm{Id} - t\chi^1)}{\det(\mathrm{Id} - t\chi^2)} \cdot \frac{\det(\mathrm{Id} - t\chi^3)}{\det(\mathrm{Id} - t\chi^4)} \cdot \dots \tag{2}$$

(Step 5) Now steps 2 and 4 combine to give that

$$\zeta_{(\Xi,f)}(t) \neq 1.$$

Remark: Observe that the proof shows that f has a periodic point, which of course corresponds to a periodic orbit in the flow. However, there is no information concerning the period (with respect to f) of the orbit. A more detailed discussion of this problem can be found in **[10]**, for the moment we restrict ourselves to the remark that by carefully choosing a compact set Ξ and map $f : \Xi \to \Xi$ it is fairly easy to construct examples where the flow resulting from the suspension has an isolating neighborhood N with the homology index of a hyperbolic periodic orbit, but such that the cross section and corresponding Poincaré section has no fixed point. Thus Theorem 3.2 is sharp.

Recall that Conley's original definition of the index was in terms of the homotopy type of a space. This is a stronger invariant than the cohomological index condition assumed in Theorem 3.2. Thus we can pose the following question:

Assume that the flow $\varphi : R^n \times R \to R^n$. Let N be an isolating neighborhood with the homotopy index of a hyperbolic periodic orbit, and with a Poincaré section Ξ and corresponding Poincaré map f. Does Inv(N) contain a periodic orbit which is a fixed point of f

In the case where the homotopy index is $(S^1, *)$, both B. Jiang and R. Geoghegan have, independently, shown the answer to be yes, provided the section Ξ is connected (the proof will appear in [**10**]). In any other case the question remains open.

Having raised the question of applicability, a natural question is whether one could expect to find an isolating neighborhood N, compute its index, and verify that it has a Poincaré section, without already knowing the existence of a periodic orbit by some other means. In a forthcoming paper [**14**] some examples based on singular perturbation techniques which demonstrate that this can be done will be presented.

A final remark, this discussion has ignored the important work of C. Conley, E. Zehnder, A. Floer, and others (see [**21**] and references therein) for variational systems where the cup length estimates of the index have been used to obtain lower bounds on the number of fixed points.

4. Chaotic Dynamics

Extensive numerical experimentation indicates that complicated or chaotic dynamics is a common feature of nonlinear dynamical systems. Theoretical work has demonstrated that such dynamics can persist over open sets of parameter space. However, beginning with a differential equation, the number of systems for which chaotic dynamics has been proven is surprising small. Furthermore, most of the problems for which rigorous results are known involve perturbations from rather special dynamics, i.e. low dimensional or integrable systems. There are two reasons for this. First, the existence of chaotic dynamics is a problem of global analysis. Second, the fine structure of the dynamics is determined in part by global bifurcations, and there are few tools for detecting such changes.

Fixed point theory stands in dramatic contrast. There are numerous results which guarentee the existence of fixed points to highly nonlinear functions defined on a wide variety of topological spaces. Its effectiveness is due to two properties: the existence of algebraic fixed point indexes, and the invariance of this index under homotopy.

In this section we shall describe a set of algebraic invariants which imply the existence of chaotic dynamics and which posses the same continuation properties as the Conley index. The archetype for our result lies in the Smale horseshoe described in Section 2. An important first observation is that the index of the total isolating neighborhood is not in general sufficient information. In Section 2 it was shown that the index of the horseshoe is trivial, yet our goal is to provide theorems which indicate that it contains chaotic dynamics.

The key to understanding the dynamics exhibited by the horseshoe is to observe that it consists of two hyperbolic fixed points each of whose stable and unstable manifolds intersect transversely. Of course, this information is much finer than that which is carried by the Conley index. Let us therefore transform

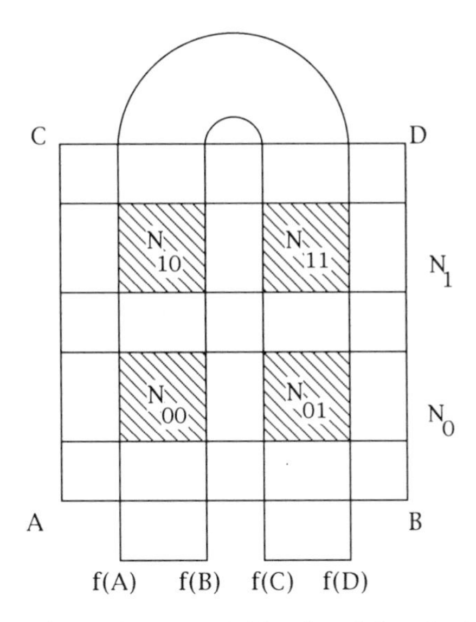

FIGURE 5. An isolating neighborhood for the horseshoe.

this information into the language of Conley. Referring to Figure 4 it is clear that if one defines

$$M = N_{00} \cup N_{01} \cup N_{10} \cup N_{11}$$

then M is an isolating neighborhood and $\text{Inv}(M) = \text{Inv}(N)$. Obviously, the fixed points of the horseshoe lie in the sets N_{00} and N_{11} which have the indexes of hyperbolic fixed points, in this case saddles. Furthermore, $N_{00} \cup N_{10} \cup N_{11}$ and $N_{00} \cup N_{01} \cup N_{11}$ are isolating neighborhoods whose invariant sets have attractor repeller pairs of the form $(\text{Inv}(N_{11}), \text{Inv}(N_{00}))$ and $(\text{Inv}(N_{00}), \text{Inv}(N_{11}))$, respectively. For the horseshoe we know that these attractor repeller pairs are related by transverse connecting orbits and that it is this transversality which leads to the chaotic structure. One of the most highly exploited aspects of the Conley index is that the algebra indicates the existence of connecting orbits between attractor repeller pairs. In particular, in a cohomological setting there is a map from the cohomological index of the attractor to the cohomological index of the repeller, and the nontriviality of this map implies the existence of a connecting orbit. In applications, the effectiveness of the index arises from the fact that transversality need not be explicitly demonstrated.

With this simple translation of information one can already give the idea of the proof of the existence of chaotic dynamics. Clearly,

$$CH^1(\text{Inv}(N_{kk}), Z_2) \approx Z_2, \qquad k = 0, 1$$

As will be shown later the existence of the connecting orbit from $\text{Inv}(N_{kk})$ to $\text{Inv}(N_{ll})$ results in a isomorphism from $CH^*(\text{Inv}(N_{ll})$ to $CH^*(\text{Inv}(N_{kk})$. Furthermore, $\chi^*(\text{Inv}(N_{kk}))$ is an isomorphism. Since these isomorphisms can be composed arbitrarily often, given any finite sequence of symbols of this set there exists an orbit with this dynamics. The closure of this set of orbits contains chaotic dynamics.

With the example of the horseshoe and this simple explanation in mind, we shall consider a slightly more general setting in which to formally state the results. This generality is employed inorder to facilitate a discussion of the limitations of the current results.

Let $N \subset X$ be an isolating neighborhood under $f : X \to X$ which decomposes into K disjoint compact subsets i.e.,

$$N = \{N_k \mid k = 1, \ldots, K\}.$$

Let

$$\mathcal{A} = \left[\alpha_{ij}\right]$$

be a $K \times K$ matrix with entries $\alpha_{ij} \in \{0, 1\}$. Define

$$\Sigma_{\mathcal{A}} \subset \prod_{-\infty}^{\infty} \{1, \ldots, K\}$$

by

$$\Sigma_{\mathcal{A}} = \{a = (\ldots, a_{-1}, a_0, a_1, \ldots) \mid \alpha_{a_n a_{n+1}} = 1\}$$

and let $\sigma : \Sigma_{\mathcal{A}} \to \Sigma_{\mathcal{A}}$ be given by

$$\sigma(a)_n = a_{n+1}.$$

The dynamical system $(\Sigma_{\mathcal{A}}, \sigma)$ is called the subshift dynamics generated by $\mathcal{A}$. Our goal will be to map the dynamics of $\mathrm{Inv}(N)$ onto the appropriate subshift dynamics.

Let $\Sigma_{\mathcal{A}}^n \subset \Sigma_{\mathcal{A}}$ denote the collection of all subsequences of length n in $\Sigma_{\mathcal{A}}$ and let Σ_K^n denote all subsequences of length n on K symbols. Given $a = (a_0, \ldots, a_n) \in \Sigma_{\mathcal{A}}^n$ set

$$N_a := \bigcap_{i=0}^{n} f^i(N_{a_i})$$

and for any $A \subset \Sigma_{\mathcal{A}}^n$ define

$$N_A := \bigcup_{a \in A} N_a$$

where $N_A = \emptyset$, if $A \subset \Sigma_K^n \setminus \Sigma_{\mathcal{A}}^n$. To simplify the notation set

$$S_A := \mathrm{Inv}(N_A).$$

It is easy to check that for any $A \subset \Sigma_K^n$ the set N_A is an isolating neighborhood under f. Thus, N_k, N_l and $N_{\{kk,kl,ll\}}$ are isolating neighborhoods. This sets up the following theorem which relates the index automorphism to the existence of connecting orbits.

THEOREM 4.1. [K Mischaikow and M. Mrozek [**13**]]

(i) (S_k, S_l) *is an attractor-repeller pair in* $S_{\{kk,kl,ll\}}$.

(ii) *The Conley indices of* S_k, S_l *and* $S_{\{kk,kl,ll\}}$ *are related by the following commutative diagram with exact rows*

$$\begin{array}{ccccccccc} 0 & \to & CH^*(S_l) & \to & CH^*(S_{kk,kl,ll}) & \to & CH^*(S_k) & \to & 0 \\ & & \downarrow \chi(S_l) & & \downarrow \chi(S_{kk,kl,ll}) & & \downarrow \chi(S_k) & & \\ 0 & \to & CH^*(S_l) & \to & CH^*(S_{kk,kl,ll}) & \to & CH^*(S_k) & \to & 0 \end{array}$$

in which vertical arrows denote the corresponding index maps.

(iii)

$$CH^*(S_{\{kk,kl,ll\}}) = CH^*(S_k) \oplus CH^*(S_l).$$

(iv) *If*

$$\chi(S_{\{kk,kl,ll\}}) \text{ and } \chi(S_k) \oplus \chi(S_l) \text{ are not conjugate}$$

then there exists a connecting orbit from S_l *to* S_k *in* $S_{\{kk,kl,ll\}}$, *i.e. an element* $x \in S_{\{kk,kl,ll\}}$ *such that* $\alpha(x) \subset S_k$ *and* $\omega(x) \subset S_l$.

Consider, now, the special case when for a fixed $m \in N$

$$\text{Con}^m(S_k) \approx (F, \text{Id}), \qquad k = 1, \dots, K. \tag{3}$$

Theorem 4.1.2 implies that in matrix form

$$\chi^m(S_{\{kk,kl,ll\}}) = \begin{bmatrix} 1 & \mu \\ 0 & 1 \end{bmatrix}, \qquad k, l = 1, \dots, K.$$

Since the Jordan form determines the conjugacy class of a matrix, the number

$$\alpha_{kl} := \begin{cases} 0 & \text{if } \mu = 0 \\ 1 & \text{otherwise} \end{cases}$$

is an invariant of the attractor-repeller pair (S_k, S_l) in $S_{\{kk,kl,ll\}}$.

In [**13**] α_{kl} was called the *connection number*, since $\alpha_{kl} \neq 0$ implies that $\chi^m(S_{\{kk,kl,ll\}})$ and $\chi^m(S_k) \oplus \chi^m(S_l)$ are not conjugate. Hence, by Theorem 4.1.4, if $\alpha_{kl} = 1$, then there exists a connection from S_l to S_k.

Observe that the horseshoe fails to satisfy (3) when the coefficients are rationals, because then $\chi^1(S_1) = -\text{Id}$. However, (3) is is satisfied using Z_2 coefficients.

The following theorem describes the dynamics on $\text{Inv}(N)$.

THEOREM 4.2. [K. Mischaikow and M. Mrozek [**13**]] *Let* N *be an isolating neighborhood. Assume that* N *is the disjoint union of compact sets* $\{N_k \mid k = 1, \dots, K\}$ *and that for* $k = 1, \dots, K$

$$Con^m(S_k) \approx Con^m(Inv(N_k)) \approx (G, Id).$$

Let α_{kl} denote the connection number obtained from $N_{\{kk,kl,ll\}}$ and set $\alpha_{kk} = 1$. Define $\mathcal{A} = \left[\alpha_{kl}\right]$ to be the resulting $K \times K$ matrix. If $\alpha_{kl} = 0$ implies that $C(S_l, S_k) = \emptyset$, then there exists $d \in N$ and a continuous surjection

$$\rho : Inv(N) \to \Sigma_{\mathcal{A}}$$

such that the following diagram commutes

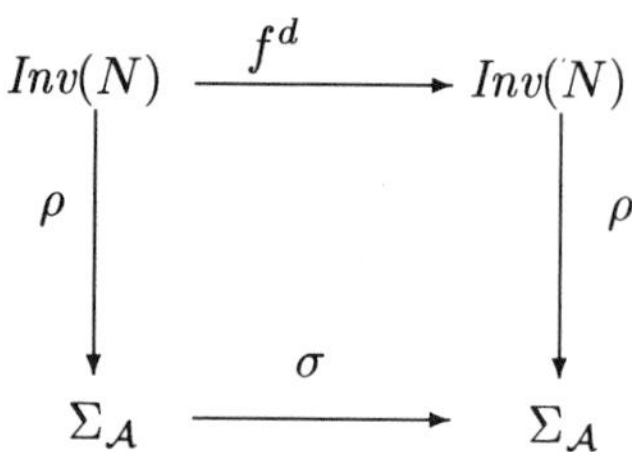

Several remarks need to be made at this point.

(i) Observe that ρ defines a semiconjugacy from $\mathrm{Inv}(N)$ onto the subshift dynamics determined by the connection numbers. Thus Theorem 4.2 provides a lower bound, rather than an exact description, of the dynamics on $\mathrm{Inv}(N)$. For those versed in fixed point theory this should come as no surprise; the fixed point index only guarentees the existence of a fixed point, other arguments are needed to obtain uniqueness.

(ii) An outline of the proof would be too lengthy to present here, however, a key point which needs to be addressed is the introduction of the connection number. In the brief idea of proof given above it was suggested that the map $\chi^*(S_{kk,kl,ll})$ could be used to map the generators of the "fixed points" around, thereby guarenteeing the existence of trajectories with the desired symbols. Unfortunately, the map $\chi^*(S_{kk,kl,ll})$ is not naturally defined. This goes back to the fundamental problem of applying the Conley index. Recall that for maps we obtain the cohomological index by quotienting by the generalized kernel of the index map acting on the index pair. In doing so we lose the topological and geometric information. The problem is that in the theorem one begins with the algebraic information and needs to recover the dynamics. It is still not understood what information is is lost in the process of obtaining the algebraic index. The connection number is introduced in order to circumvent this problem.

(iii) The caveat "$\alpha_{kl} = 0$ implies that $C(S_l, S_k) = \emptyset$" in Theorem 4.2 is essential. One need only consider the example of the horseshoe using rational coefficients. A reasonable concern is how adversely this assumption effects the applicability of the theorem. Two responses are possible. The first is that one can hope to verify this condition directly. In particular, if there is no connecting orbit between two invariant sets then a finite number of iterations of the map will indicate this fact. Of course, the finite number may be large, however, the nature of the problem is fundamentally different from proving the existence of a connecting orbit. The second response is that the condition is often vacuously true; for example in the horseshoe.

The last remark points to an important unsolved problem. Much of the interest in chaotic dynamics in the sciences and engineering is generated by strange attractors. Theorem 4.2 cannot be applied to an attractor in its entirety since the assumptions about the non-zero values of the connection numbers is essential. This forces $\mathrm{Inv}(N)$ to be an unstable invariant set. To obtain a theorem based on the Conley index which can describe strange attractors will need to make greater use of the index map χ and probably require the additional information carried by the homotopy index [**17**].

5. Global Dynamics

We now turn to global questions and as such we begin with an isolated invariant set S for which a Morse decompostion $\mathcal{M}(S)$ is given. The goal is to obtain a semi-conjugacy from S to a well understood dynamical system with the added property that the Morse decomposition is preserved.

This approach is not without precedent. In fact, the framework of C. Conley's approach to dynamical systems is the following theorem [**1**].

THEOREM 5.1. *Let S be a compact invariant set under a flow φ. Then, there exists a space X with a gradient-like flow $\psi : R \times X \to X$ and a continuous surjective map $f : S \to X$ such that*

$$f \circ \varphi = \psi \circ (id \times f) : R \times S \to X.$$

It is worth mentioning that the map f is obtained by collapsing each component of the chain recurrent set to a distinct point. The resulting quotient space is X and the flow ψ is that induced by φ. While Theorem 5.1 provides an important theoretical framework for understanding the dynamics on invariant sets it is too general to be of use in analyzing specific system. In particular:

(i) *a priori* the space X and the flow ψ are not known;

(ii) all the dynamics within the chain recurrent set is ignored.

The results described below are, in spirit, very similar to this theorem. A large class of flows is considered and a "minimal" structure for these flows is described by constructing a semi-conjugacy onto a simple "model" flow. As above, some, *but not all*, of the internal dynamics of the Morse set is collapsed out. However, and this is the crucial distinction, the space and flow that one projects onto is given *a priori*.

We begin with the simplest result. Let

$$D^P := \{z = (z_0, \dots, z_{P-1}) \mid \|z\| \leq 1\} \subset R^P$$

be the closed unit ball in R^P and $S^{P-1} = \partial D^P$ be the unit sphere. Let $\psi^P : R \times D^P \to D^P$ denote the flow generated by the following system of ordinary differential equations

$$\dot{\zeta} = Q\zeta - \langle Q\zeta, \zeta\rangle\zeta \qquad \zeta \in S^P \tag{4}$$

$$\dot{r} = r(1-r) \qquad r \in [0,1] \tag{5}$$

where

$$Q = \begin{bmatrix} 1 & 0 & \dots & 0 \\ 0 & \frac{1}{2} & & \\ \vdots & & \ddots & \\ 0 & & & \frac{1}{P} \end{bmatrix}.$$

The dynamics of ψ^P is easily understood if one realizes that (4) is obtained by projecting the linear system $\dot{z} = Qz$ onto the unit sphere. Let $e_p^{\pm} = (0, \dots, \pm 1, \dots, 0)$ be the unit vectors in the pth direction.

The assumptions on the family of dynamical systems is as follows.

A1: *S is a global compact attractor for a semi-flow Φ on a Banach space. Furthermore, if φ denotes the restriction of Φ to S then φ defines a flow on S.*

A2: *Under the flow $\varphi : R \times S \to S$*

$$\mathcal{M}(S) = \{M(p^{\pm}) \mid p = 0, \dots, P-1\} \bigcup \{M(P)\}$$

with ordering $0^{\pm} < 1^{\pm} < \dots < P-1^{\pm} < P$ is a Morse decomposition of S.

A3: *The cohomology Conley indices of the Morse sets are*

$$CH^k(M(P)) \approx \begin{cases} Z & \text{if } k = P \\ 0 & \text{otherwise} \end{cases}$$

and for $p = 0, \dots, P-1$

$$CH^k(M(p^{\pm})) \approx \begin{cases} Z & \text{if } k = p \\ 0 & \text{otherwise} \end{cases}$$

A4: *The connection matrix for $\mathcal{M}(S)$ is given by*

$$\Delta = \begin{bmatrix} 0 & 0 & 0 & \dots & 0 \\ D^0 & 0 & 0 & & \vdots \\ 0 & D^1 & & \ddots & 0 \\ \vdots & & & 0 & 0 \\ 0 & & \dots & D^{P-1} & 0 \end{bmatrix}$$

where, up to a choice of orientation, for $p = 0, \dots, P-2$
$D^p : CH^p(M(p^-)) \oplus CH^p(M(p^+)) \longrightarrow$

$$CH^{p+1}(M(p+1^-)) \oplus CH^{p+1}(M(p+1^+))$$

is given by

$$D_p = \begin{bmatrix} 1 & -1 \\ 1 & -1 \end{bmatrix}$$

and

$$D^{P-1} : CH^{P-1}(M(P-1^-))) \oplus CH^{P-1}(M(P-1^+)) \to CH^P(M(P))$$

is given by

$$D_P = \begin{bmatrix} 1 & -1 \end{bmatrix}.$$

THEOREM 5.2. *Given assumptions* A1 - A4 *there exists a continuous surjective map* $f : S \to D^P$ *and a flow* $\tilde{\varphi}$ *obtained by an order preserving time reparameterization of* φ *such that the following diagram commutes*

$$\begin{array}{ccc} R \times S & \xrightarrow{id \times f} & R \times D^P \\ \tilde{\varphi} \downarrow & & \downarrow \psi^P \\ S & \xrightarrow{f} & D^P \end{array}$$

where $M(p^{\pm})) = f^{-1}(e_p^{\pm})$ *for* $0 \leq p \leq P-1$, *and* $fM(P) = f^{-1}(0)$.

It is worth contrasting this theorem to the original result of Conley. Observe that since,

$$M(p^{\pm}) = f^{-1}(e_p^{\pm}),$$

Morse sets of S are preimages of Morse sets of the flow ψ. Furthermore, any set of connecting orbits in S is the preimage of a corresponding set of connecting orbits in D^P. As in Theorem 5.1 the Morse sets are all sent to fixed points, however, in this case we know that $X = D^P$ and the flow ψ is well understood. Thus we obtain specific information about the flow φ on S. Of course, the hypotheses on Theorem 5.2 are much stronger than those on Theorem 5.1. In particular, **A1** provides an a priori bound on the complexity of the total invariant set. While it is true that any invariant set possesses a Morse decomposition, **A2** provides a both minimal bound on the number of possible pieces in the decomposition and how these Morse sets are related by the flow. **A3** provides useful dynamic information concerning the "amount" of instability of each Morse set. Finally, **A4** implies that each Morse set is connected to each adjacent Morse set where adjacency is measured by means of the order on the indexing set of the Morse decomposition. It should be pointed out that in itself **A4** provides very little information about the structure of the connecting orbits.

The next theorem shows that one can obtain similar results without collapsing the Morse sets to fixed points, assuming, of course, that the Morse sets have sufficient structure. The model flow for this theorem is identical to that presented above except it is defined on D^{2P} and

$$Q = \begin{bmatrix} Q_1 & 0 & \dots & 0 \\ 0 & Q_2 & & \\ & & \ddots & \\ 0 & & & Q_P \end{bmatrix} \qquad \textit{where} \quad Q_p = \begin{bmatrix} p^{-1} & 2\pi \\ -2\pi & p^{-1} \end{bmatrix}, \ p = 1, 2, \dots P.$$

Again, the fact that (4) is obtained by projecting the linear system $\dot{z} = Qz$ onto the unit sphere makes it is easy to see that the following proposition holds.

PROPOSITION 5.1. *ψ is a Morse-Smale flow for which:*

(i) *The origin $0 = \Pi(P)$ is a fixed point with a $2P$ dimensional unstable manifold $W^u(0)$ and $cl(W^u(0)) = D^{2P}$.*

(ii) *For each $p = 0, \dots, P-1$, the set*

$$\Pi(p) := \{z = (z_0, \dots, z_{2P-1}) \mid z_{2p}^2 + z_{2p+1}^2 = 1\} \subset D_{2p}$$

is a periodic orbit with period 1 and

$$cl(W^u(\Pi(p)) = \{z \mid \sum_{i=0}^{2p+1} z_i^2 = 1\}$$

(iii) *$\mathcal{M}(D^{2P}) := \{\Pi(p) \mid p = 0, \dots, P\}$ is a Morse decomposition of D^{2P} with ordering $0 < 1 < \dots < P$.*

The assumptions in this case are slightly more complicated due to the fact that some of the structure of the Morse sets will be preserved.

B1: *S is a global compact attractor for a semi-flow Φ on a Banach space. Furthermore, if φ denotes the restriction of Φ to S then φ defines a flow on S.*

B2: *Under the flow $\varphi : R \times S \to S$*

$$\mathcal{M}(S) = \{M_p \mid p = 0, \dots, P\}$$

with ordering $0 < 1 < \dots < P$ is a Morse decomposition of S.

B3: *For each $p = 0, \dots, P-1$, $M(p)$ has a Poincaré section Ξ_p define on a neighborhood of $M(p)$.*

B4: *The cohomology Conley indices of the Morse sets are*

$$CH^k(M(P); Z) \approx \begin{cases} Z & \textit{if } k = 2P \\ 0 & \textit{otherwise} \end{cases}$$

and for $p = 0, \dots, P-1$

$$CH^k(M(p); Z) \approx \begin{cases} Z & \textit{if } k = 2p, 2p+1 \\ 0 & \textit{otherwise} \end{cases}$$

B5: *For each $M(p)$, $p < P$, there is a continuation of the flow in a neighborhood of $M(p)$ to an isolated invariant set which consists of the disjoint union of a hyperbolic periodic orbit and a set with trivial Conley index. The continuation preserves the Poincaré section.*

THEOREM 5.3. *Let the flow φ on S satisfy assumptions* B1 - B5. *Then there exist a continuous surjective function*

$$f : S \to D^{2P}$$

for which $f(M(p)) = \Pi(p)$ $(p = 0, \dots, P)$ and a continuous flow $\widetilde{\varphi} : R \times S \to S$ obtained via an order preserving time reparameterization of φ such that the following diagram commutes

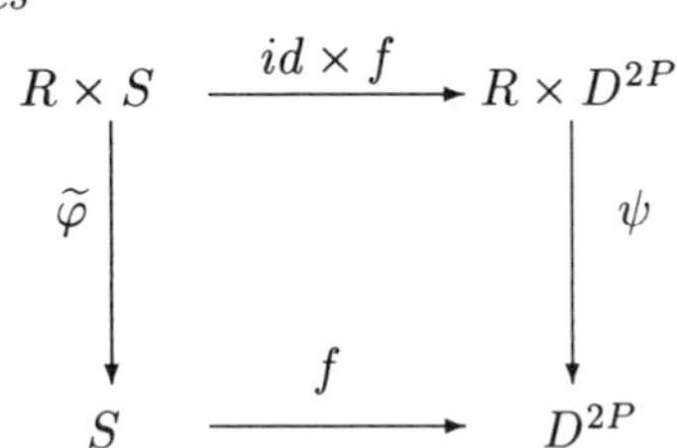

A comparison of the assumptions of this theorem and those of Theorem 5.2 are in order. **B1**, **B2** and **B4** correspond to **A1**, **A2** and **A3**. The only difference is that in this case the indexes are those of periodic orbits rather than fixed points. The new assumptions are **B3** and **B5**. **B3** forces each Morse set to have recurrent dynamics and in conjunction with **B4** implies that the Morse sets $M(p)$, $p = 0, \dots, P-1$, contain periodic orbits. **B5** plays the crucial role of determining how the unstable sets of the Morse sets intersect the isolating neighborhoods of the Morse sets. An open question is whether the assumption is necessary. It is not know whether given assumptions **B1** - **B4** it is possible for **B5** to be violated. The careful reader may be wondering why there is no global assumption of the form **A4** in this Theorem. The answer is that the connection matrix is uniquely determined given the information in assumptions **B1**, **B2** and **B4**.

Why assumptions **A1** - **A4** and **B1** - **B5**? They were motivated by concrete applications. Theorem 5.2 applies to "bistable" gradient-like systems which arise in a variety of physical and biological models [**11**]. Theorem 5.3 was motivated by and used to describe the dynamics on the global attractor for scalar delay equations with negative feedback [**9**]. It also applies to the attractor for the Ginzburg-Landau equation for an open set of parameter values [**12**]. and to attractors, under additional hypothesis, of cyclic feedback systems [**4**].

However, the similarity between these two theorems suggests that they are merely examples of a more general result. Ideally, what one wants is a theorem of the following type. The assumptions are:

(i) a Morse decompostion of an isolated invariant set S,

(ii) the Conley indices of S and the Morse sets, and
(iii) the connection matrices for the Morse decomposition.

The conclusions are:

(i) a mechanism for generating the model flow, and
(ii) a semi-conjugacy onto the model flow.

Unfortunately, at this level of generality it is fairly easy to construct examples which indicate that such theorem is not possible. Thus, the first question becomes what is the appropriate generalization of Theorems 5.2 and 5.3. The following directions of research might lead to a greater understanding of the problems involved in trying to construct model dynamics.

(i) *(Morse decompositions with Tori)* Consider a Morse decomposition which satisfies the assumptions of Theorem 5.3, except for the fact that some of the Morse sets have indices of hyperbolic invariant tori. There are two difficulties which arise in this setting, neither of which is well understood.
    (a) There is more freedom in the structure of the connection matrices. In particular, the indices of tori contain two homology generators on the same level which can be related to a variety of different generators for different Morse sets. It is not understood how these relationships correspond to the α and ω limit sets of the connecting orbits.
    (b) While it may be assumed that the Morse sets have Poincaré sections, this is not enough to determine the dynamics on the Morse sets of the model flow. In particular, one would expect to have an invariant torus with a Poincaré section in the model flow. However, the index information is clearly not sufficient to determine, for instance, the rotation number of the associated Poincaré map.

(ii) *(Gradient-Like Systems)* In Theorem 5.2 the connection matrix provides the global information upon which the dynamics of the model flow is constructed. The exact relationship between nontrivial entries in the connection matrix and connecting orbits is still unknown. In an attempt to understand this more fully, one could consider much simpler dynamics, i.e. assume that:
    (a) S is a compact global attractor in a Banach space on which the flow is gradient like;
    (b) the Conley indices of the Morse sets are those of hyperbolic fixed points; and
    (c) the connection matrix and the flow defined order for the Morse decomposition is known.

    What additional information, if any, is required to determine a model flow and a semi-conjugacy onto the model flow?

(iii) *(Topological Structures)* The problem (b) of Morse decompositions with tori, emphasizes the fact that the structure of the dynamics on the Morse set is crucial with regards to defining the semiconjugacy. For many applications, however, it is unrealistic to assume that the dynamics on the Morse sets have a uniform structure for all parameter values. This suggests further weakening what is meant by describe from the level of dynamics to the level of topology. To be more precise, as before, consider a flow $\varphi : R \times S \to S$ and assume the Morse decomposition, Conley indices, total ordering and connection matrices are known. The goal is to define a flow $\psi : R \times X \to X$ and a continuous surjective map $f : S \to X$ with the property that for every isolated invariant set Π constructed from the Morse sets in X, the pre-image $f^{-1}(\Pi)$ is an isolated invariant set in S constructible from the Morse sets of S. Notice that it is no longer required that f be a semi-conjugacy. Thus instead of preserving the dynamics of individual orbits, f merely preserves the Morse decomposition and the stable and unstable sets of the Morse sets.

REFERENCES

1. C. Conley, *Isolated Invariant Sets and the Morse Index.* CBMS Lecture Notes **38** A.M.S. Providence, R.I. 1978.
2. C. Conley, A qualitative singular perturbation theorem, *Global Theory of Dynamical Systems*, (eds. Z. Nitecki and C. Robinson) Lecture Notes in Math. **819** Springer–Verlag 1980, 65–89.
3. F. B. Fuller, Note on trajectories in a solid torus, *Ann. Math.* **56** (1952), 438–439.
4. T. Gedeon and K. Mischaikow, Global Dynamics of Cyclic Feedback Systems, in preparation.
5. J. K. Hale, *Asymptotic Behaviour of Dissipative Systems*, Math. Surveys and Monographs **25** A.M.S., 1988.
6. J. K. Hale, L. T. Magalhães, and W. M. Oliva, *An Introduction to Infinite Dimensional Dynamical Systems - Geometric Theory*, Appl. Math. Sci. **47**, Springer-Verlag 1984.
7. C. McCord. Mappings and homological properties in the homology Conley index. *Erg. Th. & Dyn. Sys.* **8*** (1988) 175 - 198.
8. C. McCord. Mappings and Morse decompositions in the homology Conley index. *Indiana Univ. Math. J.* **40** (1991) 1061 - 1082.
9. C. McCord and K. Mischaikow, On the global dynamics of attractors for scalar delay equations, preprint CDSNS92-89.
10. C. McCord, K. Mischaikow, and M. Mrozek. Zeta functions, periodicity, and the Conley index for maps and flows with transverse cross-sections. In preparation.
11. K. Mischaikow, Global asymptotic dynamics of gradient-like bistable equations, preprint CDSNS92-94.
12. K. Mischaikow and Y. Morita, Dynamics on the Global Attractor of a Gradient Flow arising in the Ginzburg-Landau Equation, preprint CDSNS92-100.
13. K. Mischaikow and M. Mrozek, Isolating Neighborhoods and Chaos, preprint CDSNS92-116.
14. K. Mischaikow, M. Mrozek, and J. Reineck, Singular Pertubations and The Conley Index, in progress.
15. M. Mrozek, Leray functor and cohomological index for discrete dynamical systems. *Trans. A. M. S.* **318** (1990) 149 - 178.
16. M. Mrozek, Open index pairs, the fixed point index and rationality of zeta functions, *Ergod. Th. & Dyn. Sys.*, (1990) **10**, 555-564.
17. M. Mrozek, Shape index and other indices of Conley type for local maps on locally compact Hausdorff spaces, preprint.

18. J.W. Robbin and D. Salamon, Dynamical systems, shape theory and the Conley index, *Ergodic Theory and Dynamical Systems* 8*(1988), 375-393.
19. K. P. Rybakowski, *The Homotopy Index and Partial Differential Equations*, Universitext, Springer-Verlag 1987.
20. D. Salamon, Connected simple systems and the Conley index of isolated invariant sets. *Trans. A. M. S.* **291** (1985) 1 - 41.
21. D. Salamon and E. Zehnder, Periodic Solutions of Periodic Hamiltonian Systems, preprint.
22. J. Smoller, *Shock Waves and Reaction Diffusion Equations*, Springer Verlag, New York, 1980.
23. P. A. Schweitzer, Counterexamples to the Seifert conjecture and opening closed leaves of foliations, *Ann. Math.* **100** (1974), 386–400.
24. R. Temam, *Infinite-Dimensional Dynamical Systems in Mechanics and Physics,* Springer-Verlag, New York, 1988.

Center for Dynamical Systems and Nonlinear Studies, School of Mathematics, Georgia Institute of Technology, Atlanta, GA 30332-0001

E-mail address: mischaik@math.gatech.edu

Contemporary Mathematics
Volume **152**, 1993

A Survey of Relative Nielsen Fixed Point Theory

HELGA SCHIRMER

ABSTRACT. The relative Nielsen number $N(f; X, A)$ was introduced in 1986. It gives a lower bound for the number of fixed points on X for maps of pairs of spaces in the homotopy class of $f: (X, A) \to (X, A)$, and in many cases it gives an optimal lower bound. In recent years several other Nielsen numbers have been introduced for maps of pairs which provide lower bounds for the number of fixed points on $X - A$ (with f on A either kept fixed or allowed to vary under a homotopy) and the closure of $X - A$. The methods used to solve these problems have been found helpful in related settings which concern fixed points of fibre maps and of equivariant maps, periodic points, coincidences and dynamical systems. This paper surveys results, applications and open problems.

1. Relative Nielsen theory - why did it start?

Nielsen fixed point theory is concerned with the determination of minimal fixed point sets, and more generally with properties of fixed point sets, of a given map $f: X \to X$. The Nielsen number $N(f)$ provides a homotopy invariant lower bound for the number of fixed points of f, and in many cases $N(f)$ is the best possible lower bound. In 1986 an extension of Nielsen theory to the setting of maps of pairs of spaces $f: (X, A) \to (X, A)$ was begun and has developed rapidly since then. This survey of the topic, which has become known as "relative Nielsen theory", contains an outline of the main results obtained so far.

1991 *Mathematics Subject Classification.* Primary 55-02 Secondary 55M20, 57M99, 57R99, 54H25.

Key words and phrases. Nielsen fixed point theory, relative Nielsen numbers, fixed points of maps of pairs, minimal fixed point sets, fixed points of homeomorphisms, fixed points of equivariant maps, fixed points of fibre maps, fixed points of maps of triads, periodic points, coincidences.

This research was partially supported by NSERC Grant A 7579

This paper is in final form and no version of it will be submitted for publication elsewhere.

There are several reasons why one wants to extend Nielsen fixed point theory to a relative setting. An obvious, but not very good, reason is that such a setting is frequently used in algebraic topology, and so one's urge for generalization can produce the wish for an extension of the definition of the classical Nielsen number $N(f)$ in order to obtain a relative one. This urge gets an additional motivation if one discovers the existence of a relative Lefschetz number [1] (see also Section 2.2). But while the definition of a Lefschetz number for maps of pairs of spaces is a fairly straightforward algebraic generalization of that of the Lefschetz number $L(f)$, this is not the case for the Nielsen number, where new concepts are needed.

The actual reason for the introduction of a relative form of the Nielsen number $N(f)$ was, however, a different and better one, namely an observed but unexplained problem concerning fixed point sets of homeomorphisms on the pants. The following example was given in 1980 by Boju Jiang in [5].

EXAMPLE 1.1. Let P be the pants, i.e. the disk with two holes removed, and let f be the homeomorphism obtained by a reflection on an axis of symmetry which interchanges the boundaries of the two holes. Then $N(f) = 1$, but any homeomorphism isotopic to f will map the outer boundary of P onto itself in an orientation-reversing manner and hence have at least two fixed points on this boundary circle. Thus $N(f)$ cannot be realized by a homeomorphism in the isotopy class of f.

The map f in this example is a homeomorphism, but the property of f which causes the difference between $N(f)$ and the minimum number of fixed points in the isotopy class of f is the fact that any map in this class must map the boundary ∂P to itself, and in particular must induce a selfmap of degree -1 on the outer boundary. Hence f is a map of pairs $f\colon (X, A) \to (X, A)$, where $X = P$ and $A = \partial P$. As Nielsen theory is better developed for continuous maps than for homeomorphisms, it was natural to study first an extension of classical Nielsen theory to continuous maps of pairs of spaces in order to explain the observed inadequacy of $N(f)$ in relation to Example 1.1. Thus relative Nielsen theory started, and the relative Nielsen number $N(f; X, A)$ was introduced.

This number is discussed first, in Section 2.1. The rest of this survey is not organized in historical order, but as follows. In Section 2, we also discuss further relative Nielsen numbers which have been defined in order to study the location of fixed points of maps of pairs, in particular numbers which give information about fixed points on the complement $X - A$ and on its closure $c\ell(X - A)$. Next, in Section 3, we describe some other relative Nielsen type numbers which have been introduced to study related topics, namely fixed points of map extensions, of maps of triads and of fibre map pairs as well as periodic points of maps of pairs. In Section 4 we outline some work which was motivated by the fact that the topics of Sections 2 and 3 have lead to a rather bewildering variety of new numbers, and in Section 5 we discuss some cases where the results and methods developed in relative Nielsen theory have helped to obtain improved solutions of problems outside this field. In particular, we describe in the final Section 5.5 an application to dynamical systems theory. Some open questions are listed in

Section 6.

We have suggested a change of existing terminology. In Section 2.2 we have replaced the term "Nielsen number of the complement" from [S4] by "Nielsen number of the closure", and written the number as $N(f;\overline{X-A})$ rather than $\tilde{N}(f;X,A)$. Thus we can replace, in Section 2.3, the awkward name "Nielsen number of the complementary space" of $N(f;X-A)$ by "Nielsen number of the complement". These names and notations should obviously have been used when these numbers were first defined, and this survey seems a suitable place for a change which has already started to creep into the literature. The changes will be pointed out carefully where they occur, and so I hope that they will help rather than hinder in the understanding.

We do not specify the precise properties needed by a space or a pair of spaces when we state definitions, and we take it for granted that a fixed point index exists in our setting. Results are usually stated only for the case where $X = M$ is a connected compact n-manifold and A a (not necessarily connected) compact submanifold, as this case is usually the most interesting one and often also the one which is easiest to describe. But many of the results are true for more general pairs of spaces, in particular for pairs of compact polyhedra or sometimes even for pairs of compact ANR's. To find precise assumptions it is necessary to consult the sources.

We assume familiarity with certain material from classical Nielsen fixed point theory which can be found in [2], [6] or [7]. References on relative Nielsen theory are as up-to-date as possible, but in such a rapidly growing area they may not be complete. We also give a few background references.

2. Nielsen type numbers for maps of pairs

In this section, we will deal with maps of pairs of spaces $f\colon (X,A) \to (X,A)$, and discuss three Nielsen type numbers which give information about the number of fixed points on the total space X, on the closure $c\ell(X-A)$ and on $X-A$. These numbers are the relative Nielsen number $N(f;X,A)$, the Nielsen number of the closure $N(f;\overline{X-A})$ and the Nielsen number of the complement $N(f;X-A)$. They will be obtained from a study of the interplay of the fixed point classes of two maps defined by $f\colon (X,A) \to (X,A)$, namely the fixed point classes F of the map $f\colon X \to X$ defined by f if the condition that $f(A) \subset A$ is immaterial, and the fixed point classes $\overline{F}$ of the restriction $\bar{f} = f|A\colon A \to A$. It is important in many of the definitions that, for any F, the set $F \cap A$ is either empty or the union of fixed point classes of $\bar{f}\colon A \to A$. All numbers are invariant under homotopies of pairs, i.e. under homotopies of the form $H\colon (X \times I, A \times I) \to (X,A)$, where I is the unit interval.

2.1. Fixed points on the total space: the relative Nielsen number.

Relative Nielsen theory has the advantage that a very simple example, which we call the Disk Example, can often be used to motivate its definitions and illustrate its results. But one should keep in mind that its simplicity is deceptive, and that everything outlined in this survey applies to much more sophisticated settings. First we use the Disk Example to show that the Nielsen number $N(f)$ can be a very poor lower bound for the number of fixed points on X for a map of pairs.

EXAMPLE 2.1 (DISK EXAMPLE). Let $X = D^2$ be the 2-disk in the complex plane and $A = S^1$ its boundary. Let $f\colon (X, A) \to (X, A)$ be a map which restricts to the map $\bar{f} = f|S^1$ given by $\bar{f}(z) = z^d$ on the boundary circle, where d is an integer $\neq 1$. As $\bar{f}$ has $|d-1|$ fixed points, the map f must have at least that many. But the Nielsen number $N(f) = 1$.

The relative Nielsen number $N(f; X, A)$, which was introduced in 1986 [S1], provides a much better lower bound for the number of fixed points on X for such maps of pairs. The crucial idea underlying its definition uses the concept of a common fixed point class which is defined as follows.

DEFINITION 2.2. Let $f\colon (X, A) \to (X, A)$ be a map of a pair of spaces. A fixed point class F is an essential *common* fixed point class of f and $\bar{f}$ if it is an essential fixed point class of $f\colon X \to X$ and contains an essential fixed point class $\overline{F}$ of $\bar{f}$.

Let $N(f, \bar{f})$ be the number of essential common fixed point classes of f and $\bar{f}$, let $N(f)$ be the Nielsen number of $f\colon X \to X$ and $N(\bar{f})$ the Nielsen number of $\bar{f}\colon A \to A$.

DEFINITION 2.3. The *relative Nielsen number* of the map $f\colon (X, A) \to (X, A)$ is

$$N(f; X, A) = N(\bar{f}) + N(f) - N(f, \bar{f}).$$

As $0 \leq N(f, \bar{f}) \leq N(f)$, the relative Nielsen number is a non-negative integer. In Example 2.1 the definition gives $N(f; X, A) = |d-1| + 1 - 1 = |d-1|$. It is not hard to see that $N(f; X, A)$ is a lower bound for the number of fixed points on the total space X for the map $f\colon (X, A) \to (X, A)$, as essential fixed point classes must contain at least one fixed point and the non-common classes which are counted by $N(f) - N(f, \bar{f})$ cannot contain a fixed point of one of the essential classes counted by $N(\bar{f})$. The proof that $N(f; X, A)$ is invariant under homotopies of pairs is obtained by an adaptation of the proof of the homotopy invariance of $N(f)$.

Two basic problems arise in classical Nielsen fixed point theory. The first is the notoriously hard computation of the Nielsen number $N(f)$. Although there is no algorithm, it can be carried out for an increasing number of spaces and maps. The second concerns the question of the optimality of $N(f)$ as a lower bound. Optimality has been established, under suitable conditions on X and/or f, by constructing a map homotopic to f with precisely $N(f)$ fixed points. Both of these problems arise in connection with the relative Nielsen number, and the other Nielsen type numbers introduced in relative Nielsen theory.

Not much work has been done so far about the computation of the relative Nielsen number. In concrete examples it is often possible to find $N(f; X, A)$ by inspection. This was true in the Disk Example 2.1. As is to be expected from classical Nielsen theory, it is also easy to find $N(f; X, A)$ in terms of the Euler characteristics of X and all components of A if f is a map in the identity class [S2]. Xuezhi Zhao [Z1] has computed $N(f; X, A)$ in terms of the induced homomorphism on the first homology groups of X and A under the assumptions that X and A are Jiang spaces and A is connected, but that last assumption is unfortunately rather restrictive. Brigitte Norton-Odenthal and Peter Wong [NW] define a relative generalized Lefschetz number for $f: (X, A) \to (X, A)$ in terms of the generalized Lefschetz numbers of f and $\bar{f}$, and show that it can be used in certain cases to compute $N(f; X, A)$ as a trace.

On the other hand, much is known about the optimality of $N(f; X, A)$. The construction of a map with a minimal fixed point set can be carried out by first minimizing the number of fixed points on A in the usual way, then uniting fixed points on $X - A$ whenever possible and finally uniting fixed points on $X - A$ which lie in common fixed point classes with fixed points of $\bar{f}$ which have been moved to the boundary ∂A of A. Hence it is necessary to assume that A is "Wecken", i.e. that it satisfies conditions which ensure that $N(\bar{f})$ is an optimal lower bound for the number of fixed points of $\bar{f}$, and that $X - A$ satisfies conditions which in the manifold case are equivalent to those which allow the Whitney trick. But a further assumption is needed which does not occur in classical Nielsen theory but has become an almost standard one in relative Nielsen theory. This is the assumption of by-passing contained in the next definition.

DEFINITION 2.4. A subspace A of a space X can be *by-passed* in X if every path in X with endpoints in $X - A$ is path-homotopic to a path in $X - A$.

The by-passing assumption is used to prove a minimum theorem which establishes optimality, and an example in [Z2] shows that it cannot be omitted. We state the next theorem for the case where X and all components of A are manifolds, but a more general form for maps of pairs of compact polyhedra can be found in [S1].

THEOREM 2.5 (MINIMUM THEOREM). *Let (X, A) be a pair of compact spaces so that X, $X - A$ and each component of A are connected triangulable manifolds of dimension ≥ 3. If A can be by-passed in X, then every map $f: (X, A) \to (X, A)$ is homotopic to a map $g: (X, A) \to (X, A)$ with $N(f; X, A)$ fixed points on X.*

The by-passing condition is clearly satisfied if $X = M^n$ is an n-dimensional manifold and $A = \partial M^n$ its boundary. If, further, $n \geq 4$, then all assumptions of Theorem 2.5 are satisfied and hence $N(f; X, A)$ can be realized, but — as is to be expected from results in the non-relative case — the conclusions of Theorem 2.5 need not hold if $n = 2$ or 3. Robert Brown and Brian Sanderson [BSa] have introduced the following terms for manifolds with non-empty boundary, where $MF[f; M, \partial M]$ is the least number of fixed points on M for all maps in the homotopy class of $f: (M, \partial M) \to (M, \partial M)$: (i) M is called *boundary-Wecken* if $N(f; M, \partial M) = MF(f; M, \partial M)$ for all maps $f: (M, \partial M) \to (M, \partial M)$, (ii) M is called *almost boundary-Wecken* if the difference $MF(f; M, \partial M) - N(f; M, \partial M)$ is bounded for all such maps, (iii) M is called *totally non-boundary-Wecken* if the difference $MF(f; M, \partial M) - N(f; M, \partial M)$ can be made arbitrarily large. Hence all n-manifolds are boundary-Wecken if $n \geq 4$. Boju Jiang [J1] has shown that a compact orientable 3-manifold with boundary is boundary-Wecken if and only if all of its boundary components are Wecken (i.e. have non-negative Euler characteristic), and otherwise it is totally non-boundary-Wecken. Not yet complete results exist for the case $n = 2$. In [BSa] Brown and Sanderson proved that the disk, annulus and Möbius band are boundary-Wecken, and their results and those of Michael Kelly [K] show that all surfaces obtained by deleting $r \geq 1$ open disks from a surface other than the 2-sphere or the projective plane are totally non-boundary-Wecken, but that the pants are almost boundary-Wecken. An outline of these results is contained in [B].

2.2. Fixed points on the closure of the complement.

Almost two decades before the introduction of the relative Nielsen number the Polish mathematician C. Bowszyc [1] introduced a relative Lefschetz number. The definition of this number is clear: induced homomorphisms of the homology groups $H_*(X)$ are replaced by induced homomorphisms of the relative homology groups $H_*(X, A)$ to obtain a relative Lefschetz number $L(f; X, A)$ for a map of pairs $N(f; X, A)$. (This notation for the relative Lefschetz number is ours.) But work is needed to find its meaning. Bowszyc proved that under suitable assumptions $L(f; X, A) \neq 0$ implies the existence of at least one fixed point on $c\ell(X - A)$, and hence the relative Lefschetz number can serve to make the location of a fixed point more precise.

The results in [S4] show that relative Nielsen theory can give better information if $L(f; X, A) = 0$, and can also detect multiple fixed points on $c\ell(X - A)$. But a new concept and a new Nielsen type number are needed. It is necessary to relate the fixed point index $\text{ind}(X, f, F)$ of a fixed point class F of $f: X \to X$ with the fixed point index $\text{ind}(A, \bar{f}, F \cap A)$ of the fixed point classes of $\bar{f}: A \to A$ which make up the set $F \cap A$. This is done in the next definition.

Definition 2.6. Let $f: (X, A) \to (X, A)$ be a map of pairs. Then a fixed point class F of $f: X \to X$ *assumes its index in* A if $\text{ind}(X, f, F) = \text{ind}(A, \bar{f}, F \cap A)$.

To understand the meaning of this definition, it helps to think of the case where $F = x_0 \in \partial A$ is an isolated fixed point of f. Then two fixed point indices are defined for x_0, namely $\text{ind}(X, f, x_0)$ and $\text{ind}(A, \bar{f}, x_0)$, and these two indices can be different. If x_0 can be moved, under a homotopy of pairs, to a fixed point x_1 of a map g so that it now lies in the interior $\text{int}A$, then $\text{ind}(X, g, x_1) = \text{ind}(A, \bar{g}, x_1)$, and if no other changes are made to the fixed point set, then the homotopy invariance of the fixed point index implies $\text{ind}(X, f, x_0) = \text{ind}(A, \bar{f}, x_0)$. This special case leads to the following general definition from [S4], where # stands for the cardinality. We have replaced the name "Nielsen number of the complement" by the more meaningful "Nielsen number of the closure", and $\tilde{N}(f; X, A)$ by the more meaningful $N(f; \overline{X - A})$.

Definition 2.7. Let $f\colon (X,A) \to (X,A)$ be a map of pairs. Then the *Nielsen number of the closure* is

$$N(f;\overline{X-A}) = \#\{F : F \text{ does not assume its index in } A\}.$$

Note that all fixed point classes of f, whether essential or not, are counted in the definition of $N(f;\overline{X-A})$. The proof that $N(f;\overline{X-A})$ is a homotopy invariant lower bound for the number of fixed points on $c\ell(X-A)$ is not hard. $N(f;\overline{X-A})$ provides more information than the relative Lefschetz number, as $L(f;X,A) \neq 0$ implies $N(f;\overline{X-A}) > 0$, and examples exist where $L(f;X,A) = 0$ but $N(f;\overline{X-A}) > 1$. Optimality can be shown if the conditions of the Minimum Theorem 2.5 are satisfied and if, in addition, it is assumed that the interior of each component of A is non-empty.

2.3. Fixed points on the complement.

Xuezhi Zhao [Z1] obtained further information about the location of the fixed point set of a map $f\colon (X,A) \to (X,A)$ in a study concerning the number of fixed points on the complement $X - A$. To see that once more a new concept is needed to define a number which provides a lower bound for the number of such fixed points, we modify the Disk Example 2.1 so that now f is a rotation around the origin 0. Then $\bar{f}$ is homotopic to the identity and is fixed point free. The fixed point set of f consists of one essential class $F = \{0\}$ which is not common, but it is easy to see that $f\colon (X,A) \to (X,A)$ is homotopic as a map of pairs to a map which is fixed point free on the interior of X.

To deal with such situations, Zhao introduced the concept of an essential fixed point class F of f which is "weakly common", which means that it "weakly contains" a fixed point class of $\bar{f}$. Such a fixed point class of $\bar{f}$ can be inessential and hence empty, as is the case in our example. The precise definition of the term "weakly contained" is given in [Z1] in terms of Jiang's fixed point class functor (see [**6, Chapter III**]) and omitted here, but the following definition is equivalent in consequence of [**Z1, Theorem 2.3**].

Definition 2.8. An essential fixed point class F of $f\colon X \to X$ is an essential *weakly common fixed point* class of f and $\bar{f}$ if there exists a path α from a point $x_0 \in F$ to a point in A so that α is homotopic to $f \circ \alpha$ under a homotopy of the form $(I,0,1) \to (X,x_0,A)$, which means that the homotopy keeps the starting point x_0 of α fixed and allows the endpoint of α to vary in A.

An essential fixed point class F of f which is common is weakly common, but not necessarily vice versa. Hence if $E(f,\bar{f})$ is the number of essential weakly common fixed point classes of f and $\bar{f}$, then $N(f,\bar{f}) \leq E(f,\bar{f}) \leq N(f)$, and so the Nielsen number given in the next definition is a positive integer. We have changed the unfortunate name "Nielsen number of the complementary space" to the simpler "Nielsen number of the complement". The original name was proposed to avoid duplication, as the name "Nielsen number of the complement" had been used in [S4] for $N(f;\overline{X-A})$.

DEFINITION 2.9. Let $f: (X, A) \to (X, A)$ be a map of pairs. Then the *Nielsen number of the complement* is

$$N(f; X - A) = N(f) - E(f, \bar{f}).$$

Zhao showed that $N(f; X - A)$ is a homotopy invariant lower bound for the number of fixed points on $X - A$, computed it in the case where A is connected and both X and A are Jiang spaces, and proved optimality for $N(f; X - A)$ if the same assumptions on (X, A) are satisfied as in the Minimum Theorem 2.5, in particular if A can be by-passed in X. He also showed that optimality is false without the by-passing condition, and defined and studied in [Z2] a "surplus number" $SN(f; X - A) \geq N(f; X - A)$ which is an optimal lower bound for the number of fixed points on $X - A$ even if the by-passing condition is not satisfied.

2.4. Fixed points on the boundary.

After characterising fixed points on X and its subspaces $c\ell(X - A)$ and $X - A$, it seems natural to consider fixed points on ∂A. A "Nielsen number of the boundary" was introduced in [S4], but it only works well in the case where $f: (X, A) \to (X, A)$ has a minimal fixed point set on X, and it has turned out to be of little interest. I doubt that the study of minimal fixed point sets on ∂A is worth-while.

3. Variations of the setting: More Nielsen type numbers

In this section we will discuss several studies which use the methods of the preceding section, but vary the setting either by restricting the class of homotopies (see 3.1), or by imposing additional conditions on the class of maps (see 3.2 and 3.3), or by studying iterates of maps (see 3.4).

3.1. Fixed points of map extensions.

So far we have discussed Nielsen type numbers for maps of pairs which are invariant under homotopies of maps of pairs, i.e. under homotopies of the form $H: (X \times I, A \times I) \to (X, A)$. But a relative homotopy in algebraic topology is usually defined as a homotopy $H: X \times I \to X$ which satisfies the condition $H(a, t) = H(a, 0)$ for every $a \in A$ and $0 \leq t \leq 1$. Under such a homotopy the fixed points on A of the original map $f = H(\cdot, 0)$ remain constant but the fixed points on $X - A$ can vary, and so the problem of minimizing their number under these changed assumptions arises. There is another, equivalent, way to look at this problem: given a map $\bar{f}: A \to A$, find the least number of fixed points on $X - A$ for all extensions of $\bar{f}$ to a map $f: X \to X$ in a given homotopy class. This topic was studied by Robert Brown, Robert Greene and Helga Schirmer in [BGS]. Surprisingly, it was discovered that its answer can be different for continuous and for smooth map extensions, a type of behaviour which had not been observed in Nielsen fixed point theory before.

The Disk Example 2.1 provides again a simple example which illustrates the situation. We consider the case where $d = 2$, and hence $\bar{f}(z) = z^2$ has one fixed point $z_1 = 1$. It is not hard to see that $\bar{f}$ can be extended to a continuous map $f: (D^2, S^1) \to (D^2, S^1)$ which has no fixed points on $X - A = \text{int}D^2$, but it follows from [BGS] that every smooth extension $f: X \to X$ of $\bar{f}$ must have at least one fixed point on $\mathrm{int}D^2$.

To deal with fixed point sets of map extensions, two extension Nielsen numbers have been introduced, namely a continuous and a smooth one. By smooth we mean C^1, but the case C^∞ is the same. The smooth extension number is only defined in the case where $X = M$ is a smooth manifold with boundary $A = \partial M$. The continuous number can be defined more generally, e.g. in the case where X is a compact polyhedron and A a subpolyhedron (see e.g. Section 4.2), but here we only give its simplified definition in the setting where the smooth number also exists. For the definition of the smooth extension Nielsen number we need a new property of an essential fixed point class F of $f: X \to X$.

DEFINITION 3.1. Let $f: (M, \partial M) \to (M, \partial M)$ be a self-map of a manifold with boundary. Then a fixed point class F of $f: M \to M$ is *representable on* ∂M if there exists a (possibly empty) subset F'_∂ of $F \cap \partial M$ such that $\mathrm{ind}(M, f, F) = \mathrm{ind}(\partial M, \bar{f}, F'_\partial)$.

DEFINITION 3.2. Let M be a smooth compact n-manifold with boundary and $\bar{f}: \partial M \to \partial M$ a smooth self-map of its boundary which has an extension to a map $f: M \to M$. Then

(1) the *extension Nielsen number* is

$$N(f|\bar{f}) = \#\{\text{essential } F : F \cap \partial M = \emptyset\},$$

(2) the *smooth extension Nielsen number* is

$$N^1(f|\bar{f}) = \#\{\text{essential } F : F \text{ is not representable on } \partial M\}.$$

It is easy to see that $N(f|\bar{f}) \le N^1(f|\bar{f}) \le N(f)$. But equality need not hold, as can be seen from the case in our Disk Example 2.1 where $\bar{f}(z) = z^2$. The map $f: D^2 \to D^2$ has one essential fixed point class F, and clearly $N(f|\bar{f}) = 0$. But $\mathrm{ind}(D^2, f, F) = L(f) = 1$ and $\mathrm{ind}(\partial D^2, \bar{f}, F \cap \partial D^2) = \mathrm{ind}(\partial D^2, \bar{f}, z_1) = L(\bar{f}) = -1$, so F is not representable on ∂D^2 and hence $N^1(f, \bar{f}) = 1$. The different behaviour of smooth and continuous extensions with respect to the fixed point set is a consequence of the following theorem, whose assumptions are satisfied in Example 2.1.

THEOREM 3.3 (INDEX THEOREM). *Let $p \in \partial M$ be an isolated fixed point of a smooth map $f: (M, \partial M) \to (M, \partial M)$. If p is a transversal fixed point of $\bar{f}$, then either $\mathrm{ind}(M, f, p) = \mathrm{ind}(\partial M, \bar{f}, p)$ or $\mathrm{ind}(M, f, p) = 0$.*

The Index Theorem does not hold for continuous maps, and not even for smooth maps if the condition that p is a transversal fixed point of $\bar{f}$ is omitted. Examples exist which show that $\mathrm{ind}(M, f, p)$ and $\mathrm{ind}(\partial M, \bar{f}, p)$ can then be arbitrary integers.

It is not hard to see from their definitions that $N(f|\bar{f})$ and $N^1(f|\bar{f})$ are lower bounds for the number of fixed points on $\mathrm{int}M$ for continuous and smooth extensions of $\bar{f}$ in the homotopy class of f, respectively, and minimum theorems exist which show that they are optimal lower bounds if the dimension of M is

≥ 3 and (in the smooth case) if $\bar{f}$ is transversally fixed. The by-passing condition is of course always satisfied if $A = \partial M$ and $X = M$.

Fixed points of smooth maps $f: (M, \partial M) \to (M, \partial M)$, where $\bar{f}$ is transversally fixed, and their behaviour under homotopies which are no longer constant on ∂M but leave the map transversally fixed on ∂M are studied in [S6]. Additional Nielsen type numbers, e.g. a "boundary transversal Nielsen number", are introduced which give homotopy invariant lower bounds for the number of fixed points for such maps, and are used to obtain further examples of transversally fixed smooth maps on the boundary of manifolds where smooth extensions over the interior of the manifold must have more fixed points than continuous ones. Such maps exist on the boundary of the n-ball B^n for all dimensions $n \geq 2$ if $(-1)^n d \geq 2$, where d is the degree of $\bar{f}$. An example where $M = S^1 \times B^2$ is the solid torus shows that the difference between $N^1(f|\bar{f}) - N(f|\bar{f})$ can be arbitrarily large. It is likely that boundary maps with a difference in the number of fixed points of smooth versus continuous extensions occur frequently, but no general results exist so far.

In [Z4] Zhao used methods from [Z2] to study continuous map extensions for compact polyhedral pairs (X, A) which do not satisfy the by-passing condition, and introduced a "surplus extension Nielsen number" $SN(f|\bar{f}) \geq N(f, \bar{f})$ which provides an optimal lower bound in such cases if X and all components of A are manifolds of dimension ≥ 3.

3.2. Fixed points of maps of a triad.

By a triad (X, A_1, A_2) we mean a space X with two subspaces A_1 and A_2 such that $A_1 \cup A_2 = X$. Hence a triad reduces to the pair of spaces (X, A_1) if $A_1 \subset A_2$, and so a map of a triad $f: (X, A_1, A_2) \to (X, A_1, A_2)$ generalizes a map of a pair of spaces. Fixed points of maps of triads are studied in [S7]. The map f defines by restriction maps $f_j = f|A_j: A_j \to A_j$ for $j = 0, 1, 2$, where $A_0 = A_1 \cap A_2$. Nielsen and relative Nielsen numbers as well as a new number occur in the definition of the *Nielsen number of the triad*

$$N(f; A_1 \cup A_2) = N(f_1; A_1, A_0) + N(f_2; A_2, A_0) - N(f_0) - IJ(f_1, f_2).$$

The new number $IJ(f_1, f_2)$ is the number of "inessentially joined pairs" of essential fixed point classes of f_1 and f_2. A precise definition of this term needs an extension of Zhao's work described in Section 2.3, in particular of the notion of weakly common fixed point classes. It can be found in [S7]. The Nielsen number of the triad is a lower bound for the number of fixed points on X and invariant under (suitably defined) homotopies of triad maps. A minimum theorem which shows that it is an optimal lower bound, and also specifies the location of minimal fixed point sets, can be proved under conditions typical of relative Nielsen theory (i.e. dimension and by-passing conditions) if A_0 is "thin", which means that it has an empty interior in X. The computation of the Nielsen number of the triad clearly reduces to the computation of other Nielsen numbers if $IJ(f_1, f_2) = 0$ or 1, and this is the case if the triad is obtained in one of the following four concrete ways: by constructing the double of a manifold with boundary, by constructing the suspension of a polyhedron, by attaching a handle to a manifold with boundary, or by taking the connected sum of two manifolds. The assumptions of the minimum theorem, in particular the by-passing and thinness assumptions, are satisfied in these four cases if the dimensions of X and its subspaces are sufficiently high.

3.3. Fixed points of fibre map pairs.

Nielsen numbers for fibre maps in the setting of maps of pairs was the topic of a thesis by Aaron Schusteff [Schu]. If $f: E \to E$ is a fibre map of a fibre space $p: E \to B$, then information about the fixed point classes of f can be obtained from the fixed point classes in the base and the fibre. In special cases the Nielsen number $N(f)$ can be computed from the "naive product formula" $N(f) = N(\bar{f}) \cdot N(\hat{f})$, where $\bar{f}$ and $\hat{f}$ are the maps induced by f on the base and fibre respectively. Under weaker hypotheses, there is a generalized product formula which contains a correcting factor called the Pak number. (See e.g. [**6, Chapter IV, §4, Theorem 4.10**].) Schusteff considered fibre space pairs (E, p, B) and (E_0, p_0, B_0), where $E_0 \subset E$ and $p_0 = p|E_0: E_0 \to B_0$, with $B_0 = p_0(E_0)$. A map $f: (E, E_0) \to (E, E_0)$ is called a *fibre map of the fibre space pair* (E, E_0) if both $f: E \to E$ and its restriction $f_0 = f|E_0: E_0 \to E_0$ are fibre maps. For the simpler case of product maps of product space pairs, [**Schu, Proposition III.1.6**] lists four conditions which each ensure the existence of a "naive product formula" $N(f; E, E_0) = N(\bar{f}; B, B_0) \cdot N(\hat{f}; F, F_0)$. These results are used to obtain various additional assumptions which yield suitably modified Pak numbers for generalized product formulas for fibre maps of a fibre space pair. Such conditions include, e.g., that f and f_0 are eventually commutative and that all spaces are Jiang spaces or nilmanifolds. The usual hypotheses about orientability can in many cases be eliminated. (See e.g. [**Schu, Theorems III.2.26 and III.2.27**].)

3.4. Periodic points of maps of pairs.

Two Nielsen type numbers exist for periodic points of a map $f: X \to X$ of a compact connected space X, namely $NP_n(f)$, the Nielsen type number of period n, and $N\Phi_n(f)$ (or $NF_n(f)$), the Nielsen type number of the nth iterate. The first is a lower bound for the number of periodic points of least period n and the second a lower bound for the number of periodic points of all periods $m|n$ (i.e. fixed points of the n-th iterate). Both numbers are homotopy invariant. (See e.g. [**6, Chapter III, §4**].) Using information about the periodic points of $f: X \to X$ and $\bar{f}: A \to A$ as well as the interplay of these data, two Nielsen numbers for periodic points of maps of pairs $f: (X, A) \to (X, A)$ were introduced by Philip Heath, Helga Schirmer and Chengye You in [HSY]. They specialize to $NP_n(f)$ and $N\Phi_n(f)$ if $A = \emptyset$ and have the expected lower bound properties. Relations between them and the ordinary Nielsen numbers for periodic points, calculations for certain cases and many examples are included in [HSY]. In preparation for this work it was first necessary to carry out a study of periodic points for maps on non-connected spaces, as the subspace A (e.g. the boundary of a manifold) need not be connected. Results are contained in [4].

4. How many Nielsen type numbers are there?

It is natural at this stage to notice that very many new Nielsen type numbers have been introduced, and to wonder whether there might be too many. But as different numbers solve different problems, it is unlikely that some of them can also serve the purpose of other ones which thus could be abolished, or that a new "super number" can be found which specializes to many of them. Nevertheless, some attempts in these directions exist.

4.1. Basic relative Nielsen numbers.

In [Z3] Xuezhi Zhao has introduced so-called basic relative Nielsen numbers $N_{ijkl}(f; X, A)$, where $i, j, k, l = 0, 1$, for a map of pairs $f: (X, A) \to (X, A)$. They can serve as building blocks for other relative Nielsen numbers. To define his numbers, Zhao considered four basic properties of a fixed point class F of $f: X \to X$, namely: (1) Is F essential? (2) Is F common? (3) Is F weakly common? (4) Does F assume its index in A? The numbers $N_{ijkl}(f; X, A)$ count the number of fixed point classes F of $f: X \to X$ which do or do not have these properties so that negative, resp. positive, answers to each of these four properties correspond to the subscript 0 or 1. Thus, for example, $N_{1010}(f; X, A)$ is defined as the number of essential fixed point classes F which are not common but are weakly common and which do not assume their index in A. Zhao shows that $R(f)$, $N(f)$ and all relative Nielsen numbers defined in Section 2 can be written as sums of basic relative Nielsen numbers. He also uses his basic numbers to strengthen some results from [S1], [S4] and [Z1] concerning the location of minimal fixed point sets on A, $X - A$ and $c\ell(X - A)$ of a map $f: (X, A) \to (X, A)$.

4.2. Relations among Nielsen type numbers.

In [W3] Peter Wong has studied relations between the extension Nielsen number $N(f|\bar{f})$ and the local Nielsen number $n(f, U)$ of E. Fadell and S. Husseini [3]. Here we need a more general definition of the extension Nielsen number than the one given in Definition 3.2. If $f: (X, A) \to (X, A)$, then the extension Nielsen number $N(f|\bar{f})$ is defined in [BGS] as the number of fixed point classes of $f: X \to X$ which do not assume their index in A and do not intersect ∂A. To define $n(f, U)$, assume that $f: U \to X$ is compactly fixed on the open subset U of X, and call two fixed points of f on U *locally Nielsen equivalent* if there exists a path α in U between them so that α is homotopic to $f \circ \alpha$ in X. Then $n(f, U)$ is defined as the number of such local equivalence classes which are essential, where essential means that the usual fixed point index is not zero. Wong related $N(f|\bar{f})$ to $n(f|X - A, X - A)$, and used his results to compute $N(f|\bar{f})$ for certain maps of pairs of nilmanifolds.

5. Some applications of relative Nielsen theory

In this section, we discuss some cases where definitions, results or methods from relative Nielsen theory have been used to solve problems outside this area.

5.1. Homeomorphisms of surfaces with boundary.

Let us return to the example which motivated the introduction of the relative Nielsen number, namely Example 1.1 with its homeomorphism of the pants obtained by reflection. More generally, let us consider a homeomorphism h of a manifold M with boundary ∂M. Such a homeomorphism must be of the form $h: (M, \partial M) \to (M, \partial M)$, and so it must be a map of pairs. Using Thurston's classification of surface homeomorphisms, Boju Jiang announced in 1981 that if M is a surface, then any homeomorphism h of M is isotopic to an embedding with $N(h)$ fixed points, and remarked that if no boundary component of M is mapped onto itself in an orientation-reversing manner, then the embedding can be chosen to be a homeomorphism. The crucial tool in this proof is the classification of surface homeomorphisms by Thurston, but relative Nielsen theory helps to clarify the last remark, as the absence of orientation-reversing maps on of the boundary circles implies $N(\bar{h}) = 0$ and hence $N(h; M, \partial M) = N(h)$. Relative Nielsen numbers are used in the following sharper result by Boju Jiang and Jianhan Guo [JG].

THEOREM 5.1. *Let M be a compact surface with boundary and $h: M \to M$ a homeomorphism. Then h is isotopic to a diffeomorphism which has $N(h; M, \partial M)$ fixed points on M and $N(h; M - \partial M)$ fixed points in int$M$.*

It is not yet known whether an extension of this result to manifolds with boundary of higher dimensions is true, but it can be expected that the relative Nielsen number will again play some role in the outcome.

5.2. Fixed point sets in a given homotopy class.

Methods developed in relative Nielsen fixed point theory have been used to obtain conditions which ensure that a subset A of a manifold M (or, more generally, a compact polyhedron X) can be realized as the fixed point set of a map in the homotopy class of a given map $f: M \to M$. Such conditions are easy if f is the identity: it is known that every closed and non-empty subset K of M can be the fixed point set of a deformation. If g is homotopic to a map f in an arbitrary homotopy class and if K is the fixed point set of g, then it is clearly necessary that the following two conditions hold: (1) $f|K: K \to M$ is homotopic to the inclusion, (2) K has $\geq N(f)$ components. In 1977 P. Strantzalos claimed that these conditions are sufficient as well as necessary for the realization of K as a fixed point set if the dimension of M is $\neq 2, 4, 5$ and if K can be by-passed in M. But this result is incorrect, and a counter-example exists in which M is a three-dimensional manifold. Although the problem here is not stated as a problem concerning a map of pairs, techniques from relative Nielsen fixed point theory, in particular the uniting of fixed points on $M - K$ with fixed points on K, are used in [S5] to replace the conditions (1) and (2) by correct, but somewhat more complicated, necessary and sufficient ones. An extension of the problem to maps of pairs $f: (M, A) \to (M, A)$ for the case of deformations is solved in [S2], but the general case is open.

5.3. Coincidence-producing maps.

The next problem which we discuss has lead to the introduction of a relative Nielsen number for coincidences, and so it also belongs to Section 3 as a case where the setting used in the definition of the relative Nielsen number has been varied to solve related problems with the help of new Nielsen numbers. But we prefer to discuss it here, because the interest in this topic does not lie so much in the definition of yet another Nielsen number as in the extension of earlier results which concern coincidence-producing maps.

A coincidence of two maps $f, g: X \to Y$ is a point $x \in X$ with $f(x) = g(x)$, and a map $g: X \to Y$ is called *coincidence-producing* (or *universal*) if g has a coincidence with every map $f: X \to Y$. In the case $X = Y$ the identity map is coincidence-producing if and only if X has the fixed point property, and so coincidence-producing maps can be considered to generalize this property. Clearly coincidence-producing maps $g: X \to Y$ can only exist if Y has the fixed point property, for a map $h: Y \to Y$ with $h(y) \neq y$ for all $y \in Y$ defines a map $h \circ g: X \to Y$ which has no coincidence with g. So the simplest choice for Y is the n-ball B^n. It was proved independently by Holsztyński in 1964 and Schirmer in 1966 that a map $g: (B^n, \partial B^n) \to (B^n, \partial B^n)$ is coincidence-producing if and only if the induced homomorphism $g_*: H_n(B^n, \partial B^n) \to H_n(B^n, \partial B^n)$ is not the zero-homomorphism. In particular, the map f in our Disk Example 2.1 is coincidence-producing if and only if $d \neq 0$.

In [BS] Robert Brown and Helga Schirmer used relative Nielsen theory to extend the range of coincidence-producing maps. Classical coincidence theory deals with maps $f, g: M \to N$, where M and N are compact triangulable orientable n-manifolds without boundary. To study coincidence-producing maps, this theory is extended to the case where M and N have a boundary and f and g are maps of the form $f: M \to N$ and $g: (M, \partial M) \to (N, \partial N)$. Hence g plays the role of the generalized identity map. A Lefschetz number for coincidences of such maps f, g was defined by M. Nakaoka [8] in 1980, and a coincidence index and a Nielsen number which have the usual properties can be found in [BS]. Lefschetz and Nielsen numbers for such maps are used to characterise coincidence-producing maps in the case where N is acyclic over the rationals and hence has the fixed point property. The proof of the following theorem uses both Lefschetz and Nielsen fixed point theory for a map of pairs.

THEOREM 5.2. *Let $g: (M, \partial M) \to (N, \partial N)$, where M and N are compact, oriented, triangulable n-manifolds with boundary.*

(1) *If $n = 1$, then g is coincidence-producing if and only if it is onto;*
(2) *if $n \geq 2$ and N is acyclic over the rationals, then g is coincidence-producing if and only if $g_*: H_n(M, \partial M) \to H_n(N, \partial N)$ is a non-zero homomorphism.*

This theorem extends the earlier results for coincidence-producing maps onto n-balls considerably, as there exist many examples of compact orientable triangulable $\mathbb{Q}$-acyclic manifolds with boundary and thus many new examples of coincidence-producing maps. Some can be found in [BS].

5.4. Periodic points, equivariant maps and fibre-preserving maps.

In [H2], and its announcement in [H1], Philip Heath "dualised", as he called it, the relative Nielsen number to propose a new Nielsen type number $N_{\mathcal{F}}(f,p)$ for a fibre preserving map f of an essentially fibre uniform fibration $p\colon E \to B$. The term "essentially fibre uniform" means that the Nielsen number $N(f_b)$ of the restriction of f to the fibre over $b \in B$ is independent of b for any point b which lies in an essential fixed point class of the induced map $\bar{f}\colon B \to B$. The definition of the number $N_{\mathcal{F}}(f,p)$ is reminiscent of the naive product formula, and Heath shows that $N_{\mathcal{F}}(f,p) = N(f;E,F_{\xi})$, where F_{ξ} is a certain subspace of E which depends on $\bar{f}$. This connection with relative Nielsen theory simplifies some of the proofs considerably, in particular the one which establishes the lower bound property of $N_{\mathcal{F}}(f,p)$ for essentially fibre uniform maps.

The method used in [S1], and in the case of a deformation in [S2], to unite a fixed point on $X - A$ with a fixed point on ∂A has helped Peter Wong in problems concerning equivariant maps. The deformation case was used in [W1] for a study of the location of fixed points of G-deformations on manifolds, and the general case in [W2] for the proofs of minimum theorems for fixed orbits and fixed points in the G-homotopy class of a G-invariant map on a manifold.

Since some techniques used in equivariant and in relative Nielsen fixed point theory are similar, relative Nielsen theory can also be used to study $NP_n(f)$, the Nielsen number of period n, after relating periodic points to fixed points of an equivariant map. More precisely, a map $f\colon X \to X$ defines a $\mathbb{Z}_n$-equivariant map g_f of the n-fold product $Y = X \times X \times \cdots \times X$ by $g_f(x_1, x_2, \ldots, x_n) = (f(x_n), f(x_1), \ldots, f(x_{n-1}))$. Let $Y_{(1)}$ be the subspace of Y on which the $\mathbb{Z}_n$-action is free. Then $B = Y - Y_{(1)}$ is a closed connected subspace of Y, and therefore $g_f\colon (Y,B) \to (Y,B)$ is a map of pairs. It is easy to see that $\mathrm{Fix} g_f = \{(x, f(x), \ldots, f^{n-1}(x)) | x \in \mathrm{Fix} f^n\}$, and therefore the fixed points of g_f on $Y_{(1)} = Y - B$ correspond to the periodic points of f with least period n. This fact was used by Peter Wong in [W4] to obtain some information about $NP_n(f)$ from the Nielsen number of the complement space $N(g_f; Y - B)$. In particular $N(g_f; Y - B) \leq NP_n(f) \leq n \cdot N(g_f; Y - B)$, and [W4] contains some algebraic conditions under which equality holds in one of the two inequalities.

5.5. Application to dynamical systems.

We finish with an application to dynamical systems theory. The asymptotic growth rate of the number of periodic orbits is an important index of complexity in dynamics. To obtain estimates for it, Boju Jiang ([J2]; see also [J3] for an exposition) considered homeomorphisms $h\colon (M,P) \to (M,P)$, where M is a closed surface and P a given finite subset which represents punctures of M. His focus is on the number of fixed points in $M - P$ under isotopies of h relative to P. Jiang's approach is to blow up each puncture, i.e. each point in P, to a circle and then to re-compactify $M - P$ into a surface $\widehat{M}$ which has one boundary component for each point in P. As $\widehat{M} - \partial\widehat{M}$ can be identified with $M - P$, it is possible (after dealing with a technical difficulty) to extend $h|M - P$ to a homeomorphism $\hat{h}\colon (\widehat{M}, \partial\widehat{M}) \to (\widehat{M}, \partial\widehat{M})$ and then study the fixed points of $\hat{h}$ on $\widehat{M} - \partial\widehat{M}$. The reason for this approach is the fact that the fundamental group of $\widehat{M}$ is much richer than that of M, and so Nielsen numbers for $\hat{h}$ can provide more information than those for h. For example, if M is the 2-sphere, then $N(h; M - P) \leq 1$ as M is simply connected. But $\pi_1(\widehat{M})$ is a free group, and so $N(\hat{h}, \widehat{M} - \partial\widehat{M})$ can be a much better lower bound for the number of fixed points of h on $M - P$. (The technical difficulty in the extension of $h|M - P$ to a homeomorphism of $\widehat{M}$ is dealt with by smoothing h near P before extending it, and then showing that $N(\hat{h}; \widehat{M} - \partial\widehat{M})$ is independent of the local smoothing.) When the zeta function technique is used to study the periodic orbits of $\hat{h}$, one can often obtain isotopy-invariant lower bounds for the asymptotic growth rate of the number of periodic orbits of h.

6. Relative Nielsen theory - where will it go?

We have seen that in the short time since its beginning, relative Nielsen theory has developed into a very active part of Nielsen fixed point theory. It started with an extension of classical Nielsen theory to maps of pairs of spaces $f\colon (X, A) \to (X, A)$ in order to explain some examples where the Nielsen number $N(f)$ cannot be realized. At first stronger results about the least number of fixed points on X for maps of pairs of spaces (X, A) were obtained, and soon further information about the location of such fixed points and their number on $X - A$ and $c\ell(X - A)$ were found. It was also noticed that the methods developed in this work can be used to obtain information about fixed points, periodic points and coincidences in related settings, and can help to solve problems in some other areas.

I did not expect this rapid development of relative Nielsen theory at its start in 1986, and needless to say I find it impossible to predict what further progress it will make. But I can finish with a list of some open problems which may help in further research. Some have already been mentioned in previous sections.

(1) Find an optimal lower bound for the number of fixed points on X for maps in the homotopy class of $f\colon (X, A) \to (X, A)$ if A cannot be by-passed in X. It is known from an example by Xuezhi Zhao [Z2] that the relative Nielsen number $N(f; X, A)$ can be too small. For the number of fixed points on $X - A$ the corresponding problem has been solved with a "surplus number" $SN(f; X - A) \geq N(f; X - A)$ in [Z2], but the approach of this paper does not seem to work for fixed points on the total space X.

(2) Find an optimal lower bound for the number of fixed points on $c\ell(X - A)$ if (X, A) satisfies one of the following two conditions:

(i) A cannot be by-passed in X,
(ii) A has a component which has an empty interior.

Nothing has been published so far which concerns these questions, and it is possible that the answer is too complicated to be enjoyable.

(3) Find "smooth" versions of the various relative Nielsen numbers discussed in this paper. This problem has only been solved for map extensions (see Section 3.1), where surprisingly a smooth and a continuous setting have lead to different answers. It is known that no difference between minimal fixed point sets on a manifold of dimension ≥ 3 exists in the non-relative case, as $N(f)$ can always

be realized smoothly [5]. But in the relative case not all steps in the proofs of the various minimum theorems which exist for continuous maps are smooth. In particular, the construction used in [S1] to unite a fixed point on $X - A$ with a fixed point on ∂A is not a smooth one.[1]

(4) Boju Jiang [J1] has shown that any orientable compact 3-manifold with boundary is boundary-Wecken if and only if all of its boundary components are Wecken. The non-orientable case is open. A similar result can be expected, but the proof from [J1] cannot not be modified to include all non-orientable cases.

(5) Only a few results exist which help to estimate or compute the various relative Nielsen numbers, and there has been only a very partial success in extending the algebraic tools of classical Nielsen fixed point theory to the various relative settings. (See e.g. [NW], [W3] and [Z1]). But it may be possible to extend more. Estimating and calculating the various new Nielsen numbers is an important and basic problem in relative Nielsen theory, but it may not be easy to get useful results.

(6) Try to obtain characterisations of classes of manifolds with boundary, or of maps on their boundary, so that smooth and continuous map extensions can have different minimal fixed point sets. The results from [BGS] and [6] described in Section 3.1 contain only specific examples.

(7) Decide whether Theorem 5.1 is true in higher dimensions. This problem needs, of course, first an extension in the non-relative case. Partial results in this case concerning embeddings for manifolds of dimension ≥ 5 are contained in a preprint by Michael Kelly. (See also [B] for an exposition.)

(8) Characterise all possible fixed point sets in the homotopy class of a relative map $f\colon (X, A) \to (X, A)$. Techniques exist to solve this problem, but the answer may be messy. If so, try and find cleaner solutions in special cases.

(9) It is of considerable interest to find more applications to other areas. Existing ones are mainly within the area of fixed point theory, but we have described an application to dynamical systems theory in Section 5.5. As fixed point theory is widely used, other applications should be possible.

Acknowledgement. I want to thank Robert Brown and Boju Jiang for their critical readings of this paper. Boju Jiang also helped me with Section 5.5.

References

(i) References on relative Nielsen theory.

[B] R.F. Brown, *Wecken properties for manifolds*, these Proceedings.

[BGS] R.F. Brown, R. E. Greene and H. Schirmer, *Fixed points of map extensions*, Topological Fixed Point Theory and Applications (Proceedings, Tianjin 1988), Lecture Notes in Math, vol. 1411, Springer-Verlag, 1989, pp. 24–45.

[BS] R.F. Brown and H. Schirmer, *Nielsen coincidence theory and coincidence-producing maps of manifolds with boundary*, Topology and its Appl. **46** (1992), 65–79.

[BSa] R.F. Brown and B. Sanderson, *Fixed points of boundary-preserving maps of surfaces*, to appear, Pacific J. Math..

[1] This problem has recently been solved. Smooth realizations of $N(f; X, A)$, $N(f; X-A)$ and $N(f; \overline{X - A})$ have been obtained by Robert E. Greene and Helga Schirmer, and are contained in a forthcoming paper "Smooth realizations of relative Nielsen numbers".

[H1] P. R. Heath, *Nielsen type numbers for fibre preserving maps, coincidences of fibre preserving maps, and for periodic points of fibre preserving maps*, C.R. Rep. Acad. Sci. Canada **14** (1992), 25–30.

[H2] ———, *A Nielsen type number for fibre preserving maps*, to appear, Topology and its Appl.

[HSY] P. R. Heath, H. Schirmer and C. You, *Nielsen type numbers for periodic points of pairs of spaces*, preprint.

[J1] B. Jiang, *Commutativity and Wecken properties for fixed points on surfaces and 3-manifolds*, to appear.

[J2] ———, *Estimation of the number of periodic orbits*, preprint.

[J3] ———, *Nielsen theory for periodic orbits and applications to dynamical systems*, these Proceedings.

[JG] B. Jiang and J. Guo, *Fixed points of surface diffeomorphisms*, to appear, Pacific J. Math.

[K] M. R. Kelly, *The relative Nielsen number and boundary-preserving surface maps*, preprint.

[NW] B. Norton-Odenthal and P. Wong, *A relative generalized Lefschetz number*, preprint.

[S1] H. Schirmer, *A relative Nielsen number*, Pacific J. Math. **122** (1986), 459–473.

[S2] ———, *Fixed point sets of deformations of pairs of spaces*, Topology and its Appl. **23** (1986), 193–205.

[S3] ———, *Some recent results on fixed point theory*, Math. Student **54** (1986), 65–72.

[S4] ———, *On the location of fixed point sets of pairs of spaces*, Topology and its Appl. **30** (1988), 253–266.

[S5] ———, *Fixed point sets in a prescribed homotopy class*, Topology and its Appl. **37** (1990), 153–162.

[S6] ———, *Nielsen theory of transversal fixed point sets (with an appendix by Robert E. Greene)*, Fund. Math. **141** (1992), 65–79.

[S7] ———, *Nielsen numbers for maps of triads*, to appear, Topology and its Appl.

[Schu] A. L. Schusteff, *Product formulas for relative Nielsen numbers of fibre map pairs*, Ph. D. Thesis, UCLA, 1990.

[W1] P. Wong, *On the location of fixed points of G-deformations*, Topology and its Appl. **39** (1991), 159–165.

[W2] ———, *Equivariant Nielsen numbers*, to appear, Pacific J. Math.

[W3] ———, *A note on the local and the extension Nielsen numbers*, to appear, Topology and its Appl.

[W4] ———, *Estimation of Nielsen type numbers for periodic points*, to appear.

[Z1] X. Z. Zhao, *A relative Nielsen number for the complement*, Topological Fixed Point Theory and Applications (Proceedings, Tianjin 1988), Lecture Notes in Math., vol. 1411, Springer Verlag, 1989, pp. 189–199.

[Z2] ———, *Estimation of the number of fixed points on the complement*, Topology and its Appl. **37** (1990), 257–265.

[Z3] ———, *Basic relative Nielsen numbers*, Topology – Hawaii, World Scientific, Singapore, 1992, pp. 215–222.

[Z4] ———, *Estimation of the number of fixed points of map extension*, to appear, Acta Math. Sinica B.

(ii) Some other references.

[1] C. Bowszyk, *Fixed point theorems for the pairs of spaces*, Bull. Acad. Polon. Sci. **16** (1968), 845–850.

[2] R. F. Brown, *The Lefschetz Fixed Point Theorem*, Foresman and Co., Glenview, IL, 1971.

[3] E. Fadell and S. Husseini, *Local fixed point index theory for non-simply-connected manifolds*, Ill. J. Math. **25** (1981), 673–699.

[4] P. R. Heath, H. Schirmer and C. You, *Nielsen type numbers for non-connected spaces*, preprint.

[5] B. Jiang, *Fixed point classes from a differentiable viewpoint*, Fixed Point Theory (Proceedings, Sherbrooke, Quebec, 1980), Lecture Notes in Math., vol. 886, Springer-Verlag, 1981, pp. 163–170.

[6] ———, *Lectures on Nielsen Fixed Point Theory*, Contemporary Mathematics, vol. 14, Amer. Math. Soc., Providence, RI, 1983.

[7] T. H. Kiang, *The Theory of Fixed Point Classes*, Springer-Verlag, Berlin, 1989.

[8] M. Nakaoka, *Coincidence Lefschetz numbers for a pair of fibre-preserving maps*, J. Math. Soc. Japan **32** (1980), 751–779.

CARLETON UNIVERSITY, OTTAWA, ONT., K1S 5B6, CANADA

E-mail address: schirmer@carleton.ca

Contemporary Mathematics
Volume 152, 1993

CLASSIFICATION of LIFTS of AUTOMORPHISMS of SURFACES to the UNIT DISK

LEV SLUTSKIN

1. Introduction

The goal of this paper is to prove Nielsen's results [7] on the fixed points of lifts of automorphisms of a compact surface S with boundary of genus p (p>1) to its universal covering space (the unit disk *U*), on its boundary ∂U based on Thurston's classification theorem [10] for automorphisms of surfaces. Therefore, our work can be seen as complementary to the results of Miller [13] and Handel and Thurston [12] who showed that Thurston's theorem follows from the Nielsen theory. Besides the fact that in the light of the Thurston theory Nielsen's results get a simple geometric interpretation as a combination of lifts of irreducible pieces of S, it, also, gives us a convenient tool for constructing new examples of different boundary behavior of lifts of automorphisms to U. In particular, we give the example, asked by Nielsen [7], of g*, a lift of an automorphism g, with fixed points on intervals of regularity of the group of hyperbolic transformations fixed by g*.

First, we study the lifts of pseudo-Anosov automorphisms of *S* to *U*. Here we treat simultaneously the compact case and that of a surface with punctures. The corresponding theorem for compact surfaces was stated by Thurston in [10] and proved, later, by Fathi and Laudenbach [3]. The case of a punctured surface

1991 Mathematics Subject Classification. Primary 32G15, 57S30. The final version of this paper will be submitted for publication elsewhere.

was studied by Marden and Strebel [5] under the assumption that a lift of a pseudo-Anosov automorphism has either a fixed point inside U, or its fixed point is a parabolic fixed point. We show that the only remaining case is that of a lift with exactly two fixed points on the boundary of U. Then we consider the general case of a surface with boundary by shrinking the boundary curves of S into punctures. In the last paragraph we study arbitrary automorphisms of S by dividing it into irreducible pieces.

Finally I would like to thank Professor Frederick Gardiner with whom I discussed the original version (the compact case) of theorem 3. I, also, want to thank Professor Jane Gilman for her advice and interest in my work.

2. Lifts of pseudo-Anosov automorphisms

Our treatment of pseudo-Anosov automorphisms is based largely on the work of Marden and Strebel [5] where they studied pseudo-Anosov automorphisms by applying their earlier results on geometric properties of lifts to U of trajectories of quadratic differentials on a Riemann surface. Their approach to studying pseudo-Anosov automorphisms through the associated quadratic differentials originates in the observation of Bers [1] that a pseudo-Anosov automorphism is a Teichmuller self-mapping with the same initial and terminal quadratic differential.

2.1. Compact case. Let S be a compact oriented surface of genus $p>1$, and $\Phi=(\Phi_1,\Phi_2)$ a pair of transverse measured foliations on S. Then there exists a conformal structure $\tau: S \to X$ on S, where X is a Riemann surface of the same genus, and q, a quadratic differential on X, such that Φ coincides with the pair $\Phi_q=(\Phi_h,\Phi_v)$, where Φ_h denotes the foliation of S composed by the horizontal geodesics of q and, Φ_v by the vertical ones, respectively. Indeed, the open rectangles with the horizontal sides on leaves of Φ_1, and the vertical sides on leaves of Φ_2, which do not have inside singularities of Φ, define local coordinates on S outside critical points. If we add to them the interiors of the unions of the closed rectangles around singular points, we obtain a Riemann surface X. Now, we define a holomorphic quadratic differential q on X. $q=dz^2$ in any rectangle which does not contain a singular point, and $q = 4^{-1}n^2z^{n-2}dz^2$ in a neighborhood of a singular point, where n is equal to the number of the leaves of Φ_h (Φ_v) stemming from the singular point. We note, here, that a linear element $ds=\int|q^{1/2}|$ defines transverse measures for Φ_h and Φ_v identical with corresponding transverse measures for Φ_1 and Φ_2. When g is a pseudo-Anosov automorphism of S with the associated pair of transverse measured foliations $(\Phi_u,\Phi_s)=(\Phi_1,\Phi_2)$ then $\tau g\tau^{-1}$ is a Teichmuller self-mapping of X with the same initial and terminal quadratic differential equal q.

Now, we recall some properties of lifts of quadratic differentials to the unit disk U. Consider q^*, the lift of q to U, the universal covering space of X. Then $q^*dz\geq 0$, $q^*dz\leq 0$ determine, respectively, the horizontal and vertical geodesics of q^* in U. The following results were proved by Marden and Strebel [4].

a) Each geodesic γ from a point in U tends to its end point γ_e, a uniquely determined point on ∂U.

b) Let γ_1, γ_2 be two geodesics from the same point in U. Then γ_1, γ_2 have different end points on ∂U.

Let g^* be a lift of g to U.

Theorem 1. (Thurston [10])

1. g^* does not have fixed points in U.
Then g^* has exactly two fixed points on ∂U. One of them is attracting and another is repelling fixed point. Positive iterates of g^* of any points of U converge to the attracting fixed point.

2. $g^*(z)=z$, $z\in U$.
Then z is the only fixed point of g^* in U. g^* has $2n$ periodic points on ∂U, where n is the number of horizontal (vertical) geodesics from z. The end points of horizontal geodesics from z are the periodic attracting points of g^*, and the end points of vertical geodesics from z are, respectively, the periodic repelling points of g^*, and g^* permutes the fixed points on ∂U in the same manner as g permutes horizontal and vertical geodesics from z.

2.2. **Non-compact case.** Let S be a compact oriented surface of genus p with m $(m\geq 0)$ points removed. We always assume that $2p-2+m>0$. We say that an orientation preserving automorphism of S is a *pseudo-Anosov automorphism* if there exists a conformal structure $\tau: S\rightarrow X$ on S, where X is a Riemann surface of the same genus with m punctures on it, and q, a quadratic differential on X, with, at most, simple poles at punctures, such that $\tau g\tau^{-1}$ is a Teichmuller self-mapping of X with the same initial and terminal quadratic differential equal q.

We treat simultaneously the compact and non-compact cases. The treatment of the non-compact case is similar to that of the compact one, if we notice that, geometrically, there is a similarity between parabolic fixed points on ∂U and fixed points inside U. Let z be a singular point (a zero or a pole) of q. Then $ord_q(z)$ denote the order of z. It follows that $-1 \le ord_q(z) \le 4p-4+m$. The number of horizontal (vertical) geodesics from z is equal $2(ord_q(z)+2)$.

Theorem 2. (Marden & Strebel [5]) *Let g^* be a lift of a pseudo-Anosov automorphism g of a surface S with a finite number of punctures, and let $g^*(z)=z$, $z \in U$ is a singularity of q^*; or $g^*(z)=z$, $z \in \partial U$ is a parabolic fixed point which corresponds to a puncture of S, and g^* preserves the directions from z. Then the end points of horizontal geodesics from z are the attracting fixed points, and the end points of vertical geodesics from z are, respectively, the repelling fixed points of g^*. Moreover, let β, β' be two adjacent vertical geodesics from z. Then the positive iterates of any point inside the sector of $\bar{U}$ bounded by β and β' which contains exactly one horizontal geodesic from z, say α, converge to α_e. When $z \in \partial U$, there are a finite number N, $N \le 2(4p-2+m)$, of λ-orbits of fixed points of g^* on ∂U, different from z, where λ is a parabolic element of the Fuchsian group of S which fixes z.*

NOTE 1. z is neither attracting nor repelling fixed point of g^*.

NOTE 2. In theorem 2 the same proof may be carried out

1. For any $z \in U$, such that $g^*(z)=z$.
2. When g^* permutes directions from z.

It implies the following theorem.

Theorem 2'. *Let g^* be a lift of a pseudo-Anosov automorphism g of a surface S with a finite number of punctures, and let $g^*(z)=z$, $z \in U$, or $g^*(z)=z$, $z \in \partial U$, is a parabolic fixed point, and let g not preserve the directions from z. Then if*

1. $z \in U$.

Then the end points of horizontal and vertical geodesics from z are the periodic points of g^* with the same period P, $P \le 2^{-1}(4p-2+m)$, which is determined by the angle of rotation of directions about z. z is the only fixed point of g^*. It is neither attracting nor repelling. Moreover, let β, β' be two adjacent vertical geodesics from z. Then the positive iterates of

any point inside the sector bounded by β and β' in $\bar{U}$ which contains exactly one horizontal geodesic from z, say α, converge to the $g*$-orbit of α_e. Points on β and β' converge to z.
2. $z \in \partial U$.

Then the iterates of any point of $\bar{U}$ converge to z. z is neither attracting nor repelling fixed point of $g*$.

We describe, now, the procedure that will allow us to complete the classification of fixed points of $g*$. First, we recall the divergence principle for the end points of geodesics of quadratic differentials.

Theorem. Divergence principle for the end points. (Marden & Strebel [4]) *Let α be a horizontal straight segment in U, and β, β' be two vertical geodesics from the end points of α which projections on S are not closed loops. Then β, β' have different end points on ∂U.*

We assume, first, that S is compact. Let Φ_h be the collection of all horizontal geodesics of $q*$. We show how to make a partition of Φ_h into an infinite countable number of layers. This partition is uniquely determined by an arbitrary singular point $z_0 \in U$. The collection of finitely many horizontal geodesics from z_0 forms Φ_0, *the zero layer of Φ_h with respect to z_0*. Let k_0 ($k_0 \geq 1$) be the order of z_0. Then there exist k_0+2 critical vertical geodesics from z_0: $\beta_1, \beta_2, \ldots, \beta_{k+2}$. The collection of all horizontal geodesics through points on $\beta_1, \beta_2, \ldots, \beta_{k+2}$, except those of Φ_0, forms Φ_1, *the first layer of Φ_h with respect to z_0*. Consider $\gamma \in \Phi_1$. Then there are two possibilities: either γ is not a critical geodesic or there is a singular point z of order $k>0$ which lies on γ. In the latter case there are $k+1$ horizontal geodesics from z which do not intersect any of $\beta_1, \beta_2, \ldots, \beta_{k+2}$. They comprise k sectors around z. No point inside these sectors belongs to a geodesic from Φ_1. In each of these sectors we consider the geodesics of the first layer with respect to z. We repeat the same procedure at each singular point of Φ_1. The collection of all geodesics obtained in this manner form Φ_2. When we iterate the process we obtain Φ_3, Φ_4, ..., Φ_n, ... On each step of iteration Φ_n ($n>0$) is composed of geodesics of the first layer with respect to singular points of Φ_{n-1}.

Theorem. (Slutskin [8]) *Φ_0, Φ_1, ..., Φ_n, ... form a partition of Φ_h, i. e., $\Phi_h = \cup\Phi_n$ and $\Phi_i \cap \Phi_j = \emptyset$, where i, j = 0, 1, 2, ..., and $i \neq j$.*

NOTE. The partition of Φ_h into layers could be carried out, in a similar manner, when z_0 is a parabolic fixed point (S is a surface with punctures). In this case we have to consider a countable infinite number of sectors about z_0 or any parabolic fixed point which we may encounter in the process of the partition.

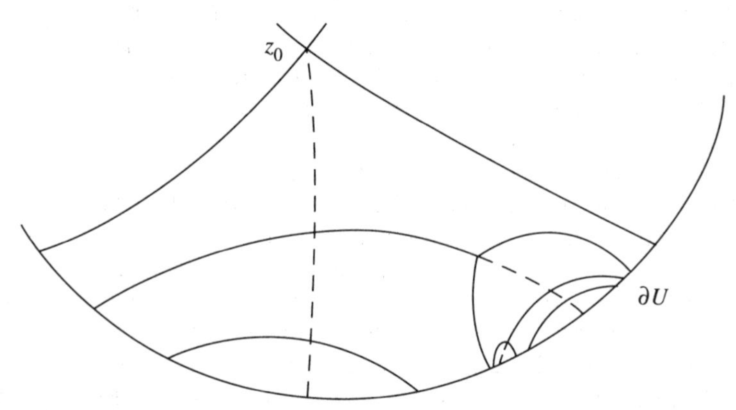

Fig. 1. The partition of Φ_h into layers

Theorem 3. *If g^* has three or more fixed points on ∂U, then g^* has either a fixed point inside U, or one of its fixed points is a parabolic fixed point.*

Proof. Assume that g^* has neither a fixed point inside U, nor any of its fixed points is a parabolic fixed point.

Lemma 1. *Let g^* has neither a fixed point inside U, nor any of its fixed points is a parabolic fixed point. Let $x \in \partial U$ be a fixed point of g^*. Then x is not the end point of any horizontal or vertical geodesic.*

Proof. Indeed, let x be the end point of a horizontal or vertical geodesic γ. Then since x is not a parabolic fixed point, it follows that $g^*(\gamma)=\gamma$. g^* stretches or contracts γ. It implies that g^* has, at least, one fixed point on γ. Contradiction. ∎

Let z be a singular point of q^* or a parabolic fixed point. Then horizontal geodesics from z form, respectively, a finite or an infinite number of sectors about z.

Lemma 2. *Let z be a singular point of q^* or a parabolic fixed point, such that $g^*(z)\neq z$. Then the fixed points of g^* on ∂U may belong, at most, to two sectors about z.*

Proof. g^* sends the sectors about z onto the corresponding sectors about $g^*(z)$. Let S be the sector about z, such that $g^*(z)\in S$; and S', the sector about $g^*(z)$ which contains z. It follows that g^* sends all sectors about z, except $(g^*)^{-1}(S')$, into S. It implies that all fixed points of g^* must be either in S or in $S_1=(g^*)^{-1}(S')$. In order to prove the theorem it is enough to find z, a singular point of q^* or a parabolic fixed point, such that there are fixed points of g^* in three different sectors about z. First of all let us find z, a singular point of q^* or a parabolic fixed point, such that there are fixed points of g^* in two different sectors about z. Let z_0 be a singular point of q^*. If there are two sectors about z_0 which contain fixed points of g^* we are done. If not, let S_0 be the sector about z_0 which contains the fixed points of g^*. Consider the first layer of Φ_h with respect to z. By lemma 1 there is no geodesic of the first layer having a fixed point of g^* as its end point. It implies that there exists z_1, a singular point of q^* or its parabolic fixed point, such that the fixed points of g^* are in the sectors about z_1. If there are two sectors about z_1 which contain the fixed points of g^* we are done. If not, let S_1 be the sector about z_1 which contains the fixed points of g^*. In this way we obtain the descending sequence of sectors $S_0 \supset S_1 \supset \ldots \supset S_n \ldots$, such that each of them contains all fixed points of g^*. Then there are two possibilities - either it ends after a finite number of steps, or not. In the former case it follows that there exists a singular point of $q*$ or its parabolic fixed point, such that the fixed points of $g*$ are in two different sectors about it. In the latter case $\cap S_n$ is a point on ∂U (see [8]). Contradiction.

Let z be as above, i. e., the fixed points of $g*$ belong to two different sectors about z. It follows that, at least, one of them, S_1, contains x_1, x_2, fixed points of $g*$. Let y be a fixed point of $g*$ which does not belong to S_1. Again, we can find the finite descending sequence of sectors $S_0 \supset S_1 \supset \dots \supset S_n$, such that each of them contains x_1, x_2. Let S_n be the last sector in this sequence. We have the following two cases.
1. For some $z_1 \in S_n$, there are two different sectors about it, S' and S'', containing x_1 and x_2, respectively. Then the sector about z_1 which contains geodesics of the n-th order with respect to z contains y, and it is different from S' and S''. Contradiction.
2. There are two points inside S_n, z_1 and z_2, such that one of the sectors about z_1 contains x_1, and one of the sectors about z_2 contains x_2. Let α_1, α_2 be horizontal geodesics from z_1 and z_2, respectively, which intersect β, the vertical geodesic which bisects S_n. and let b_1, b_2 be the respective points of intersection. Consider z', the vertex of S_n. Assume that b_1 precedes b_2 when we move from z' along β. Then there are three different sectors about z_1 which contain x_1, x_2, and y, respectively. Contradiction. ∎

The following theorem completes the classification of lifts of pseudo-Anosov automorphisms.

Theorem 4. *Let $g*$ not have fixed points in U nor on ∂U among its parabolic fixed points. Then*
1. $g*$ has exactly two fixed points on ∂U. These fixed points are not the end points of any horizontal or vertical geodesic.
2. The positive iterates of any point of $\bar{U}$ different of the two
fixed points of $g*$ converge to one of them.

Proof. *1*. It follows from theorem 3 that $g*$ may

have only one or two fixed points on ∂U. Let us show that the first case cannot happen. Indeed, let $x \in \partial U$ be the only fixed point of g^*. Take z, an arbitrary singularity inside U. Let S be a sector about z which contains x, and x_1 and x_2 the end points of geodesics from z which compose S. Consider S_1, $S_1 \subsetneq S$, a sector about a singular point z_1 which contains x and contains neither $(g^*)^{-1}(x_1)$ nor $(g^*)^{-1}(x_2)$. Since g^* is monotone on ∂U, it follows that the end points of geodesics from $g^*(z_1)$ on ∂U are separated by x in S. It implies that the geodesics from $g^*(z_1)$ intersect the geodesics from z_1. Contradiction.

By lemma 1 of theorem 3 the fixed points of g^* are not the end points of horizontal or vertical geodesics. ∎

2. Let x be the attracting and y the repelling fixed points of $g^*|_{\partial U}$.
Consider $x' \in \bar{U}-\{x,y\}$. Pick a singularity $z \in U$, such that the sector S about z contains y, but not x or x'. Let $B=\bar{U}-S$. Consider the descending sequence of g*-iterates of B: $B \supset g^*(B) \supset \ldots \supset (g^*)^n(B) \supset \ldots$.

Lemma. $\cap (g^*)^n(B)=x$.

Proof. Let $c=B \cap \partial U$. Then $c \supset g^*(c) \supset \ldots \supset (g^*)^n(c) \supset \ldots$ is a descending sequence of arcs on ∂U, such that $\cap (g^*)^n(c)=x$. Let $z \in U \cap (g^*)^n(B)$. Consider γ, a horizontal geodesic through z. It follows that $\gamma \in \cap (g^*)^n(B)$, and it implies that $\gamma_e=x$. But this contradicts lemma 1 of theorem 3. ∎

2.3. Surface with boundary. Assume, now, that S is an oriented surface with boundary, i. e., S may be obtained from a compact surface of genus p by removing k open disks. We always assume that $2p-2+k>0$. It means that the projective map $\pi: U \to S$ sends the maximal open intervals of ∂U which do not contain limit points of the corresponding Fuchsian group, onto boundary curves of S. These maximal intervals are called *the intervals of regularity*. We show that the study of lifts of automorphisms of S is conducive to the study of lifts of surfaces with punctures. Let $Int(S)$ denote S without its boundary curves.

Theorem 5. *Let $f*$, $f*: U \to U$, be a lift of a homeomorphism f. $f: Int(S) \to S'$, where S' is a surface with k punctures, to the respective universal covering spaces. Then $f*$ can be extended continuously onto ∂U in such a way that it sends intervals on ∂U corresponding to the boundary curves of S to the points on ∂U corresponding to the respective punctures of S'.*

Proof. First, we notice that $f*$ can be extended to the intervals of regularity on ∂U. Indeed, let U' denote the union of U and the intervals of regularity. Choose any point $z \in U$. If x is a point belonging to an interval of regularity, we define $f*(x)$ by the formula $f*(x) = lim\ f*(a)$, where a is a path in U from z to x. It is easy to see that this definition does not depend on a and $f*$ is continuous on U'. We note here that $f*$ sends the intervals of regularity to the points on ∂U corresponding to the respective punctures of S'. Next, we extend $f*|_{\partial U'}$ continuously to all ∂U. To show that $f*$ is continuous on ∂U, we consider $G(S)$, the Fuchsian group corresponding to the projective map $\pi: U \to S$. Let $x \in \partial U$ and x does not belong to any interval of regularity. We have two cases.

1. x is not the end point of any interval of regularity.

We consider a sequence of non-Euclidean lines $c_1, c_2, \ldots, c_l, \ldots$, with the boundary points at the end points of intervals of regularity, from the different sides of x and converging to x. Their images under $f*$ are arcs in U with boundary points at parabolic fixed points. Since $f*$ is a homeomorphism of U it follows that all $\{f*(c_l)\}$, but a finite number, are outside the disk of radius $1-\rho$, $\rho > 0$. We note that none of the non-Euclidean lines $\{c_l\}$, $l = 1, 2, \ldots$, intersects the hyperbolic axes of the elements of the Fuchsian group corresponding to the intervals of regularity. It implies that $\{f*(c_l)\}$, $l = 1, 2, \ldots$, do not intersect the images of the hyperbolic axes under $f*$ which are loops coming out of parabolic fixed points. It follows that $\{f*(c_l)\}$, $l = 1, 2, \ldots$, converge to $f*(x)$. ∎

2. x is the end point of an interval of

regularity I.

We consider a sequence of non-Euclidean lines converging to x, each of them having one of its boundary points at the end point of an interval of regularity different from I and another inside I. Now we repeat the argument in 1. ∎

Let $G(Int(S))$, $G(S')$ be the Fuchsian groups of $Int(S)$, and $G(S')$, respectively. Then $f*$ induces an automorphism μ: $G(Int(S)) \to G(S')$.

Corollary. *μ sends the hyperbolic elements of $G(IntS)$ with the fixed points at the end points of interval of regularity onto the parabolic elements of $G(S')$.*

DEFINITION. g, an automorphism of $Int(S)$, is called a *pseudo-Anosov automorphism of* $Int(S)$ if there exists a homeomorphism f, f: $Int(S) \to S'$, where S' is a surface with k punctures, such that fgf^{-1} is a pseudo-Anosov automorphism of S'.

Lemma 3. *Let g be a pseudo-Anosov automorphism of $Int(S)$. Then there exists g_1, a pseudo-Anosov automorphism of $Int(S)$ homotopic to g, which may be extended homeomorphically to the boundary of S.*

Proof. Let τ, τ: $Int(S) \to X$, be a conformal structure on $Int(S)$, and $g'=\tau g\tau^{-1}$ a Teichmuller self-mapping of X with the same initial and terminal quadratic differential. Let $\bar{S}$, $S\subset\bar{S}$, be a closed Riemann surface, such that the boundary curves of S are the boundaries of disks on $\bar{S}$. In other words, $\bar{S}$ is obtained from S by closing its holes. If we puncture $\bar{S}$ in the centers of these discs we obtain a punctured surface $\bar{S}'$ with k punctures. Now, we contract an annulus around each boundary curve of S to a disk on $\bar{S}$ in such a way that the outer circle remains fixed and the inner one is mapped to the center. We assume that the contraction occurs uniformly along radii towards the center. Let J: $S \to \bar{S}$ denote the contracting map. It follows that J sends $Int(S)$ onto $\bar{S}'$. Consider f, f: $\bar{S}' \to X$ a Teichmuller mapping homotopic to τJ^{-1}. Then $g_1=(fJ)^{-1}g'fJ$ is an automorphism of $Int(S)$. Now, we show how to extend g_1 to the boundary of S.

Let B be a boundary curve of S and $x \in B$. Consider $\bar{u}$, a radial direction from the center of the disk to x. Since a Teichmuller mapping has directional derivatives in all points, including singularities, it follows that $f^{-1}g'f$ determines a direction from the center of the disk on $\bar{S}$ which depends continuously on $\bar{u}$. Let $g_1(x)$ denote the point on the boundary curve of S corresponding to this direction. By substituting a small angle with the vertex at x for $\bar{u}$ the same argument shows that g_1 is continuous at x. Since we could consider $(g_1)^{-1}$ instead of g_1 it follows that g_1 is an automorphism of S. ∎

Corollary 1. *Let a be a critical direction from the puncture of X corresponding to a boundary curve of S. Then $\tau_1^{-1}(a)$ converges to a point on the boundary curve of S.*

Corollary 2. *Let $g_1(B)=B$, where B is a boundary curve of S. Then, either g_1 does not have fixed points on B, or the fixed points of g_1 on B are the end points of critical directions from B.*

Proof. Let $\bar{u}_1$ be a radial direction from the puncture $x_1 = fJ(B)$ on X. In local coordinates in a neighborhood of x_1, g' is the composition of the map $w = z^{n+2/2}$, where n $(n>-2)$ is the order of x_1, and the affine map with the dilatation $K>1$, $\sqrt{K}$ being the stretching factor of g_1. g' either rotates the critical directions from x_1, or fixes them. It follows that g' may preserve only the critical directions from x_1. Since there is a one-to-one correspondence between the directions from x_1 and the points on B, it follows that the only possible fixed points of g' on B are those that correspond to the critical directions from x. ∎

DEFINITION. Let g_1 be an automorphism of S. Then g_1 is called a *pseudo-Anosov automorphism of S*, if $g_1|_{Int(S)}$ is a pseudo-Anosov automorphism of $Int(S)$.

Let $G(S)$ be a Fuchsian group of S, and $\Lambda(G(S))$ the limit set of $G(S)$.

Theorem 6. (Dynamics of lifts of pseudo-Anosov automorphisms on ∂U). *Let S be a surface with boundary, and g a pseudo-Anosov automorphism of S. Then one of the following occurs.*

a) g^* does not have fixed points on ∂U. Then there are $2N$, $1<N\le 4p-2+k$, alternately attracting and repelling periodic points of g^* on ∂U, all with the same period, and none of them is the fixed point of an element of $G(S)$. The g^*-iterates of any other point on ∂U converge to the g^*-orbit of the corresponding attracting periodic point.
b) g^* has $2N$, $1<N\le 4p-2+k$, fixed points on $\boldsymbol{\Lambda}(G(S))$. Then they are alternately attracting and repelling fixed points of g^* on ∂U. None of them is the fixed point of an element of $G(S)$. The g^*-iterates of any other point on ∂U converge to the corresponding attracting fixed point.
c) g^* has two fixed points on $\boldsymbol{\Lambda}(G(S))$, and none of them is the fixed point of an element of $G(S)$. Then they are the fixed points of g^* on ∂U. One of them is attracting and another repelling.
d) g^* has two fixed points on $\boldsymbol{\Lambda}(G(S))$ which are the end points of an interval of regularity $\boldsymbol{\Delta}$. Then they are the fixed points of g^* on $\partial U-\boldsymbol{\Delta}$. One of them is the attracting and another is the repelling fixed point of g^* on $\partial U-\boldsymbol{\Delta}$.
e) g^* has an infinite countable number of fixed points on $\boldsymbol{\Lambda}(G(S))$, and two of them, x and y, are the end points of an interval of regularity $\boldsymbol{\Delta}$. Then the fixed points of g^* on $\partial U-\boldsymbol{\Delta}$, except x and y, are alternately attracting and repelling fixed points of g^* on $\partial U-\boldsymbol{\Delta}$, and none of them is the fixed point of an element of $G(S)$. x and y are neither attracting nor repelling fixed points of g^*. The g^*-iterates of any other point on $\partial U-\boldsymbol{\Delta}$ converge to the corresponding attracting fixed point. Let $\boldsymbol{\lambda}$ be the element of $G(S)$ with the fixed points at x and y. Then the fixed points of g^* on $\partial U-\boldsymbol{\Delta}$, different from x and y, split into $2N$, $N\le 4p-2+k$, of $\boldsymbol{\lambda}$-orbits.

Proof. We have the following commutative diagram

$$\begin{array}{ccc} U & \xrightarrow{f^*} & U \\ \pi\big\downarrow & & \big\downarrow\pi_1 \\ Int(S) & \xrightarrow{f} & S' \end{array}$$

where we keep the notations of theorem 5. It

follows from theorem 5 and lemma 3 that the dynamical behavior of $g*$ on $\Lambda(G(S))$ is determined by the dynamical behavior of $f*g*(f*)^{-1}$ on ∂U. Now, the theorem follows from theorem 5 and results in 2.2, if we notice that the transition from the statements about the fixed points of $g*$ on $\Lambda(G(S))$ to the corresponding statements on ∂U follow from the fact that $g*$ is monotone on ∂U.

The only fact that remains to be proved is that $g*$ does not fix fixed points of elements of $G(S))$ except, maybe, the end points of an interval of regularity. It follows from the corollary to theorem 5 that we can assume that S is a surface with punctures. Then we can make use of the result of Marden and Strebel [5], [6] stating that $g(\gamma)$ is not freely homotopic to γ for any closed curve γ on S which cannot be continuously deformed into a boundary component of S. ∎

NOTE. Actually, Marden and Strebel [5], [6] proved their result in the case when the end points of γ are fixed, but the same proof can be carried out for freely homotopic curves.

3. Classification theorems

We assume that S is an oriented compact surface of genus p with k boundary curves, such that 2p-2+k>0. Let $M(S)$ be the mapping class group of orientation-preserving automorphisms of S. We allow a free homotopy along the boundary curves of S. Consider $g: S \to S$, an element of finite order of $M(S)$ ($g^r \sim 1_S$, $n>0$). We can assume that $g^r = 1_S$. Then S can be provided with a conformal structure σ in such a way that g becomes a conformal automorphism of S. Then $g*$, a lift of g to U, is a Möbius transformation of U. Let $G(S)$ be the Fuchsian group corresponding to S. The dynamical behavior of $g*$ in U is well-known:

a) $g*$ is the identity map.

b) $g*$ is elliptic. Then there exists one fixed point inside U, and all other points of $\bar{U}$ are periodic with the same period equal r.

c) $g*$ is hyperbolic, and the fixed points of $g*$ are not the end points of an interval of regularity;. Then there exists one attracting and one repelling fixed point on ∂U which are the

fixed points of a hyperbolic element of $G(S)$.
d) g^* is hyperbolic, and the fixed points of g^* are the end points of an interval of regularity. Then one of them is an attracting and another a repelling fixed point.

The following is the equivalent statement of Thurston's classification theorem due to Bers [1].

Theorem. (Thurston) *Let g be an automorphism of S. Then there exist g_1 isotopic to g and a finite number of closed, mutually disjoint Jordan curves on S - C_1, C_2, ..., C_n, $0 \leq n \leq 3p-3+k$, interchanged by g_1, with no two of them being freely homotopic and none of them can be shrunk to a point or to a boundary component of S, and, such that on each component of $S-\{C_1 \cup C_2 \cup \ldots \cup C_n\}$ g_1^d is either homotopic to a pseudo-Anosov or to a periodic automorphism, for some $d>0$.*

NOTE 1. We assume that the set of curves C_1, C_2, ..., C_n is *minimal* in the sense that there is no subset with the same property.

NOTE 2. The automorphisms of different components may not agree on their common boundary curves.

Let S_1, S_2, ..., S_j be the set of components of $S-\cup C_i$. Then the preimages of C_1, C_2, ..., C_n in U divide it in a countable infinite number of pieces, each one being the universal covering space for one of the components S_i, $i=1,2,\ldots,j$. We can assume that C_1, C_2, ..., C_n are geodesics on S. Let $A(S_i)$ be a component of U projected on S_i. It is bounded by non-Euclidean lines which are projected on those curves among C_1, C_2, ..., C_n which are boundary curves of S_i. By using the argument similar to that of theorem 5 we obtain the following result.

Lemma 4. *Let E^*, $E^*: A(S_i) \to U$, be a lift of E, the identity automorphism of S_i, to the respective universal covering spaces. Then E^* can be extended homeomorphically onto $\partial A(S_i)$ in such a way that it sends non-Euclidean lines on $\partial A(S_i)$ onto the intervals of regularity of ∂U corresponding to the respective boundaries of S_i.*

Let $G(A(S_i))$ be the maximal subgroup of $G(S)$

which leaves invariant $A(S_i)$. Then $G(A(S_i))$ is the group of transformations of $A(S_i)$, and it is a finitely generated subgroup of $G(S)$ of second kind. If $G(S_i)$ is the Fuchsian groups of S_i, then E^* induces an automorphism $\mu_{E^*}: G(A(S_i)) \to G(S_i)$.

Corollary 1. *μ_{E^*} sends the elements of $G(A(S_i))$ with the fixed points at the end points of intervals of regularity and non-Euclidean lines on $\partial A(S_i)$ onto the hyperbolic elements of $G(S_i)$ with the fixed points at the end points of intervals of regularity.*

It follows from lemma 4 that the dynamical behavior of a lift of an automorphism of S_i on $\partial A(S_i)$ is the same as on ∂U. We also note, here, that since we allowed a free homotopy along boundary curves, it follows that the action of g^* is determined neither on intervals of regularity on ∂U, nor on non-Euclidean lines on $\partial A(S_i)$. This implies the following corollary.

Corollary 2. *Let g_1, g_2 be automorphisms of S_i. Then there are g_1^*, g_2^*, lifts of g_1, g_2 to U, which coincide on $\partial A(S_i) \cap \partial U$ if and only if $g_1 \sim g_2$.*

Let G be a Fuchsian group. Then $\Lambda(G)$ denote the limit set of $G(S)$. Consider $\lambda \in G(A(S_i))$. Then the end points of λ lie on $\Lambda(G(A(S_i)))$. From the other side, any point on $\Lambda(G(A(S_i)))$ which is not the end point of an element of $G(A(S_i))$, is not, either, the end point of an element of $G(S)$. The second statement follows from the fact that if $\lambda_1 \in G(S) \setminus G(A(S_i))$, then $\lambda_1 A(S_i)) \cap A(S_i) = \emptyset$ or $\{x,y\}$, where x and y are the end points of a non-Euclidean line on $\partial A(S_i)$.

Theorem 7. *Let g be an automorphism of S, and for some $A(S_i)$, $1 \le i \le j$, $g_1^*(A(S_i) = A(S_i)$.*
1. $g|_{S_i}$ is homotopic to a periodic automorphism of S_i of finite order ($g^r \sim 1_{S_i}$, $r>0$). Then one of the following occurs.
a) g^* is the identity on $\Lambda(G(A(S_i)))$.
b) g^* has no fixed points on ∂U. Then each point of $\Lambda(G(A(S_i)))$ is a periodic point of g^* with the same period equal r.

c) g^* has two fixed points on $\Lambda(G(A(S_i)))$ which are not the end points either of a non-Euclidean line on $\partial A(S_i)$, or of an interval of regularity. Then they are the fixed points of g^* on ∂U. One of them is the attracting and another is the repelling fixed point. They are, also, the fixed points of an element of $G(A(S_i))$.

d) g^* has two fixed points on $\Lambda(G(A(S_i)))$ which are either the end points of a non-Euclidean line on $\partial A(S_i)$, or of an interval of regularity. Then they are the fixed points of g^* on $\partial U-\Delta$, where Δ is the interval between them which has the empty intersection with $\Lambda(G(A(S_i)))$. One of them is the attracting and another is the repelling fixed point of g^* on $\partial U-\Delta$.

2. $g|_{S_i}$ is homotopic to a pseudo-Anosov automorphism of S_i. Then one of the following occurs.

a) g^* does not have fixed points on ∂U. Then there are $2N$, $1<N\leq 4p-2+k$, alternately attracting and repelling periodic points of g^* on ∂U, all with the same period, and none of them is the fixed point of an element of $G(S)$. The g^*-iterates of any other point on ∂U converge to the g^*-orbit of the corresponding attracting periodic point.

b) g^* has $2N$, $1<N\leq 4p-2+k$, fixed points on $\Lambda(G(A(S_i)))$. Then they are alternately attracting and repelling fixed points of g^* on ∂U. None of them is the fixed point of an element of $G(S)$. The g^*-iterates of any other point on ∂U converge to the corresponding attracting fixed point.

c) g^* has two fixed points on $\Lambda(G(A(S_i)))$, and none of them is the fixed point of an element of $G(S)$. Then they are the fixed points of g^* on ∂U. One of them is attracting and another repelling.

d) g^* has two fixed points on $\Lambda(G(A(S_i)))$ which are the end points either of a non-Euclidean line on $\partial A(S_i)$, or of an interval of regularity. Then they are the fixed points of g^* on $\partial U-\Delta$, where Δ is the interval between them which has the empty intersection with $\Lambda(G(A(S_i)))$. One of them is the attracting and another is the repelling fixed point of g^* on $\partial U-\Delta$.

e) g^* has an infinite countable number of fixed points on $\Lambda(G(A(S_i)))$, and two of them, x and y, *are either the end points of a non-Euclidean line on $\partial A(S_i)$, or the end points of an interval of regularity. Let Δ be the interval between them which has the empty intersection with $\Lambda(G(A(S_i)))$. Then the fixed points of $g*$ on $\partial U-\Delta$, except x and y, are alternately attracting and repelling fixed points of $g*$ on $\partial U-\Delta$, and none of them is the fixed point of an element of $G(S)$. x and y are neither attracting nor repelling fixed points of $g*$. The $g*$-iterates of any other point on $\partial U-\Delta$ converge to the corresponding attracting fixed point. Let λ be the element of $G(S)$ with the fixed points at x and y. Then the fixed points of $g*$ on $\partial U-\Delta$, different from x and y, split into $2N$, $N \leq 4p-2+k$, of λ-orbits.*

Proof. *1.* a) and b) follow from lemma 4. What concerns c) and d), it follows from lemma 4 that the points of $\Lambda(G(A(S_i)))$ converge to the respective fixed point of $g*$, say x. Since $g*$ is monotone on ∂U, it implies that all points of ∂U converge to x, except, maybe, points on the interval between x and another fixed point of $g*$ which has the empty intersection with $\Lambda(G(A(S_i)))$, when x is the end points of a non-Euclidean line on $\partial A(S_i)$, or of an interval of regularity. ∎
2. a) - e) follow from the corresponding sections of theorem 6. ∎

Lemma 5. *Let curves C_1, C_2, ..., C_n have been chosen in Thurston's classification theorem, such that g_1, $g_1 \sim g$, interchanges C1, C_2, ..., C_n, and on each component of $S-\{C_1 \cup C_2 \cup ... \cup C_n\}$ $g_1{}^k$ is either homotopic to a pseudo-Anosov or to a periodic automorphism, for some $k>0$. Let S_i be a component of $S-\{C_1 \cup C_2 \cup ... \cup C_n\}$, such that $g_1(S_i)=S_i$. Then $g_1|_{Si}$ is either homotopic to a pseudo-Anosov or to a periodic automorphism of S_i.*

Proof. Let $g_1|_{Si}$ be a reducible automorphism of S_i. Then $g_1^m(\gamma)$, for some $m>0$, is freely homotopic to γ, where γ is a closed curve on S_i. It follows that $g_1^{mk}(\gamma) \sim \gamma$. It implies that g_1 is a periodic automorphism of S_i. Contradiction. ∎

Let U' denote U without the preimages of the curves C_1, C_2, ..., C_n.

Theorem 8. *Let S be a surface with boundary, and*

g an automorphism of S. Let g_1* not leave invariant any component of U'. Then one of the following occurs.

1. There is a non-Euclidean line γ in U, such that $\pi(\gamma)=C_i$, $1\le i\le n$, and $g_1*(\gamma)=\gamma$. Then $g*$ does not have fixed points on ∂U. The end points of γ are periodic points of $g*$ of order two. Let $A(S_i)$ be a component of U' which borders γ. Then $(g_1*)^{(2)}(A(S_i))=A(S_i)$. $(g_1*)^{(2)}|_{A(Si)}$ and $(g_1*)^{(2)}|_{g1*(A(Si))}$ are conjugated by g_1*. There are two cases.

a) $g*$-iterates of all other points on ∂U converge to two periodic points of $g*$.

b) $g*$ has an infinite countable number of periodic points of period two on ∂U. Then the periodic points of $g*$, except the end points of γ which are neither attracting nor repelling, are alternately attracting and repelling periodic points of $g*$ on ∂U, and none of them is the end point of an element of $G(S)$. $g*$-iterates of any other point on ∂U converge to the $g*$-orbit of the corresponding attracting periodic point.

2. $g*$ has two fixed points on ∂U. None of them belongs to an interval of regularity.

Proof. *1.* g_1* interchanges two components of $\bar{U}-\gamma$. Since g is an orientation-preserving automorphism, it follows that g_1* interchanges the end points of γ. It implies that $g*$ does not have fixed points on ∂U, and interchanges $A(S_i)$ with $g_1*(A(Si))$.

Lemma 1. *$(g_1*)^{(2)}|_{A(Si)}$ and $(g_1*)^{(2)}|_{g1*(A(Si))}$ are conjugated by g_1*.*

Proof. $(g_1*)^{(2)}|_{A(Si)} = g_1*|_{g1*(A(Si))}g_1*|_{A(Si)}$. By the same token

$(g_1*)^{(2)}|_{g1*(A(Si))} = g_1*|_{A(Si)}g_1*|_{g1*(A(Si))}$. It follows that

$$(g_1*)^{(2)}|_{g1*(A(Si))} = g_1*|_{A(Si)}(g_1*)^{(2)}|_{A(Si)}(g_1*)^{(-1)}|_{A(Si)}. \quad ■$$

$(g_1*)^{(2)}$ leaves invariant $A(Si)$ and fixes the end points of a non-Euclidean line on $\partial A(Si)$ not corresponding to a boundary curve on S. It follows from theorem 7 that, either $(g_1*)^{(2)}$-iterates of any point on ∂U, except those on the interval Δ between the end points of γ, converge to one of the end points of γ, or there are an infinite countable number of alternately attracting and repelling fixed points of $(g_1{*})^{(2)}$ on $\partial U-\Delta$. Now, a) and b) follow from the lemma.

2. Let $g_1{*}(\gamma_1)\neq\gamma_1$ for any non-Euclidean line γ_1 in U which is projected onto one of the curves C_1, C_2, ..., C_n. Let a non-Euclidean line γ be a preimage of one of the curves C_1, C_2, ..., C_n. Consider $g_1{*}(\gamma)$. Then $g_1{*}(\gamma)\cap\gamma=\emptyset$. We have two possibilities:

a) $g_1{*}$ preserves the direction of γ in U, i. e. in the quadrangle Π composed by γ, $g_1{*}(\gamma)$ and two arcs on ∂U, γ and $g_1{*}(\gamma)$ have the same direction. Let A be the component of $U-\gamma$ which contains $g_1{*}(\gamma)$. It follows that $g_1{*}(A)\subset A$. Consider the decreasing sequence of sets

$$\ldots\supset(g_1{*})^{(-2)}(A)\supset(g_1{*})^{(-1)}(A)\supset A\supset g_1{*}(A)\supset(g_1{*})^{(2)}(A)\supset\ldots$$

Since $(g_1{*})^{(m)}(\gamma)$ converges to a point $x\in\partial U$ when $m\to\infty$, it follows that x is the attracting point of $g_1{*}$. By the same token y, the limit point of $(g_1{*})^{(-m)}(\gamma)$, $m\to\infty$, is the repelling point of $g_1{*}$. It is easy to see that neither x nor y belongs to any interval of regularity, or is a parabolic fixed point. ∎

b) $g_1{*}$ changes the direction of γ in U, i. e. in the quadrangle Π composed by γ, $g_1{*}(\gamma)$ and two arcs on ∂U, γ and $g_1{*}(\gamma)$ have the opposite directions.

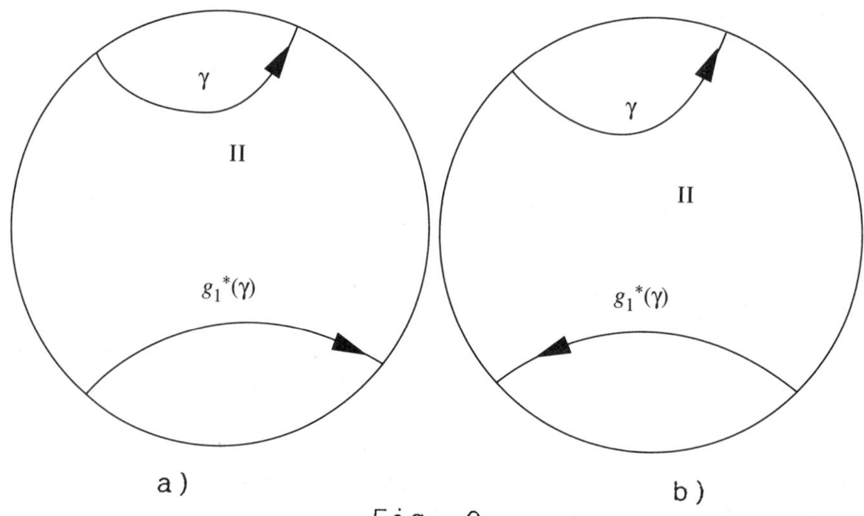

Fig. 2.

Lemma 2. $(g_1{*})^{(2)}(\gamma)\cap\gamma=\emptyset$.

Proof. Assume that $(g_1{*})^{(2)}(\gamma)=\gamma$. It follows that $g_1{*}(\Pi)=\Pi$. Let z be a fixed point of $g_1{*}/_{\Pi}$. Since $g_1{*}$ interchanges the sides of Π on ∂U, it follows that $z\in U$. It implies that $g_1{*}$ preserves either a component of U' or a preimage of one of the curves $C_1, C_2, \ldots, C_n$. Contradiction. ∎

$(g_1{*})^{(2)}$ preserves the direction of γ in U. By the same argument as in a) we can conclude that $(g_1{*})^{(2)}$ has two fixed points on ∂U, and none of them belongs to an interval of regularity. $g_1{*}$ has at most two fixed points on ∂U. If $g_1{*}$ has less than two fixed points on ∂U, we have two possibilities.

1. $g_1{*}$ does not have fixed points on ∂U. Then $g_1{*}$ has a fixed point inside U. It implies that either $g_1{*}$ preserves a component of U' or a preimage of one of the curves C_1, C_2, $\ldots$, C_n. Contradiction.
2. $g_1{*}$ has one fixed point z on ∂U. It implies that for any $z_1\in\partial U$, $(g_1{*})^{(m)}(z_1)$ converges to z when $m\to\infty$. This contradicts the fact that $(g_1{*})^{(2)}$ has two fixed points on ∂U.

It follows that $g_1{*}$ has two fixed points on ∂U. ∎

Theorem 9. *Let S be a surface with boundary, g an automorphism of S, and $g_1{*}(A(S_i))=A(S_i)$, where $A(S_i)$ is a component of U' projected on S_i. Then one of the following occurs.*
1. $g{*}$ does not fix the end points of any non-Euclidean line on $\partial A(S_i)$. Then $A(S_i)$ is the only component of U' left invariant by g_1.
2. $g{*}$ fixes the end points of a non-Euclidean line a on $\partial A(S_i)$, and it is not the identity on $\Lambda(G(A'))$ for any component A' of U'. Then the only components of U' left invariant by $g_1{*}$ are the two ones which border a.
3. $g{*}$ is the identity on $\Lambda(G(A'))$, where A' is a component of U'. Then A' is the only component of U' having this property. The components of U' left invariant by $g_1{*}$ are those which border the non-Euclidean lines on $\partial A'$.

Proof. 1. Let A be a component of U' different

from $A(S_i)$. Then there exists a non-Euclidean line γ on $\partial A(S_i)$ which separates A from $A(S_i)$. $g_i{*}(\gamma)$ is a non-Euclidean line on $\partial A(S_i)$ different from γ. It follows that $g_i{*}(A) \cap A = \emptyset$. ∎

2. Let A be a component of U' which borders $A(S_i)$. Assume that there exists B, a component of U', different from A and $A(S_i)$, and, such that $g_i{*}(B)=B$. Let γ be a non-Euclidean line on $\partial A(S_i)$ which separates $A(S_i)$ from B. It follows from theorem 7 that we have one of the following cases: 1.d), 2.d), or 2.e). In any case the end points of a are, respectively, the attracting and repelling points of $g_i{*}$ on ∂U without the interval between them. It follows that $\gamma = a$. By the same token B is separated from A by a. It implies that, either $B=A$, or $B=A(S_i)$. Contradiction. ∎

3. We can assume that $A'=A(S_i)$. Let A be a component of U' which borders a non-Euclidean line γ on $\partial A(S_i)$. Then $g_i{*}(A)=A$.

Lemma. *Let A, $\lambda(A)$ be two different components of $U' \backslash A(S_i))$ which border non-Euclidean lines on $\partial A(S_i)$, where $\lambda \in G(A(S_i))$. Then*

1. $g_i{*}|_A$ and $g_i{*}|_{\lambda(A)}$ are conjugated by λ.

2. $g_i{*}$ is not the identity on $\Lambda(G(A))$.

Proof. 1. $g_i{*}\lambda(g_i{*})^{-1}=\lambda$. It follows that $\lambda^{-1} g_i{*}\lambda = g_i{*}$. ∎

2. Indeed, otherwise $g_i{*}$, by 1. of the lemma, must be the identity on any component of U' which is a preimage of $\pi(A)$ and borders a non-Euclidean line on $\partial A(S_i)$. It follows that g is homotopic to the identity automorphism of $S_i \cup \pi(A)$. But, this contradicts the fact that the collection of the curves $\{C_1, C_2, \ldots, C_n\}$ is minimal. By theorem 7 it follows that we have one of the cases: 1-d, 2-d, or 2-e. In each of them $g_i{*}$ interchanges non-Euclidean curves on ∂A except γ. It follows that $g_i{*}$ does not leave invariant any component of U', except A, which is separated from $A(S_i)$ by γ. ∎

Theorem 10. **(Classification theorem)** *Let S be a surface with boundary, and g an automorphism of S. Then one of the following occurs.*

1. $g{*}$ does not have fixed points on ∂U. Then five cases are possible.

a) $g{*}$ has two periodic points on ∂U which are the *fixed points of an element of $G(S)$ not corresponding to a boundary curve. g^*-iterates of all other points on ∂U converge to the periodic points of g^*. In this case g is a reducible automorphism.*

b) g^* has an infinite countable number of periodic points of period two on ∂U. Two of them, x and y, are the fixed points of λ, an element of $G(S)$ not corresponding to a boundary curve. Then the periodic points of g^*, except x and y, are alternately attracting and repelling periodic points of g^* on ∂U, and none of them is the end point of an element of $G(S)$. x and y are neither attracting nor repelling periodic points of g^*. g^*-iterates of any other point on ∂U converge to the g^*-orbit of the corresponding attracting periodic point. The periodic points of g^*, different from x and y, split into $4N$, $N \leq 4p-3+k$, of λ-orbits. In this case g is a reducible automorphism.

c) The points of $\Lambda(G(S))$ are the periodic points of g^* with the same period equal r. In this case g is homotopic to a periodic automorphism of period r.

d) There is G_1, a finitely generated proper subgroup of $G(S)$ of second kind, such that $\Lambda(G_1) \subset M1$, where M_1 is the set of periodic points of g^* on $\Lambda(G(S))$. All points of M_1 have the same period equal P. No point of $\Lambda(G_1)$ is attracting or repelling periodic point. There are two possibilities: either, no point of any interval of regularity of G_1 belongs to M_1, or there exist a finite number of points $x_1, x_2, \ldots, x_{2s}$, $s \leq 4p-3+k$, belonging to intervals of regularity of G_1, such that $M_1 \setminus \Lambda(G_1) = \cup G_1(x_i)$, where $G_1(x_i)$, $1 \leq i \leq s$, is the G_1-orbit of x_i. In the former case any point x on an interval of regularity of G_1 not corresponding to a boundary curve of S converges to the g^*-orbit of its end point which does not depend on x. In the latter case each interval of regularity of G_1 which contains, at least, one point belonging to an orbit of $G_1(x_i)$, $1 \leq i \leq s$, contains infinite countable number of alternately attracting and repelling periodic points of g^*. None of them is the end point of an element of $G(S)$. g^*-iterates of any other point in such an interval converge to the g^*-orbit of the corresponding attracting periodic point. In both cases g is a reducible automorphism.

e) There are $2N$, $1<N\leq 4p-2+k$, alternately attracting and repelling periodic points of g^* on ∂U, all with the same period P, and none of them is the fixed point of an element of $G(S)$. The g^*-iterates of any other point on ∂U converge to the g^*-orbit of the corresponding attracting periodic point. In this case g is a pseudo-Anosov or reducible automorphism.

2. g^* has two fixed points on $\Lambda(G(S))$. Three cases are possible.

a) The fixed points of g^* are the end points of an interval of regularity Δ. Each point on $\partial U-\Delta$ converges to one of them.

b) g^* has two fixed points on ∂U. One of them is attracting and another is repelling. If, at least, one of the fixed points of g^* is the fixed point of an element of $G(S)$, then g in not a pseudo-Anosov automorphism. If, at least, one of the fixed points of g^* is not the fixed point of an element of $G(S)$, then g is not a periodic automorphism.

c) g^* has two fixed points on ∂U. None of them is, either, attracting or repelling. In this case g is a reducible automorphism.

3. g^* has $2N$, $1<N\leq 4p-2+k$, fixed points on ∂U. They are alternately attracting and repelling fixed points of g^*. None of them is the fixed point of an element of $G(S)$. The g^*-iterates of any other point on ∂U converge to the corresponding attracting fixed point. In this case g is a pseudo-Anosov or reducible automorphism.

4. g^* has an infinite countable number of fixed points on $\Lambda(G(S))$. Three cases are possible.

a) Two fixed points of g^*, x and y, are the end points of an element of $G(S)$ not corresponding to a boundary curve. The fixed points of g^*, except x and y, are alternately attracting and repelling fixed points, and none of them is the end point of an element of $G(S)$. x and y are neither attracting nor repelling fixed points. g^*-iterates of any other point on ∂U converge, either to the corresponding attracting fixed point, or to one of the points $\{x,y\}$. Let λ be the element of $G(S)$ with the fixed points at x and y. Then the fixed points of g^* on ∂U, different from x and y, split into $2N$, $N\leq 2(4p-$

$3+k$), λ-orbits. In this case g is a reducible automorphism.
b) Two fixed points of g^*, x and y, are the end points of an interval of regularity Δ. The fixed points of g^* on $\partial U-\Delta$, except x and y, are alternately attracting and repelling fixed points, and none of them is the end point of an element of $G(S)$. x and y are neither attracting nor repelling fixed points. g^*-iterates of any other point on $\partial U-\Delta$ converge to the corresponding attracting fixed point. Let λ be the element of $G(S)$ with the fixed points at x and y. Then the fixed points of g^*, different from x and y, on $\partial U-\Delta$ split into $2N$, $N\leq 4p-2+k$, of λ-orbits. In this case g is a pseudo-Anosov or reducible automorphism.
5. g^* has a continuum set of fixed points on $\Lambda(G(S))$. Then two cases are possible.
a) Each point of $\Lambda(G(S))$ is the fixed point of g^*. In this case g is homotopic to the identity automorphism.
b) There is G_1, a finitely generated proper subgroup of $G(S)$ of second kind, such that $\Lambda(G_1)\subset M$, where M is the set of fixed points of g^* on $\Lambda(G(S))$. No point of $\Lambda(G_1)$ is attracting or repelling. There are two possibilities: either, no point of any interval of regularity of G_1 belongs to M, or there exist a finite number of points x_1, x_2, ..., x_{2s}, $s\leq 4p-3+k$, belonging to intervals of regularity of G_1, such that $M\backslash\Lambda(G_1) = \cup G_1(x_i)$, where $G_1(x_i)$, $1\leq i\leq s$, is the G_1-orbit of x_i. In the former case any point x on an interval of regularity of G_1 not corresponding to a boundary curve of S converges to its end point which does not depend on x. In the latter case each interval of regularity of G_1 which contains, at least, one point belonging to an orbit $G_1(x_i)$, $1\leq i\leq s$, contains infinite countable number of alternately attracting and repelling fixed points of g^*. None of them is the end point of an element of $G(S)$. g^*-iterates of any other point in such an interval converge to the corresponding attracting fixed point. In both cases g is a reducible automorphism.

Proof. We consider the following four cases.
1. g_1^* does not leave invariant any component of U'.

There are three cases which were considered in theorem 8. The cases 1-a and 1-b of theorem 8 correspond to respective cases, here. The case 2 of theorem 8 corresponds to 2-b or to 2-c.

2. g_1* leaves invariant, exactly, one component of U'.

Let $A(S_i)$ be the component of U' left invariant by g_1*. By theorem 9 g_1* does not fix the end points of any non-Euclidean line on $\partial A(S_i)$. All these cases were considered in theorem 7. The following is the table which shows the correspondence between different cases.

Theorem 7	Theorem 10
1-b	1-c or 1-d
1-c	2-b
1-d	2-a
2-a	1-e
2-b	3
2-c	2-b
2-d	2-a
2-e	4-b

The only non trivial fact, here, is the case 1-b. We can distinguish between two cases:

a) g* is homotopic to a periodic automorphism of S. It corresponds to 1-c of the theorem.

b) g* is a reducible automorphism. Look at $g*^{(p)}$. It is the identity on $\Lambda(G(A'))$, where A' is a component of U'. This case corresponds to 1-a of theorem 7 and will be discussed later.

3. g_1* leaves invariant, exactly, two components of U'. Let A, B be the components of U' left invariant by g_1*. By theorem 9 g_1* fixes the end points of a non-Euclidean line a which borders A and B, and is the identity, neither on $\Lambda(G(A))$, nor on $\Lambda(G(B))$. Then on each component A and B we have, by theorem 7, one of the cases: 1-d, 2-d, or 2-e. We have the following correspondence table.

A	B	Theorem 10
1-d	1-d	2-b or 2-c
1-d	2-d	2-b or 2-c
1-d	2-e	4-a
2-d	2-d	2-b or 2-c
2-d	2-e	4-a
2-e	2-e	4-a

The difference between the cases (1-d; 2-e) or (2-d; 2-e) and the case (2-e; 2-e) is that the fixed points of g^* are not separated by the end points of a in the first two cases.

4. g^* is the identity on $\Lambda(G(A'))$, where A' is a component of U'. If A' is the only component of U' then g^* is homotopic to the identity on S. It corresponds to 5-a of the theorem. Otherwise, it follows from theorem 9 that A' is the only component of U', such that all points on its common boundary with ∂U, except, may be, those on intervals of regularity, are fixed by g^*. Let a be a non-Euclidean line on $\partial A'$. Consider A, $A \neq A'$, a component of U' which borders a. Let Δ be an interval on ∂U between the end points of a, such that $\Delta \cap \Lambda(G(A'))=\emptyset$. On $\Lambda(G(A))$ we have, by theorem 7, one of the cases: 1-d, 2-d, or 2-e. In the first two cases the end points of a are, respectively, the attracting and repelling fixed points of g^* on Δ. In the last case there are a finite number of points $x_1, x_2, \ldots, x_{s1}$ belonging to Δ, such that $M_1 = \cup(\lambda_1)_{xi}$, where M_1 is the set of the fixed points of g^* on Δ, λ_1 is the element of $G(A')$ which fixes the end points of a, and $(\lambda_1)_{xi}$, $1 \leq i \leq s_1$, is the λ_1-orbit of x_i. By the lemma of theorem 9, g_1^* is conjugated by the elements of $G(A')$ on the components of U' which border the non-Euclidean lines on $\partial A'$. It means that the orbits $G(A')(x_i)$, $i=1,2,\ldots,s_1$, yield the set of fixed points of g^* on $G(A')(\Delta)$. 5-b follows from the fact that the non-Euclidean lines on $\partial A'$ split into a finite number of different $G(A')$-orbits which correspond to different curves among $C_1, C_2, \ldots, C_n$ which border $\pi(A')$. ∎

The respective statement in the case when g^* is periodic on $\Lambda(G(A'))$ follows easily from 4.

Finally, we give the upper bounds for the number of λ-orbits in 1-b, 1-d, 4-a, and 5-b. In order to do it we introduce the number $N(S)$, $N(S)=4p-2+k$, which represents the maximal number of horizontal (vertical) trajectories stemming from a point on S. It follows that $N(S)>0$.

Lemma 1. *Let $S=S_1 \cup S_2$, and $S_1 \cap S_2=C_1 \cup C_2 \cup \ldots \cup C_r$, where $C_1, C_2, \ldots, C_r$ are mutually disjoint simple closed curves. Then $N(S) \geq N(S_1)+N(S_2)$.*

Proof. Let S_1 be a surface of genus p_1 with k_1

boundary curves, and S_1 a surface of genus p_2 with k_2 boundary curves, respectively. Then S is a surface of genus $p=p_1+p_2+r-1$ with $k=k_1+k_2-2r$ boundary curves. It follows that $N(S)-(N(S_1)+N(S_2))=2r-2\geq 0$. ∎

Lemma 2. *Let $S'=S\setminus C$, where C is a simple closed curve on S which does not separate S. Then $N(S)-N(S')=2$.*

Proof. Let S be a surface of genus p with k boundary curves. Then S' is of genus $p-1$ with $k+2$ boundary curves. It follows that

$$N(S)-N(S')=(4p-2+k)-(4(p-1)-2+k+2)=2 \quad ∎$$

Corollary 1. *Let S_1, S_2, ..., S_j be the set of components of $S-\cup C_i$. Then*

$$N(S)\geq N(S_1)+N(S_2)+...+N(S_j).$$

Proof. S is obtained from S_1, S_2, ..., S_j by gluing either different components along their common boundary curves, or two boundary curves within the same component. In either case it follows from lemmas 1 and 2 that $N(S)\geq N(S_1)+N(S_2)+...+N(S_j)$. ∎

Corollary 2. *Let S', $S'\neq S$, be a component of $S-\cup C_i$. Then $N(S)\geq N(S')+1$.*

Proof. It follows from corollary 1. ∎

In the cases 1-b and 4-a there are two components of U', A and A', left invariant by either g_1* or $(g_1*)^2$. On each of them the number of λ-orbits is less or equal than $2(N(S'))$, where $S'=\pi(A')$. The upper bound for λ-orbits follows, now, from corollary 2.

The case 1-d, as we have already noticed above, can be reduced to the case 5-b. Let A be a component of U' left invariant by G_1. Then, if S' is a component of $S-\pi(A)$ adjacent to $\pi(A)$, it may contribute at most $N(S')$ G_1-orbits on $M\setminus\Lambda(G_1)$. It follows from corollary 1 that the number of different G_1-orbits on $M\setminus\Lambda(G_1)$ is less or equal than $N(S)-N(\pi(A))<N(S)$. ∎

Finally, we construct the example, asked by Nielsen [7], of a lift of an automorphism g of S with fixed points on intervals of regularity of the subgroup of the elements of G(S) fixed by μ_{g*}, not corresponding to boundary curves of S (the case 5-b of theorem 10).

Theorem 11. *Let p_1, s be positive integers, such that $4p_1+2s\leq 4p-2$. Then there exists $g*$, a lift to U of an automorphism g of S, such that there exist exactly 4s inequivalent fixed points of $g*$ on intervals of regularity not corresponding to boundary curves of S of G_1, the subgroup of G(S) of the rank $2p_1$, fixed by μ_{g*}.*

Proof. Let C be a curve on S which divide it into two pieces S_1 and S_2 of genus p_1 and $p-p_1$, respectively. We define $g|_{S1}=1|_{S1}$. Let S_2' be a compact surface of genus $p-p_1$. There exists g_2', a pseudo-Anosov diffeomorphism of S_2', which has a singularity z of order 2s-2 (2s critical horizontal and 2s critical vertical directions from z) (see Veech [11]). Now, we map S_2 onto S_2' in a way, that C is sent to z. Then g_2' induces g_2, a pseudo-Anosov diffeomorphism of S_2. We can assume that g_2 is the identity on C. Now, we put $g|_{S2}=g_2$. Consider $g*$, the lift of g to U which is the identity on $A(S_1)$, a component of U corresponding to S_1. Let $A(S_2)$, be the component of U corresponding to S_2 which has the common boundary with $A(S_1)$ along $C*$, a lift of C. If $g*$ preserves the critical directions from $C*$, we are done. Otherwise, there exists n, $|n|>1$, such that $g*\lambda^n$ preserves the critical directions from $C*$, where λ is the element of G(S) which fixes the end points of $C*$ on ∂U. Now, we consider the n-th power of the Dehn's twist of S along C done in the corresponding direction. ∎

REFERENCES:

[1] Bers, L., An extremal problem for quasiconformal mappings and a theorem by Thurston, Acta Math. 141 (1978), 73-98.
[2] Gilman, J., On the Nielsen type and the classification for the mapping class group, Advances in Math. 40 (1981), 68-96.
[3] Fathi, A., Laudenbach F., The dynamics of the lift of

a pseudo-Anosov diffeomorphism to the Poincare disk, Publications Mathematique d'Orsay (1983).
[4] Marden, A., Strebel, K., On the ends of trajectories, In Differential Geometry and Complex Analysis (I. Chavel and H. Farcas, eds.), Springer-Verlag, Berlin, 1985, 195-204.
[5] Marden, A., Strebel, K., Pseudo-Anosov Teichmuller mappings, Journal d'Analyse Mathematique 46, (1986), 194-220.
[6] Marden, A., Strebel, K., Geodesics for quadratic differentials on punctured surfaces, Complex Variables 5, (1986), 271-280.
[7] Nielsen, J., Topology of closed orientable surfaces, II, Collected Mathematical Papers, vol. 1, Birkhauser, 1986.
[8] Slutskin, L., Thesis, Columbia University, 1987.
[9] Strebel, K., Quadratic Differentials, Springer, 1984.
[10] Thurston, W., On the geometry and dynamics of diffeomorphisms of surfaces, I, Bulletin of the Amer. Math. Society 19 (1988), 417-431.
[11] Veech, W., Gauss measures for transformations on the space of interval exchange maps, Ann. of Math. (2) 115, 1982.
[12] Handel, M., Thurston W., New proofs of some results of Nielsen, Advances in Math. 56 (1985), 173-191.
[13] Miller, R., Nielsen's viewpoint on geodesic laminations, Advances in Math. 45 (1982), 189-212.

Contemporary Mathematics
Volume **152**, 1993

Equivariant Nielsen fixed point theory and periodic points

PETER WONG

ABSTRACT. The purpose of this article is to give a brief account of Nielsen fixed point theory in the presence of a group action and its application to the Nielsen theory of periodic points.

1. Introduction

In classical topological fixed point theory, the Lefschetz number $L(f)$ of a selfmap $f : X \to X$ on a compact polyhedron X gives an algebraic count of the indices of fixed points of f while its Nielsen number $N(f)$ measures the size of the essential fixed point set $Fixf$. It is well known that $N(f) = 0$ implies that f is deformable to be fixed point free if X satisfies the so-called Wecken condition [**Br**] (e.g. when X is a manifold of $dimX \geq 3$). If, in addition, X is a Jiang space [**Br**] [**J1**], then the converse of the Lefschetz fixed point theorem holds true. In the presence of a group action requiring all maps to respect the action, the vanishing of $N(f)$ or $L(f)$ is not sufficient to deform the map equivariantly to be fixed point free.

Let $G = \mathbb{Z}_2$ be the cyclic group of order 2 acting on the two torus $T^2 = \{(z_1, z_2) \in \mathbb{C}^2 || z_i| = 1, i = 1, 2\}$ via the action $\zeta(z_1, z_2) = (z_2, z_1)$ where ζ is the generator of $\mathbb{Z}_2$. Consider the G-map $f : T^2 \to T^2$ (i.e., $f\zeta = \zeta f$) given by $(z_1, z_2) \mapsto (\bar{z}_2, \bar{z}_1)$ where $\bar{z}$ denotes the complex conjugate of z. One can easily show that $L(f) = 0$ and thus $N(f) = 0$. The fixed point set of the G-action is given by $(T^2)^G = \{(z_1, z_2)|\zeta(z_1, z_2) = (z_1, z_2)\} = \{(z, z)||z| = 1\} \approx S^1$. The restriction $f^G = f|(T^2)^G$ sends (z, z) to $(\bar{z}, \bar{z})$ and has $L(f^G) = 2$. Since $(T^2)^G$

1991 *Mathematics Subject Classification.* Primary 55M20; Secondary 57S99, 58F99.

Key words and phrases. equivariant fixed point theory, Nielsen fixed point theory, periodic points.

Travel supported by a grant from Bates College
This paper is in final form and no version of it will be submitted for publication elsewhere

is G-invariant, we conclude that every map G-homotopic to f must have a fixed point in $(T^2)^G \subset T^2$. This shows that we need to consider a sequence of Nielsen numbers which depend upon the isotropy types of the space.

Let us take a closer look at the example above. If we consider the involution $\varphi : S^1(\subset \mathbb{C}) \to S^1$ given by $\varphi(z) = \bar{z}$, then the map f is simply defined by $f(z_1, z_2) = (\varphi(z_2), \varphi(z_1))$. Moreover, $Fixf = \{(z, \varphi(z)) | z \in Fix\varphi^2\}$ and $Fixf^G = \{(z, z) | z = \varphi(z)\} \approx Fix\varphi$. It is easy to see that φ can be deformed to a map ψ such that $Fix\psi$ consists of two points and $Fix\psi = Fix\psi^2$. In other words, φ is deformable to a map with two fixed points and no periodic points with least period 2. This homotopy of φ induces an equivariant homotopy of f to a G-map f' with exactly two fixed points which lie in $(T^2)^G$. This example shows that the periodic points of φ of period two are precisely the fixed points of the G-map f. In fact the fixed point indices and hence the Lefschetz numbers of f and φ^2 coincide. Moreover, there is a one to one correspondence between their fixed point classes and thus $N(\varphi^2) = N(f)$.

Throughout G will denote a compact Lie group and M a compact smooth G-manifold upon which the G-action is smooth. For any $x \in M$, the *isotropy subgroup at* x is given by $G_x = \{g \in G | gx = x\}$. Let $Iso(M)$ be the isotropy types of M, i.e., the set of conjugacy classes of closed subgroups of G which appear as isotropy subgroups. Denote by $\mathcal{F} = \{(H) \in Iso(M) || WH| < \infty\}$ where $WH = NH/H$ is the *Weyl group* of H in G and NH is the normalizer of H in G. For any closed subgroup $H \leq G$, $M^H = \{x \in M | hx = x, \forall h \in H\}, M_H = \{x \in M | G_x = H\}$. If $(H), (K) \in Iso(M)$ and (H) is subconjugate to (K), we write $(H) \leq (K)$. We can choose an *admissible* ordering on $Iso(M)$ so that $(H_j) \leq (H_i)$ implies $i \leq j$. We then have a filtration of G-subspaces $M_1 \subset M_2 \subset ... \subset M_k = M$, where $M_i = \{x \in M | (G_x) = (H_j), j \leq i\} = GM^{H_i}$. Since M^H is a WH-space so for any G-map $f : M \to M$, the restriction $f^H = f|M^H : M^H \to M^H$ is a WH-map. By a G-homotopy, we mean a G-map $F : M \times [0,1] \to M$ where the G-action on $[0,1]$ is trivial. Minimizing $Fixf$ through equivariant deformations is accomplished inductively on each open subspace M_{H_i} via WH_i-deformations which are extended to G-deformations **[FW]** in a G-invariant neighborhood of M_i and hence to M_{i+1}.

For any G-map $f : M \to M$, there is an induced map $\bar{f} : M/G \to M/G$ such that $pf = \bar{f}p$ where $p : M \to M/G$ is the orbit map and M/G is the orbit space. In the case where G acts freely on M, $p : M \to M/G$ is a principal G-bundle. If $dimG \geq 1$, then every G-map $f : M \to M$ is G-deformable to be fixed point free **[Wi]**. Consider the real projective space $\mathbb{R}P^{2n}$ as the orbit space of the $2n$-sphere S^{2n} under the free antipodal action of $\mathbb{Z}_2$. For any $\mathbb{Z}_2$-map $f : S^{2n} \to S^{2n}$ of degree -1, f is equivariantly homotopic to a fixed point free map whose induced map on the orbit space must have an essential fixed point since $\mathbb{R}P^{2n}$ has the fixed point property. Therefore, we only need to focus on finite group actions or on the isotropy subgroups with finite Weyl groups and we have to work on the G-space M but not the orbit space M/G.

The purpose of this paper is to survey some of the ideas behind the fixed point theory for equivariant maps. In particular, we shall discuss how techniques from various Nielsen type theories and equivariant topology are employed. We shall also establish a connection between the equivariant Nielsen theory and the Nielsen theory for periodic points. For further background in topological fixed point theory, we refer the reader to [**Br**] and [**J1**]. The basic references for equivariant topology are [**B**] and [**tD**].

I would like to thank the referee for a number of helpful suggestions.

2. Equivariant Nielsen theory

2.1 Equivariant Wecken Theorem.

When M is simply connected, $L(f) = 0$ implies $N(f) = 0$ for any selfmap $f : M \to M$. It then follows from the Wecken theorem that the converse of the Lefschetz Fixed Point Theorem holds true. In the presence of a group action, we have seen from §1 that a necessary condition to deform a G-map f equivariantly to be fixed point free is the vanishing of the ordinary Lefschetz numbers $\{L(f^H)\}$ for all $(H) \in \mathcal{F}$. If all the M^H are simply connected then this condition is also sufficient. The following is due to D. Wilczyński.

THEOREM 1 ([**Wi**]). *Let G be a compact Lie group acting smoothly on a compact smooth manifold M. Suppose that each M^H is connected of $dim M^H \geq 3$ and is simply connected for $(H) \in \mathcal{F}$. For any G-map $f : M \to M, f$ is G-homotopic to a fixed point free map if , and only if, $L(f^H) = 0, \forall (H) \in \mathcal{F}$.*

In [**Wi**], a slightly more general result is proven with simple connectivity assumptions on *each* connected component of M^H. A special case of this result was also obtained independently by A. Vidal [**V**] using Bredon's equivariant obstructions. In fact, Theorem 1 can be generalized by replacing simple connectivity with Jiang condition on M^H. In order to do this, we must have an equivariant analog of the classical Wecken theorem.

THEOREM 2 ([**FW**]). *Let G be a compact Lie group acting smoothly on a compact smooth manifold M. Suppose that for $(H) \in \mathcal{F}$, M^H is connected of $dim M^H \geq 3$ and $dim M^H - dim(M^H - M_H) \geq 2$. For any G-map $f : M \to M, f$ is G-homotopic to a fixed point free map if , and only if, $N(f^H) = 0, \forall (H) \in \mathcal{F}$.*

The basic strategy is to remove the inessential fixed points by a stepwise induction on the M_i as described above. First, we choose an admissible ordering $(H_1), ..., (H_k)$ on $Iso(M)$ and the associated filtration $M_1 \subset M_2 \subset ... \subset M_k = M$. By an equivariant version of the Hopf construction, we may assume without loss of generality that $Fix f$ is finite [**Wi**][**W2**]. The crucial step is to coalesce fixed points of the same class equivariantly. If $x \in Fix f$, then by equivariance, the orbit Gx is pointwise fixed under f. Moreover, if $G_x = H$, then the WH-orbit of x has exactly $|WH|$ points since WH acts freely on the open subspace M_H. Given $x, y \in Fix f \cap M_H$, if x and y are *locally Nielsen equivalent* [**FH2**], i.e., there exists a path α in M_H from x to y such that $\alpha \sim f^H\alpha$ relative to the endpoints in M^H, then the orbits WHx and WHy (hence Gx and Gy) are said to be *Nielsen equivalent.* Equivalent orbits can be coalesced equivariantly via the classical Wecken method as in [**Br**]. Now we begin the induction and let $k = 1$. In this case, each fixed orbit has $|WH_1|$ points each of which has the same fixed point index. Having coalesced fixed orbits that are equivalent, we arrive at a WH_1-map for which no points from different orbits are Nielsen equivalent in the ordinary sense. Thus, $N(f^{H_1}) = 0$ implies that all the fixed orbits are of index zero and hence can be removed locally and equivariantly. We then extend it to a fixed point free G-map. For the inductive step, we may assume that $Fixf \subset M_k - M_{k-1}$. Under the codimension hypothesis, the ordinary Nielsen equivalence and the local Nielsen equivalence (on M_{H_k}) coincide. We apply the same coalescing procedure to unite fixed orbits in M_{H_k} and hence in $GM_{H_k} = M_k - M_{k-1}$.

Theorem 2 can be proven using equivariant obstruction theory [**BG**] similar to the non-equivariant approach to Wecken theorem presented in [**FH1**]. There is also a different obstruction theoretic approach given in [**W2**] using the local obstruction theory developed in [**FH2**].

The codimension condition in Theorem 2 is similar to the so-called *gap hypothesis* in equivariant surgery theory. For instance, for any finite group G of odd order acting smoothly on a compact smooth manifold M, the fixed point set of the action M^G has codimension at least two. Note that the purpose of the codimension hypothesis is to ensure that local Nielsen classes coincide with the ordinary ones. This condition can be relaxed [**W2**] by assuming that $M^H - M_H$ can be *by-passed* in M^H, i.e., $\iota_\# : \pi_1(M_H) \to \pi_1(M^H)$ is surjective.

We now state a stronger version of Theorem 1 which is essentially a consequence of Theorem 2.

THEOREM 3. *Let G be a compact Lie group acting smoothly on a compact smooth manifold M. Suppose for all $(H) \in \mathcal{F}$ that M^H is connected of $dim M^H \geq 3$, satisfies the Jiang condition and $M^H - M_H$ can be by-passed in M^H. For any G-map $f : M \to M$, f is G-deformable to be fixed point free if, and only if, $L(f^H) = 0, \forall (H) \in \mathcal{F}$.*

2.2 Equivariant Nielsen type numbers.

As seen in §2.1, one obvious choice for an equivariant Nielsen type number is the k-tuple given by

$$N_G^*(f) = (N(f^{H_1}), ..., N(f^{H_k}))$$

where $(H_i) \in \mathcal{F}$. One can show that $N_G^*(f)$ has the usual properties such as G-homotopy and G-homotopy type invariance and commutativity [**W2**]. It is, however, not clear how the components of $N_G^*(f)$ give a sharp lower bound for $Fixf$ in the G-homotopy class of f.

In view of the coalescing of fixed orbits in Theorem 2 ([**FW**][**W2**]), we must also consider uniting fixed orbits of different isotropy types especially when

$N_G^*(f) \neq 0$. This is similar to the basic idea in relative Nielsen theory [**S1**] [**S2**] [**Z**]. Since inessential fixed point classes play an important role in [**Z**], we must therefore define our new invariants via the covering space approach.

For every $(H) \in \mathcal{F}$, WH is a finite group and M^H is a WH-space. For any lift $\tilde{f^H}$ of f^H to the universal cover $\eta_H : \tilde{M^H} \to M^H$, the WH-invariant set $WH(\eta_H Fix \tilde{f^H})$ is called a *WH-fixed point class* (WH-fpc). If $x, y \in Fixf^H \neq \emptyset$, then x and y belong to the same WH-fpc if, and only if, (1) $y = \sigma x$ for some $\sigma \in WH$ or (2) there exists a path $\alpha : [0,1] \to Y$ such that $\alpha(0) = x, \alpha(1) = \sigma' y$ for some $\sigma' \in WH$ and $\alpha \sim f^H \alpha$ (rel endpoints) [**W4**]. By assigning the usual fixed point index to each WH-fpc, we define $N_{WH}(f^H)$ to be the number of *essential* WH-fpcs of f^H. We also define $N_G(f_H)$ to be $|WH|$ times the number of esssential WH-fpcs that do not intersect with any WK-fpc for any $(K) > (H)$. Roughly speaking, $N_G(f_H)/|WH|$ measures the number of fixed orbits in M_H. Note that $N_G(f_H)$ is defined in the same spirit as the relative Nielsen number on the complement defined by X. Zhao [**Z**].

The number $N_G(f_H)$ is invariant under G-homotopy and serves as a lower bound for the number of fixed points in M_H. Moreover, we have the following minimality result.

THEOREM 4 ([**W4**]). *Let G and M be as in Theorem 2. For any $(H) \in \mathcal{F}$,*

$$N_G(f_H) = min\{\#Fix\varphi_H | \varphi \sim_G f\}.$$

To see how the techniques from relative Nielsen theory are employed in the equivariant theory, let us review briefly some of the background material. Let $f : (X, A) \to (X, A)$ be a map of a finite polyhedral pair. Define $N(f; X - A)$ to be the number of essential fixed point classes of f that do not *contain* any fixed point class of $f|A$. When $X - A$ is not a surface, has no local cut points and A can be by-passed, $N(f; X - A)$ is a sharp lower bound for the number of fixed points in $X - A$ in the relative homotopy class of f [**Z**]. There are two basic steps in the procedure of coalescing fixed points. One is to coalesce fixed points in $X - A$ of the same class *along a path in* $X - A$. This is guaranteed by the by-passed condition (compare the codimension hypothesis in Theorem 2). The other is to move the fixed points in $X - A$ to the boundary of A in X if they are Nielsen equivalent to some fixed points in A. Therefore, in order to prove Theorem 4, we need to carry out these two steps WH-equivariantly with $X = M^H$ and $X - A = M_H$.

Let $x, y \in Fixf^H$. Suppose that $x \in M_H$ and x and y belong to the same WH-fpc but $y \neq \sigma x, \forall \sigma \in WH$. Then there exists a path α in M^H from x to some translate y' of y such that $\alpha \sim f^H \alpha$ (rel endpoints). Under the hypotheses of Theorem 4, we may assume that $\alpha([0,1)) \subset M_H$. We alter f^H in a small contractible neighborhood N_α of $\alpha([0,1))$ so that we can move x to y' along α. Taking the WH-translates of N_α yields a WH-homotopy which is constant outside $WH(N_\alpha)$ (see [**W4**]).

We should point out that in general, we may not be able to find a G-homotopy $h \sim_G f$ such that the map h realizes $N_G(f_H)$ for *all* $(H) \in \mathcal{F}$ *simultaneously.* The difficulty here lies in the fact that a WH-fpc may contain fixed points of different isotropy types. There is also no simple formula for the minimal number of fixed points of maps in the G-homotopy class (see [**W1**][**W4**]).

3. Periodic Points

We now relate the equivariant Nielsen theory to the Nielsen theory of periodic points developed in [**HPY**] and [**J1**].

3.1 Nielsen type numbers for periodic points.

Let X be a compact connected smooth manifold and $f : X \to X$ be a self map. For any positive integer n, let

$$Y_n = X \times ... \times X \qquad \text{(n-fold product).}$$

The cyclic group $\mathbb{Z}_n =< \zeta >$ acts on Y_n via

$$\zeta \cdot (x_1, ..., x_n) = (x_n, x_1, ..., x_{n-1}) \qquad x_i \in X.$$

We associate to f a $\mathbb{Z}_n$-map $g_f : Y_n \to Y_n$ defined by

$$g_f(x_1, ..., x_n) = (f(x_n), f(x_1), ..., f(x_{n-1})).$$

For any positive integer m with $m|n$,

$$Y_n^{\mathbb{Z}_m} \approx X \times ... \times X \qquad \text{(n/m-fold product)}$$

and

$$Fixg_f^{\mathbb{Z}_m} = \{(x, f(x), ..., f^{n/m-1}(x)) | x \in Fixf^{n/m}\}.$$

It is easy to see that there is a 1-1 correspondence between $Fixg_f$ and $Fixf^n$. An *f-orbit* of a periodic point x of period n (i.e., $\{x, f(x), ..., f^{n-1}(x)\}, x \in Fixf^n$) corresponds to the $\mathbb{Z}_n$-fixed orbit of the fixed point $(x, f(x), ..., f^{n-1}(x))$ of g_f. Recall from [**J1**] a periodic point class of period n which is simply a fixed point class of f^n, is *irreducible* if it does not contain any periodic point class of period $m < n$. A set of periodic point classes (of diverse periods) is said to be *f-invariant* if it is a union of f-orbits. The *height* of an f-invariant set of periodic point classes is the sum of the periods of the f-orbits in the set. The *Nielsen type number of period* n is given by

$NP_n(f) =$ the height of the set of irreducible essential periodic point classes of period n.

Let $\tilde{f} : \tilde{X} \to \tilde{X}$ be a lift of f where $p : \tilde{X} \to X$ is the universal cover. Let $\tilde{f^n} = \tilde{f} \circ ... \circ \tilde{f}$ (n-copies) be the lift of f^n so that every periodic point class of period n is of the form $pFix\alpha\tilde{f^n}$ for some $\alpha \in \pi$ where $\pi = Cov(p) \equiv \pi_1(X)$. Consider the lift $\tilde{g}_f$ of g_f given by

$$\tilde{g}_f(\tilde{x}_1, ..., \tilde{x}_n) = (\tilde{f}(\tilde{x}_n), \tilde{f}(\tilde{x}_1), ..., \tilde{f}(\tilde{x}_{n-1})), \qquad \tilde{x}_i \in \tilde{X}.$$

We have the following commutative diagrams of liftings

$$\begin{array}{ccc} \tilde{X} & \xrightarrow{\tilde{f}^n} & \tilde{X} \\ p\downarrow & & \downarrow p \\ X & \xrightarrow[f^n]{} & X \end{array}$$

and

$$\begin{array}{ccc} \tilde{X} \times \ldots \times \tilde{X} = \tilde{Y}_n & \xrightarrow{\tilde{g}_f} & \tilde{Y}_n \\ \eta\downarrow & & \downarrow \eta = p \times \ldots \times p \\ X \times \ldots \times X = Y_n & \xrightarrow[g_f]{} & Y_n \end{array}$$

If S is the f-orbit of $pFix\alpha\tilde{f}^n$ for some $\alpha \in \pi$, we associate to it the $\mathbb{Z}_n$-fpc $\mathbb{Z}_n(\eta Fix\hat{\alpha}\tilde{g}_f)$ where $\hat{\alpha} = 1 \times \ldots \times 1 \times \alpha \in Cov(\eta) \equiv \pi \times \ldots \times \pi$. This is a bijection between the set of nonempty f-orbits of periodic point classes and the set of nonempty $\mathbb{Z}_n$-fpcs of g_f.

Furthermore, there is a 1-1 correspondence between the set of f-orbits of irreducible essential periodic point classes of period n and the set of essential $\mathbb{Z}_n$-fpcs of g_f. We have the following identification.

THEOREM 5 ([**W4**]).

$$NP_n(f) = N_{\mathbb{Z}_n}((g_f)_{(1)})$$

where (1) *is the trivial subgroup of* $\mathbb{Z}_n$.

The definition of $NP_n(f)$ was first given by B. Halpern but we choose the one given in [**J1**]. This number was shown to be a Nielsen type number satisfying all the usual properties of the classical Nielsen number. For a compact connected smooth manifold X of $dimX \geq 5$, $NP_n(f)$ is a sharp lower bound for the number of periodic points of least period n in the homotopy class of f [**J1**][**W4**]. The equality in Theorem 5 can be thought of as a new interpretation of $NP_n(f)$. The notion of Nielsen equivalence on the orbits of fixed points has also been employed in [**J3**].

One can estimate $NP_n(f)$ using relative Nielsen theory and the map g_f. More precisely, if $N(g_f; Y_{n_{(1)}})$ denotes the relative Nielsen number of g_f on the complement $Y_n - Y_n^{>(1)} = Y_{n_{(1)}}$ where $Y_n^{>(1)}$ is the subspace of points whose isotropy subgroups are not trivial, then

THEROEM 6 ([**W6**]).

$$N(g_f; Y_{n_{(1)}}) \leq NP_n(f) \leq n \cdot N(g_f; Y_{n_{(1)}}).$$

Under certain algebraic conditions, $NP_n(f)$ can be calculated in terms of $N(g_f; Y_{n_{(1)}})$. The key connection here is that g_f and f^n have the same fixed point indices [**K**] and there is a one to one correspondence between their fixed point classes and hence $N(g_f) = N(f^n)$ [**W6**].

3.2 Congruence relations.

In [**D1**], A. Dold established a congruence relation among the fixed point indices $\{I(f^k)\}$ of the iterates. More precisely, for any positive integer n,

THEOREM 7 ([**D1**]).

$$\sum_{d|n} \mu(n/d) I(f^d) \equiv 0 \quad mod\ n$$

where μ is the Möbius function.

Here $I(f^d)$ is the fixed point index of f^d on the open set V_d where $V_1 = V$ is an open subset of an euclidean neighorhood retract X and $V_n = f^{-1}(V_{n-1})$ provided that $Fix f^d$ is compact in V_d for all $d \leq n$. In the case where $V = X$ and X is compact, $I(f^d)$ coincides with the Lefschetz number $L(f^d)$.

P. Heath, R. Piccinini and C. You showed a similar congruence relation replacing $I(f^d)$ by $N(f^d)$, under the so-called *n-toral* condition which is an algebraic condition roughly meaning that for all divisors m of n no distinct fixed point classes of f^m are combined as classes of f^n. Moreover, they showed the following

THEOREM 8 ([**HPY**]). *Let $f : X \to X$ be an n-toral map such that for every $m|n, L(f^m) \neq 0$ and $J(f) = \pi$. Then,*

$$NP_n(f) = \sum_{d|n} \mu(n/d) N(f^d).$$

Using the $\mathbb{Z}_n$ action and the map g_f described in §3.1, K. Komiya [**K**] deduced Dold's result and thus obtained a more general congruence relation for the fixed point indices of equivariant maps. A similar generalization of Theorem 8 is given in [**W4**] for G a finite abelian group.

4. Concluding Remarks

4.1 Equivariant vectorfields.

It is well known that a compact connected smooth manifold M admits a nonsingular vectorfield if and only if its Euler characteristc $\chi(M)$ vanishes. This is also equivalent to the existence of a fixed point free deformation (i.e., a map homotopic to the identity map). An equivariant analog of this result was proven in [**Wi**] (see also [**J2**]). There is also a similar result on the existence of equivariant nonsingular *pathfields* on finite G-complexes [**W3**]. The use of pathfields leads to the necessary and sufficient conditions for a nonempty invariant subset to be the fixed point set of a G-deformation of M [**W1**]. This is the equivariant analog of the so-called *Complete Invariance Property* (CIP) with respect to deformations.

4.2 Equivariant fixed point indices.

Let G be a compact Lie group. In [**D2**], A. Dold established an isomorphism between the equivariant fixed point situations over B, denoted by $G\text{-}FIX_B$, and $\omega_G^0(B)$, the equivariant 0-th stable cohomotopy group of B. Furthermore, when B is a point, $G\text{-}FIX_{pt}$ is isomorphic to $A(G)$, the Burnside ring of G which organizes the lattice of conjugacy classes of subgroups of G. This ring $A(G)$ is an important object of study in transformation groups (see [**tD**]). For a more detailed treatment of fixed point indices of equivariant maps, see [**U**] by H. Ulrich. The approach to periodic points via the $\mathbb{Z}_n$ action and the map g_f described in §3 has led to a new proof [**EGK**] of a result of J. Franks on period doubling cascades. This describes Morse index type conditions when, for a smooth map $f : D^n \to intD^n$, the existence of a periodic orbit of least period p implies that for all $k \geq 0$ there are periodic orbits of least period $2^k p$ whose differentials Df^k reverse the orientation on their unstable generalized eigenspaces.

4.3 Other Nielsen type theories.

We see from the equivariant Wecken theorem in §2 that one can employ the local Nielsen theory [**FH1**] as well as the relative Nielsen theory [**S1**][**Z**]. This suggests a connection between the two. In fact, under certain circumstances, the local theory is related to a variant of the relative theory[**W5**], namely, the relative Nielsen theory for map extensions [**BGS**]. We believe that relative Nielsen theory is a natural tool to study the fixed point theory of equivariant maps and iterates of maps.

References

[**A**] Aigner, A., *Combinatorial Theory*, Springer-Verlag, Heidelberg, 1979.

[**BG**] Bosari, L. and Gonçalves, D., *G-deformation to fixed point free maps via obstruction theory*, preprint.

[**B**] Bredon, G., *Introduction to Compact Transformation Groups*, Academic Press, New York, 1972.

[**Br**] Brown, R.F., *The Lefschetz Fixed Point Theorem*, Scott Foresman, Illinois, 1971.

[**BGS**] ______, Greene, R. and Schirmer, H., *Fixed points of map extensions*, Springer LNM **1411** (1989), 24–45.

[**tD**] tomDieck, T., *Transformation Groups*, de Gruyter, Berlin, New york, 1987.

[**D1**] Dold, A., *Fixed point indices of iterated maps*, Invent. Math. **74** (1983), 419–435.

[**D2**] ______, *Fixed point theory and homotopy theory*, Contemp. Math. AMS **12** (1982), 105–115.

[**EGK**] Erbe, L. Gęba, K. and Krawcewicz, W., *Equivariant fixed point index and the period-doubling cascades*, Can. J. Math. **43** (1991), 738–747.

[**FH1**] Fadell, E. and Husseini, S., *Fixed point theory for non-simply connected manifolds*, Topology **20** (1981), 53–92.

[**FH2**] ______, *Local fixed point index theory for non-simply connected manifolds*, Ill. J. Math. **25** (1981), 673–699.

[**FW**] ______and Wong, P., *On deforming G-maps to be fixed point free*, Pacific J. Math. **132** (1988), 277–281.

[**HPY**] Heath, P., Piccinini, R. and You, C., *Nielsen type numbers for periodic points,I*, Springer LNM **1411** (1989), 88–106.

[**J1**] Jiang, B., *Lectures on Nielsen Fixed Point Theory*, Contemp. Math. v.14, AMS, 1983.

[**J2**] ______, *A note on equivariant vector fields*, Acta Math. Sci. **11** (1991), 274–282.

[J3] ———, *Nielsen theory for periodic orbits and applications to dynamical systems*, these proceedings.

[K] Komiya, K., *Fixed point indices of equivariant maps and Möbius inversion*, Invent. Math. **91** (1988), 129–135.

[S1] Schirmer, H., *A relative Nielsen number*, Pacific J. Math. **122** (1986), 459–473.

[S2] ———, *A survey of relative Nielsen fixed point theory*, these proceedings.

[U] Ulrich, H., *Fixed Point Theory of Parametrized Equivariant Maps*, Springer LNM 1343, 1988.

[V] Vidal, A., *Äquivariante Hindernistheorie für G-Deformationen*, Dissertation, Universität Heidelberg, 1985.

[Wi] Wilczyński, D., *Fixed point free equivariant homotopy classes*, Fund. Math. **123** (1984), 47–60.

[W1] Wong, P., *On the location of fixed points of G-deformations*, Top. Appl. **39** (1991), 159–165.

[W2] ———, *Equivariant Nielsen fixed point theory for G-maps*, Pacific J. Math. **150** (1991), 179–200.

[W3] ———, *Equivariant path fields on G-complexes*, Rocky Mtn. J. Math. **22** (1992), 1139–1145.

[W4] ———, *Equivariant Nielsen numbers*, Pacific J. Math., to appear.

[W5] ———, *A note on the local and the extension Nielsen numbers*, Top. Appl. **48** (1992), 207–213.

[W6] ———, *Estimation of Nielsen type numbers for periodic points*, preprint.

[Z] Zhao, X., *A relative Nielsen number for the complement*, Springer LNM **1411** (1989), 189–199.

Department of Mathematics, Bates College, Lewiston, ME 04240

E-mail address: pwong@abacus.bates.edu

Recent Titles in This Series

(*Continued from the front of this publication*)

(See the AMS catalog for earlier titles)